BIRKHÄUSER

Atmospheric and Oceanic
Mesoscale Processes

Edited by
Maithili Sharan
Sethu Raman

Birkhäuser
Basel · Boston · Berlin

Reprint from Pure and Applied Geophysics
(PAGEOPH), Volume 164 (2007) No. 8 - 9

Editors:

Maithili Sharan
Professor, Centre for Atmospheric Sciences
Indian Institute of Technology Delhi
Hauz Khas, New Delhi 110016
India
mathilis@cas.iitd.ernet.in

Sethu Raman
Professor, Department of Marine, Earth, and Atmospheric Sciences
North Carolina State University
Raleigh, North Carolina 27695
USA
sraman@ncsu.edu

Library of Congress Control Number: 2007928186

Bibliographic information published by Die Deutsche Bibliothek:
Die Deutsche Bibliothek lists this publication in the Deutsche Nationalbibliografie; detailed bibliographic data is available in the Internet at <http://dnb.ddb.de>

ISBN 978-3-7643-8492-0 Birkhäuser Verlag AG, Basel · Boston · Berlin

Basel · Boston · Berlin
P.O. Box 133, CH-4010 Basel, Switzerland
Part of Springer Science+Business Media
Printed on acid-free paper produced from chlorine-free pulp. TCF ∞

ISBN 978-3-7643-8492-0
9 8 7 6 5 4 3 2 1

ISBN 978-3-7643-8493-7 (eBook)
www.birkhauser.ch

Contents

C. Seasonal Monsoon

D. Land Surface Processes

Pure appl. geophys. 164 (2007) 1423–1425
0033–4553/07/091423–3
DOI 10.1007/s00024-007-0226-3

Pure and Applied Geophysics

Preface

Prediction of mesoscale high impact weather events like tropical storms, very heavy monsoon rainfall, thunderstorms, excessive snowfall, etc. pose great challenges to operational and research meteorologists. Similarly, air-pollution episodes in urban areas, accidental release of poisonous gases, and deteriorating air quality are also influenced by mesoscale circulation patterns in the atmospheric boundary layer.

Understanding, modeling and prediction of mesoscale processes in the atmosphere, ocean and environmental systems, have gained importance in the last decade or so. This is because of the availability of more sophisticated observational systems, provided by technological innovations and more realistic simulations using advanced dynamical models.

An International Conference on Mesoscale Processes in Atmosphere Ocean and Environmental Systems (IMPA2006) was held at the Indian Institute of Technology, Delhi during 14–17 February, 2006 with the aim to bring together scientists and engineers from industry, research organizations, government and academia involved in monitoring, modeling, simulation and management of the various events related to mesoscale weather.

Numerous scientific papers were presented at IMPA2006 on the theme of mesoscale processes. In this volume, peer-reviewed papers related to observational analysis, numerical simulations and atmospheric and oceanic processes are included to gain insight into the mesoscale processes related to weather and climate.

KRISHNAMURTI *et al.* in their paper addressed hurricane intensity through angular momentum and scale energetics approaches. The angular momentum of a parcel of air that arrives at the region of the maximum winds of a hurricane is generally depleted along its path. They have noted that the cloud torques are the major contributor to this depletion. SANDEEP *et al.* investigate the relative merits of utilizing Holland and Rankine wind profiles together with the NCAR-AFWA synthetic vortex scheme on the structure, precipitation and track of a tropical cyclone, which formed over the Bay of Bengal during 10–12 November, 2002 using a mesoscale model. The performance of the Planetary Boundary Layer and convective parameterization on the simulated features of the Andhra 2003 cyclone is studied by SRINIVAS *et al.* using a numerical model.

The study by ABHILASH *et al.* focuses on the impact of Doppler Weather Radar and other conventional and non-conventional data in simulating the intensity and propagation of rainbands associated with mesoscale intense convective events over Kolkata. Observational studies of sea breeze front about its onset, depth of circulation and induced precipitation over Chennai in southern peninsular India have been discussed in the paper by SURESH using a S-band Doppler Weather Radar Chennai, India. PATTANAIK in his paper examines the variability of convective activity over the north Indian Ocean on an interannual and longer time scale and its association with the rainfall activity over the four different homogeneous regions of India during the monsoon season. BOYLES *et al.* investigate the mesoscale circulations caused by the transition from clay to sandy soils over the Carolinas region of the southeastern United States. The next three papers are concerned with the use of MM5, WRF and ETA models. In the paper by HUANG *et al.* the WRF-model along with variational data assimilation was utilized to study a heavy rainfall event along the west coast of India with and without the assimilation of GPS occultation refractivity soundings during the 2002 monsoon. RAMA RAO *et al.* in their paper have used WRF and Eta models to investigate heavy rainfall events that occurred (i) over parts of Maharastra in west India during July 26 to 27, 2005 and (ii) over coastal Tamilnadu and south coastal Andhra Pradesh during October 24 to 28, 2005. RAKESH *et al.* study the sensitivity of weather forecasts produced using the MM5 model during the launch of an Indian satellite on 5th May, 2005 with different cumulus parameterization schemes.

The paper by RATNAM *et al.* presents an ensemble forecast for the monsoon season of 2005 using the NCEP's T170L42 AGCM on C-DAC's supercomputer PARAM Padma. A new analytical formulation of entrainment and detrainment in the Tiedtke's mass flux cumulus parameterization is presented by DEB *et al.* The proposed analytical profiles of entrainment and detrainment are tested in GCM for Indian monsoon rainfall. An objective approach is presented by MOHAPATRA and MOHANTY for prediction of daily summer monsoon rainfall over Orissa (India) due to the interaction of mesoscale and large-scale synoptic systems. SANJEEVA RAO and SIKKA address interactive aspects of the Indian and the African summer monsoon systems to understand the possible mutual interactions of subseasonal variability of the two neighboring regional monsoon systems through data analysis. One of the aims of the paper by STRAUS and KRISHNAMURTHY is to show that by combining the two approaches (EOF analysis of a single level and consideration of vertical shear), one can obtain a single, multi-level seasonal mean circulation pattern (or mode) that is closely associated with the seasonal mean all-India monsoon rainfall. Another purpose of their study is to examine the boundary-forced hypothesis by discussing the geographic distribution of observed seasonal mean Indian rainfall associated with this mode, as well as the correlation with global SST. DIMRI and GANJU in their paper studied the mean climatic conditions and the intraseasonal variability of temperature and precipitation during a particular winter season, Oct. 1999—Mar. 2000, over the western Himalaya using a regional climate model. A statistical evaluation of the Coupled Ocean/Atmosphere Mesoscale Prediction System performed over the Arabian Gulf region for the period, 1 August to 5 October, 2004 is described by

EAGER *et al.* Influence of the southern Pacific waves on the Indian Ocean Rossby waves is discussed (VAID *et al.*) by examining the Rossby wave propagations in the southern Pacific and Indian Oceans. The Pacific to Indian Ocean connection is investigated with respect to its effect on southern tropical Indian Ocean Rossby waves.

The land surface processes and surface characteristics play a dominant role in the development of the boundary layer and in turn to proper mesoscale simulation and prediction. ALFIERI *et al.* present in their paper a reclassification approach using a logical model for improving the accuracy of land use/land cover data sets when the focus is on a local scale. The observations from the Land Surface Processes Experiment conducted at Anand (Gujarat, India) during the year 1997–1998 are examined (ADITI and SHARAN) to study the characteristics of surface layer under weak wind stable conditions. The magnitude of surface fluxes computed from the Monin-Obukhov similarity theory is shown to be smaller in comparison to those based on the observations in weak wind stable conditions. The diurnal structure of the boundary layer during the Indian summer monsoon period is studied (RAMAKRISHNA and SHARAN) using a one-dimensional meteorological boundary layer model and the observations from the Monsoon Trough Boundary-Layer Experiment conducted in 1990 at Jodhpur, India. NAGAR *et al.* examine the evolutionary features of the dynamic and thermodynamic characteristics of marine atmospheric boundary layer over the Arabian Sea and the onset of monsoon over Kerala during the Arabian Sea Monsoon Experiment in 2003.

We would like to place on record our gratitude to all the authors who have contributed to this special issue. Their contributions to this special issue are very much appreciated. We wish to thank all the reviewers for their valuable comments/suggestions. We also thank Dr. Renata Dmowska, Topical Editor, Pure and Applied Geophysics for her assistance in bringing out this special issue.

Finally, we express our sincere thanks to the editorial staff of Pure and Applied Geophysics for the meticulous care they have taken in preparing this special edition.

Maithili Sharan
Professor, Centre for Atmospheric Sciences
Indian Institute of Technology, Delhi,
Hauz Khas, New Delhi 110016, India.

Sethu Raman
Professor, Department of Marine, Earth,
and Atmospheric Sciences, North Carolina
State University, Raleigh, North Carolina
27695, USA.

Published Online First: June 19, 2007

A. Tropical Cyclone

Pure appl. geophys. 164 (2007) 1429–1441
0033–4553/07/091429–13
DOI 10.1007/s00024-007-0238-z

Pure and Applied Geophysics

Addressing Hurricane Intensity through Angular Momentum and Scale Energetics Approaches

T. N. Krishnamurti, L. Stefanova, L. Watson, and S. Pattnaik

Abstract—This paper addresses two avenues for gaining insight into the hurricane intensity issue—the angular momentum approach and the scale interaction approach. In the angular momentum framework, the torques acting on a parcel's angular momentum are considered along an inflowing trajectory in order to construct the angular momentum budget. These torques are separable into three components: The pressure torque, the surface friction torque, and the cloud torque. All torques are found to diminish the angular momentum of an inflowing parcel, with the cloud torques having the most important role. In the scale interaction approach, energy exchanges among different scales within a hurricane are considered as a means of understanding hurricane intensity. It is found that the majority of kinetic energy contribution to the hurricane scales originates from potential-to-kinetic in-scale energy conversions. The contribution of mean-wave interactions in the kinetic energy varies with distance from the center and with the life stage of a storm. In the early stages, as the disorganized convection becomes organized on the hurricane scales, upscale energy transfers (*i.e.*, from small to large scale) are found to take place in the outer radii of the storm. In a mature storm, the kinetic energy transfers are downscale, except for the inner radii.

Key words: Hurricane intensity, angular momentum, scale interactions, hurricane energetics.

1. Introduction

Addressing the intensity forecasts of hurricanes entails a suite of complex multifaceted issues. Many factors seem to have a role in the final shaping of the strongest winds in a hurricane. There are both external and internal forcings that deserve to be addressed. The eventual framework that one wishes to arrive at must have the flexibility to include all possible candidates that might contribute to the final intensity. That is a difficult task since some of the data and modeling constraints may not fully accommodate all possibilities and may then have a somewhat limited scope.

There are a number of issues that need to be addressed for a full understanding of the intensity problem. One of these is how does the outer angular momentum configure the intensity as parcels of air arrive at the storm's intensity? Do we have an adequate

Department of Meteorology, Florida State University, Tallahassee, Florida, USA.
E-mail: tnk@io.met.fsu.edu

modeling capacity to estimate the torques that modify the outer angular momentum—torques arising from friction, cloud turbulence, and pressure asymmetries? (HOLLAND, 1983; MOLINARI and VOLLARO, 1989; KRISHNAMURTI *et al.*, 2005). Another aspect of the intensity problem is: If energy released from deep convective clouds drives a hurricane, how do cloud scales (of the order of a few km) pass energy to the hurricane scale (hundreds of km)? The formal scale interactions among these diverse scales need to be understood especially in the context of the growth of the mean and eddy kinetic energy that controls the eventual intensity of a hurricane. The mathematical framework for such scale interactions was first developed by SALTZMAN (1970), and was applied in a hurricane context in KRISHNAMURTI *et al.* (2005). There is increasing evidence that the internal forcing, i.e., cloud microphysics, exhibits sensitivity to the hurricane's intensity. Modeling must accommodate research on sensitivities such as growth of ice, graupel, snow, liquid water, and fall velocities on the intensity of the storm. This calls for a fairly high-resolution microphysical non-hydrostatic modeling of a hurricane (LIN *et al.*, 1983; TAO and MCCUMBER, 1989; TAO and SIMPSON, 1993; BRAUN, 2002; MCFARQUHAR and BLACK, 2004; ROGERS *et al.*, 2004; PATTNAIK and KRISHNAMURTI, 2007a,b).

A number of issues are generally commented upon in tropical storms discussion groups when a current storm exhibits an unusual intensity change. These include the passage of a hurricane over warm or cold pockets of SST and oceanic heat content anomalies, dry air intrusion, effects of TUTT (tropical upper tropospheric troughs) and effects of potential vorticity intrusions at upper levels, and issues related to vertical shear of the horizontal wind. Many of these can, in principle, be accommodated by detailed post-analysis of model forecasts, provided the model is initialized with adequate data sets, and has reasonable data assimilation, high resolution, and comprehensive dynamics and physics. If a reasonable forecast of intensity is accomplished in such a framework, such diagnostic inquiries can be enlightening.

The phenomenon of eye wall replacement relates to the intensity changes of a hurricane. As an outer eye wall forms and the original inner eye collapses, a sudden drop in the intensity has been noted in many instances. That alone is a research topic in its own right. Modeling issues of this type require adequate observations on the scale of mesoconvective rain bands. Having just ten or so dropsondes and a few levels of flight legs does not seem adequate to define initial states that may lead to an eye wall replacement cycle. Carefully planned field experiments, such as the current RAINEX (http://www.joss.ucar.edu/rainex/) might address such data needs.

At the end of the 2004 hurricane season Dr. James Franklin of the National Hurricane Center provided a summary on the performance of the suite of operational models for real time hurricane forecasts (FRANKLIN, 2004). According to this summary the FSU multimodel superensemble had the least errors among the member models for the position and intensity forecasts of the 2004 season. Those were mostly large-scale operational models. Even though the superensemble provided the best guidance for hurricane intensity forecasts among the various models, there is still much to be desired for providing a useful product. Given the need for improved intensity forecasts, a suite of mesoscale models that can adequately simulate some of the aforementioned features may be useful for the improvement and understanding of hurricane intensity. Rapid intensification (as in Hurricane Charley of 2004) or rapid weakening (as in Hurricane Lili of 2002) is a major forecast issue. We have heard of accounts such as "there was a flare-up of convection followed by rapid intensification" in Hurricane Opal of 1995.

All of the above are major unsolved problems that relate to intensity forecasting. Analysis of simulations by mesoscale models with microphysical details may provide some insight on these issues.

2. The Angular Momentum Approach

The intensity issue of hurricanes was addressed in a recent paper by Krishnamurti *et al.* (2005) using the angular momentum budget of a hurricane in storm relative cylindrical coordinates, and a scale interaction approach in a storm centered frame of reference. Within the angular momentum budget in storm relative coordinates, a large outer angular momentum of the hurricane is depleted continually along inflowing trajectories. This depletion occurs via surface and planetary boundary layer friction, model diffusion, and 'cloud torques'; the latter is a principal contributor to the diminution of outer angular momentum. The eventual angular momentum of the parcel near the storm center determines the storm's final intensity.

Figure 1, taken from Krishnamurti *et al.* (2005), illustrates a typical inflowing three-dimensional trajectory into a hurricane (Hurricane Bonnie, August 1998). The starting point of this trajectory was on August 22 at the 336 hPa level. This trajectory is based on forecasts from an MM5 model that provided a reasonable forecast of the storm track and intensity. This trajectory was ended on August 25 at the 850 hPa level near the region of the hurricane's maximum winds. The isotachs of the predicted wind for August 25 are illustrated by thin black lines in units of ms^{-1}. It is possible to tabulate the angular momentum history of this inflowing parcel. This is shown in Table 1 (also from Krishnamurti *et al.*, 2005). Here we show a number of torques, namely the pressure torque, the surface frictional torque, and the cloud torque. The latter relates to the parcel moving through clouds, and can be expressed as $\left(\frac{dM}{dt}\right)_{CL} = -r\frac{\partial}{\partial z}\overline{W'V'_\theta}$, where r is the radial distance from the storm center, W' is the vertical velocity and V_θ is the azimuthal

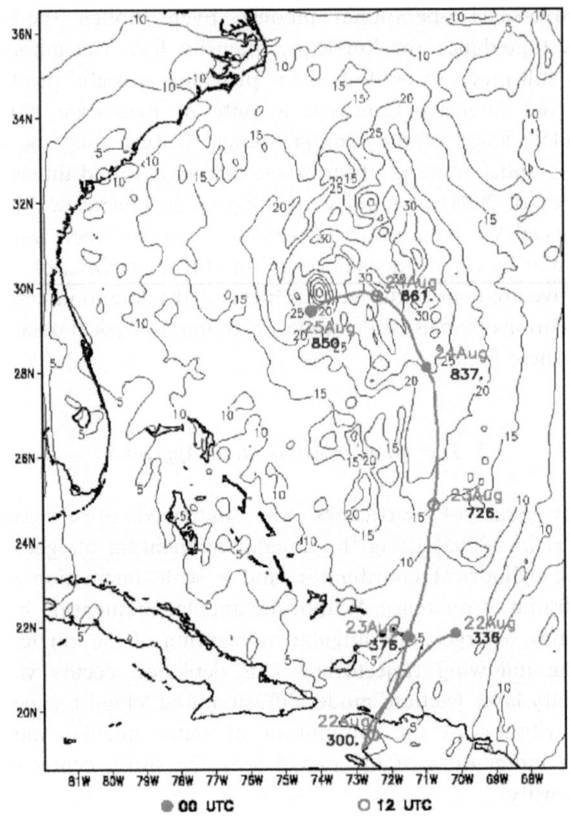

Figure 1
72-hr 3-D backward trajectory of maximum wind at 850 hPa (ms^{-1}) for Hurricane Bonnie of 1998. Shown here is a trajectory terminating at the wind maxima in the vicinity of the center of the hurricane at the end of the 72-hr forecast. Isopleths indicate the maximum wind distribution at 72-hr forecast time (from KRISHNAMURTI *et al.*, 2005).

wind. Given a cloud resolving model, this term can be calculated explicitly. The salient finding here is that the parcels lose angular momentum as they move into the interior of the hurricane largely from the cloud torques they encounter. The final intensity of the hurricane is determined by the prorated angular momentum. Thus it is possible to relate the final intensity of the hurricane to the angular momentum an inflowing parcel acquires.

Table 1

Values of the contribution of different torques to the angular momentum change (m^2s^{-1}) along the path of the 3-D trajectory of wind maximum at 850 hPa for Hurricane Bonnie of 1998 (from KRISHNAMURTI et al., 2005)

Forecast hours	00	12	24	36	48	60	72
Position of V_{max} at 850 hPa	20.74N 68.32W	22.58N 69.95W	25.56N 68.91W	25.17N 70.81W	26.60N 71.37W	27.65N 72.36W	29.48N 72.12W
Pressure of the Parcel (hPa)	336	300	376	726	837	861	850
Angular Momentum M ($\times 10^6$)	74.54	65.31	53.84	45.10	34.86	21.05	12.41
Total angular momentum change Δ M ($\times 10^6$)	–	−9.23	−11.47	−8.14	−10.24	−13.81	−8.64
Pressure torque angular momentum change ΔM_{PT} ($\times 10^6$)	–	−2.47	−3.17	−3.46	−2.69	−3.20	−3.66
Frictional torque angular momentum change ΔM_{FT} ($\times 10^6$)	–	−0.8	−1.35	−1.59	−3.39	−4.9	−3.2
Cloud torque angular momentum change ΔM_{CT} ($\times 10^6$)	–	−5.96	−6.95	−3.09	−4.16	−5.71	−1.78

3. The Scale Interaction Approach

Scale interactions can be assessed using the familiar energetics in the wave number domain. Such an approach is useful for answering questions relating to the energy exchanges that may be important for the scales of motion that define a hurricane. The mathematical framework for the calculation of such energy exchanges is shown in the appendix. Fourier decomposition of data sets from a hurricane analysis in the azimuthal direction suggests that wavenumbers 0, 1 and 2 are the dominant components determining the hurricane scale. The scale interaction approach is the familiar energetics in the wave number domain where the eddy and azimuthally averaged kinetic energy of a hurricane offer some insights on its intensity (KRISHNAMURTI *et al.*, 2005). The dynamical and thermodynamical equations of a complete model are cast in storm-centered cylindrical coordinates. The energy equations for kinetic and potential energy are then cast in the azimuthal wavenumber domain. The wave numbers include the azimuthally averaged wavenumber 0, the principal hurricane scale asymmetries (wavenumbers 1 and 2), and other (cloud) scales. The main questions asked here relate to the role of separate scales in supplying energy to the hurricane circulation, thus contributing to its intensity. A principal finding is that clouds carry most of their variance, via organized convection, directly on the scales of the hurricane. This is best seen from the azimuthal distribution of cloud or rain water mixing ratios after a model has been spun up for a few hours. This organization of convection goes with the generation of available potential energy on the hurricane scales. Large transformations of available potential to kinetic energy via vertical overturning occur at these large scales. This suggests that once a storm is formed these larger scales play a central role in the energetics and hence for the intensity issue as well. Less of the kinetic

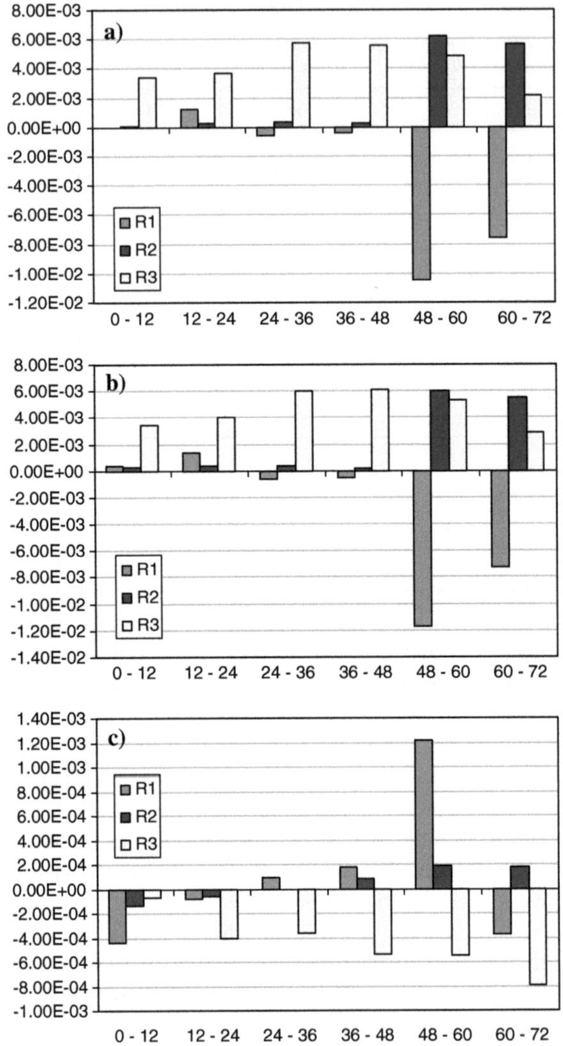

Figure 2

Vertically and radially integrated mean to eddy kinetic energy exchanges from a) all wavenumbers; b) wavenumbers 1 and 2; c) wavenumbers 3 through 180. The different color bars represent the three areas: The inner radius (R1 = 0–33 km), the area of maximum winds (R2 = 33–125 km) and outer radius (R3 = 125–334 km) for Hurricane Felix of 2002. Forecasts averaged over 12-hr periods, units of m^2s^{-3}.

energy is generated on the scales of individual clouds that are of the order of a few km. The other major components of the energetics are the kinetic-to-kinetic energy exchange and available potential-to-available potential energy exchange among different scales. These occur via triad interactions among scales and were noted to be essentially downscale transfer, i.e., a cascading process. It is the balance among all these processes that seems to dictate the final kinetic energy of the hurricane scales and hence of the storm intensity.

Figure 2 displays the wave-mean interactions for kinetic energy exchanges during a 72-hour forecast of the genesis stage of Hurricane Felix of 2002. These are vertically and radially averaged values for three areas: The inner area, encompassing the center of the storm between 0 and 33 km radii, indicated as R1; the area of the maximum winds, between 33 and 125 km, indicated as R2, and the outer area, from 125 to 334 km, indicated as R3.

Throughout the 72-hour simulation, the mean flow gains kinetic energy from the totality of wave-mean interactions (Fig. 2a) in the inner radius, except for a modest loss between 24 and 48 h. In the middle and outer radii, the mean flow loses kinetic energy as a result of the totality of wave-mean interactions. The wave-mean interactions can be separated into two parts: Interactions of the mean flow with the storm asymmetries at wave numbers 1 and 2, and interactions of the mean flow with the cloud scales. Among the kinetic-kinetic energy exchanges, the mean-asymmetry interactions (Fig. 2b) are responsible for most of the changes in kinetic energy of the mean flow, as evidenced by the similarity of Figures 2a and 2b. Figure 2c displays the exchange of kinetic between the mean flow and the cloud scales (wavenumbers 3 to 180). The mean flow in the inner and middle radius shows initially a slight gain followed by increasing loss to the cloud scales. In the outer radius however, the mean flow shows a gain of kinetic energy from the cloud scales and the gain increases throughout the simulation, along with the intensification of the storm. The order of magnitude of the mean-cloud exchanges is smaller by an order of magnitude than the exchanges involving the mean-asymmetry interactions. During the formation of the storm at all radii the largest transfers of kinetic energy occur between the mean flow and the asymmetries. Overall, the total kinetic energy lost at all radii to the scales of the asymmetries is larger by an order of magnitude than the amount of kinetic energy gained from the cloud scales, indicating that, in terms of kinetic energy transfer, the role of disorganized convection is secondary.

The transfer of energy from small to large scales seen at the outer radii in the formative stages of a hurricane is associated with the process of organization of convection on the larger scale of the mean circulation. In a mature storm, on the other hand, the wave-mean interactions in the kinetic energy domain proceed in a

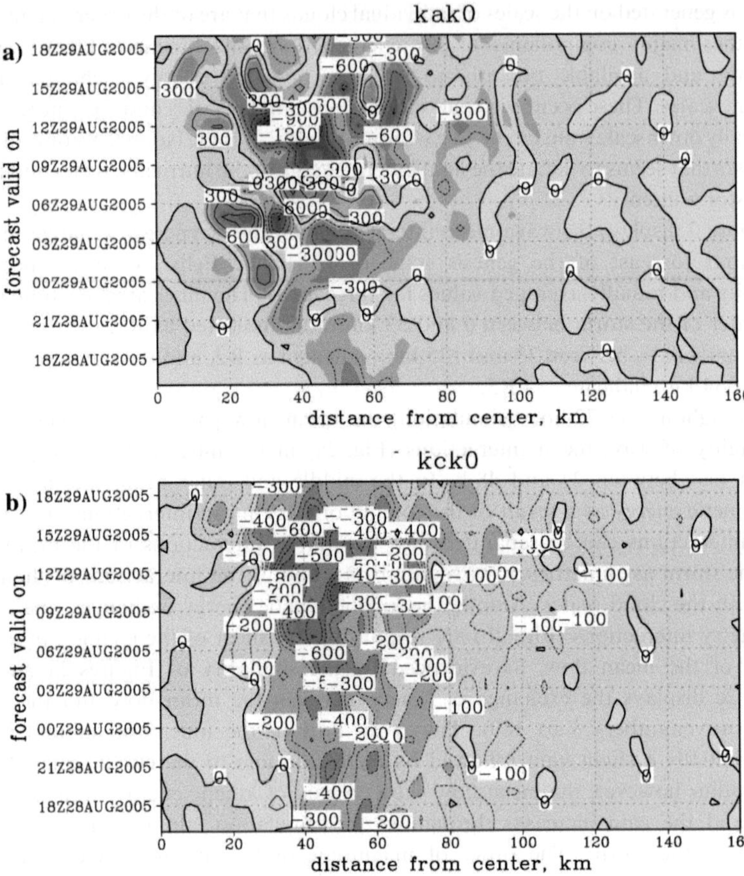

Figure 3
Radius-forecast time Hovmüller diagram of vertically averaged kinetic energy transfer a) from the asymmetries to the azimuthal mean and b) from the cloud scales to the azimuthal mean for Hurricane Katrina of 2005. Units of $10^{-4}m^2s^{-3}$.

downscale manner. This is illustrated in Figure 3, on the example of an analysis of a simulation of hurricane Katrina, 2005. Here we see the vertically averaged exchanges between wavenumber zero and the principal asymmetries (Fig. 3a) and the vertically averaged exchanges between wavenumber zero and the cloud scales (Fig. 3b) as a radius-forecast time Hovmüller diagram. Outside of the storm's eye, which extends to about 30 km, both the asymmetry and the cloud scales draw energy from the mean flow as can be seen by the negative values in Figure 3. The rate of loss of kinetic energy of the mean flow is largest near the area of maximum winds, and at the time of peak storm intensity. Inside the eye, the asymmetries, and to a lesser degree the cloud scales, act as a source of kinetic energy to the mean flow.

4. Summary and Conclusions

The angular momentum of a parcel of air that arrives at the region of the maximum winds of a hurricane has generally been depleted along its path. We have noted that the cloud torques are the major contributor to this depletion. In a high resolution nonhydrostatic mesoscale model with explicit clouds and cloud microphysics it is possible to calculate these cloud torques directly. Over the regions of active convection the cloud torques are negative, implying a net divergence of flux of angular momentum along the path of large-scale parcel trajectories. If ΔM is the diminution of angular momentum M for an inflowing parcel between radii r_1 and r_2, we can write

$$\Delta M = M_1 - M_2 = V_{\theta_1} r_1 + \frac{f r_1^2}{2} - V_{\theta_2} r_2 - \frac{f r_2^2}{2},$$

where V_{θ_1} and V_{θ_2} are the tangential velocities at r_1 and r_2 respectively. If the outer tangential velocity V_{θ_1} is known then the inner tangential velocity V_{θ_2} can be explicitly determined once ΔM is known. Since the angular momentum and the tangential velocity of a parcel relate as $V_\theta = M/r - fr^2/2$, a depletion of angular momentum of an inflowing parcel implies a reduced contribution to the tangential wind (and thus, intensity) of a hurricane.

The hurricane intensity is thus impacted by the outer angular momentum and by the inner forcings of the clouds and the related cloud torques. In the modeling of cloud microphysics, cloud growth is sensitive to a number of parameters, PATTNAIK and KRISHNAMURTI (2007a,b). Future work must address the sensitivity of the cloud torques and the hurricane intensity to the microphysics of the model.

The scale interactions approach reveals that the kinetic energy of hurricane scales, and thus hurricane intensity, is dictated by the net effect of generation of kinetic energy due to vertical overturning and the wave-mean exchanges of kinetic energy. The vertical overturning constitutes conversion of potential to kinetic energy (see e.g., WIIN-NIELSEN and CHEN, 1993). This in-scale process is the dominant contributor to the kinetic energy at the hurricane scales. The wave-mean kinetic energy exchanges can be separated into asymmetry-mean and cloudscale-mean interactions. The former are found to be larger than the latter by up to an order of magnitude. Generally, the wave-mean interactions act as a sink to the kinetic energy of the mean flow, with the exception of a) the inner radii, where the asymmetries tend to transfer energy towards the mean, and b) the outer radii of a storm in its formative stages, where the organization of convection leads to transfer of energy from the cloud scales to the scale of the mean. Future work on scale interactions requires an observational thrust. The US fleet of reconnaissance aircraft flying into hurricanes does provide 3 to 4 vertical levels of data coverage from dropwindsonde and flight level winds. Such data sets are available as soon as a flight is completed. Several groups also carry single levels of data analysis from these. It is important to ask which of the salient components of the scale interactions can be quickly computed as these data become available. Data of such type would have great practical utility if some guidance on the short-term tendencies for the storm's maximum winds can be produced. Even a 6-hour guidance on the rapid intensification or weakening of a storm would be a very important contribution if it were possible to produce it via an analysis of the scale interactions.

Appendix

In storm-centered coordinate system the equations of motion are given by

$$\frac{\partial v_\theta}{\partial t} = -v_\theta \frac{\partial v_\theta}{r\partial\theta} - v_r \frac{\partial v_\theta}{\partial r} - \omega \frac{\partial v_\theta}{\partial p} - \frac{v_r v_\theta}{r} - f v_r - g \frac{\partial z}{r\partial\theta} - F_\theta, \tag{1}$$

$$\frac{\partial v_r}{\partial t} = -v_\theta \frac{\partial v_r}{r\partial\theta} - v_r \frac{\partial v_r}{\partial r} - \omega \frac{\partial v_r}{\partial p} - \frac{v_\theta^2}{r} + f v_\theta - g \frac{\partial z}{\partial r} - F_r, \tag{2}$$

and the continuity equation is

$$0 = -\frac{\partial v_\theta}{r\partial\theta} - \frac{\partial v_r}{\partial r} - \frac{v_r}{r} - \frac{\partial \omega}{\partial p}. \tag{3}$$

The independent variables in this coordinate system are the azimuthal angle θ, the radial distance from the center r, and the pressure p. The tangential and radial winds are v_θ (positive anticlockwise) and v_r (positive outward), respectively. The vertical velocity in pressure coordinates is ω, f is the Coriolis parameter, gz is the geopotential height, and F_θ and F_r are the tangential and radial components of the frictional force per unit mass.

A hydrostatic framework has been used for the analysis here, since the scales of interest are sufficiently large for this to be a reasonable approximation. The Fourier transform of any variable $q(\theta, r, p)$ along the azimuthal (θ) direction is given by:

$$q(\theta, r, p) = \sum_{n=-\infty}^{\infty} Q(n, r, p) e^{in\theta}, \tag{4}$$

where the complex Fourier coefficients $Q(n, r, p)$ are given by

$$Q(n, r, p) = \frac{1}{2\pi} \int_0^{2\pi} q(\theta, r, p) e^{-in\theta} \, d\theta. \tag{5}$$

The Fourier transform of the product of two functions q_1 and q_2 is given by

$$\frac{1}{2\pi} \int_0^{2\pi} q_1(\theta, r, p) q_2(\theta, r, p) e^{-in\theta} d\theta = \sum_{m=-\infty}^{\infty} Q_1(n, r, p) Q_2(n - m, r, p). \tag{6}$$

Using this, equations (1)–(3) can be shown to lead to the following equation for the kinetic energy change of wavenumber n due to nonlinear interactions with all other wavenumbers excluding 0:

$$\begin{aligned}
\frac{\partial K(n)}{\partial t} = \sum_{\substack{m=-\infty \\ m \neq 0}}^{\infty} & V_\theta(m) \left(\frac{1}{r} \Psi_{v_\theta \frac{\partial v_\theta}{\partial \theta}}(m, n) + \Psi_{v_r \frac{\partial v_\theta}{\partial r}}(m, n) + \Psi_{\omega \frac{\partial v_\theta}{\partial p}}(m, n) + \frac{1}{r} \Psi_{v_\theta v_r}(m, n) \right) \\
& + V_r(m) \left(\frac{1}{r} \Psi_{v_\theta \frac{\partial v_r}{\partial \theta}}(m, n) + \Psi_{v_r \frac{\partial v_r}{\partial r}}(m, n) + \Psi_{\omega \frac{\partial v_r}{\partial p}}(m, n) - \frac{1}{r} \Psi_{v_\theta v_\theta}(m, n) \right) \\
& - \frac{1}{r} \frac{\partial}{\partial r} r [V_\theta(m) \Psi_{v_r v_\theta}(m, n) + V_r(m) \Psi_{v_r v_r}(m, n)] \\
& - \frac{\partial}{\partial p} [V_\theta(m) \Psi_{\omega v_\theta}(m, n) + V_r(m) \Psi_{\omega v_r}(m, n)],
\end{aligned} \tag{7}$$

where $\Psi_{ab}(m, n) = A(n - m)B(-n) + A(-n - m)B(n)$.

The last term in (7) vanishes upon integration from top to bottom of the atmosphere, provided $\omega(p_{top}) = \omega(p_{bottom}) = 0$.

The rate of change of potential energy due to nonlinear interactions in a cylindrical coordinate system is calculated starting from the equation for the local change in temperature given by

$$\frac{\partial T}{\partial t} = -v_\theta \frac{\partial T}{r \partial \theta} - v_r \frac{\partial T}{\partial r} - \omega \frac{\partial T}{\partial p} + \frac{RT}{C_p p} \omega + \frac{Q}{C_p}. \tag{8}$$

Defining the available potential energy as $P(n) = C_p \eta |B(n)|^2$, where $\eta = -\frac{R^2}{C_p p^{R/C_p}} \left(\frac{\partial \Theta}{\partial p} \right)^{-1}$ is the static stability factor, one can obtain

Pure appl. geophys. 164 (2007) 1443–1463
0033–4553/07/091443–21
DOI 10.1007/s00024-007-0231-6

Pure and Applied Geophysics

On the Prediction of Tropical Cyclones over the Indian Region Using a Synthetic Vortex Scheme in a Mesoscale Model

S. Sandeep,[1] A. Chandrasekar,[1] and S. K. Dash[2]

Abstract—The tropical cyclones form over the oceanic regions where conventional meteorological observations are not available. This contributes to a poor initial analysis of the cyclonic vortex and hence inadequate forecast. One way of overcoming the above problem is to modify the initial analysis by replacing the weak and ill-defined vortex in the initial analysis with a synthetic vortex having the correct size and intensity at the correct location. In this study we are investigating the effect of inclusion of a synthetic vortex based on Rankine as well as on Holland wind profiles, using NCAR-AFWA bogussing scheme for the prediction of four tropical cyclones, which formed over the Bay of Bengal during November 2002 and 2005, December 2005 and over the Arabian Sea during May 2004, using the MM5 model. Two numerical experiments are designed in this study for each of the above four cyclones. In the first experiment the model is integrated with a synthetic vortex based on Rankine wind profile while in the second experiment we utilize the Holland wind profile. For the November 2002 cyclone, in both the experiments the model is integrated from 10 November 2002 18 UTC to 12 November, 2002 12 UTC with the synthetic vortex inserted at the initial time. The results of the study for the November 2002 cyclone show that the model simulation with the Holland vortex has produced a stronger cyclone in terms of minimum sea-level pressure and maximum wind speed. Also, the results for the November 2002 cyclone with the Holland vortex showed a better longitudinal height section of the horizontal wind speed across the center of the cyclone. The track error of the cyclone for the November 2002 cyclone is less in the model simulation with the Holland vortex at the initial time and at 24 hours of forecast. The results for the November 2002 cyclone with the Rankine vortex showed greater vertical wind speed as compared to the Holland vortex. However, for the November 2002 cyclone there were no significant differences in the spatial distribution of precipitation for both the experiments. In order to provide an adequate number of case studies for a good statistical sample, the present study is extended for three additional cyclones over the Indian region. All four cyclones studied here show that the Holland vortex has produced a stronger cyclone in terms of the minimum sea-level pressure and maximum wind speed. The Holland vortex showed a better vertical structure of wind speed in the longitudinal height section at 24 hours of forecast for the November 2005 cyclone while the structure was better for the Rankine vortex for the remaining two cyclones. There were no significant differences in the spatial distribution of precipitation for the two experiments corresponding to all four cyclones. Some statistical results pertaining to all four cyclones are provided such as the average track error as well as the average difference between the observed and the model minimum sea-level pressure and the maximum wind speed. The statistical results corresponding to the average of all the four cyclones are at only a slight variance with the results corresponding to the November 2002 cyclone.

Key words: Synthetic vortex, NCAR, AFWA, tropical cyclone, MM5, Holland, Rankine.

[1] Department of Physics and Meteorology, IIT Kharagpur, India. E-mail: chand@phy.iitkgp.ernet.in
[2] Centre for Atmospheric Sciences, IIT Delhi, India.

1. Introduction

The performance of a numerical model greatly depends on the initial conditions. It is obvious that inaccurate initial conditions will lead to a poor forecast by the model. The mesoscale models utilize three-dimensional analysis (which are usually model outputs obtained from a coarse resolution global model) for the initial and lateral boundary conditions. As the tropical cyclones form over data sparse oceanic areas, the initial analysis obtained from the global coarse resolution model may be inadequate and incorrect with an ill-defined center. This inadequate representation of a cyclonic vortex in the initial analysis will deteriorate the forecast of tropical cyclones by the mesoscale models. One way of overcoming the above is to replace the weak and diffused cyclonic vortex in the initial analysis with a synthetic vortex, which is closer to the observed cyclonic vortex in terms of size, intensity and location. Earlier studies with inclusion of a synthetic vortex (MATHUR, 1991; TRIVEDI *et al.*, 2002; SANDEEP *et al.*, 2006), have shown improved tropical cyclone prediction. Alternate methods of hurricane initialization by KURIHARA *et al.* (1993) and KRISHNAMURTI *et al.* (1993) are also available in the literature. Davis and Low-Nam (2001) have proposed the National Center for Atmospheric Research (NCAR) – Air Force Weather Agency (AFWA) synthetic vortex scheme which is now available as part of the Pennsylvania State University (PSU) – NCAR fifth generation mesoscale model (MM5) (GRELL *et al.*, 1994). TRIVEDI *et al.* (2002) studied a super-cyclone formed over the Bay of Bengal during October 1999 using the National Meteorological Center's vortex initialization scheme (MATHUR, 1991) and found that the initial synthetic vortex greatly effected the prediction of the track and intensity of the cyclone. MOHANTY *et al.* (2004) investigated the above-mentioned cyclone using the MM5 model. In order to obtain small-scale features associated with the cyclone, MOHANTY *et al.* (2004) integrated the MM5 model for 12 hours with analysis nudging during the pre-forecast period and found that the simulated track was in reasonable agreement with the observed track. SANDEEP *et al.* (2006) studied the effect of inclusion of a synthetic vortex on the prediction of a Bay of Bengal tropical cyclone and found improved prediction in the structure and track of the cyclone. ROY ABRAHAM *et al.* (1995) studied the effect of the Rankine and the Holland wind profiles on two tropical cyclones over India and found that the simulated cyclone moved faster when the model was initialized with the Holland wind profile.

The NCAR-AFWA tropical cyclone synthetic vortex scheme has two main components (DAVIS and LOW-NAM, 2001). The first component detects and extracts the tropical cyclone from the first guess (analysis) field while the second component calculates the synthetic vortex and incorporates the same with the modified background field to form the final analysis. Identification of the center of the vortex is the first step in the removal of the initial weak vortex, and this is done by searching for the maximum vorticity in the analysis pressure level nearest the surface within a prescribed radial distance from the observed cyclonic center. DAVIS and LOW-NAM

(2001) modified the vorticity, geostrophic vorticity and divergence fields and calculated the change in the non-divergent stream function, geopotential and velocity potential and using the above computed a modified velocity field. The main features of NCAR-AFWA synthetic vortex scheme are axi-symmetry, fixed radius of maximum wind, mass and wind fields in nonlinear balance, nearly saturated (with respect to water or ice), no eye, and maximum winds of the synthetic vortex as a predetermined fraction of the observed maximum winds. This scheme assumes Rankine specification of wind distribution in the tropical cyclones as follows

$$V_m = A(z)F(r);$$
$$F(r) = (V_m/r_m)r; \ (r \le r_m),$$
$$F(r) = (V_m/{r_m}^{\alpha})r^{\alpha}; \ (r > r_m),$$

where $F(r)$ is the radial dependence and $A(z)$ is the amplitude and vertical dependence, r_m and v_m are the radii of maximum wind and the maximum wind at r_m and α is a parameter to yield an approximately correct functional relationship near the storm and reduce the influence of the storm at large radii. The vertical weight function $A(z)$ is assigned a value of unity from the surface through 850 hPa, 0.95 at 700 hPa, 0.9 at 500 hPa, 0.7 at 300 hPa, 0.6 at 200 hPa and 0.1 at 100 hPa. This scheme assumes zero vorticity outside the radius of maximum wind in order to define the nondivergent wind associated with the first guess storm. HOLLAND (1980) proposed an analytic model for wind and pressure profiles in tropical cyclones as follows.

$$r^B \ln[(P_n - P_c)/(P - P_c)] = A,$$

where P is the pressure at the radius r, P_n is the ambient pressure, P_c is the central pressure, and A and B are scaling parameters.

On taking antilogarithms, one derives

$$P = P_c + (P_n - P_c)\exp(-A/r^B).$$

Hence using the gradient wind equations, the wind profile is

$$V_g = \left[AB(P_n - P_c)\exp(-A/r^B)/\rho r^B + r^2 f^2/4\right]^{1/2} - rf/2,$$

where f is the Coriolis parameter and ρ is the density of the air.

The chief objective of this work is to investigate the relative merits of utilizing Holland and Rankine wind profiles together with the NCAR-AFWA synthetic vortex scheme on the structure, precipitation and track of a tropical cyclone, which formed over the Bay of Bengal during 10–12 November, 2002 using the MM5 model. Section 2 describes the synoptic features of the four tropical cyclones while Section 3 outlines the model and design of numerical experiments. The results and discussions are presented in Section 4 and the conclusions in Section 5.

2. Synoptic Features of the Tropical Cyclones

2.1. November 2002 Cyclone over the Bay of Bengal

The tropical cyclone which is being investigated in this study was first identified as a depression centered about 235 km east southeast of Chennai, India on 10 November, 2002 06 UTC. The depression moved steadily northeastward and intensified into a tropical cyclone at 12 UTC of 11 November and was positioned at 750 km south southwest of Kolkata, India. The intensity of the tropical cyclone did not change until 12 November 06 UTC after which it accelerated in the northeast direction and was located at 195 km south southwest of Kolkata with a maximum wind speed of 28 m s^{-1} and a minimum central pressure of 988 hPa. The above system crossed Sagar Island at 09 UTC on 12 November, 2002 and subsequently weakened into a depression by 12 UTC. About 50 casualties and severe property damage were caused by the above tropical cyclone.

2.2. November 2005 Cyclone over the Bay of Bengal

On 26 November, 2005 an area of persistent convection had developed over the Bay of Bengal, approximately 555 km northwest of Penang, Malaysia. A tropical cyclone formation alert was issued at 12 UTC on the same day and the first cyclone warning was issued at 18 UTC. The system gradually moved westward and also intensified reaching its maximum intensity at 18 UTC on 28 November, 2005 and was located at 787 km east southeast of Chennai, India. The system slowed down and its motion became erratic as it approached the Indian coastline on 30 November. There were no reports of loss of life.

2.3. December 2005 cyclone over the Bay of Bengal

On 4 December, 2005 an area of persistent convection was observed over the Bay of Bengal, approximately 1574 km, east southeast of Chennai, India. A tropical cyclone formation alert was issued at 1430 UTC on 5 December and the first cyclone warning was issued at 00 UTC on 6 December when the system was located 1018 km east of Chennai. The system intensified to a deep depression on 6 December at 09 UTC and intensified to a cyclonic storm at 03 UTC the next day. The cyclone intensified over the next 2 days and crossed land near Vedaranyam at 03 UTC on 10 December.

2.4. May 2004 Cyclone over the Arabian Sea

By 0400 UTC on 4 May, an area of convection lay overland approximately 500 km west of Chennai, India. At the same time a new area of convection had developed in the Arabian Sea, roughly 277 km northwest of Cochin, India. A tropical cyclone formation alert was issued at 13 UTC when the system was located

426 km west-northwest of Cochin and the first cyclone warning was issued at 00 UTC on 5 May when the system was experiencing winds of 15 m s^{-1} and was located at 370 km west-northwest of Cochin. The cyclone wandered aimlessly for the next three days and on 9 May 00 UTC started weakening over the sea itself. There were no reports of loss of life.

3. Model Description and Numerical Experiments

The present study utilized the MM5 version 3.6 (GRELL *et al.*, 1994). The model was configured with twenty-three vertical layers (centered at σ = 0.995, 0.985, 0.97, 0.945, 0.91, 0.87, 0.825, 0.775, 0.725, 0.675, 0.625, 0.575, 0.525, 0.475, 0.425, 0.375, 0.325, 0.275, 0.225, 0.175, 0.125, 0.075, 0.025) and two nested domains (Outer Domain: 90 km grid spacing with 85 $\times$ 75 grid cells in east-west and north-south directions; Inner Domain: 30 km grid spacing with 130 $\times$ 118 grid cells in the east-west and north-south directions). The physics options used are the MRF PBL scheme, the Grell cumulus scheme, a mixed phase Reisner scheme for explicit moisture, a cloud radiation scheme and a multi-level soil model. The NCEP reanalysis data available at a horizontal resolution of 2.5° $\times$ 2.5° and a time resolution of 6 hours were used to develop the initial and lateral boundary conditions. A one-way nesting option was employed. Two numerical experiments were designed to study the impact of the inclusion of synthetic vortices based on Rankine and Holland wind profiles on the prediction of the track of the tropical cyclone for the November 2002 cyclone. The first experiment (called Rankine) utilized the NCEP reanalysis for the initial and lateral boundary conditions and the model integrations were performed from 10 November, 2002 18 UTC to 12 November, 2002 12 UTC with a synthetic vortex (based on Rankine wind profile) inserted in the initial analysis at 18 UTC of 10 November, 2002. This experiment used the following values for the parameters $\alpha = -0.75, r_m = 90\,\text{km}, v_m = 25\,\text{m}\,\text{s}^{-1}$, vortex size = 300 km and search radius = 400 km. The second experiment (called Holland) utilized a synthetic vortex based on the Holland wind profile at 18 UTC of 10 November, 2002 and the model integrations for the Holland run were the same as for the Rankine run. For the Holland run, while the parameter B was assigned a value of 1.11, all the other parameter values, such as r_m, v_m, vortex size, and search radius remained unchanged. The results of the MM5 simulation corresponding to the two numerical experiments were then compared with NCEP reanalysis as well as with observations for the time period between 10 November, 2002 18 UTC to 12 November, 2002 12 UTC. Similar additional experiments were performed for three additional cyclones (5 May, 2004 00UTC to 9 May, 2004 00UTC for the May 2004 cyclone, 28 November, 2005 18 UTC to 1 December, 2005 18UTC for the November 2005 cyclone and 7 December, 2005 00UTC to 10 December, 2005 00UTC for the December 2005 cyclone). In addition to the changes in the initial location of the

Roy Abraham *et al.* (1995). Figure 1d shows that the MWS simulated by the Holland run is closer (after one day forecast) with the observations (28 m s^{-1}) while the corresponding MWS for the Rankine simulation has a lower value of 25 m s^{-1}. Roy Abraham *et al.* (1995) found that the minimum 24 hour SLP simulated by the Holland vortex was lower by 4 hPa while in our study the maximum difference of lowest SLP is 3 hPa. Figure 2 depicts the longitude-height cross section of the horizontal wind speed across the center of the cyclone. Although both Rankine and Holland runs could simulate a well-defined cyclonic structure, it can be seen that at 18 hours of forecast, when the system had already attained cyclonic intensity, and beyond, the Holland run simulated a system with a larger horizontal extent. Also, Figure 2 shows that the Holland run at 30 hours of forecast has simulated a stronger cyclonic system as compared to the Rankine run. However, the Rankine vortex at the end of 42 hours of forecast reveals that the cyclone had weakened considerably with a loss of vertical structure to the west of the cyclone center, consistent with the observation. The Holland run however shows no loss of vertical structure at the end of 42 hours of forecast, at variance with the observations. Figure 3 depicts the longitude-height cross section of vertical velocity and potential temperature close to the center of the cyclone. It is seen from Figure 3 that the Rankine run simulated a stronger vertical velocity at and after 18 hours of forecast as compared to the Holland run. Figure 4 depicts the sea-level pressure patterns for both runs. It is clear that the Holland simulation produced a stronger cyclone in terms of lower central sea-level pressure values. Also, the Holland run shows that the cyclone crossed the land at 12 November, 2002 12 UTC. This compares more closely with the observed time at landfall, which is 12 November, 2002 09 UTC. Figure 4 at the end of 42 hours of forecast for the Rankine run shows that the cyclone has only partially crossed land. Figure 5 depicts the lower tropospheric wind pattern simulated by the Rankine and Holland runs. It is seen from Figure 5 that both runs simulated more or less the same cyclonic wind magnitudes. Figure 6a shows the 24-hour accumulated precipitation obtained from Tropical Rainfall Measuring Mission (TRMM) at 24 hours of forecast. Figures 6b and 6c depict the corresponding 24-hour accumulated precipitation produced by Rankine and Holland simulations, respectively at the same time. The comparison of simulated precipitations of both runs with the TRMM precipitation indicate minor differences between the model simulated spatial precipitation pattern except that the meridional extent of the precipitation is more northward for the Rankine run. Roy Abraham *et al.* (1995) also obtained a more northward shift of the rainfall pattern in the Rankine run as compared to the Holland run. Roy Abraham *et al.* (1995) had obtained more rainfall amounts in the Rankine run as compared to the Holland run. However, the results of this study do not show appreciable differences in the rainfall amounts between the two runs. Figure 7 depicts the model predicted track due to the Rankine and Holland runs together with the observed track as well as the NCEP reanalysis track. It is seen that the Holland run cyclone track is closer to the observed track at the initial time and at

24 hours of forecast. The Rankine run track is closer to the observed track at other times. Table 1 provides the vector displacement of track errors (with respect to observed track) in kilometers for both the Rankine and the Holland runs for the

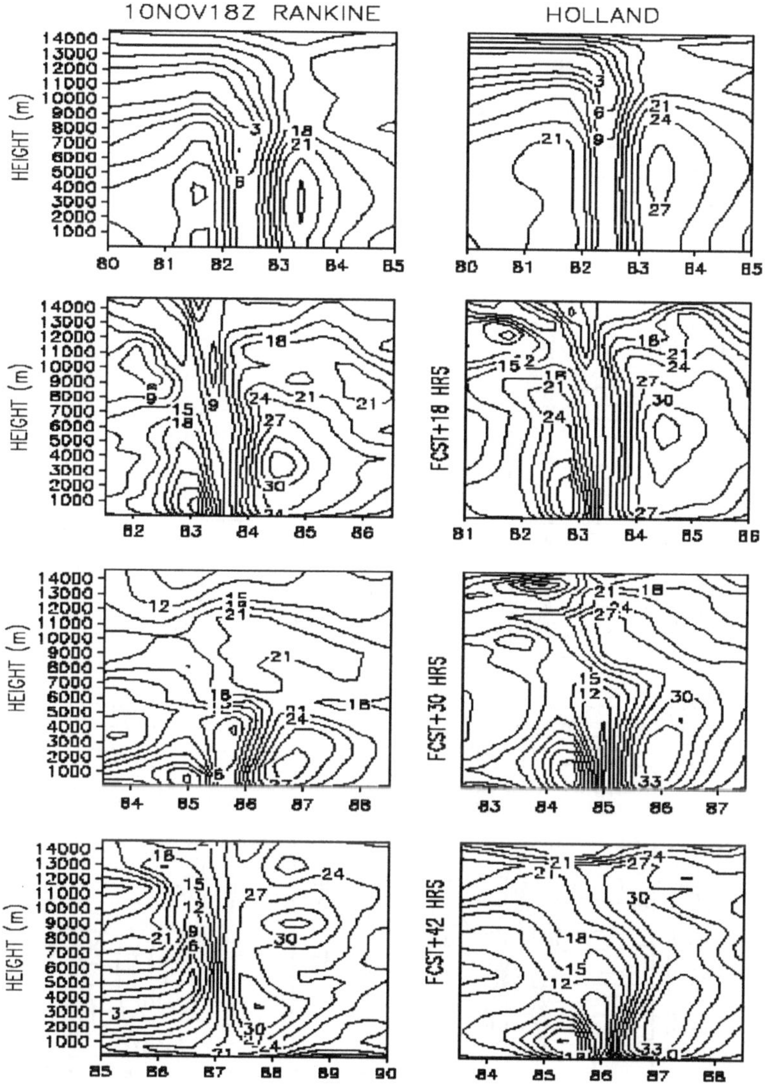

Figure 2
Longitude-height cross section of horizontal wind speed (m s^{-1}) across the center for the November 2002 cyclone.

November 2002 cyclone. Also, Table 1 provides the difference between the observed and the model minimum SLP and the MWS for the November 2002 cyclone. The results of the Rankine run yield reduced track errors for most of the forecast.

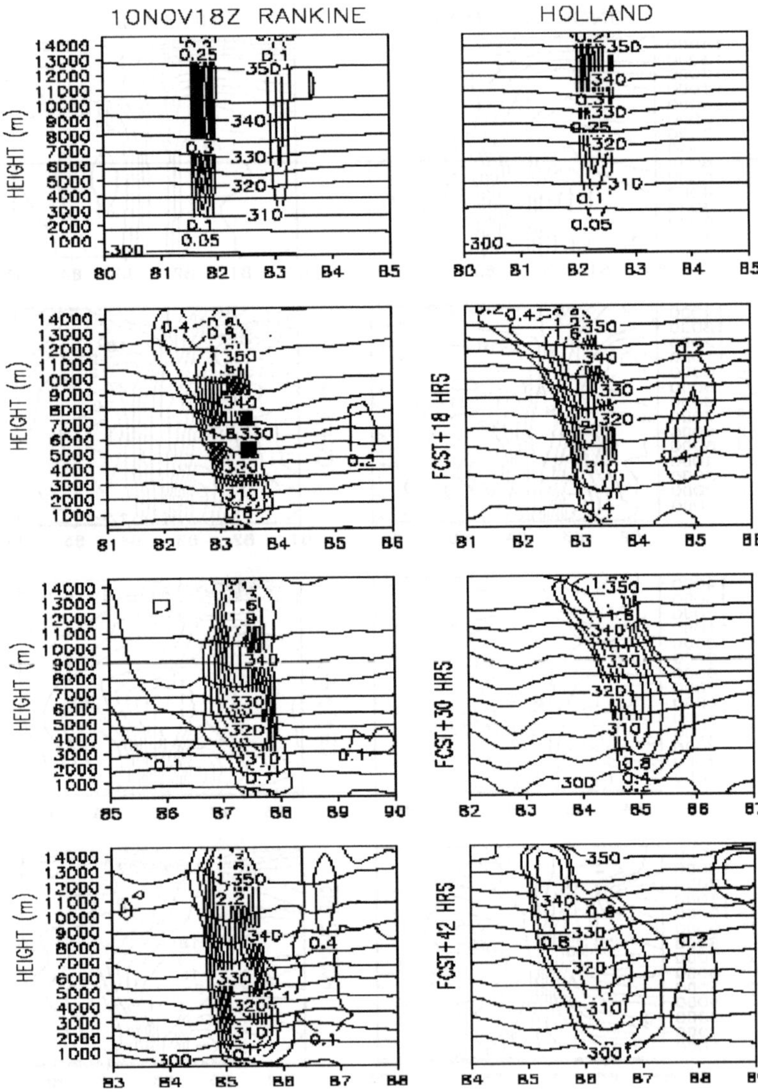

Figure 3
Longitude-height cross section of vertical velocity (m s^{-1}) and potential temperature (K) near the center for the November 2002 cyclone.

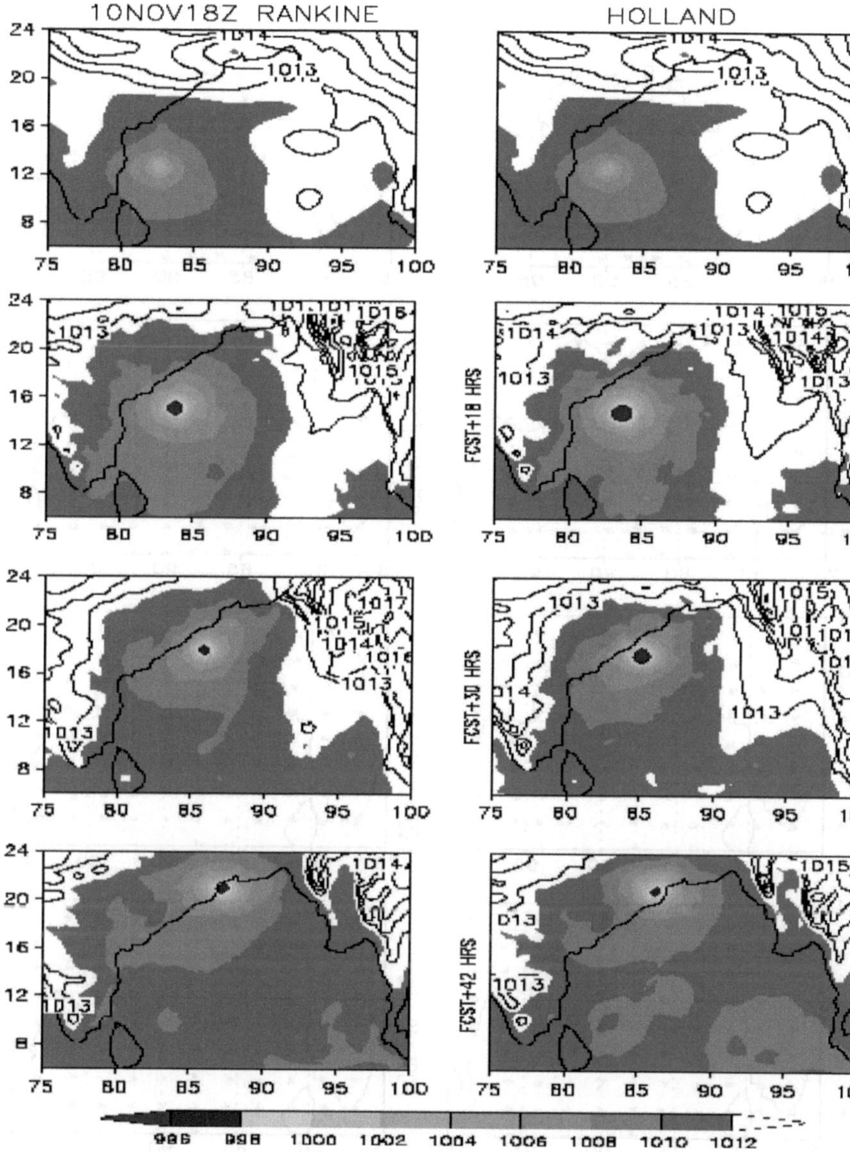

Figure 4
Sea-level pressure pattern (hPa) for the November 2002 cyclone.

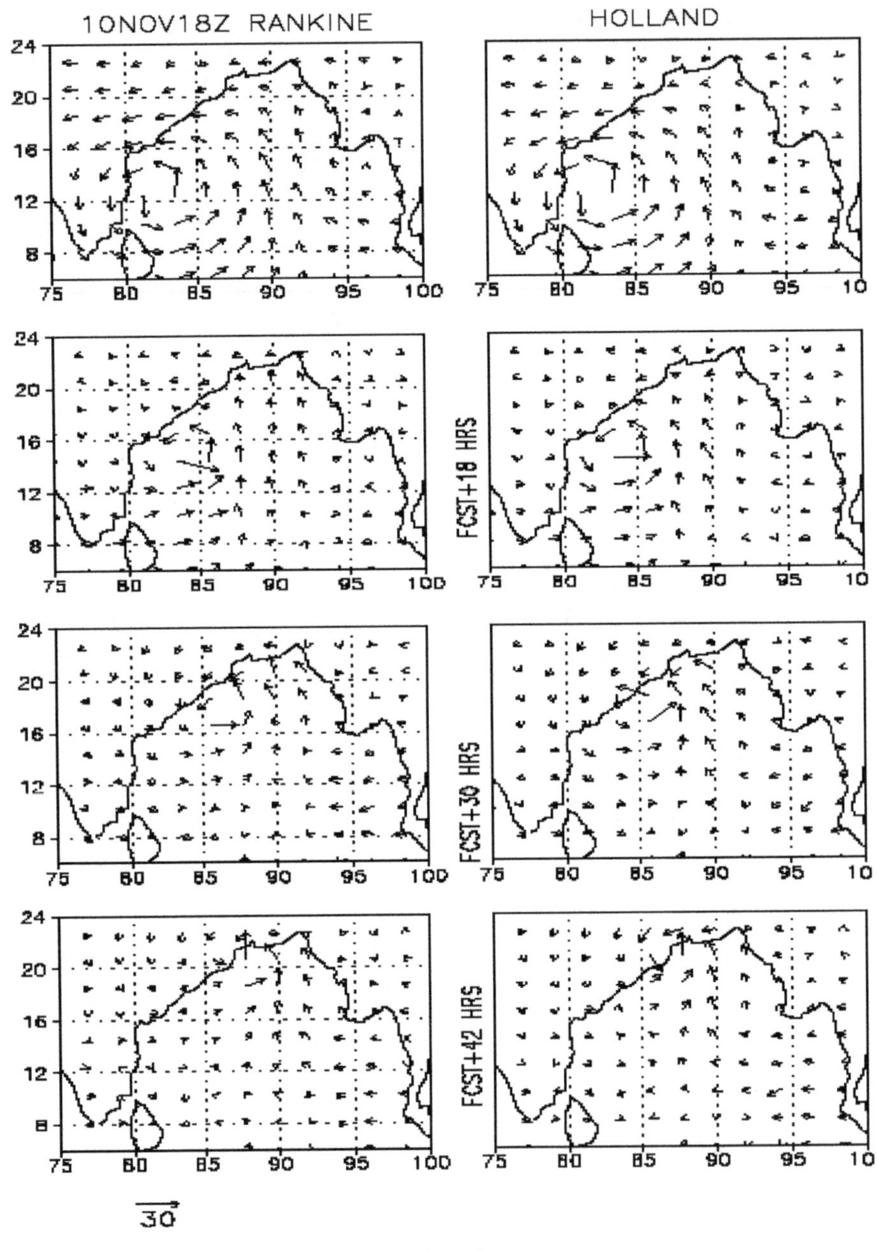

Figure 5
Lower tropospheric winds (m s^{-1}) at ó = 0.995 for the November 2002 cyclone.

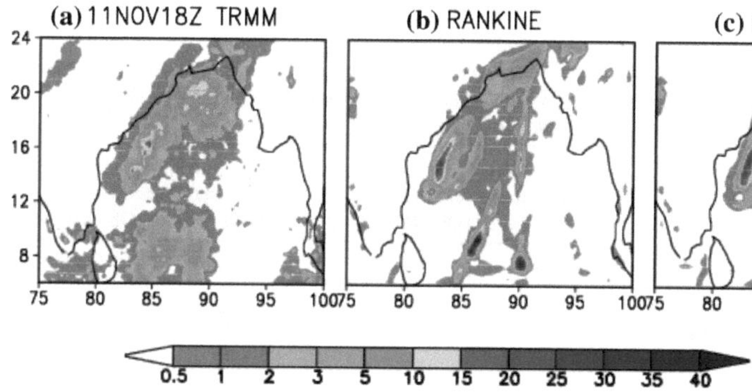

Figure 6
(a–c) 24-hour accumulated precipitation (in cm) at 11 November 18 UTC, 2002 from TRMM, Rankine and Holland runs.

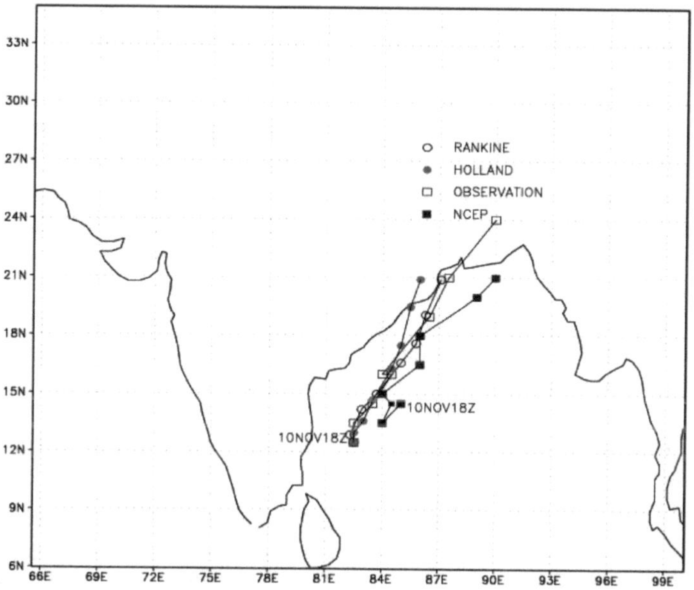

Figure 7
Track of the November 2000 tropical cyclone.

ROY ABRAHAM *et al.* (1995) had obtained faster movement of the cyclone in the Holland run. Both runs appear to lag behind the observed track. The difference of the model-simulated minimum SLP with respect to the observed is less for the

2002 cyclone. Figures 8c–8d, 9c–9d and 10c–10d provide the longitudinal height section of the horizontal wind speed across the center of all three cyclones corresponding to 24 hours of forecast for both the Rankine and the Holland profiles. The longitudinal height section of the wind speed across the center of the cyclone for the November 2005 cyclone was consistent with the results of the November 2002 cyclone. However, for the other two cyclones (December 2005 and May 2004) the results of the longitudinal height section of the wind speed across the center of the cyclone had a better structure for the Rankine profile. Figures 8e-–8f, 9e–9f and 10e–10f depict the mean SLP at 24 hours of forecast for all three cyclones corresponding to the Rankine and Holland profiles. The results of the three additional cyclones in terms of mean SLP are consistent with the results of the November 2002 cyclone in that the Holland profile simulates a stronger vortex as compared to the Rankine profile. Figures 8g–8h, 9g–9h and 10g–10h depict the lower tropospheric wind ($\sigma = 0.995$) as well as the 24-hour accumulated precipitation corresponding to the 24 hours of forecast for all three cyclones corresponding to the Rankine and the Holland profiles. Figures 8i, 9i and 10i show the 24-hour accumulated precipitation from TRMM observations corresponding to the 24 hours of forecast. The results of the three additional cyclones show slight difference in the spatial distribution of simulated rainfall corresponding to the Rankine and the Holland profiles, a result which is consistent with the November 2002 cyclone. The track errors (not shown for brevity) for the three additional cyclones show lower errors for the Holland profile during the first 24 hours of forecast, while the track errors are less for the Rankine case at later forecast times. Table 2 provides the average track displacement error for all four tropical cyclones. Also, Table 2 provides the average differences between the observed and the model minimum SLP and the MWS for all four cyclones. In effect, Table 2 provides the average picture of four tropical cyclones as compared to the results of Table 1, which provides the individual picture corresponding to a single cyclone. The operation of averaging is bound to reduce the magnitudes (in an absolute sense) and it is thus not surprising to find that the average track error as well as the averaged differences in the minimum SLP and MWS in Table 2 have lower absolute magnitudes compared with the corresponding values in Table 1. The average track error is found to be lower for the Rankine profile for most times as compared with the Holland profile, which is consistent with the results shown for the November 2002 cyclone. The averaged differences between the observed and the model simulated minimum SLP also show that the differences (in an absolute sense) are lower for the Rankine case until 24 hours of forecast, while they are higher after the first day of forecast. This is quite similar to the individual picture seen in Table 1 corresponding to the November 2002 cyclone, except that in the latter the differences in the Rankine case were lower until 30 hours of forecast. The averaged differences between the observed and the model simulated MWS show that the differences (in an absolute sense) are lower for the Rankine case for all

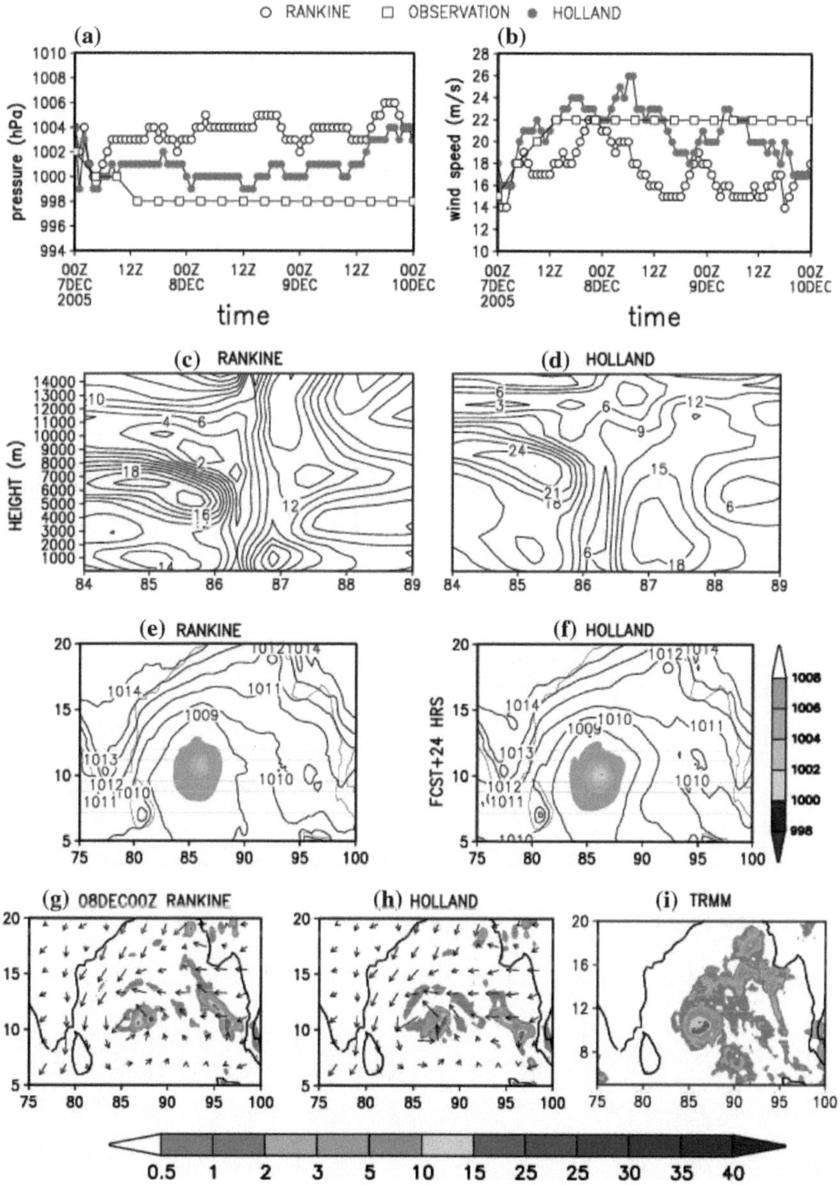

Figure 9
(a–i) Similar to Figure 8 except for the December 2005 cyclone.

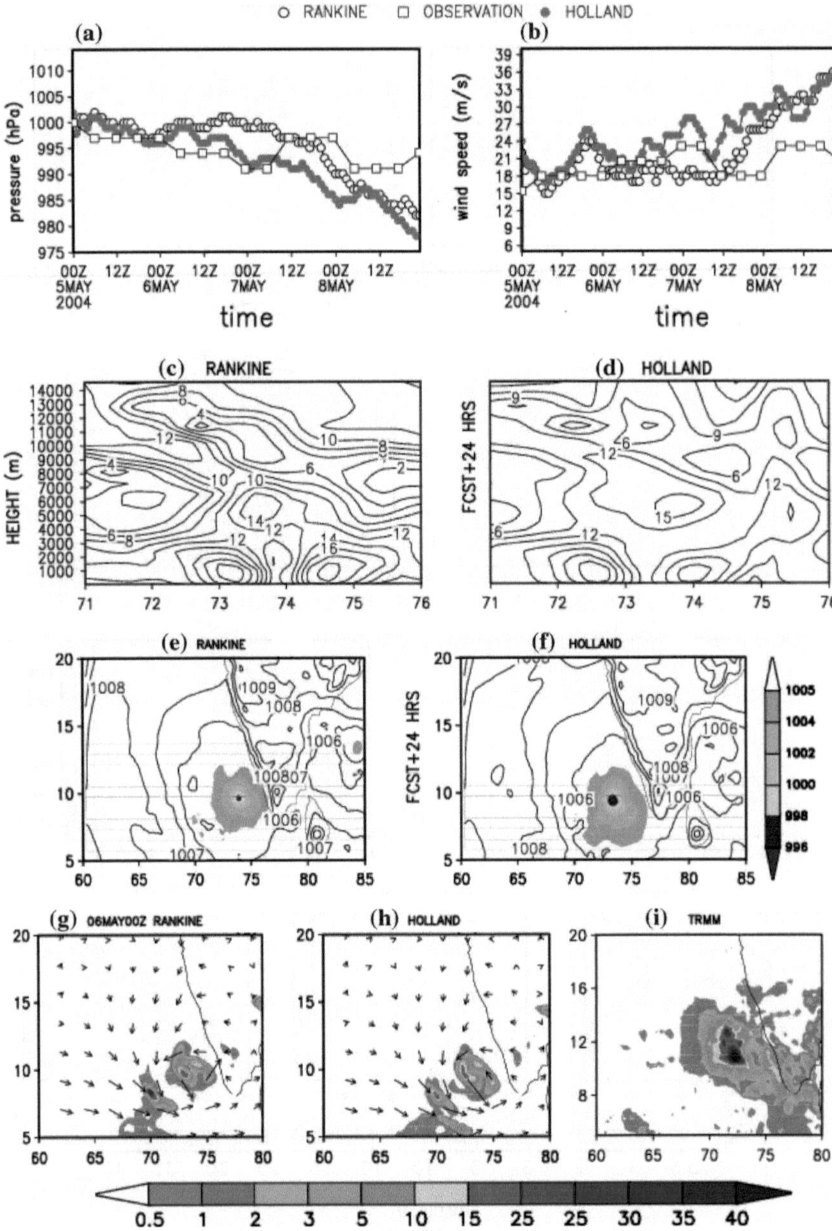

Figure 10
(a–i) Similar to Figure 8 except for the May 2004 cyclone.

Table 2

Average vector displacement of track errors with respect to observed track (km) and difference between the observed and the model minimum sea-level pressure (hPa) and maximum wind speed (m s^{-1}) *for both Rankine and Holland profiles for four cyclones*

Time of forecast (HRS)	Average track error (km) for four cyclones		Average difference in minimum sea-level pressure (hPa) (Obs-Model) for four cyclones		Average difference in maximum wind speed (m s^{-1}) (Obs-Model) for four cyclones	
	Rankine	Holland	Rankine	Holland	Rankine	Holland
06	77	66	−0.75	0.0	−2.75	−4.5
12	110	123	−0.25	2.0	−4.25	−6.5
18	138	150	−1.0	1.0	−4.50	−7.5
24	123	113	−1.25	2.0	−4.75	−5.75
30	111	169	−4.50	−1.25	−1.0	−3.50
36	143	193	−6.0	−2.25	0.25	−4.50
42	236	279	−7.0	−3.0	0.50	−1.50

times. This result is somewhat different from the individual picture seen in Table 1 corresponding to the November 2002 cyclone, where the differences between the two profiles were not very different.

5. *Conclusions*

This study investigated the impact of inclusion of synthetic vortices based on Rankine and Holland wind profiles using NCAR-AFWA synthetic vortex scheme on the prediction of a few tropical cyclones, which formed over the Indian region. The results of the study show that model run with the Holland vortex could simulate a stronger system in terms of minimum SLP, and MWS. Also, at and after 30 hours of forecast, the Holland run reflected a stronger vertical structure in the horizontal wind speed. The cyclone simulated by the Holland run moved faster (except at and near 24 hours of forecast) when compared to the Rankine run. The above result is in agreement with the results obtained by ROY ABRAHAM *et al.* (1995). Both the runs could simulate the structure of the precipitation pattern normally associated with cyclones for the November 2002 cyclone. The Rankine run shows greater meridional extent of the precipitation pattern as compared to the Holland run. However, the rainfall amounts were not appreciably different between the two runs for all four cyclones. Earlier results in the literature (ROY ABRAHAM *et al.*, 1995) seem to agree with some of the above results especially with respect to the minimum SLP, MWS, and larger meridional extent of the rainfall pattern. However, the earlier study showed less rainfall amounts and faster movement in the Holland run as compared to the Rankine run. Overall, the results of this study seem to agree broadly with the earlier results related to the differences between the simulations using both Holland and Rankine wind profiles. In order to provide an adequate number of statistical samples, the present study was extended to include three additional cyclones, which formed over the Indian region. The average track error as well as the averaged differences between the observed and the simulated minimum SLP corresponding to the four cyclones had a similar behavior with the results of the November 2002 cyclone, except that the magnitudes in the averaged results were lower in absolute terms.

Acknowledgements

MM5 model, NCAR-AFWA synthetic vortex scheme and NCEP reanalysis data were obtained from NCAR and NCEP, respectively. IMD provided observed data for all cyclones over the Indian region. TRMM data were obtained from NASA. This work is supported under a research project funded by the Department of Science and Technology (DST), India, and two of the authors (AC and SKD) thank DST for its support. S. Sandeep acknowledges the Council of Scientific and Industrial Research (CSIR), India for providing a research fellowship to carry out this work.

References

Davis, C.A. and Low-Nam, S. (2001), *The NCAR-AFWA tropical cyclone bogussing scheme*, Air Force Weather Agency Report, NCAR Boulder, CO., 13 p.

Grell, G.A., Dudhia, J., and Stauffer, D.R. (1994), *A description of the fifth-generation Penn State/NCAR mesoscale model (MM5)*, NCAR Technical Note TN-398+STR National Center for Atmospheric Research Boulder, CO., 122 p.

Holland, G. J. (1980), *An analytic model of wind and pressure profiles in hurricanes*, Mon. Wea. Rev. *108*, 1212–1218.

Krishnamurti, T.N., Bedi, H.S., and Ingles, K. (1993), *Physical initialization using SSM/I rain rates*, Tellus *45A*, 247–269.

Kurihara, Y., Bender, M.A., and Ross, R.J. (1993), *An initialization scheme of hurricane models by vortex specification*, Mon. Wea. Rev. *121*, 2030–2045.

Liu, Y., Zhang, D.L., and Yau, M.K. (1997), *A multi-scale numerical simulation of hurricane Andrew (1992). Part I. Explicit simulation and verification*, Mon. Wea. Rev. *125*, 3073–3093.

Mathur, M.B. (1991), *The National Meteorological Center's quasi-Lagrangian model for hurricane prediction*, Mon. Wea. Rev. *109*, 1419–1447.

Mishra, D.K. and Gupta, G.R. (1976), *Estimation of maximum wind speeds in tropical cyclones occuring in India seas, Indian Journal of Meteorology*, Hydrolosy and Geophysics, *27*, 285–290.

Mohanty, U.C., Mandal, M., and Raman, S. (2004), *Simulation of Orissa Super-Cyclone (1999) using PSU/NCAR mesoscale model*, Natural Hazards *31*, 373–390.

Roy Abraham, K., Mohanty, U.C., and Dash, S.K. (1995), *Simulation of cyclones using synthetic data*, Proc. Ind. Acad. Sci. (Earth and Planetary Sciences). *104*, 635–666.

SANDEEP, S., CHANDRASEKAR, A., and DASH, S.K. (2006), *Effect of inclusion of a synthetic vortex on the prediction of a tropical cyclone over the Bay of Bengal using a mesoscale model*, Int. J. Ecol. Develop. *4*, 35–51.

TRIVEDI, D.K., SANJAY, J., and SINGH, S.S. (2002), *Numerical simulation of a super-cyclone storm, Orissa 1999: Impact of initial conditions*, Met. Appl. *9*, 367–376.

(Received February 9, 2006, accepted September 11, 2006)
Published Online First: June 23, 2007

Pure appl. geophys. 164 (2007) 1465–1487
0033–4553/07/091465–23
DOI 10.1007/s00024-007-0228-1

Pure and Applied Geophysics

Numerical Simulation of Andhra Severe Cyclone (2003): Model Sensitivity to the Boundary Layer and Convection Parameterization

C. V. Srinivas,[1] R. Venkatesan,[1] D. V. Bhaskar Rao,[2] and D. Hari Prasad[2]

Abstract—The Andhra severe cyclonic storm (2003) is simulated to study its evolution, structure, intensity and movement using the Penn State/NCAR non-hydrostatic mesoscale atmospheric model MM5. The model is used with three interactive nested domains at 81, 27 and 9 km resolutions covering the Bay of Bengal and adjoining Indian Peninsula. The performance of the Planetary Boundary Layer (PBL) and convective parameterization on the simulated features of the cyclone is studied by conducting sensitivity experiments. Results indicate that while the boundary layer processes play a significant role in determining both the intensity and movement, the convective processes especially control the movement of the model storm. The Mellor-Yamada scheme is found to yield the most intensive cyclone. While the combination of Mellor-Yamada (MY) PBL and Kain-Fritsch 2 (KF2) convection schemes gives the most intensive storm, the MRF PBL with KF2 convection scheme produces the best simulation in terms of intensity and track. Results of the simulation with the combination of MRF scheme for PBL and KF2 for convection show the evolution and major features of a mature tropical storm. The model has very nearly simulated the intensity of the storm though slightly overpredicted. Simulated core vertical temperature structure, winds at different heights, vertical winds in and around the core, vorticity and divergence fields at the lower and upper levels—all support the characteristics of a mature storm. The model storm has moved towards the west of the observed track during the development phase although the location of the storm in the initial and final phases agreed with the observations. The simulated rainfall distribution associated with the storm agreed reasonably with observations.

Key words: Tropical cyclone, simulation, physical processes, intensity, track.

1. Introduction

Tropical Cyclones (TC) cause enormous damage to the life and property at the place of their landfall in the tropical coastal regions. Heavy winds, torrential rains and storm surge cause the damage due to the cyclones. On an average about five tropical cyclones occur annually over the Bay of Bengal (Bhaskar Rao *et al.*, 2001)

[1] Radiological Safety Division, Safety Group, Indira Gandhi Centre for Atomic Research, Kalpakkam, India.
[2] Department of Meteorology & Oceanography, Andhra University, Visakhapatnam, India.
E-mail: cvsri@igcar.gov.in

in the North Indian Ocean which contributes to 6 percent of the global annual frequency (RAGHAVAN and SEN SARMA, 2000). Based on the maximum sustained winds associated with the system and geographical region of their occurrence, they are classified into a depression, a tropical storm, severe cyclone or hurricane (ASNANI, 1983). Tropical cyclones form over warm oceans which supply energy to the atmosphere in the form of latent heat and sensible heat. Under favorable atmospheric thermo-dynamical conditions associated with low-level convergence a surface low develops into a cyclonic storm. The movement of the tropical cyclones is generally known from the knowledge of the upper atmospheric conditions and the prevailing circulations. However, prediction of the location of the landfall and intensity a few days in advance would be highly desirable for planning and implementation of the mitigation measures effectively.

Limited observations are available over oceans to explore the active region of cyclone development and limit their prediction. Observations from satellites provide useful information about their location, intensity and movement (DVORAK, 1975). Synoptic methods and satellite remote-sensing techniques are usually applied to forecast the storm movement. Recent advances in numerical modeling and availability of computer resources have led to the application of numerical models. The synoptic forecasting methods have limitations of subjectivity whereas simulations from dynamical models are dependent on representative initial/boundary conditions and incorporation of necessary physics. Because of recent advances in numerical modeling with improved physics, resolution and availability of high spatio-temporal resolution observations, hydro-dynamical models are being used to understand the structure and predict the tropical cyclones.

The early modeling studies of tropical cyclones were made with axi symmetric 2-D models to understand the physical and dynamical mechanisms and the energetics of tropical cyclone development (ROSENTHAL, 1971; ANTHES, 1977; YAMASAKI, 1968; BHASKAR RAO and ASHOK, 1999 among several others). These studies helped to investigate the role of boundary layer and convection in the tropical storm development. In order to study the asymmetric effects, interaction with the environment and to resolve the different scales of motion associated with the tropical storms, nested 3-D models are used. Three-dimensional models are employed to issue more realistic operational forecasts of tropical storms by many national weather forecasting centres (MATHUR, 1991; CHEN et al., 1995; KURIHARA et al., 1993, 1995). The India Meteorological Department uses high-resolution limited area models for forecasting tropical cyclones over the Indian Ocean region with assimilation of synthetic observations (PRASAD et al., 1997; PRASAD and RAMA RAO, 2003).

Application of numerical models for tropical cyclone prediction necessitates a study of the suitable physics. The chief physical process for the intensification of low pressure into a cyclonic storm is known to be the conditional instability of second kind (CISK) mechanism and the cooperative interaction between the cloud scale and synoptic scale circulations. The exchange of energy at the ocean-atmosphere interface and its supply through the planetary boundary layer to the free atmosphere play an important role in the intensification of the system. Thus the PBL and convection processes, which play an important role in the development of tropical storms, need to be carefully represented in the numerical models for realistic predictions.

The community model NCAR MM5 has been widely used for tropical cyclone simulation and sensitivity studies. LIU *et al.* (1997) reported an improvement in the simulation of hurricane Andrew with triple-nested MM5 using horizontal grid spacing less than 5 km. BRAUN and TAO (2000) reported improvement in simulated cyclone intensity with Burk-Thomson and Bulk Aerodynamic PBL parameterization. In a simulation study of hurricane Diana, DAVIS and BOSSART (2001) reported that model physics play an important role during the transformation from marginal to hurricane intensity. MANDAL *et al.* (2004) studied the impact of various physical parameterization schemes available with NCAR MM5 model on the prediction of two tropical cyclones formed during November 1995 over the Bay of Bengal. They used two PBL and four convection schemes and found that the combination of MRF PBL scheme with Grell cumulus scheme performed better than the other combinations.

MOHANTY *et al.* (2004) used a single domain MM5 with 30-km horizontal resolution to simulate Orissa super-cyclone and reported that the model underpredicted the cyclone intensification after 48 hours. RAO and BHASKAR RAO (2003) simulated Orissa super-cyclone using Grell, MRF and Simple Ice schemes for the physical processes and reported a good prediction of track but an underestimation of intensity. BHASKAR RAO *et al.* (2006) conducted a sensitivity study on the simulation of 1999 Orissa super-cyclone with different schemes for the parameterization of PBL, convection and explicit moisture. They reported that the combination of KF2 convection scheme with Mellor-Yamada PBL scheme provides the best results for intensification while KF2 convection scheme with MRF PBL scheme provides the best combination for the least track error. With domain and physics experiments using MM5 model, MANDAL and MOHANTY (2006) found the combination of Grell convection, MRF PBL and CCM2 radiation schemes gave the best simulation for the Orissa super-cyclone. In a simulation study of the Orissa super-cyclone, TRIVEDI *et al.* (2006) reported that SLP, horizontal, vertical wind structure, temperature anomalies and rainband characteristics are best simulated with Kain-Fritisch and MRF schemes. In a modeling study of tropical cyclones SHIVHARE and SRINIVAS (2006) found that the Grell convection scheme gave better cyclone track as compared to KF2. Many of these studies mainly focused on the simulation of the Orissa super-cyclone which is reported with hurricane intensity. Very few sensitivity studies exist for moderate intensity cyclones. In the present study the PSU/NCAR MM5 mesoscale model is used to simulate the development of the recent Andhra severe cyclone (2003), which caused heavy damage to some parts of coastal Andhra

Table 1

Details of the grids and the physics options used in the NCAR MM5 model

Dynamics	Primitive equation, Non-hydrostatic		
Vertical resolution	23 sigma levels		
Horizontal resolution	81 km	27 km	9 km
Domains of integration	74.50 E – 105.50 E 1.00 N – 32.00 N	72.643 E – 94.504 E 3.5315 N – 24.377 N	76.20 E – 88.84 E 5.595 N – 18.81 N
Radiation	Dudhia scheme for short wave radiation Rapid Radiation Transfer Model (RRTM) for longwave radiation		
Planetary boundary layer	Medium Range Forecast (MRF); Eta Mellor-Yamada (MY), Blackadar (BL); Burk-Thomson (BT)		
Sea-surface temperature	Real Sea Surface temperatures		
Cumulus parameterization	Grell (GR); Kain-Fritisch2 (KF2)		
Explicit moisture	Simple-Ice (SI) scheme		
Surface processes	Multi-layer soil diffusion model		
Sea-surface temperature	Real SST		

environmental air, except at the top and bottom of the circulations. It uses the quasi-equilibrium closure assumption, as proposed by ARAKAWA and SCHUBERT (1974). The precipitation efficiency is assumed to be a function of the mean wind shear in the lower troposphere. The effects of the convective-scale downdrafts are parameterized. The Kain-Fritisch scheme considers a Lagrangian Parcel method along with vertical momentum dynamics to estimate the properties of cumulus convection. It incorporates a trigger function, a mass flux formulation and closure assumption. The trigger function identifies the potential updraft source layers associated with convection, whereas the mass flux formulation calculates the updraft, down draft and environmental mass flux associated with that. The closure assumption for the scheme is that the convective effects remove convective available potential energy (CAPE) in a grid element with an advective time period by rearranging mass in a column using updrafts, downdraft and environment mass flux until at least 90% of the CAPE is removed. The CAPE is calculated based on the path of an entraining diluted parcel. The cloud radius which controls the maximum possible entrainment rate is specified as a function of sub-cloud layer convergence. A minimum cloud depth required for activation of deep convection is allowed to vary as a function cloud base temperature. The scheme assumes conservation of mass, thermal energy, total moisture and momentum, and since it represents different processes associated with convection it is chosen for the present study. The other physics used in the model are identical in each case of numerical experiments. These are the Dudhia simple Ice scheme for grid scale resolved precipitation, the DUDHIA (1989) radiation parameterization, and a five-layer soil model.

5. Data and Model Initialization

The data for the terrain elevation, land-use and soil types for the 81, 27 and 9 km resolution domains are obtained from the USGS topography data at 30', 10' and 5' resolutions. The initial conditions for the model domains have been interpolated from NCEP final analysis data available at 1° × 1° degree resolution corresponding to 00 UTC of 12 December, 2003. The time-varying lateral boundary conditions are derived at every 6 h interval during the period 00 UTC of 12 December, 2003 to 00 UTC of 16 December, 2003 from NCEP FNL data. The time-varying sea-surface temperature data are also given at 6 h intervals from NCEP data interpolated for the three model domains.

The data on the intensity and the position of the December cyclone are taken from the reports of the India Meteorological Department for comparison with the model results. The observations of rainfall recorded at the coastal stations are collected from IMD reports for comparison with model-derived rainfall. The TRMM satellite rainfall observations have also been used for comparison with the model derived precipitation.

6. Results

In this study results from several numerical experiments of the Andhra-2003 cyclone conducted using MM5 are presented to evaluate the performance of the model for the intensification, movement and structure. The experiments are categorized into two groups to study the sensitivity of the simulated features of the storm to the parameterization of PBL and convection processes.

6.1. Experiments with PBL Schemes

In the first set of experiments the PBL schemes of Medium Range Forecast (MRF), Blackadar (BL), Eta Mellor-Yamada (MY) and Burk-Thompson (BT) are selected with the combination of Grell (GR) for convection and Simple-Ice for explicit moisture. The simulated minimum central sea-level pressure (CSLP) and the maximum surface winds at every 6 h interval are shown in Figure 2.

It is noted that the intensity of the simulated storm is underpredicted in all the cases as seen from the time series of the CSLP values. The deepening period was about 60 hours in all the cases except MRF where it was about 66 hours. The experiments with MY and MRF schemes produced the strongest cyclone with a CSLP of 992 hPa. The mature stage of the storm was about 6 hours for BL and BT schemes whereas it was about 12 hours for both MY and MRF schemes. The observed lowest CSLP of the storm was 990 hPa attained at the 84th hour (12Z on 15 Dec., 2003). The lowest CSLP in all the four experiments was attained about 18 hours earlier than the observations.

The maximum wind speed simulated in different cases varies from 28 to 34 ms^{-1}. Excepting the BL scheme, all schemes show a maximum wind speed above 30 ms^{-1} which is an over prediction to the IMD observation of 27 m s^{-1}. Further, in all experiments the maximum winds are underpredicted after the landfall. The maximum winds associated with the storm weaken more rapidly in the experiments with BT and BL schemes than in the case of MY and MRF. The maximum wind of the storm is 34 m s^{-1} in the case of MY scheme, 32 m s^{-1} in the case of BT scheme. According to the IMD reported observations, the maximum pressure drop associated with the storm was 17 hPa. The experiments with MY, MRF schemes produced the highest pressure drop of 14 hPa followed by BL and BT schemes. Thus the MY scheme offers the most intensive cyclone in terms of maximum winds and the pressure drop.

The track positions of the simulated cyclone from experiments with different PBL schemes along with IMD observations are shown in Figure 3. It is noted that all experiments show deviation in track positions from the initial stage of the system. The simulated track positions are to the west of the observed track. The observed movement of the system was northward in the first 24 hours and northwestward subsequently. While the experiments with BT, BL and MY schemes show wide

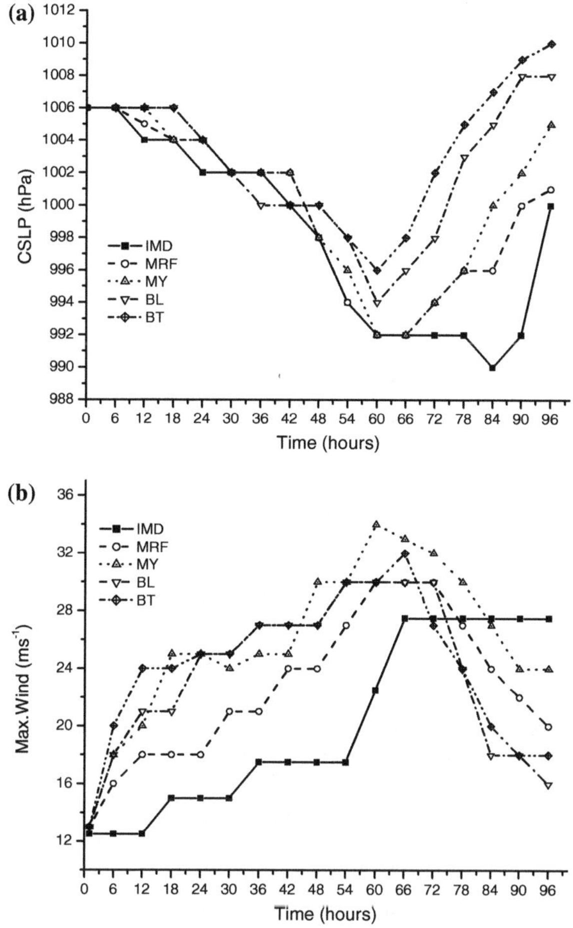

Figure 2

Time variation of simulated a) central sea-level pressure (hPa) and b) maximum wind (m s^{-1}) for the experiments with different PBL schemes along with IMD estimates.

dispersion in the simulated track, the MRF scheme gives the nearest track positions. The deviation in track positions increases with MY, BL and BT schemes. The error range in track position was least for the case with MRF (99 to 206 km) (Table 2) and progressively increases for MY, BL and BT schemes. The BT scheme shows the maximum error range (243–555 km) for the track position. Correspondingly the land fall positions in different experiments were near Ongole, Chennai, Pondicherry and Nagapattinam with MRF, MY, BL and BT schemes, respectively. Thus the results

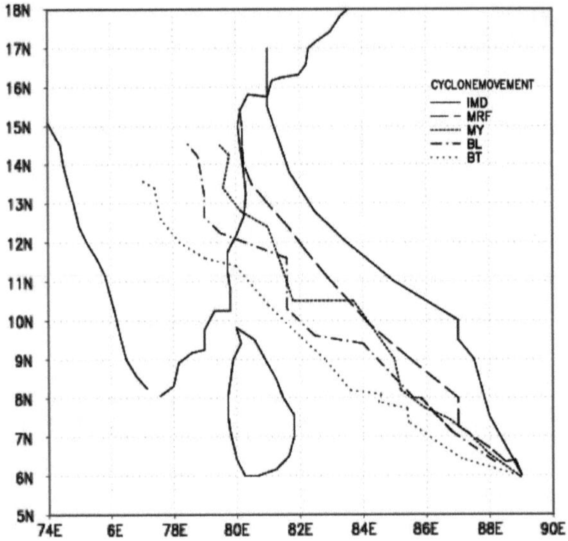

Figure 3
Model simulated track positions of the Andhra severe cyclone for the experiments with different PBL schemes along with IMD estimates.

Table 2

Errors of track positions (km) for different sensitivity experiments

Experiments with PBL schemes				
	Hours of simulation			
PBL Scheme	24	48	72	96
MRF	99	152	206	187
MY	219	252	289	324
BL	239	232	310	388
BT	243	324	432	555
Experiments with Convection schemes				
	Hours of simulation			
Convection Scheme	24	48	72	96
KF2	55	150	63	21
GR	99	152	206	187
KF2 with MY PBL	126	291	164	168

indicate that the MRF scheme provides the best simulated track with minimum errors in track positions, while also reasonably predicting the intensity of the system in terms of the time series of maximum winds and the CSLP.

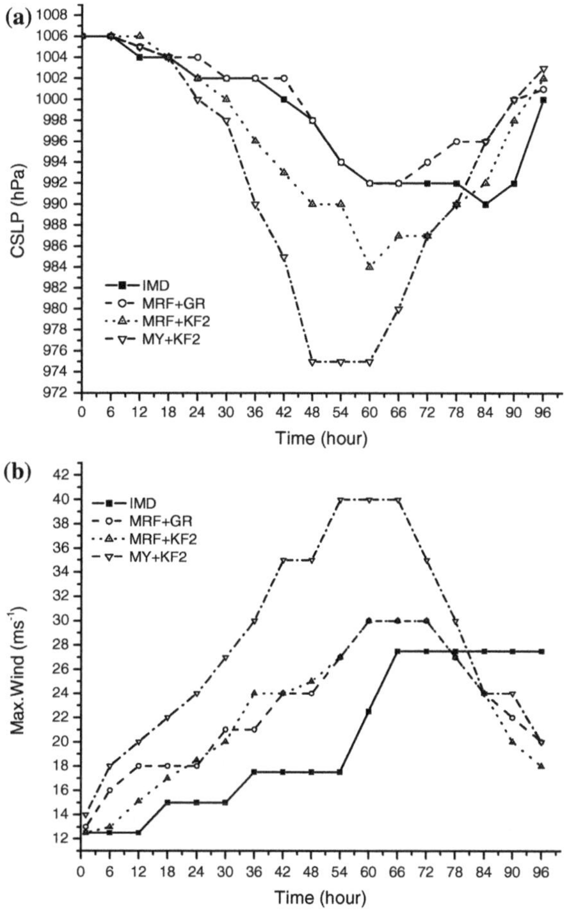

Figure 4
Time variation of simulated a) central sea-level pressure (hPa) and b) maximum wind (m s^{-1}) for the experiments with different convection schemes along with IMD estimates.

6.2. Experiments with Convection Schemes

From the results noted in the previous section, the MRF PBL scheme produced the best simulation with reasonable intensity estimation of the Andhra severe cyclone with a CSLP of 992 hPa and corresponding maximum winds of 30 m s^{-1}. Hence a set of model experiments was conducted with the MRF scheme for PBL processes, Simple Ice scheme for explicit moisture varying the convection schemes with Grell (GR) and Kain-Fritisch 2 (KF2). Since the MY scheme offered the

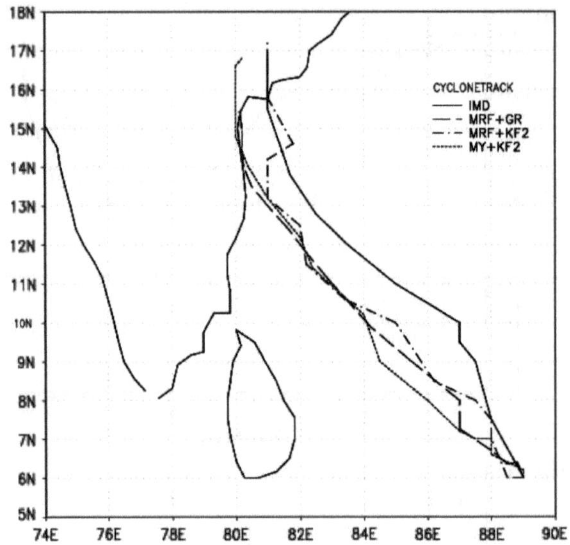

Figure 5
Simulated track positions of the Andhra severe cyclone for the experiments with different convection schemes along with IMD estimates.

highest intensive cyclone in the previous set of experiments, a separate experiment is also conducted to evaluate the performance of the model with the combination of KF2 for convection and MY scheme for PBL processes. The model derived CSLP and maximum winds for all these experiments are shown in Figure 4.

The results from the experiments using MRF PBL with a variation of convection schemes indicate that the KF2 scheme gives maximum storm intensification with a CSLP of 984 hPa and maximum winds of 30 m s^{-1}. However, this is a slight overprediction to the observations. The CSLP distribution in the experiment with the Grell scheme almost followed the observations up to 66 hr and was underpredicted thereafter. Both the GR and KF2 schemes have a deepening period of 60 hours agreeing with the observations. Observations indicate the CSLP at the mature stage is 900 hPa and the maximum winds are 27 m s^{-1}. Though the time series of CSLP is better simulated by MRF and Grell schemes up to 66 hr, a rapid weakening (or rise in CSLP) was noticed thereafter in contrast to the observations. The time series of maximum winds is almost similarly simulated by both the KF2 and Grell schemes throughout the life cycle of the storm. The KF2-MRF combination yields a higher fall in CSLP than the KF2-Grell combination during the deepening stage (from t = 33 hr to t = 75 hr) but gives closer values to the observations after 72 hr when the storm was approaching the coast. Figure 4 also shows that the experiment with KF2 for convection and MY scheme for PBL produced the most intensive storm of all experiments with a CSLP of 975 hPa, maximum winds of 40 ms^{-1} and with a deepening period of just 48 hours, which is considerably an overestimation of the observations. Mention must be made here that the MY scheme (along with the Grell convection scheme) produced the most intensive cyclone in the first set of experiments while testing different PBL schemes.

The track positions of the simulated storm with respect to GR, KF2 convection schemes (with MRF PBL) and the combination of KF2 and MY schemes are shown in Figure 5 along with observations. There is a wide deviation of track from observations after 24 hours in all the cases. The simulated cyclone moved to the west of the observed track in all three cases. The track positions in the three experiments closely follow each other up to 66 hr. The model cyclone in the experiment with KF2, MRF schemes turns towards the observed track at 66 hr at 00 UTC 15 Dec., 2003 and moves thereafter in line with observations. This trend in the sudden turn of the cyclone to the northeast may be due to better data from the land region. The location of the land fall of the model cyclone with MRF, KF2 schemes is identical to the observation. It is to be noted that the model cyclone in the experiments with the combinations of GR, MRF and KF, MY schemes moved considerably to the west of the observed track, almost along the coastline with corresponding errors in land fall positions south of Machilipatnam. The experiment with KF2, MRF schemes gives minimum track errors, followed by the cases with GR, MRF and KF2, MY schemes (Table 2). Results from all the seven model experiments indicate that the combination of KF2 for cumulus convection and MY for PBL processes produces the most intensive cyclone in terms of the CSLP and the maximum winds associated with the storm. The simulation with the combination of KF2 for convection and MRF PBL schemes produced least track errors apart from representing the maximum winds similar to Grell, MRF combination throughout the simulation and closer CSLP values in the mature and decay stages of the storm. Considering the overall characteristics, the KF2, MRF combination gives the best simulation of the cyclone.

6.3. Structure of the Simulated Cyclone

Results discussed in the previous sections clearly show that the model experiment with the combination of parameterization schemes of KF2 for convection and MRF scheme for PBL produced the best simulation of the Andhra severe cyclone as regards its movement as well as intensity. The structure of the simulated cyclone for this experiment is analyzed and discussed. The CSLP distribution, geopotential at 500 hPa and 200 hPa along with the wind field during the mature stage corresponding to 12 UTC of 15 December are shown in Figure 6a.

The model cyclone has a CSLP of 990 hPa and maximum winds associated with the storm are 25 m s^{-1}. Cyclone has attained highest intensity and its location is near

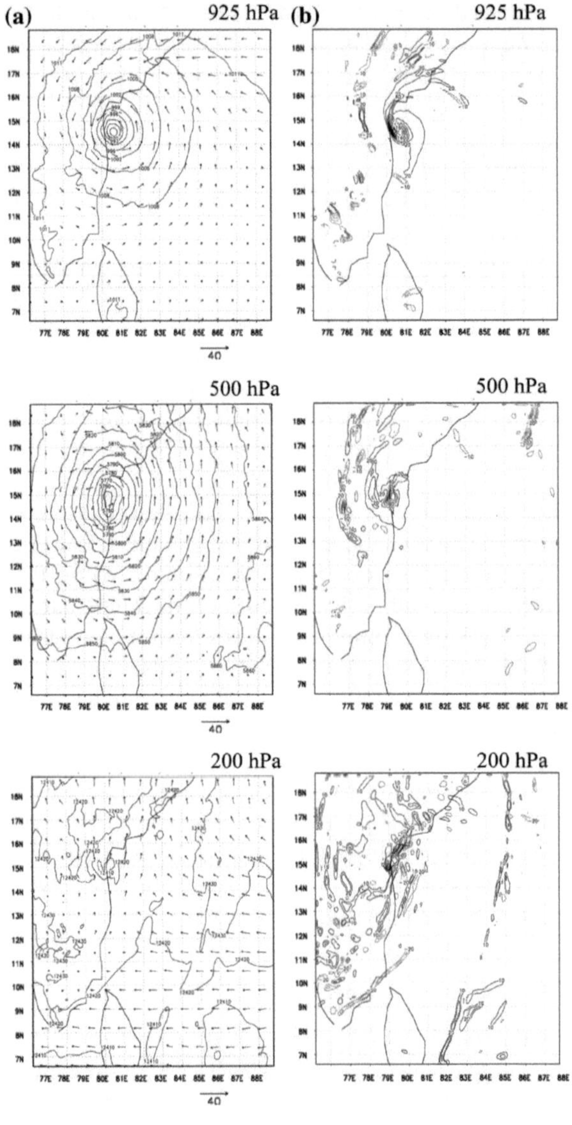

Figure 6
Distribution of CSLP (hPa)/geopotential (gpm) along with wind field and b) vorticity (x1.0e + 5 s^{-1}) corresponding to 12 UTC of 15 December, 2003. Upper, middle and lower panels correspond to 925 hPa, 500 hPa and 200 hPa levels. Cyclonic voriticity is drawn in solid lines and anticyclonic vorticity is drawn in dotted lines.

the coast at this time. The geopotential gradient is strong up to 500 hPa and is noted to weaken at 200 hPa, indicating the prevalence of cyclonic circulation reaching 500 hPa and anticyclonic motion above it.

The vorticity field at 925 hPa, 500 hPa and 200 hPa is shown in Figure 6b corresponding to 12 UTC of 15 December. The figure shows cyclonic vorticity (positive values) exceeding 20×10^{-5} sec^{-1} and anticyclonic vorticity (negative values) of -10×10^{-5} sec^{-1}. A positive cyclonic vorticity distribution is observed at the lower levels and a negative (anticyclonic) vorticity is seen at the upper levels. Cyclonic vorticity concentration is recognized around the location of the storm with its position in the south Bay of Bengal near the Andhra Coast. The low level cyclonic vorticity ($>20 \times 10^{-5}$ sec^{-1}) shows strong cyclonic circulation in a radius of about 150 km, indicating a mature storm. The anticyclonic vorticity is observed to concentrate in the northwest and southeast quadrants around the cyclonic vorticity maximum. The same trend of vorticity distribution is noticed to continue at the 500 hPa level also. At the upper level (200 hPa) a negative vorticity pattern is observed all around the cyclonic region indicating an anticylonic circulation. This represents the existence of an outflow region around the 200 hPa level towards the southeast and northeast sectors.

Analysis of the divergence pattern shows low-level convergence and high-level divergence associated with the storm. The low-level convergence pattern at 925 hPa level (Fig. 7) represents strong spiraling inflow regions towards the storm center, the maximum inflow ($\sim 60 \times 10^{-5}$sec^{-1}) concentrated on the west side of the storm. Divergence in the upper level (200 hPa) flow indicates outflow regions. The maximum outflow is found concentrated in the forward-right and rear-left quadrants around the region of low-level convergence.

The east-west vertical crosssections of temperature anomaly, horizontal wind and vertical wind at the mature stage of the storm are presented in Figure 8. The vertical distribution of equivalent potential temperature (not shown) indicates the development of a warm core region in the cyclone, and is seen descending to the lower levels at its mature stage. The maximum core temperatures (360°K) are seen at a height of roughly 1 km and seen to gradually decrease to 354°K. The temperatures are seen to radially fall in the inner wall and outer wall regions. The temperature anomaly estimated from the temperature values at the mature stage of the storm and the mean atmospheric temperatures before the development of the storm (Fig. 8a) indicate a maximum warming of 9°C to occur in the 500 to 300 hPa layer, i.e., roughly in the middle troposphere. This is consistent with the earlier observational studies. The warm core region is seen to reduce conically at lower levels. The temperatures at the middle tropospheric levels are seen about 5°C higher than the corresponding values at lower levels. Temperature gradients in the lower atmospheric levels, including the boundary layer, are seen weak.

The distribution of wind flow (Fig. 8b) indicates the presence of gusty/cyclonic winds vertically throughout the troposphere and horizontally in a radius of about

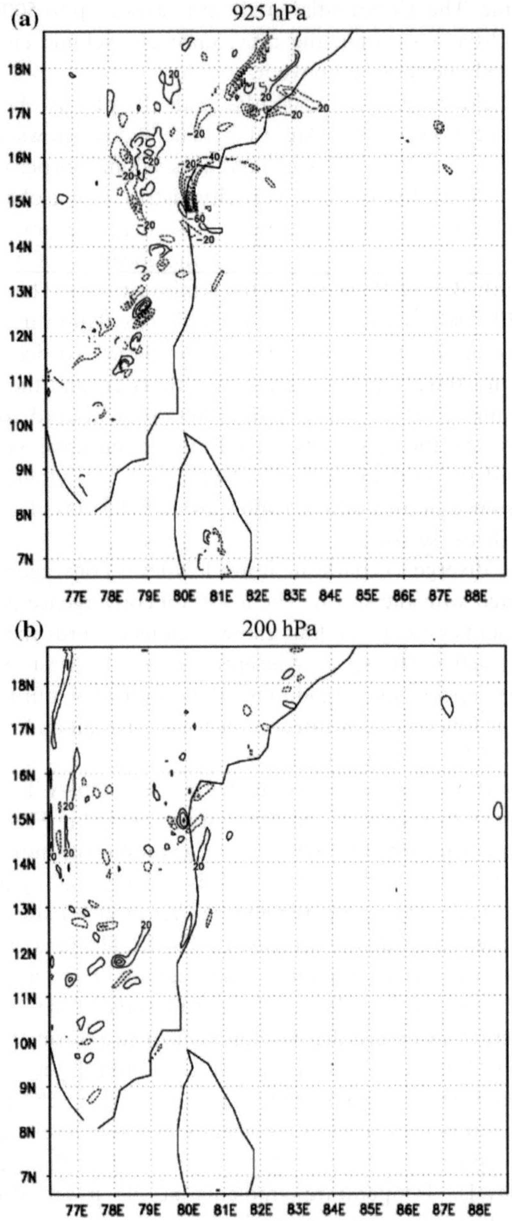

Figure 7
Convergence/ divergence (1.0e + 5 s^{-1}) at a) 925 hPa and b) 200 hPa corresponding to 12 UTC of 15 December, 2003.

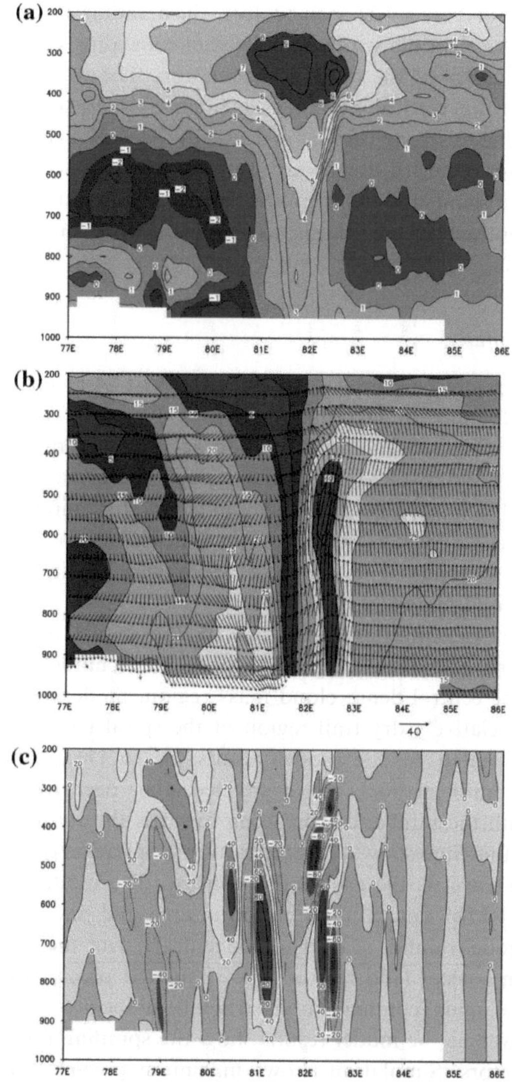

Figure 8
Vertical cross section of simulated a) temperature anomaly, b) wind (m s^{-1}) and c) vertical winds (cm s^{-1}) at mature stage of the cyclone corresponding to 12 UTC of 15 December, 2003.

500 km. This essentially shows contrasting horizontal winds around the storm. Severe cyclone winds (24–32 m s^{-1}) are noted up to 400 hPa level towards the eastern side and reaching 700 hPa level on the western side of the storm. A calm central

region with winds of 5 m s^{-1} is seen to form in a radius of about 0.2°. Gradual upward expansion of the calm wind region is seen coinciding with the outflow region and decreasing intensity of the cyclonic circulation. The vertical wind distribution (Fig. 8c) shows strong vertical motion (20–80 cm s^{-1}) up to 500 hPa level in the 20–50 km radius due to strong vertical updrafts in the wall region. The core region is essentially characterized with downward motion, which is seen to expand upward conforming to the wind and temperature distributions. Around the wall region downdrafts are noticed which is due to the subsidence on the rear side of the convective updrafts.

6.4. Simulated Rainfall

The simulated rainfall accumulated during the past 24-hr period is presented for 00 UTC of 14, 15 and 16 December, 2003 in Figure 9. The model rainfall is compared with the IMD observations of 24-hr accumulated rainfall of some coastal stations as well as with the TRMM satellite derived rainfall rates. It is known that in tropical cyclones the clouds align themselves in bands that spiral around and into the storm with a dense cloud mass at the center. The eye region due to subsidence is relatively free from clouds. The strongest winds and heaviest precipitation occur around the wall cloud region of the cyclone.

The pattern of precipitation simulated by the model agrees with the cloud distribution pattern associated with the storm. The characteristics of maximum precipitation in the central dense cloud mass region, moderate precipitation in the spiral bands and relatively dry trail region of the spiral bands in the outer sectors are all well represented by the model results on 15 Dec., 2003. The simulated maximum precipitation at 00 UTC of 14 December occurred over the southeast Bay of Bengal with the rainbands spreading towards the south and southeast. The TRMM data for this time show a large area of precipitation over the central Bay of Bengal spreading towards the east and southeast. The model rainfall distribution agrees with the TRMM rainfall distribution in area and pattern. At 00 UTC of 15 December, the model rainfall is seen to concentrate over the north Bay of Bengal and the adjoining coastal land region. The simulated precipitation pattern clearly indicates the developing comma cloud structure associated with the storm. The TRMM data as well as the model results show the spiraling rainbands in the north and northeast sectors. Simulation shows maximum precipitation in the northern direction over the north Tamilnadu and south Andhra Pradesh coastal regions. TRMM observations indicate concentration of maximum precipitation over the Bay of Bengal and adjoining land regions with extending rainbands towards the northeast. The model rainfall agrees with TRMM-derived rainfall distribution of 5 to 25 cm/day. Over the ocean the model has underpredicted the rainfall where a rainfall up to 38 cm/day is found in the TRMM data. The simulated rainfall exceeding 20 cm/day near Madras and exceeding 15 cm/day near Pondicherry is

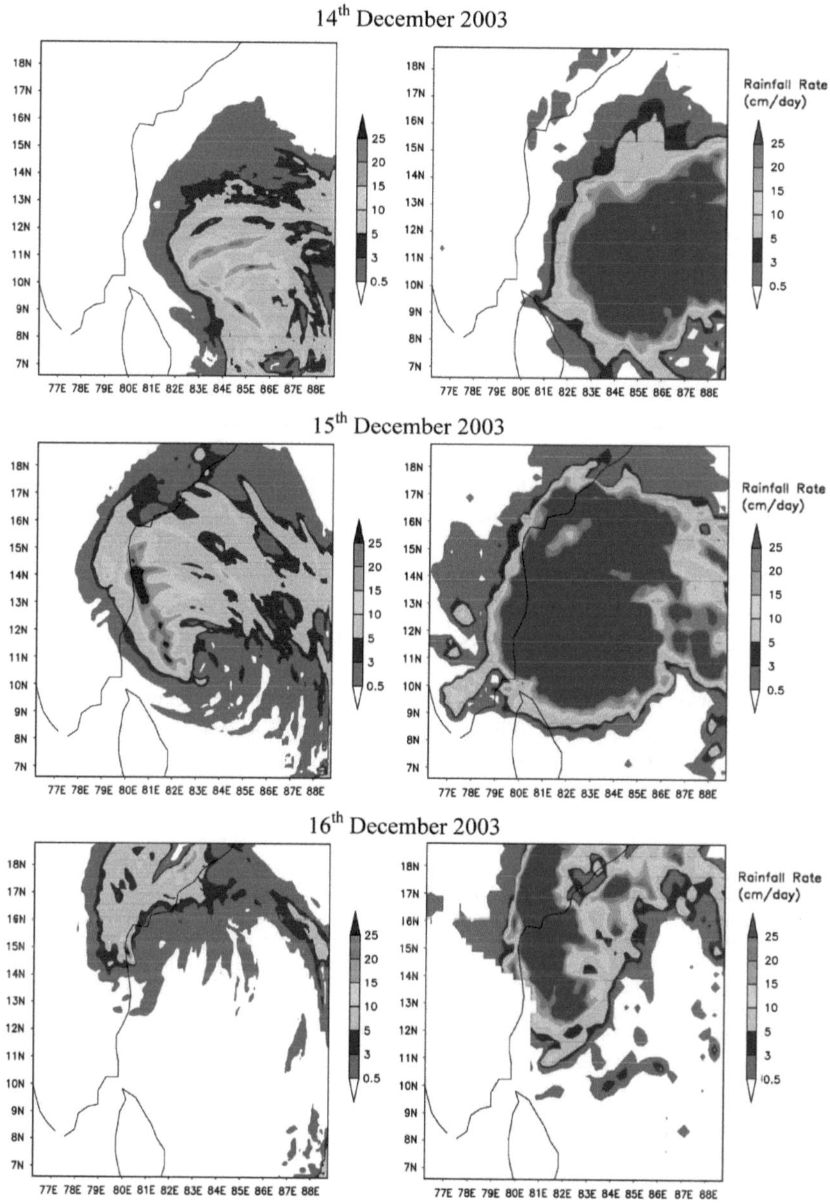

Figure 9
Model simulated total rainfall (left panel) (cm/day) and TRMM-TMI rainfall (cm/day) (right panel) corresponding to 00 UTC of a) 14 Dec., b) 15 Dec., and c) 16 Dec., 2003.

comparable with the observations reported by IMD (Table 3). At 00 UTC of 16 December, the simulated precipitation shows the movement of the maximum rainfall interior of the coast and the weakening of the system indicating a maximum of 10 to 15 cm/day agreeing with the pattern of TRMM precipitation. The inclination of the rainbands towards the north and northeast is clearly identified in both the simulation and the TRMM rainfall pattern. The model predicted rainfall on 15 December closely agrees with the observations at the locations Karaikal, Nagapattinam, Pondicherry, Madras, Nellore, Vijayawada and Kakinada, respectively. Thus the model simulated rainfall is reasonable.

7. Summary and Conclusions

A numerical study was conducted to investigate the effect of various parameterizations of the planetary boundary layer and convection processes on the intensification and movement of a tropical cyclone. The case of Andhra cyclone during December 2003, one of the severe cyclones over the Bay of Bengal, is chosen for this study. A high-resolution nested-grid mesoscale model, PSU/NCAR MM5, has been used to simulate the development of the cyclone. Four schemes for the PBL parameterization (MRF, Mellor-Yamada, Blackadar and Burk-Thomson) and two schemes for the cumulus convection (Grell, Kain-Fritisch 2) are used to evaluate the model performance.

The first set of model experiments was carried out to study the effect of different PBL parameterization schemes. These experiments indicate that the model produces the intense cyclone with a central surface low pressure of 992 hPa, maximum winds

Table 3

Observed accumulated rainfall (in 24 hours) of a few coastal stations for 15 December, 2003 on the east coast

Location	Latitude/longitude	Rainfall (mm/day)
Kallakkurichchi	11.73 78.97	39.06 mm
Vedaranniyam	10.37 79.85	43.95 mm
Vijayawada/Gannavar	16.53 80.80	43.95 mm
Thanjavur	10.78 79.13	48.83 mm
Kakinada	16.95 82.23	53.71 mm
Nellore	14.45 79.98	58.59 mm
Tirupathi	13.67 79.58	58.59 mm
Machilipatnam	16.20 81.15	68.36 mm
Nagappattinam	10.77 79.85	68.36 mm
Karaikal	10.92 79.83	73.24 mm
Cuddalore	11.77 79.77	126.95 mm
Pondicherry	11.92 79.83	126.95 mm
Madras	13.07 80.25	200.20 mm
Madras/Minambakkam	13.00 80.18	200.20 mm

of 34 m s^{-1} with the MY PBL scheme and Grell scheme for convection processes. However, the combination of MRF PBL scheme and the Grell convection schemes is found to yield reasonable simulation with the actual observed intensity. The experiments with different PBL schemes also simulated widely varying tracks and the MRF scheme gave the best simulated track with minimum errors. These results suggest that the PBL processes play an important role in the intensification as well as the motion of the cyclone.

In the second set of experiments two different convection parameterization schemes were used with MRF scheme for PBL and Simple-Ice scheme for explicit moisture. It is found that the Grell convection scheme with MRF PBL produces a better time series of CSLP and maximum winds in the deepening stage while the experiment with KF2 scheme predicted the best track of the cyclone with the least errors in the track position while also giving better horizontal wind characteristics throughout the simulation and CSLP in the mature and decay stages. These results indicate that the convection processes play an important role in modulating the track of the cyclone. The experiment with KF2 for convection and MY PBL produced the most intensive cyclone of all the experiments in terms of CSLP and winds. These results support the earlier results by BHASKAR RAO *et al.* (2006), in which it was found that the KF2, MY schemes provided the best simulation for intensity and KF2, MRF schemes offered the best combination for the least track error for the Orissa super-cyclone (1999).

The structure of the cyclone at the mature stage from the experiment with KF2 for cumulus convection and MRF for planetary boundary layer is examined. Simulated features of the cyclone are compared with observed features of winds, pressure and rainfall. The model-produced structure of the cyclone at the mature stage conforms to the previous findings. The simulated wind distributions at the lower and upper atmospheres, the vertical velocity in the middle atmosphere and vorticity fields at the lower and upper levels all support the intensification of the system. Converging cyclonic winds around the storm, diverging winds at the upper troposphere and the central warm core show the typical characteristics of a mature tropical storm. The model storm moved to the west of the actual observed track during the developing phase and closely followed the observed track in the final intensification and land-crossing time. The overall error in track positions ranges from 21 to 150 km.

The model simulated rainfall agrees with the observed rainfall distribution over the Bay of Bengal and the Tamilnadu-Andhra Pradesh coasts. The model could predict the maximum rainfall of >20 cm/day distributed along the coastal stations near the landfall, agreeing with the reports from IMD. Results indicate that the model could reasonably simulate the intensification, movement and rainfall although little difference is seen as noted from the observations. The present results support the earlier studies on the model sensitivity to PBL and convection parameterization in the case of the Orissa Super-cyclone. However in order to arrive at reasonable conclusions about the proper choice of physics in the model of a larger number of cyclone cases (severe, very severe, hurricane etc., cyclones in different years and seasons) need to be studied.

Acknowledgement

The first two authors are thankful to Dr. Baldev Raj, Director IGCAR, for giving permission to present this work at IMPA2006. The India Meteorological Department is acknowledged for its observations on cyclone track positions and rainfall used in the study.

References

ANTHES, R.A. (1977), *Hurricane model experiments with a new cumulus parameterization scheme,* Mon. Wea. Rev. *105,* 287–300.

ARAWAKA, A., and SCHUBERT, W.H. (1974), *Interaction of a cumulus cloud ensemble with the large scale environment.* Part I. J. Atmos. Sci., *31,* 674–701.

ASNANI, G.C. (1993), *Tropical Meteorology,* vols. 1 and 2., published by Prof. G.C. Asnani, c/o Indian Institute of Tropical Meteorology, Dr. Homi Bhabha Road, Pashan, Pune- 411008, India.

BHASKAR RAO, D.V. NAIDU, C.V., and SRINIVASA RAO, B. (2001), *Trends and fluctuations of the cyclonic systems over North Indian Ocean,* Mausam, *52,* 1, 1–8.

BHASKAR RAO, D.V. and ASHOK, K. (1999), Simulation of tropical cyclone circulation over the Bay of Bengal. Part I. Description of the model, initial data and results of the control experiment, Pure Appl. Geophys. *156,* 3, 525–542.

BHASKAR RAO, D.V. and DASARI HARI PRASAD (2006), *Numerical prediction of the Orissa super-cyclone (1999): Sensitivity to the parameterization of convection, boundary layer and explicit moisture processes,* Mausam *57,* 1, 61–78.

BLACKADAR, A.K., *Modeling the nocturnal boundary layer.* Preprints. Third Symp. On *Atmospheric Turbulence, Diffusion, and Air Quality* (Rayleigh, NC, Am. Meteor. Soc. (1976), pp. 46–49.

BRAUN, S.A. and TAO, W-K. (2000), *Sensitivity of high resolution simulations of hurricane Bob* (1991) *to planetary boundary layer parameterizations,* Mon. Wea. Rev. *128,* 3941–3961.

BURK, S.D. and THOMPSON, W.T. (1989), *A vertically nested regional numerical prediction model with second-order closure physics,* Mon. Wea. Rev. *117,* 2305–2324.

CHEN, D.R., YEH, T.C., HAUNG, K.N., PENG, M.S., and CHANG, S.W. (1995), *A new operational typoon track prediction system at the central weather Bureau in Taiwan.* Preprints 21 Conf. Hurr. Trop. Meteor. Soc., Boston, MA pp. 50–51.

DAVIS, C.A. and BOSSART, L.F. (2001), *Numerical simulations of the genesis of hurricane Diana* (1984). *Part I: Control Simulation,* Mon. Wea. Rev. *129,* 1859–1881.

DUDHIA, J. (1989), *Numerical study of convection observed during winter monsoon experiment using a mesoscale two-dimensional model,* J. Atmos. Sci. *46,* 3077–3107.

DVORAK, V.F. (1975), *Tropical cyclone intensity analysis and forecasting from satellite imagery,* Mon. Wea. Rev. *103,* 420–430.

GRELL, G.A. (1993), *Prognostic evaluation of assumptions used by cumulus parameterization,* Mon. Wea. Rev. *121,* 764–787.

GRELL. G.A., DUDHIA, J., and STAUFFER, D.R. (1994), *A description of the fifth-generation Penn State/NCAR mesoscale model* (MM5), NCAR Technical Note, NCAR/TN-398+STR, 117 pp.

HONG, S.Y. and PAN, H.U. (1996), *Nonlocal boundary layer vertical diffusion in a medium-range forecast model.* Mon. Wea. Rev. *124,* 2322–2339.

JANJIC, Z.A. (1990), *The step-mountain coordinate: Physical package*, Mon. Wea. Rev. *118*, 1429–1443.
KAIN, J.S. and FRITSCH, J.M., *Convective Parameterization for Mesoscale Models: The Kain- Fritsch Scheme. The Representation of Cumulus Convection in Numerical Models* (K.A. Emanuel and D.J. Raymond, eds.) (Amer. Meteor. Soc. 1993) 246 pp.
KURIHARA, Y., BENDER, M.A., TULEYA, R.E., and ROSS, R.J. (1995), *Improvements in the GFDL hurricane prediction system*, Mon. Wea. Rev. *123*, 2791–2801.
KURIHARA, Y., BENDER, M.A., TULEYA, R.E., and ROSS, R.J. (1993), *Hurricane forecasting with GFDL hurricane prediction system*, Preprints 21 Conf. Hurr. Trop. Meteor. Soc., Boston, MA pp. 323–326.
LIU, Y., ZHANG, D.-L., and YAU, M.K. (1997), *A multi-scale numerical simulation of hurricane Andrew* (1992). *Part I. Explicit simulation and verification*, Mon. Wea. Rev. *125*, 3073–3093.
MANDAL, M., MOHANTY, U.C., and RAMAN, S. (2004), *A study on the impact of parameterization of physical processes on prediction of tropical cyclones over the Bay of Bengal with NCAR/PSU mesoscale model*, Natural Hazards *31*, 2, 391–414.
MANDAL, M. and MOHANTY, U.C. (2006), *Numerical experiments for improvement in mesoscale simulation of Orissa super-cyclone*, Mausam, *57*, 1, 79–96.
MATHUR, M.B. (1991), *The National Meteorological Center's quasi-Lagrangian model for hurricane prediction*, Mon. Wea. Rev. *109*, 1419–1447.
MOHANTY, U.C., MANDAL, M., and RAMAN, S. (2004), *Simulation of Orissa super-cyclone (1999) using PSU/NCAR mesoscale model*, Natural Hazards *31*, 373–390.
PRASAD, K. and RAMA RAO, Y.V. (2003), *Cyclone track prediction by a quasi-Lagrangian model*, Meteor. Atmos. Phys. *83*, 173–185.
PRASAD, K., RAMA RAO, Y.V., and SANJIB, SEN. (1997), *Tropical cyclone track prediction by a high resolution limited area model using synthetic observation*, Mausam *46*(3), 351–366.
RAGHAVAN, S. and SEN SARMA, A.K., *Tropical cyclone impacts in India and neighbourhood.* In *Storms*, vol. 1. (eds. Roger Pielke, Jr. and Roger Pielke, Sr.) (Routledge, London 2000) pp. 339–356.
RAO, G.V. and BHASKAR RAO, D.V. (2003), *A review of some observed mesoscale characteristics of tropical cyclones and some preliminary numerical simulations of their kinematic features*. Proc. of Ind. Nat. Sci. Acad. *69*A, 5, 523–541.
ROSENTHAL, S.L. (1971), *The response of a tropical cyclone model to variations in boundary layer parameters, initial conditions, lateral boundary conditions, and domain size*, Mon. Wea. Rev. *99*, 767–777.
SHIVHARE, R.P. and SRINIVAS, V.S. (2006), *Tropical clone storm – Modeling studies in Indian Air Force*, Mausam *57*, 1, 135–140.
TRIVEDI, P., MUKHOPADHYAY and VAIDYA, S.S. (2006), *Impact of physical parameterization schemes on the numerical simulation of Orissa super cyclone (1999)*, Mausam *57*, 1, 97–110.
YAMASAKI, M. (1968), Detailed analysis of a tropical cyclone simulated with a 13 layer model, Met. and Geophys. *19*, 559–585.

(Received February 9, 2006, accepted November 13, 2006)
Published Online First: June 19, 2007

B. Mesoscale Variability and Modelling

Pure appl. geophys. 164 (2007) 1491–1509
0033–4553/07/091491–19
DOI 10.1007/s00024-007-0235-2

Pure and Applied Geophysics

Impact of Doppler Radar Wind in Simulating the Intensity and Propagation of Rainbands Associated with Mesoscale Convective Complexes Using MM5-3DVAR System

S. Abhilash,[1] Someshwar Das,[2] S. R. Kalsi,[3] M. Das Gupta,[2] K. Mohankumar,[1] John P. George,[2] S. K. Banerjee,[3] S. B. Thampi,[4] and D. Pradhan[5]

Abstract—Pre-monsoon rainfall around Kolkata (northeastern part of India) is mostly of convective origin as 80% of the seasonal rainfall is produced by Mesoscale Convective Systems (MCS). Accurate prediction of the intensity and structure of these convective cloud clusters becomes challenging, mostly because the convective clouds within these clusters are short lived and the inaccuracy in the models initial state to represent the mesoscale details of the true atmospheric state. Besides the role in observing the internal structure of the precipitating systems, Doppler Weather Radar (DWR) provides an important data source for mesoscale and microscale weather analysis and forecasting. An attempt has been made to initialize the storm-scale numerical model using retrieved wind fields from single Doppler radar. In the present study, Doppler wind velocities from the Kolkata Doppler weather radar are assimilated into a mesoscale model, MM5 model using the three-dimensional variational data assimilation (3DVAR) system for the prediction of intense convective events that occurred during 0600 UTC on 5 May and 0000 UTC on 7 May, 2005. In order to evaluate the impact of the DWR wind data in simulating these severe storms, three experiments were carried out. The results show that assimilation of Doppler radar wind data has a positive impact on the prediction of intensity, organization and propagation of rain bands associated with these mesoscale convective systems. The assimilation system has to be modified further to incorporate the radar reflectivity data so that simulation of the microphysical and thermodynamic structure of these convective storms can be improved.

Key words: Tropical mesoscale convective systems, Doppler weather radar, data assimilation, composite reflectivity, hydrometeors.

1. Introduction

Mesoscale convective cloud clusters produce large amounts of rain in the tropics and consist of numerous deep cells. The structure and life cycle of Tropical

[1] Department of Atmospheric Sciences, Cochin University of Science and Technology, Cochin 682 016, India.
[2] National Center for Medium Range Weather Forecasting, A - 50, Sector - 62, Noida, 201 307, India. E-mail: somesh@ncmrwf.gov.in
[3] India Meteorological Department, New Delhi, India.
[4] Regional Meteorological Center, IMD, Chennai, India.
[5] Regional Meteorological Center, IMD, Kolkata, India.

Mesoscale Convective Systems (TMCS) are complex. In their mature stage, the clusters consist partly of convective towers, which contain buoyant updrafts, negatively buoyant downdrafts and heavy showers of rain. The precipitation bands associated with these TMCS extend over a horizontal distance of more than 100 km. Based on the radar reflectivity and satellite images obtained during GATE and MONEX, HOUZE (1982) summarized the detailed structure of these oceanic TMCS. Several important characteristics of TMCS using satellite data have been well documented (MADDOX, 1980).

The prediction of the timing, location, organization and structure of the MCS, especially over the tropics, is recognized as one of the biggest challenges in mesoscale modeling. Obtaining an accurate initial state is important in the accurate model prediction of these MCS. The variational data assimilation approach is one of the most promising techniques available to directly assimilate heterogeneous mesoscale observations in order to improve the estimate of the models initial state. DAS GUPTA *et al.* (2005) made the first attempt to investigate the application of the MM5-3DAVR assimilation system over the Indian region. The impact of various satellite data such as SSM/I, Q-SCAT and ATOVS has been evaluated within this study for the prediction of monsoon depressions over India.

Doppler radar observation is an important data source for mesoscale and microscale weather analysis and forecasting. Radar data assimilation could be promising for short-range numerical weather prediction. Despite the importance of the DWR data for use in warning of the heavy rainfall associated with the convective systems, there remains no effort to include analyzed Doppler radar data in the assimilation cycle of the operational weather prediction models in India (DAS *et al.*, 2006). The present work is a first attempt in utilizing India Meteorological Department (IMD) Doppler radar data into a numerical model for prediction of mesoscale convective complexes. There are two major objectives of this study; one is to understand the impact of the radar wind data on the prediction of precipitation associated with TMCS and the other is to comprehend the detailed thermodynamic and microphysical structure of the convective system.

2. *Synoptic Overview of the Intense Convective Events*

Mesoscale convective cloud clusters over Kolkata region during 0600 UTC of 5 May and 1800 UTC of 6 May, 2005 are selected for the present study. The radar reflectivity and observed rainfall are used to select the convection days. The synoptic charts showed a north-south trough extending to 0.9 km above sea level from the northern states to southern peninsular India. Maximum temperature was found to be above normal by 2–3°C at many places over the region. Thunderstorms were reported over Gangetic West Bengal and Sub-Himalayan West Bengal during 5–7 May, 2005. Jalpaiguri, Cooch Behar and Malda reported 1 cm rainfall each on 5th

May. Kolkata (Alipore) reported 5 cm and Kolkata (Dum Dum) reported 2 cm on 6th May.

Figure 1 presents the IR satellite cloud images for 1200 and 1800 UTC on 5 and 6 May, 2005. The pictures show scattered convective clouds at many places over West Bengal and its environs during the period. Figure 2 shows the radar reflectivity MAX (Z) and PPI (Z) at 1108 UTC of 5 May and 1408 UTC of 6 May, 2005. The radar reflectivity pattern clearly shows the occurrence of two major cloud bands during the period, which initially formed northwest of Kolkata and propagated southeastwards.

3. Overview of the 3DVAR System in MM5

The mesoscale model, MM5 (GRELL *et al.*, 1994) has been used for the present study. The MM5 is a non-hydrostatic model with terrain following sigma coordinate system developed at the National Center for Atmospheric Research (NCAR) designed to simulate or predict mesoscale and regional scale atmospheric circulation (DUDHIA *et al.*, 2002). The three-dimensional variational data assimilation (3DVAR) system developed for MM5 is flexible enough to allow a variety of research studies apart from its operational utilization. The 3DVAR system developed by BARKER *et al.* (2003, 2004) is used for the present assimilation experiments. The basic goal of the 3DVAR system is to produce an optimal estimate of the true atmospheric state at any desired analysis time through iterative solution of a prescribed cost function (IDE *et al.*, 1997).

$$J(x) = J^b + J^o = \frac{1}{2}(x - x^b)^T B^{-1}(x - x^b) + \frac{1}{2}(y - y^o)^T (E + F)^{-1}(y - y^o) \quad (1)$$

The variational (VAR) problem can be summarized as the iterative solution of Eq. (1) to find the analysis state $\boldsymbol{x}$ that minimizes $J(\mathbf{x})$. This solution represents the *a posteriori* maximum likelihood (minimum variance) estimate of the true state of the atmosphere, given the two sources of the *a priori* data: The background (previous forecast) $\boldsymbol{x}^b$ and observations $\boldsymbol{y}^o$ (LORENC, 1986). The fit to individual data points is weighted by estimates of their errors: $\boldsymbol{B}$, $\boldsymbol{E}$ and $\boldsymbol{F}$ are the background, observation (instrumental) and representativity error covariance matrices, respectively. Representativity error is an estimate of inaccuracies introduced in the observation operator $\mathbf{H}$ used to transform the gridded analysis $\boldsymbol{x}$ to observation space $\boldsymbol{y} = \mathrm{H}\boldsymbol{x}$. Further details about the components of the 3DVAR system can be found in BARKER *et al.* (2003). Applications of the 3DVAR system have been reported in real-time analysis and forecasting (BARKER *et al.*, 2004).

The statistics of the differences between the 24-hr and 12-hr forecasts are used to estimate the background error covariances *via* the National Meteorological Center (NMC) method (PARISH and DERBER, 1992). DAS GUPTA *et al.* (2005) computed the background error ($\boldsymbol{B}$) of the MM5 model using the NMC method and studied the impact of various conventional and non-conventional data sets over the Indian region. They conducted several experiments including assimilation of temperature and moisture profiles obtained from ATOVS and GPS Radio Occultation, total precipitable water from SSMI, surface wind fields from QSCAT and showed that the assimilation of these observations using the MM5-3DVAR system improved the short-range prediction of severe weather systems over the Indian region.

4. *Single Doppler Velocity Retrieval Method*

Radial velocity measurements from single-site radar can be used to obtain wind profiles under the assumption of a linear wind model. In this method, the wind in the vicinity of the radar is expressed as;

$$U(x,y,z) = u_0 + x\frac{\partial u}{\partial x} + y\frac{\partial u}{\partial y} + (z - z_0)\frac{\partial u}{\partial z} \tag{2}$$

and likewise for $V(x,y,z)$ and $W(x,y,z)$. Using this linear wind field, the radial wind can be calculated as a function of range, azimuth and elevation. For a uniform wind field this results in

$$V_r = V_h \cos\theta \cos(\varphi - \varphi_0) - V_f \sin\theta, \tag{3}$$

where V_h and φ_0 are the horizontal wind speed and direction, respectively, φ is the azimuth angle, θ is the antenna elevation angle and V_f is the fall velocity of the particle. The horizontal wind field retrieved by this method is also called the Uniform Wind Technique (UWT). The wind field derived by the UWT technique is valid at constant or small elevation angles close to the radar site whether it is during precipitation or clear weather, but in uniform wind condition. The uniform wind is assumed only in volume scan.

When Doppler radar data are displayed at constant range and elevation θ, the radial wind is a function of azimuth φ, will have the form of sine. The wind speed and direction can be determined from the amplitude and phase of the sine, respectively. This technique is called Velocity-Azimuth Display (VAD) and it was introduced by LHERMITTE and ATLAS (1961) and BROWNING and WEXLER (1968).

Currently, Doppler radars are recording volume scans of reflectivity and radial wind data as a function of range, azimuth and elevation. Different VADs form volume scans as a function of height, and a wind profile at the radar site, can thus be obtained.

Instead of processing for each height, a single VAD or a series of VADs, one can also process all available volume data in a certain height layer at once. This so-called Volume Velocity Processing (VVP) technique has been introduced by WALDTEUFEL and CORBIN (1979). Using equation (2) of the linear wind model, the radial wind can be calculated for all points within a layer centered at height; z_0 via multi-dimensional and multi-parameter linear fit, the parameters of the linear wind can be extracted.

The algorithm developed by PERSSON and ANDERSON (1979) has been used to retrieve horizontal wind fields from measured Doppler velocity. The radial velocity can be expressed as

$$V_{r\varphi} = u_{r\varphi} \sin(\varphi) + v_{r\varphi} \cos(\varphi), \tag{4}$$

where $u_{r\varphi}$ and $v_{r\varphi}$ are the U and V components of the horizontal wind vector at range (r) and azimuth φ. The U and V components of the horizontal wind are estimated using the least-squares regression technique to minimize the function $\sum_{\varphi} \left[u_{r\varphi} \sin\varphi + v_{r\varphi} \cos\varphi - V_{r\varphi}\right]^2$ for each elevation range r. The regression is performed for each cell (any value from 4 to 100 km/cell) defined in the resolution of the grid. Details of this technique are available in RAO *et al.* (2004).

ANDERSON (1998) has published a verification of the VAD winds against radiosonde winds and winds from a NWP model. In this study, the availability and accuracy of the VAD winds of Swedish radar have been investigated. The availability of VAD winds is about 80% at 925 hPa, and it drops to about 15% at 400 hPa. The vector difference between the VAD winds and the radiosonde winds has an average magnitude of about 2 ms^{-1}.

RAO *et al.*, (2004) have compared the horizontal wind fields obtained from the DWR with Rawindsonde and Pilot balloon observations and found good agreement between them. Further, BANERJEE *et al.* (2004) found that these wind fields compared well with synoptic and upper air observations in a cyclonic situation of the 12 Nov., 2002 cyclonic storm over the Bay of Bengal. DAS *et al.* (2006) also verified the retrieved wind fields with Pilot balloon observations and wind analysis from NCMRWF T80 model interpolated to a mesoscale domain. Figure 5(a) presents the wind field retrieved from DWR Kolkata at a height of 1.0 km at 0000 UTC 5th May, 2005 and, Figure 5 (b) shows the wind analysis at 900 hPa obtained from the MM5 domain valid at the same time. The retrieved wind from DWR shows good agreement with the model wind analysis.

5. Observation Preprocessing and Quality Control

The MM5-3DVAR has a module, "3DVAR_OBSPROC" for pre-processing and quality control of observations. It packs the observations in a suitable format for ingest into 3DVAR. The main goal of the package is to (i) perform spatial and temporal checks to select the only observations located within the target domain and a specified time-window and then (ii) merge duplicate observations (same location and type) and (iii) choose observations nearest the analysis time. It estimates and assigns the error for each observation and performs a variety of quality control checks (e.g., check for negative wind speed, spike in wind profile, spike in temperature profile, superadiabatic lapse rate, height above/below the model lid/surface, etc.). Finally it packs the observations in 3DVAR format. The DWR wind fields are packed in the same way as the Pilot wind observations in the data file. Presently, no separate instrumental error profile is used for the DWR wind fields. Thus DWR winds are treated similar to the Pilot winds for the six vertical levels from 0.5 to 3.0 km.

XIAO *et al.* (2004, 2005) evaluated the impact of Doppler radar observations on the forecast of heavy rainfall associated with a squall line and a typhoon using the KMA model. They included analysis increments for the vertical velocity, cloud water mixing ratio and rain water mixing ratio in their WRF/MM5 3DVAR system. Because total water mixing ratio is used as a control variable, partitioning the moisture and hydrometeor increments is necessary in the 3DVAR system. They found that assimilation of radial velocity and reflectivity data improved the Quantitative Precipitation Forecast (QPF) skills for squall line, mesoscale cyclone and tropical cyclone cases. Also the QPF skills are further improved with 3DVAR cycling mode. Real-time application with the KMA operational model also indicates a statistically significant positive impact of the Doppler radar data assimilation on the short-range forecast of QPF.

6. *Experimental Design*

Figure 3 shows the domain configuration for the MM5 experiments. Two-way nesting is employed with a horizontal resolution of 30 km for the coarse outer grid (4.6–38.7°N, 65.3–99.0°E) and 10 km for the inner fine grid (19.5–25.2°N, 85.2–93.5°E). The location of the Kolkata radar site is marked as a circle and its maximum range is 250 km. The model physics include the Grell scheme for Cumulus parameterization (GRELL *et al.*, 1994), boundary layer parameterization of HONG and PAN (1996) as used in the Medium Range Forecast (MRF) model, explicit treatment of cloud water, rain water, snow, ice and graupel based on GSFC (NASA/Goddard Space Flight Center) microphysical scheme (TAO *et al.*, 1989; 1993). Cloud radiation interaction is allowed between explicit cloud and clear air.

To assess the impact of assimilating the DWR wind data into the MM5 3DVAR system on the prediction of mesoscale convective events, three experiments were carried out. The initial and lateral boundary conditions obtained from the global analysis of National Center for Medium Range Weather Forecasting (NCMRWF) global model (T80–resolution) are used to initialize the first (CTRL/GSFC) experiment. The offline experiments have been carried out using 3DVAR data assimilation system of MM5. In the second experiment (3DVAR/NoDWR), global data received through Global Telecommunication System (GTS)/ Internet is ingested in the system at the analysis time with ± 3-hr time window. In the 6-hr assimilation cycle, cold-start is at 0600 UTC of 4th May, 2005 so that four cycles are used to initialize the model. Modules have been developed for reading the decoded observed data from NCMRWFs operational data sets and for packing in LITTLE_R format, required for ingesting in 3DVAR observation preprocessor. In this six-hour 3DVAR update cycle, the 6-hour forecast from the previous cycle serves as the background for the next cycle. The third experiment (3DVAR/DWR) is the same as the second experiment, except that the horizontal wind derived from the DWR is used along with other conventional and non-conventional data in the assimilation system. Table 1 details the data sets used for the assimilation experiments. A summary of the experiments is presented in Table 2. Figure 4 shows the overview of the experiment design for radar data assimilation.

7. *Results*

7.1. Simulated Rainfall

The main purpose of assimilating the high resolution DWR wind data is to help improve the heavy rain forecast associated with quasi-stationary MCS. Figure 5(a) shows the 6-hr accumulated rainfall from the merged satellite product of TRMM (the details of the 3B42 data set can be found on http://lake.nascom.nasa.gov/data/dataset/TRMM). Two rainbands were observed around Kolkata; one over the north and the other over the southwest of Kolkata during 0600–1800 UTC on 5 May. The major rainbands, initially originating from the northwest, propagate southeastwards during 0600 UTC of 5 May and 0000 UTC of 6 May, which is a typical characteristic of *Norwesters* (*the severe thunderstorms that move from Northwest to Southeast direction over the West Bengal region during the pre-monsoon season*). A small patch of rainband is also observed over northwest Kolkata at 1200 UTC on 6 May, which

Table 1

Description of data used in the assimilation experiments

Data	Description
SYNOP	Surface observations from land stations
SHIP	Voluntary observation from sea
BUOY	Drifting and moored buoy observations
TEMP	Upper air profiles of temperature, humidity and wind from radiosonde
PILOT	Wind profiles from optical theodolite
AIREP/ AMDAR	Upper level wind and temperature reported by aircrafts
SATOB	Satellite observed cloud motion vectors from INSAT, METEOSAT-6, GMS and GOES
SATEM	Satellite observed wind and total precipitable water from NOAA series of satellites

Table 2

Summary of MM5 forecast experiments

Experiment	Description
CTRL_GSFC	Using global analysis of NCMRWF T80 model without data assimilation.
3DVAR_NoDWR	3DVAR 6-hr assimilation cycle using GTS data.
3DVAR_DWR	3DVAR 6-hr assimilation cycle using GTS and DWR wind observations.

moved southward and centered around the south and southeast of Kolkata at 1800 UTC on 6 May. Though the intensity of precipitation is less, this precipitation band also shows a southeastward movement.

Figures 5 (b), (c) and (d) present simulated rainfall by the three experiments, namely CTRL_GSFC, 3DVAR_NoDWR and 3DVAR_DWR, respectively. The CTRL-GSFC simulation shows 6–8 cm of widespread rainfall between 0600–1800 UTC of 6 May and 3DVAR_NoDWR predicts the similar association of rain bands with intensity little reduced as compared to the CTRL_GSFC experiment, but the TRMM does not show much rainfall at these times. The rainfall observed by TRMM may be underestimated as it depends on the pass of the satellite and the rain gauge observations included in its analysis. The observed rainfall amount from TRMM is more during 0600–1800 UTC of 5 May. Both of the experiments show a southeastward propagation of the precipitation bands as observed by TRMM. However, the Kalpana satellite cloud images (Fig. 1) and radar reflectivity (Fig. 2) do show clouds between 1200–1800 UTC of 6 May.

The simulated rainfall from 3DVAR_DWR indicates that assimilation of DWR data has produced rainfall during 1200–1800 UTC of both 5th and 6th May, south of Kolkata as seen from TRMM observations and satellite images. The CTRL_GSFC experiment overestimates the rainfall. While overall rainfall from 3DVAR_NoDWR is reduced as compared to the control simulation and further reduced in the 3DVAR_DWR experiment, making it closer to observations. The aerial extent of the precipitation band is also reduced in the 3DVAR_DWR experiment. The CTRL_GSFC and 3DVAR_NoDWR did not produce realistic rainbands on 5 May. An important result from the 3DVAR_DWR experiment is the prediction of the rain bands on both the days and the generation of new rain bands northeast of Kolkata on 5 May, 2005. While the CTRL_GSFC and 3DVAR_NoDWR experiment predicted widespread rainfall over the entire domain on 6th May, 2005.

7.2. Horizontal and Vertical Wind Fields

In this section we examine the horizontal and vertical wind fields at 1800 UTC on 6 May, 2005. Figures 7(a) and (b) show the horizontal wind vectors and vertical velocity of the three experiments at 850 and 700 hPa, respectively. Southwesterly winds are present south of Kolkata and northwesterly winds are seen north of Kolkata at 850 hPa. Both the assimilation experiments produced strong winds at

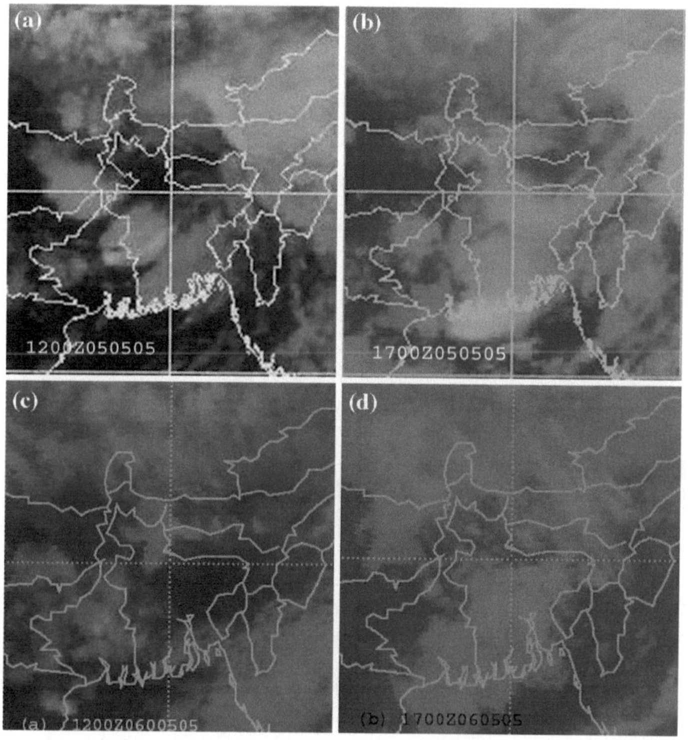

Figure 1
(a, b, c, d): Satellite Cloud images from Kalpana Satellite at a) 1200 UTC, b) 1700 UTC of 5 May and, c) 1200 UTC, d) 1700 UTC of 6 May 2005.

850 hPa as compared to the CTRL_GSFC experiment. The maximum vertical velocity is also found near the region where the horizontal wind convergence occurs. The 3DVAR_DWR experiment predicts two positive vertical velocity cores (solid circles) and a negative core (dashed circle) in between at 850 hPa, while scattered positive regions are simulated by two other experiments. The 700 hPa horizontal wind is nearly westerly to northwesterly. The horizontal wind speed at 700 hPa is reduced in the assimilation experiments, while the CTRL_GSFC predicted strong winds at this level. At the 700-hPa level, the 3DVAR_DWR experiment simulated three positive cores (solid circle) of vertical velocity whereas the other two experiments show similar vertical velocity structure as at 850 hPa. The three upward motion bands are 50–100 km apart. The maximum vertical velocity of each band reaches more than 50 cm s^{-1}. This analysis also confirms the evidence of elevated convection and is well simulated by the 3DVAR_DWR experiment.

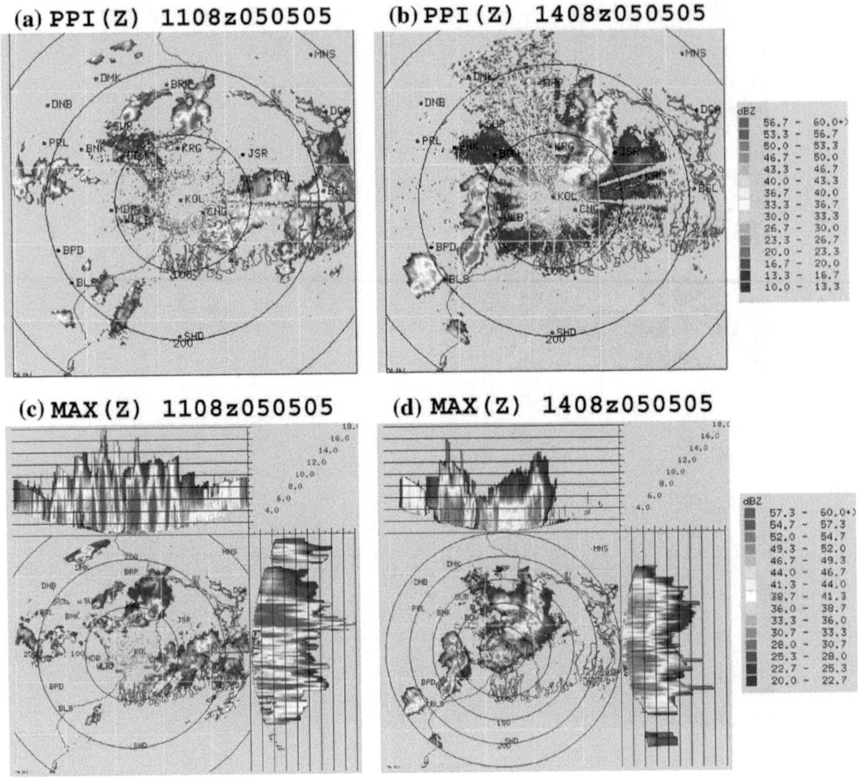

Figure 2

(a, b, c, d): Reflectivity PPI (Z) in dBZ at a) 1108 UTC, b) 1408 UTC and, MAX(Z) in dBZ at c) 1108 UTC, and d) 1408 UTC of 5 May 2005.

7.3. Composite Reflectivity and Integrated Liquid Water

Figure 8(a) shows the simulated composite radar reflectivity (shading) and surface pressure (dotted contours) of the three experiments valid at 1800 UTC on 6 May, 2005. The model-derived echoes from the CTRL_GSFC experiment are distributed over a large area with maximum reflectivity of more than 50 dBZ over the Kolkata region, while that from the 3DVAR_NoDWR experiment shows two individual maxima in radar reflectivity over and west of Kolkata. The 3DVAR_DWR experiment simulates radar echoes extending from the southwest to northeast of Kolkata with three distinct maxima of reflectivities (> 50 dBZ) one over Kolkata, the second one over the southwest and third over the northeast of Kolkata. The radar reflectivity predicted by the 3DVAR_DWR experiment is almost close to observed

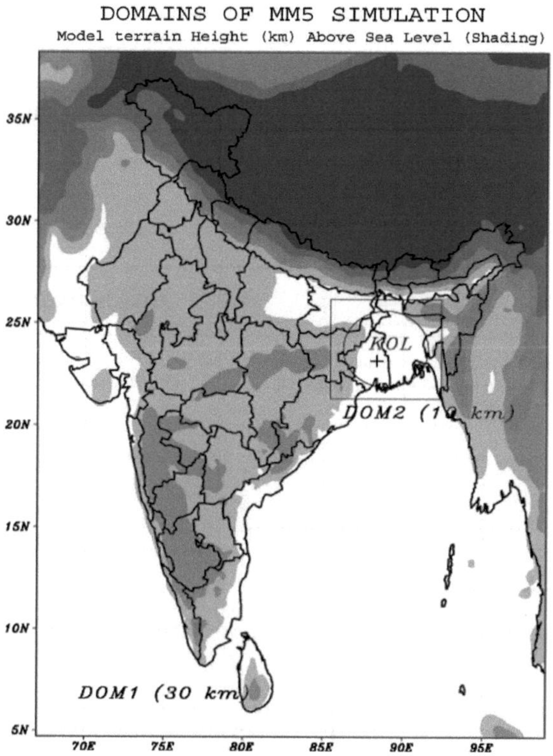

Figure 3
Domain configuration for MM5-3DVAR experiments.

low-level (elevation angle of 0.5 degree) radar reflectivity from the DWR (Fig. 2). The main result from this analysis is that the intensity, location and organization of MCS are better simulated by the 3DVAR_DWR experiment. Though the simulated reflectivity does not match exactly with observations in terms of location, its magnitudes are of the same order and range from 10–50 dBZ. The integrated cloud water represents the vertically integrated cloud water given as the depth of the water substance over the area of the grid cell in millimeters (mm) contained within the entire MM5 grid column from surface to the model top. Figure 8(b) shows the vertically integrated liquid water (shading) and potential temperature (dotted contours). All three experiments simulated the vertically integrated liquid water (VIL) reasonably well during 1200–1800 UTC of 6 May, 2005. However, the 3DVAR_DWR experiment produced better VIL at 1200 UTC on 5 May, 2005, corresponding to the rainfall observed by TRMM and radar reflectivity.

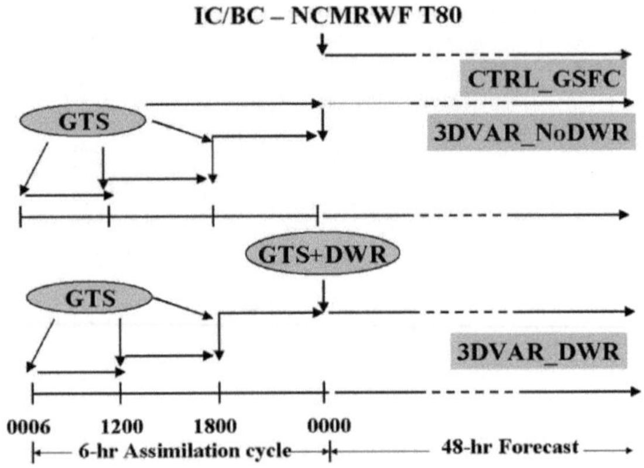

Figure 4
Experiment design for radar data assimilation.

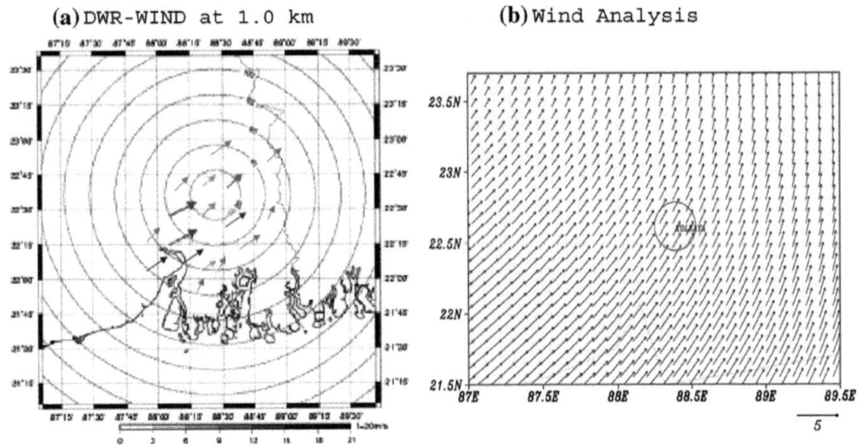

Figure 5
(a) Wind field retrieved from DWR at 1.0 km, (b) wind analysis at 900 hPa interpolated to MM5 domain at 0000 UTC 5th May.

7.4. *Hydrometeor Fields*

The physics options of the simulation experiments are able to predict the hydrometeor species like mixing ratios of cloud liquid water, rainwater, ice, snow and graupel. Figure 9 shows the latitude-height section of the sum of these five species, called total cloud hydrometeor (TCH) averaged over 87–89°E and centered around

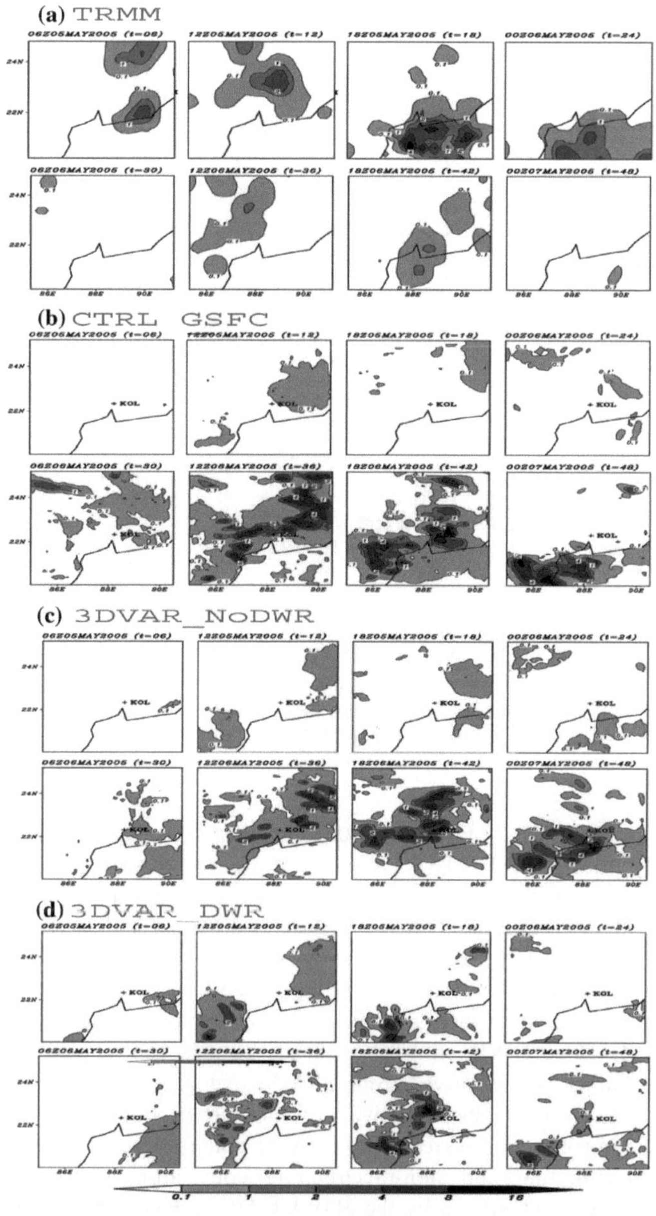

Figure 6

(a,b,c,d): Six hourly accumulated precipitation (cm) from the (a) TRMM, (b) CTRL_GSFC, (c) 3DVAR_NoDWR, (d) 3DVAR_DWR experiment based on initial condition of 00 UTC, 05 May, 2005.

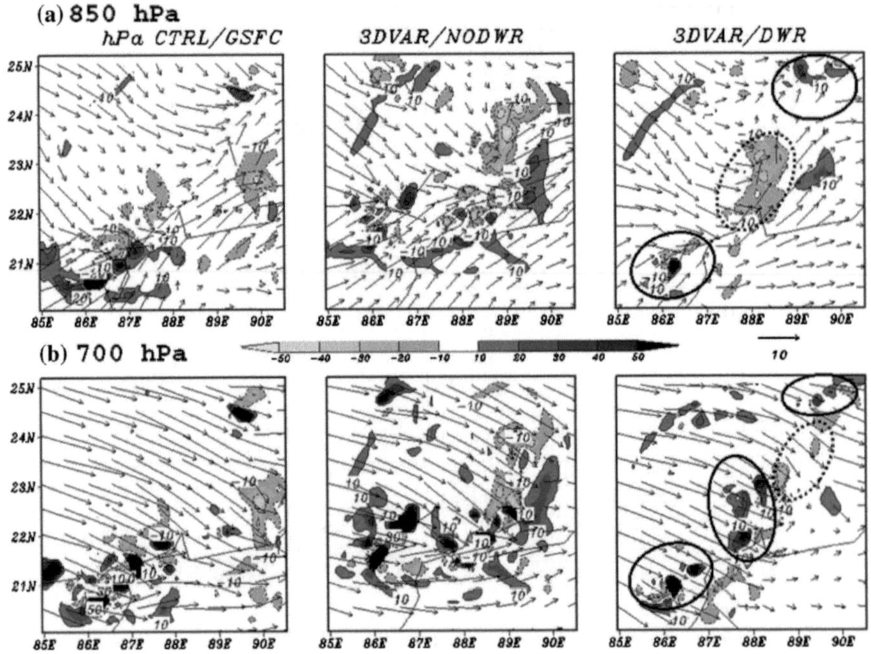

Figure 7
(a,b): Horizontal velocity vector and vertical velocity in cm s^{-1} (shading) at a) 850 hPa and b) 700 hPa valid for 1800 UTC 06 May, 2005.

88°E valid at 1800 UTC of 6 May, 2005 for the three experiments. The TCH simulated by the 3DVAR_DWR experiment clearly shows evidence of three individual convective cloud bands and such profiles are not available from DWR for comparison. An interesting feature observed in Figure 9 is that the hydrometeor structure shows a strong precipitating convection over the south and a small patch over the northeast of Kolkata and is well supported by TRMM observations. The 3DVAR_DWR experiment also simulated a non-precipitating patch over Kolkata and MAX (Z) from DWR showing evidence of this central cell, however no precipitation bands were observed from TRMM over the area. The PPI (Z) from DWR does not show strong reflectivity at the surface during the period. This supports the occurrence of elevated convection. The major outcome of this analysis is that both precipitating and non-precipitating convection as well as elevated convection are better simulated by the 3DVAR_DWR experiment as compared to the CTRL_GSFC and 3DVAR_NoDWR. Latitude-height section of radar reflectivity derived from the model also shows similar structure (Fig. 9(b)). Time series of

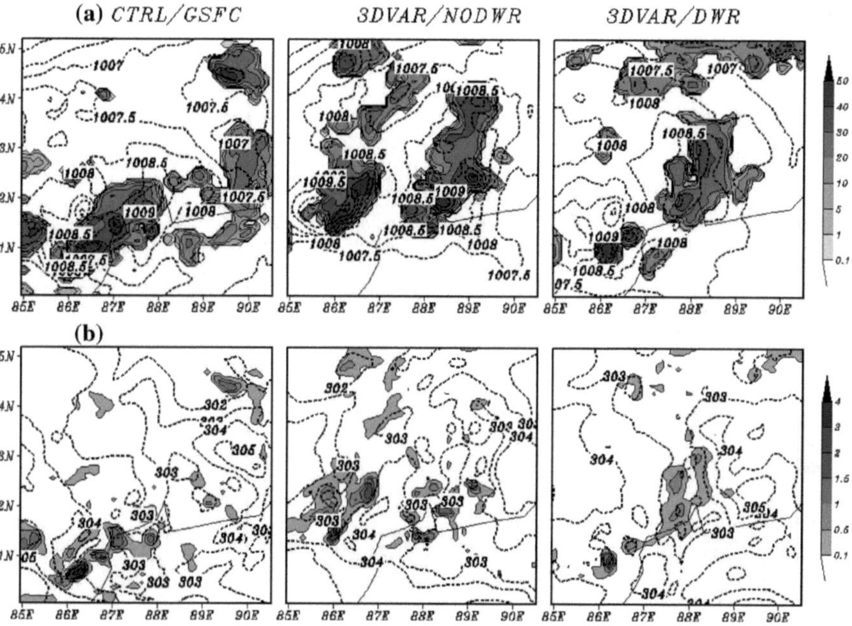

Figure 8

(a,b): Simulated a) Composite reflectivity in dBZ (shading) and surface pressure (dotted contours at 3 hPa intervals), b) vertically integrated liquid water (VIL) in mm and potential temperature (dotted contours at 3°K intervals) from the three experiments valid at 1800 UTC 06 May, 2005 based on initial condition at 0000UTC 05 May, 2005.

the cloud water mixing ratio and rainwater mixing ratio is shown in Figure 10. The 3DVAR_DWR experiment predicts the occurrence of clouds during 1200–1800 UTC on both 5 and 6, May, while CTRL_GSFC and 3DVAR_NoDWR predict the occurrence of convection only on 6 May, 2005.

8. *Summary and Concluding Remarks*

Obtaining an accurate initial state is recognized as one of the biggest challenges in accurate model prediction of convective events. This study focuses on the impact of Doppler Weather Radar (DWR) and other conventional and non-conventional data in simulating intense convective events over Kolkata. The retrieved wind field from DWR has been used in the three-dimensional data assimilation system of MM5. The convection event has been selected based on the radar reflectivity and observed rainfall.

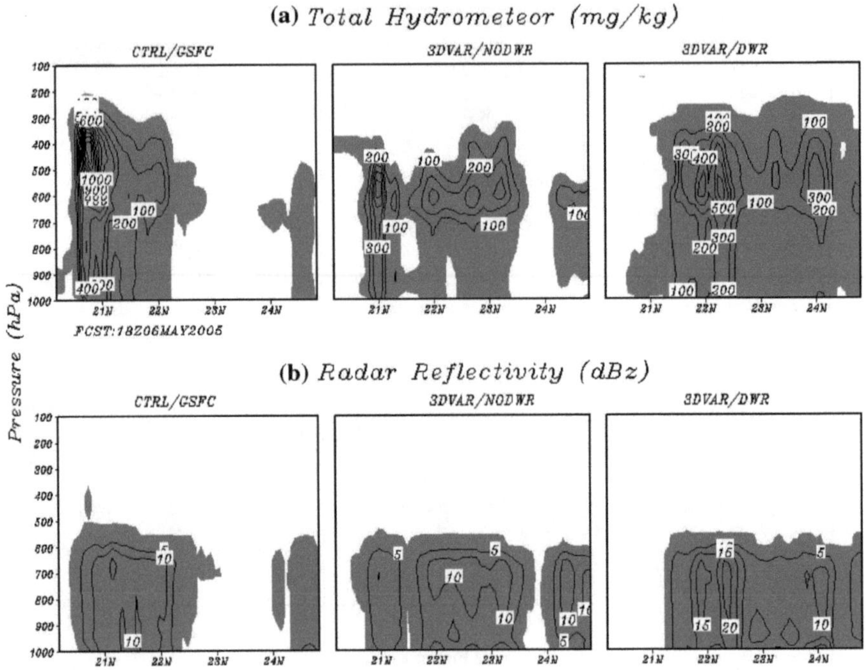

Figure 9
(a,b); Latitudinal cross section of a) total cloud hydrometeor (mg kg^{-1}) and b) radar reflectivity averaged along 2° longitude strip centered at 88°E for the three experiments valid at 1800 UTC 06 May, 2005.

The main purpose of assimilating the Doppler radar data is to enhance the short-range prediction of precipitation associated with the mesoscale convective complexes. The simulated rainfall has been compared with observed rainfall from TRMM. The 3DVAR experiment, especially the 3DVAR_DWR simulated rainfall, is closer to the observed precipitation both in intensity and location. A detailed analysis of the microphysical structure of the convective events has been made. The total cloud hydrometeor, liquid water mixing ratio and the rainwater-mixing ratio have been simulated well in the 3DVAR_DWR experiment and a comparison has been made with the reflectivity obtained from the DWR. The vertically integrated liquid and the composite reflectivity simulated by the experiments are comparable in magnitude with that derived from the radar. The intensity and structure of the system were better simulated by the 3DVAR experiments.

The attempt in utilizing the IMD Doppler Weather Radar data into a numerical mesoscale model has been accomplished with this study. In this regard, the

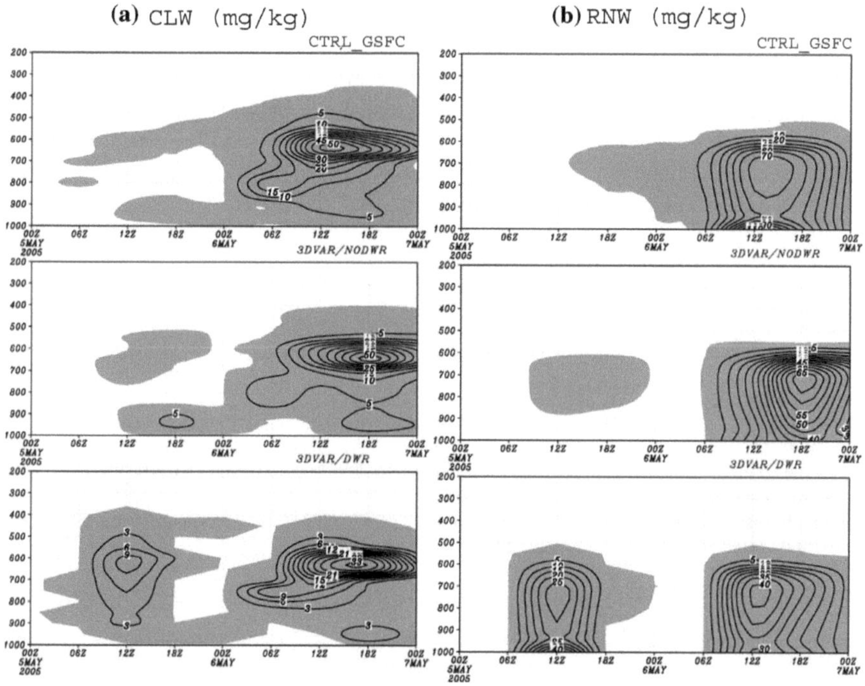

Figure 10

(a, b): Time section of a) cloud water mixing ratio (mg kg^{-1}) and b) rain water mixing ratio (mg kg^{-1}) averaged over 2° latitude/longitude around Kolkata for the three experiments based on the initial condition at 0000 UTC 05 May, 2005.

preliminary results are quite encouraging. Although the results show some positive impact of radar data assimilation on the short-range prediction of convective systems, these results are preliminary and further investigation is being carried out to assimilate the radar reflectivity and radial velocity fields in the assimilation system.

Acknowledgements

Results presented in this paper are based on the collaborative works between the National Center for Medium Range Weather Forecasting (NCMRWF), India Meteorological Department (IMD) and Cochin University of Science and Technology. We would like to thank the National Center for Atmospheric Research (NCAR), USA, in particular, Dr. Dale Barker for his immense assistance provided on the Mesoscale 3DVAR assimilation system, and to Dr. Mitch Moncrief for many stimulating discussions. We thank Mr. Prashant Mali and Dr. L. Harendu Prakash for their initial aid in reading the DWR data and graphic display of wind fields. Appreciation is also due Dr. B. Lal (Director General of Meteorology, IMD), Dr. R. C. Bhatia (Additional Director General of Meteorology, IMD) for their support. The first author gratefully acknowledges the Council of Scientific and Industrial Research (CSIR) for providing Junior Research Fellowship. We are also grateful to the referees for their valuable comments that enhanced the paper.

References

ANDERSON, T. (1998), *VAD winds from C band Ericsson Doppler Weather Radars,* Meteor. Zeitschrift *7*, 309–319.

BANERJEE, S. K., KUNDU, S. K., and SINGH, H. A. K. (2004), *A study of 12th November 2002 Cyclonic storm in the Bay of Bengal using Doppler Weather Radar*, Mausam, *55*, 4, 671–680.

BARKER, D. M., HUANG, W., GUO, -Y. R., and BOURGEOIS, Al (2003), A *Three-Dimensional Variational (3DVAR) Data Assimilation System for Use With MM5*, NACR Technical Note, NCAR/TN-453 + STR, February 2003, 68 pp.

BARKER, D. M., HUANG, W., GUO, -Y. R., and XIO, Q. N. (2004), A *three dimensional Variational (3DVAR) Data Assimilation System with MM5: Implementation and Initial results,* Mon. Wea. Rev. *132*, 897–914.

BROWNING, K. A. and WEXLER, R. (1968), Thc *determination of kinematic properties of a wind field using Doppler radar*, J. Appl. Meteor. *7*, 105–113.

DAS SOMESHWAR., ABHILASH, S., DAS GUPTA, M., GEORGE, J. P., KALSI, S. R., BANERJEE, S. K., THAMPI, S. B., PRADHAN, D., and MOHANKUMAR, K. (2006), *Assimilation of Doppler Weather Radar wind in a mesoscale model and their impact on simulation of thunderstorms and severe weather systems,* Report no. NMRF/RR/01/2006, 120 pages, Published by NCMRWF, A-50, Institutional Area, Sector-62, NOIDA, India 201 307.

DAS GUPTA, M., GEORGE, J. P., and DAS, S. (2005), *Performance of MM5-3DVAR system over Indian Subcontinent,* International Conf. on MONEX and its Legacy, 3–7 Feb. 2005, Delhi.

DUDHIA, J., GILL, D., K., MANNING, K., WANG, W., and BRUYERE, C. (2002), *PSU/NCAR Mesoscale modeling System (MM5 version 3) tutorial class notes and user's guide*. Available from National Center for Atmospheric Research, Boulder, Colorado, USA, June 2002.

GRELL, G. A., DUDHIA, J., and STAUFFER, D. R. (1994), A *description of the 5th generation Penn State/ NCAR Mesoscale model (MM5),* NCAR Technical Note, NCAR/TN-398 + STR, pp 117.

HOUZE, R. A. Jr. (1982), *Cloud clusters and large-scale vertical motions in the Tropics,* J. Met. Soc. Japan *60*, 396–410.

HONG, S. -Y. and PAN, H.-L. (1996), *Nonlocal boundary layer vertical diffusion in a medium-range forecast model,* Mon. Weather. Rev. *124*, 2322–2339.

IDE, K., COURTIER, P., GHIL, M., and LORENC, A.C. (1997), *Unified notation for data assimilation: Operational, sequential and variational,* J. Met. Soc. Japan *75*, 181–0189.

LHERMITTE, R. M. and ATLAS, D. (1961), *Precipitation motion by pulse Doppler radar*. 9th Conf. Radar Meteorology, Kansas City, Amer. Meteorol. Soc. pp. 218–223.

LORENC, A. C. (1986), *Analysis methods for numerical weather prediction,* Q. J. R. Met. Soc. *112*, 1177–1194.

MADDOX, R. A. (1980), *Mesoscale convective complexes,* Bull. Amer. Meteorol. Soc. *108*, 1374–1387.

PARISH, D. F. and DERBER, J.C. (1992), *The National Meteorological Centre's Spectral Statistical Interpolation analysis system,* Mon. Weather. Rev. *120*, 1747–1763.

PERSSON, P. O. G. and ANDERSON, T. (1987), *A Real Time System: Automatic wind field interpretation of Doppler radar wind components,* Proc. Symp. Mesoscale Analysis and Forecasting, Vancouver, European Space Agency, SP282, 61–66.

RAO, R. P., KALYANA SUNDARAM, S., THAMPI, S. B., SURESH, R., and GUPTA, J. P. (2004), *An overview of first Doppler Weather Radar inducted in the cyclone detection network of India Meteorological Department,* Mausam *55*, 155–176.

TAO, W. -K., SIMPSON, J., and MCCUMBER, M. (1989), *Ice-water saturation adjustment,* Mon. Wea. Rev. *117*, 231–235.

TAO, W.-K, SUI, C,-H., FERRIER, B., LANG, S., SCALA, J., CHOU, M.-D., and PICKERING, -K. (1993), *Heating, moisture and water budgets of tropical and midlatitude squall lines: Comparisons and sensitivity to longwave radiation,* J. Atmos. Sci. *50*, 673–690.

WALDTEUFEL, P., and CORBIN, H. (1979), *On the analysis of single Doppler radar data*, J. Appl. Meteor. *18*, 532–542.

XIAO, Q., KUO, Y.-H., SUN, J., WEN-CHAU LEE., EUNHA LIM., GUO, Y.-R., and BARKER, D. M. (2005), *Assimilation of Doppler radar observations with a regional 3-DVar system: Impact of Doppler velocities on forecasts of a heavy rainfall case,* J. Appl. Meteor. *44*, 768–788.

XIAO, Q., KUO, Y.-H., SUN, J., LEE, W.-C., BARKER, D. M., and EUNHA LIM. (2004), *Assimilation of Doppler radar observations and its impacts on forecasting of the landfalling typhoon Rusa (2002),* Proc. Third European Conf. on Radar in Meteorology and Hydrology (ERAD), Vol. 2, pp. 178–182.

(Received April 10, 2006, accepted October 17, 2006)
Published Online First: June 19, 2007

Pure appl. geophys. 164 (2007) 1511–1525
0033–4553/07/091511–15
DOI 10.1007/s00024-007-0234-3

Pure and Applied Geophysics

Observation of Sea Breeze Front and its Induced Convection over Chennai in Southern Peninsular India Using Doppler Weather Radar

R. Suresh

Abstract— Sea breeze, the onshore wind over a coastal belt during daytime, is a welcoming weather phenomenon as it modulates the weather condition by moderating the scorching temperature and acts as a favourable mechanism to trigger convection and induce precipitation over coastal and interior locations. Sea breeze aids dispersal of pollutants as well. Observational studies about its onset, depth of circulation and induced precipitation have been carried out in this paper for the period April to September, 2004–2005 using a S-band Doppler Weather Radar functioning at Cyclone Detection Radar Station, India Meteorological Department, Chennai, India. The onset of sea breeze has been observed to be between 0900 and 1000 UTC with the earliest onset at 0508 UTC and late onset at 1138 UTC. The frequency is greater during the southwest monsoon season, viz., June – September and the frequency of initial onset is greater in north Chennai. The modal length of sea breeze is between 20 and 50 km with extreme length as high as 100 km also having been observed. Though the inland penetration is on average 10 to 20 km, penetration reaching 100 km was also observed on a number of cases. The induced convection could be seen in the range 50–100 km in more than 53% of the cases. The mean depth of sea breeze circulation is 300–600 m but may go well beyond 1000 m on conducive atmospheric conditions.

Key words: Sea breeze, Doppler Weather Radar, convection, wind shear, CAPE.

1. Introduction

The differential heating due to contrasting specific heat capacity of land and water causes circulations that vary in spatial and temporal scales. While the sea breeze and land breeze are observed in spatial scales of a few tens of km, the gigantic monsoon circulations extend to several hundreds of km. Regarding temporal scale, the sea breeze/land breeze last for a few hours while that of monsoonal circulation persists for a few months. The onshore surface winds during daytime, the sea breeze, and the nocturnal offshore surface winds, i.e., the land breeze, have been studied as early as the 1890s, albeit the monsoonal circulations have been studied time immemorially since 4 B.C (Asnani, 1993). Sea breeze plays a dominant role in

Aerodrome Meteorological Office, IMD, Chennai 600 027, India. E-mail: suresh.imd@gmail.com; dr_r_suresh@yahoo.co.in; mochennai@imdmail.gov.in modulating weather conditions over coastal areas (BLANCHARD and LOPEZ, 1985) and dispersion of pollutants along coastal regions (RHOME *et al.*, 2003). Detailed descriptions and reviews of earlier works on sea/land breeze circulations can be seen in ATKINSON (1981), PIELKE (1984), SIMPSON (1994), among others.

The atmospheric conditions prevailing in the boundary layer, surface friction, marine and land pollution besides synoptic flow decide the strength and spatial distribution of sea breeze. Laboratory experiments and numerical modeling studies have confirmed that gradient flow and local terrain affect the development of sea breeze (PIELKE 1984; ARRITT, 1993). Weather radars have been used to identify the signatures of sea breeze which have risen out of inhomogeneities in refractive index caused by variations in moisture field when the moist and humid sea air interacts with the prevailing dry continental air (WILSON *et al.*, 1994; SAUVAGEOT, 1996; RINEHART, 1999). ACHTEMEIER (1991) has documented that weather radar can detect echoes from clear air, also when the insects are lifted by the convergence-caused wind-discontinuity, like sea breeze, and such echoes can be used for boundary layer studies. Use of weather radars for the boundary layer studies have been excellently dealt with in celebrated books on radar meteorology by ATLAS (1990), SAUVAGEOT (1992), DOVIAK and ZRNIC (1993) and RAGHAVAN (2003), to name a few.

Hitherto the studies on sea breeze over Chennai on the East Coast of India have been carried out based on observations made from conventional meteorological observatories at Nungambakkam, Meenambakkam airport, and Indian Air Force at Tambaram, besides a few temporary observatories set-up for the study purpose at some interior locations. However, these studies are of limited scope since the precise lateral and vertical extents to which the sea breeze propagates and the location at which the associated convection triggers cannot be determined. However, with the induction of a 10-cm Doppler Weather Radar (DWR) by replacing the outlived S-band analogue radar meant for cyclone detection purpose at Chennai, it has now become possible to study in depth, the characteristics of sea breeze front and validate the numerical modeling results.

2. *Observations*

The DWR has been in operation 24 hrs a day since its commissioning on 20.2.2002 at Cyclone Detection Radar (CDR) station, Chennai (13.08°N/80.29°E). The technical details of the DWR supplied by M/s Gematronik GmbH, Neuss, Germany have been tabulated in Appendix 1. Being an operational weather radar for National meteorological and hydrological services, a number of scan strategies for weather surveillance are adopted in a cyclic loop of 30-minutes interval throughout the year. Those scan strategies used in this paper, have been summarized in Appendix 2. The radar echoes from the sea breeze front appear as a 'thin line of enhanced reflectivity' due to inhomogeneities in the refractive index.

In view of range-height limitations of the radar, it is desirable to have a number of PPI displays of different elevations to mentally visualize the mesoscale sea-breeze circulation. In other words, one needs to consult many plan position indicator (PPI) or constant altitude PPI (CAPPI) type displays to understand the height of the sea-breeze circulations. To quickly view the data collected from a volume scan with a number of elevations, one form of display, viz.: maximum projection display is used to identify the signature of sea breeze at different ranges from the radar. Hence the maximum projection type display, which is possible from a digital radar, is quite helpful in monitoring the thin line echo of enhanced reflectivity.

Figure 1 shows maximum plot of radar reflectivity factor (dBZ) at 1015 UTC on 27 June, 2003. The plan view shows the maximum value of dBZ over each pixel from the bottom to top layer probed by the radar. In the present case, data were collected from a scan range of 120 km and displayed in 100 km range. The top plot is the maximum value of dBZ as viewed from south to north and the right-hand side plot is the maximum value of dBZ as viewed from west to east from a volume probed by the DWR in 360° azimuth through a number of elevation angles. This type of display will help to understand the propagation of the sea breeze front from coast to inland and to estimate the length of the front which is not feasible from the ground-based

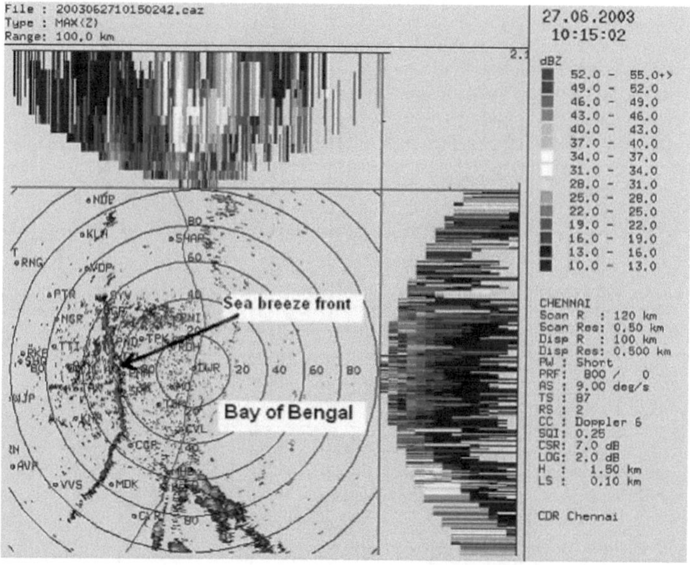

Figure 1
Maximum plot of radar reflectivity factor (dBZ) at 1015 UTC on 27 June, 2003. The plan view shows the maximum value at each pixel and the top graph denotes the maximum value as viewed from south to north and the right-hand side graph denotes the maximum value as viewed from west to east.

conventional surface observations. Figure 1 shows that the sea breeze has almost advanced, though not perfectly, parallel to the coast at a distance ranging between 40 and 60 km from the coast.

Moreover, the radar observation opens a new horizon in our understanding of the characteristics of sea breeze besides the convection induced by it. In one case of moderate wind shear reported by the aviation community, the shear was induced by sea breeze convergence (SURESH, 2004). Since the zonal flow has a westerly component during mid-March to mid-October, the sea breeze front is clearly discernible during this period over the east coast of peninsular India while during the rest of the year the sea breeze front is not fully discernible as they are superposed with the prevailing easterly flows. Horizontal wind can be estimated from the radial wind measured by DWR over a small circle not exceeding 30 km radius, from the radar using volume velocity processing (VVP) technique proposed by WALDTEUFEL and CORBIN (1978), which is an extension of the celebrated BROWNING and WEXLER (1968) algorithm on velocity—azimuth display (VAD). Figure 2 shows the vertical time section plot of horizontal wind estimated from the radial wind measured by the DWR within 30 km radius from the radar on 28 June, 2003. The DWR data have

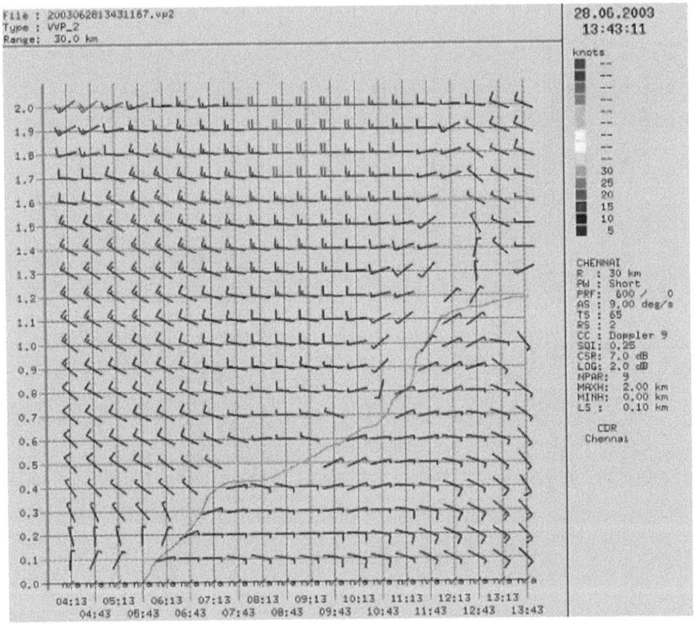

Figure 2
Vertical time section plot of horizontal wind estimated from Doppler Weather Radar at Chennai on 28 June, 2003. The sea breeze regime has been delineated.

been obtained from the same scan repeated at 30-min. intervals and hence the abscissa shows the time-step of 30 minutes and the ordinate is the height in km. The changeover from westerly to easterly and from surface to upper heights has been marked in the figure which characterizes the extent of sea-breeze front.

As the development of sea-breeze front depends on many factors and mostly on the temperature contrast between land and ocean, the sea breeze phenomenon is not observed most days during the hot weather season. For example, a quick look at the plot of horizontal wind estimated from DWR (see Fig. 3) reveals that there was no change in the wind direction throughout the day and hence it could be interpreted from this figure that there was no sea breeze on 29 June, 2003. The frequencies of occurrence/non-occurrence of sea breeze over Chennai during April to September have been documented elsewhere (see for example, ATKINSON, 1981; RAJ *et al.*, 2002).

3. Data

The onset and propagation of sea breeze over Chennai is clearly discernible from mid-April to September since westerly winds prevail at a low level during this period (RAJ *et al.*, 2002). However, persistent westerlies over Chennai on the east coast could be observed only during June – September, when the southwest

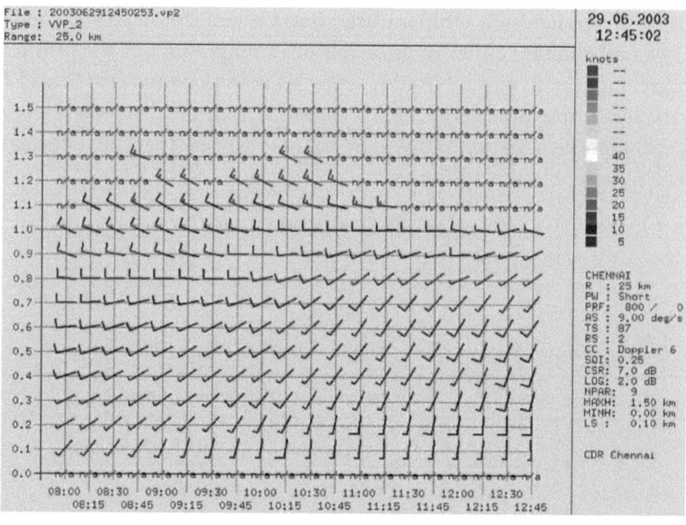

Figure 3
Vertical time section plot of horizontal wind estimated from Doppler Weather Radar at Chennai on 29 June, 2003.

monsoon current after giving copious rainfall over the west coast pass through the interior peninsular India and thereafter the east coast as dry westerlies. The interaction of onshore winds due to local effects with the relatively dry westerlies causes radar echo of enhanced reflectivity. DWR data for the period April–September, 2004–2005 have been analyzed in this study to determine the time of onset and spatial extent of propagation of sea breeze. Sea-breeze induced convection around Chennai have also been analyzed. In all 248 days data have been used in this study. In addition DWR and surface rain gauge data of 28 and 29 June, 2003 have also been utilized in this study for the purpose of analyzing sea-breeze induced convection around Chennai.

4. *Characteristics of Sea Breeze Front*

From the DWR data, the characteristics of the sea breeze front such as frequency, length, intensity, inland penetration etc have been critically analyzed.

4.1. *Frequency of Sea Breeze*

During the study period, April–September, 2004–2005, DWR data were available for analysis for 353 days. During the month of April, sea breeze could not be detected for nearly 50% of the days (see Table 1). This is so because on those days the prevailing wind already had an easterly component and the local circulation originating from the land-sea temperature contrast could not be clearly detected in view of very small changes either in direction or in magnitude of the prevailing wind. This is in agreement with earlier studies reported in ATKINSON (1981) and RAJ *et al.* (2002). In the remaining months on nearly one-fourth of the days the sea breeze front could not be identified presumably due to insufficient thermal contrast for the development of solenoidal circulation.

Table 1

Doppler Weather Radar data used for analysis of sea-breeze front over Chennai

2004–2005	No. of days	Data used	Sea breeze observed	No sea breeze
April	60	58	28	30
May	62	61	49	12
June	60	60	40	20
July	62	60	45	15
August	62	54	42	12
September	60	60	44	16
Total	366	353	248	105

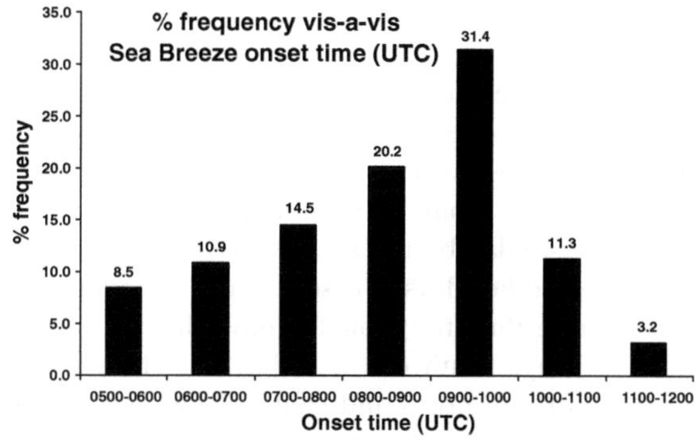

Figure 4
Sea breeze onset time over Chennai, April – September, 2004–2005.

4.2. *Time of Onset of Sea Breeze*

The onset of sea breeze depends on the strength of land-sea temperature contrast and the prevailing wind. Figure 4 depicts the frequencies of onset of sea breeze at different times. As one normally expects, the most favorable time of onset of the sea breeze is between 0800 and 1000 UTC (1330 and 1530 hrs local time) at which the maximum temperature is recorded. Nonetheless, early onset at 0508 UTC (observed on 17.7.2004) and late onset at 1138 UTC (on 6.7.2005) are not at all uncommon.

4.3. *Inland Penetration of Sea-Breeze Front*

The inland penetration of sea-breeze front has been studied from the DWR data and summarized in Table 2. The modal frequency is 10–20 km. The inland

Table 2

Inland penetration of sea-breeze front over Chennai, April – September, 2004—2005

Year 2004–2005	Inland penetration (km)						Total	% Frequency
	<10.0	10.0–20.0	20.0–30.0	30.0–40.0	40.0–50.0	> 50.0		
April	13	12	2	1	0	0	28	11.3
May	17	21	6	3	2	0	49	19.8
June	3	12	5	7	6	8	40	16.1
July	4	14	4	8	4	11	45	18.1
August	2	13	6	6	5	10	42	17.0
September	1	14	5	8	6	10	44	17.7
Total	40	86	28	33	23	39	248	100.0
% Frequency	16.0	34.6	11.2	13.3	9.2	15.7	100.0	

penetration could be seen beyond 50 km as well as in 39 out of 248 cases studied. The penetrations as far as 100 km have also been observed during the study period. Intense convection was observed at ranges farther than 50 km radius only. Though the penetration of sea-breeze front is almost parallel to the coast, there are deviations in day-to-day observations. Moreover it has been seen that the penetration of the sea-breeze front is not simultaneous north and south of the radar but varies on a day-to-day basis. On most days the penetration is seen north of the radar first while on other days it is seen south of radar. Out of 248 days of sea breeze detection during this study, on 142 days the sea breeze appeared north of the radar and on 35 days south of the radar. On 71 days the sea breeze had simultaneous penetration (both north and south of the radar). The reason for the high incidence of early northward penetration has been analyzed by studying the land-use pattern. Concentration of steel industries and location of a thermal power plant, oil refinery, etc. in north Chennai could be the reason for the high land-surface temperature triggering early sea breeze onset in north Chennai in comparison to green and thick social forest/vegetation/park over south Chennai which do not heat up as quickly as the factories and barren land dominating northern Chennai.

4.4. Length of Sea Breeze Front

Despite the range-height limitations of radar, the extent of the sea-breeze front has been worked out from the thin line echo characteristics of the front. The north–south extent of the front less than 20 km in length was observed on 22 days amounting to 8.9% frequency while 20–50 km length was observed on 164 days (66.1%). Lengths in the range 50–100 km were noticed on 37 (14.9%) and longer than 100 km was observed on 25 days (10.1%). As such the modal length of the sea breeze front is 20–50 km.

4.5. Depth of the Sea Breeze Circulation

In the earlier studies, depth of sea-breeze circulation was computed using radio wind and pilot balloon ascents taken at synoptic hours and also using a few special pilot balloon ascents taken at auxiliary synoptic hours for a few days. It has been concluded with very limited upper air ascents by BHASKARA RAO *et al.* (1984) that the depth of sea-breeze circulation around Chennai is approximately 500 m, though the depth may go beyond 1000 m as well. RAJ *et al.* (2002) has documented, based on 15 years of synoptic hours pilot balloon ascents data, that the mean depth of sea breeze varies between 490 and 765 m. In the present study, the depth has been estimated based on a westerly reversal at the upper level of the circulation. For this purpose the horizontal wind has been estimated from the measured radial wind within 30 km radius circle from the DWR by VVP algorithm as mentioned earlier. The vertical layer spacing (LS) has been chosen as 100 m for this estimation.

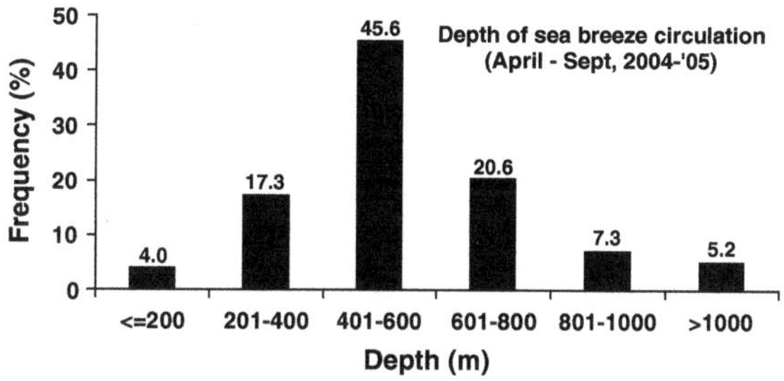

Figure 5
Depth of sea breeze circulation around Chennai, April – September, 2004–2005.

The variability of the depth of sea breeze has been shown in Figure 5. The modal depth of the circulation is between 400 and 600 m. Depths as high as 1100 m were also noticed during the study period and they were seen during July–September.

4.6. Propagation of Sea Breeze and its Speed at Surface

As the sea breeze opposes the prevailing strong zonal wind due to the southwest monsoon, the speed of the sea breeze at the surface is relatively low whereas the speed reaches its maximum (around 7.7 m s^{-1}) at heights between 300 and 600 m. The north–south oriented Chennai coast line separates the Bay of Bengal in the east from land area in the west. Hence sea-breeze circulation is expected to be from the east at the surface level and the return flow from the west at the upper level. However, observation reveals that the sea-breeze circulation is not zonal (east – west circulation) always. As has been discussed in section 4.3, the sea-breeze circulation was seen on many days from the south at the surface level and the return flow is from the north at the upper level. Hence meridional circulation (north–south circulation) has been observed on many occasions. This could perhaps be the reason as to why the penetration of sea breeze is not always parallel to the coast and maximum onset frequency is seen towards the north of the radar. The mean wind during sea-breeze occurrences has been computed and depicted in Figure 6. The propagation of sea-breeze front over inland is quite slow (about 4 km h^{-1}) close to the coast up to 30 km or so but its movement is fast between 30 and 80 km at a rate of approximately 12–15 km h^{-1}.

4.7. Sea Breeze Induced Convection

As the sea-breeze front propagates inland, the convergence at the leading edge of the front along with the pumping of moisture from the sea is favorable for triggering convection. During summer and the southwest monsoon season, the occurrence of rainfall over the east coast and little interior is often attributed to sea breeze since this

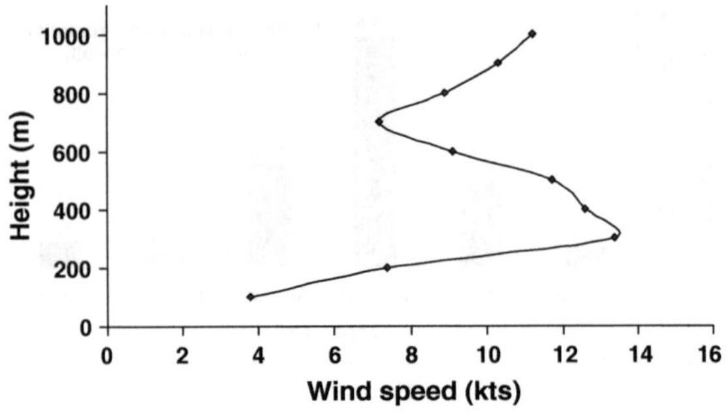

Figure 6
Variation of speed of sea breeze with height over Chennai, April–September, 2004–2005.

area is the rain-shadow area of the southwest monsoon. Figure 7 shows the induced convection observed at different ranges from the radar. Convection at ranges between 50 and 100 km was observed in 31.4% cases while in 24.2% cases the induced convection was noticed at ranges farther than 100 km. However, in 37.1% cases no convection was observed which may indirectly reveal that either the strength of the front was very weak and/or convergence was insufficient to cause vertical motion. Interestingly, in only 7.3% cases was induced convection observed in short ranges of less than 50 km.

The strength of convection has been analyzed from the radar reflectivity factor which is the summation of the 6th power of the diameters of the droplets in unit volume probed. The radar reflectivity (z) – rain rate (R) relationship that is being

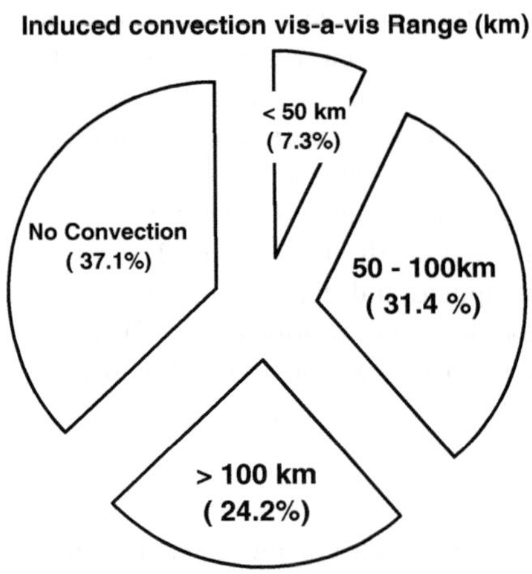

Figure 7
Percentage frequency of sea-breeze induced convection around Chennai during April – September, 2004–2005.

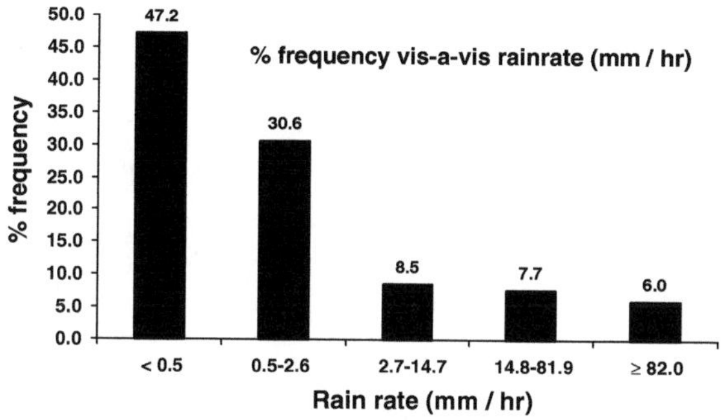

Figure 8
Rain rate from the sea-breeze induced convection over Chennai, April – September, 2004–2005.

operationally used at Chennai, viz., $z = 267R^{1.345}$ (SURESH *et al.*, 2005) has been used to estimate the rain rate. Figure 8 shows the rain rate of the induced sea-breeze convection around Chennai within 100-km radius from the radar. In about 78% of the cases, the rain rate was less than 3 mm/hr and more specifically less than 0.5 mm/hr. Intense convection exceeding 82 mm/hr was also observed in 6% of the cases analyzed.

5. *Sea Breeze-Induced Convection – A Case Study*

On a hot sunny afternoon, 28 June, 2003, the propagation of sea breeze was clearly discernible from 0645 UTC. The atmosphere was unstable at 0000 UTC, based on thermodynamical indices such as Total Total Index (TTI), George's K index, Lifted index and Showalter's index. However, the convective available potential energy (CAPE) was as low as 300 J/kg and convective inhibition energy (CINE) was 156 J/kg which are not conducive to convective cloud development since a minimum of 1000 J/kg is needed for moderate to intense convection, based on the literature survey we made. Nonetheless, there are cases documented by SURESH and BHATNAGAR (2005) to the effect that the development of severe convection does not necessarily depend on high values of CAPE in the morning hours, and a synoptic condition may aid increased CAPE and trigger convection in the afternoon. The 1200 UTC thermodynamic parameters also revealed that the atmosphere was unstable and that the CAPE was as high as 2960 J/kg. Convective activity commenced from 0930 UTC and intense convection was noted from 1030 UTC at a range of 40 to 80 km west to northwest of the radar. This reconfirms the earlier finding by BHAT *et al.* (1996) that the tropical convection is not governed by the local influence alone but depends on large scale processes as well. Chennai and its suburbs experienced significant rainfall. Though rain gauges have recorded 10 to

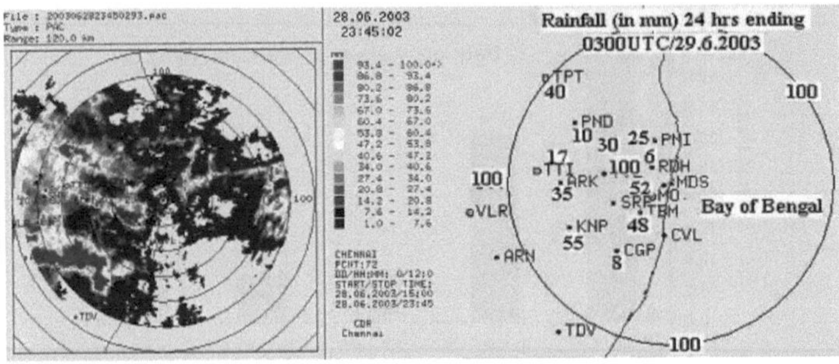

Figure 9
Precipitation accumulation as estimated from radar compared with rain gauge measured values on 28 June, 2003 around Chennai.

90 mm rainfall, estimates based on radar measured reflectivity have indicated precipitation exceeding 100 mm was quite probable in places where no rain gauges are available. Figure 9 shows precipitation accumulation based on the operational $z - R$ relationship, viz., $z = 267R^{1.345}$ from 1500 to 2345 UTC on 28 June, 2003 which is compared with the measured rainfall through rain gauges for a 24-hr period ending 0300 UTC on 29 June, 2003. The radar estimation was quite comparable with ground truths. The rainfall was attributed to sea-breeze convergence and associated moisture supply. Numerical model simulation was attempted using MM5, RAMS and ARPS. Though the sea breeze was captured well by all these models, the quantitative precipitation could not be accurately and precisely predicted. Further research is being undertaken to simulate the rainfall using 4DVAR and adopting nudging techniques which are beyond the scope of this paper.

6. Summary and Conclusions

From the Doppler weather radar based study on the sea-breeze front over Chennai, the following results have been obtained:

(i) The modal onset time of sea breeze is between 0800 and 1000 UTC with the earliest onset at 0508 UTC and late onset at 1138 UTC.

(ii) The frequency of sea breeze increases during southwest monsoon months rather than the pre-monsoon season.

(iii) Though the modal penetration of sea breeze is in the range 10–20 km, there are a number of occasions at which it has penetrated beyond 50 km and on a few occasions well beyond 100 km as well.

(iv) Whenever the penetration of sea breeze was not simultaneously observed in both north and south of the radar, the initial penetration was seen towards north of the radar in comparison to south of the radar.

(v) The modal length of the sea-breeze front was of the order 20–50 km but lengths as long as 100 km are not uncommon.
(vi) The mean depth of sea breeze circulation is around 600 m.
(vii) The speed maximum has been observed between 300 and 600 m.
(viii) The inland penetration speed of the sea breeze front is from 4 kmph (near the coast) to 15 kmph at ranges between 30 and 80 km inland.
(ix) While no convection was observed in 37% of the cases, convection was seen in about 53% of cases in ranges beyond 50 km.

Appendix 1

System specification of Doppler Weather Radar (DWR) functioning at Cyclone Detection Radar station, India Meteorological Department (IMD), Chennai

Transmitter	
Type	Klystron Amplifier
Peak power	750 kWatts
Modulator	Hard switched, switch array, solid state
Frequency	2875 to 2878 MHz
Pulse width	1μs(short pulse) & 2μs (long pulse)
Pulse Repetition Frequency	250–1200 in short pulse & 250–550 in long pulse mode.
Receiver	
Type	Double super heterodyne
Stable Local Oscillator / First Local Oscillator	2400 MHz
Second Local Oscillator	465,466,467, 468 MHz
Intermediate Frequency	10 MHz
Noise figure	better than 1.5 dB
Minimum Digitally Detectable Signal	−114 dBm in long pulse and −112 dBm in short pulse.
Digital part of the receiver	
Band width	1 MHz in reflectivity & 0.5 MHz in velocity mode
A/D conversion	40 MHz, 12 bits
Signal processing	10 DSP chips of 120MFLOPS/sec each
Simultaneous output	Radar Reflectivity, Radial Velocity and Spectrum width.
Minimum range bin spacing	75 m
Maximum number of range gates	2000
Dynamic range	Better than 95 dB.
Antenna	
Reflector type and diameter	Prime focus feed, 8.5 m dia.
Polarisation	Linear, Horizontal
Scan rate	3 to 36 °/s (0.5 to 6 r.p.m)
Beam width	~1°
Gain	44.5dBi
Height of the focal point of antenna	52 m above mean sea level.
Radome	Epoxy-foam sandwich; 11.6 m dia
Computers and peripherals	Two SUN ULTRA 10 workstations Five Pentium PCs 4 mm DAT and one DLT drive.
Un-interruptible Power Supply	60KVA

Appendix 2

Scan strategies adopted at DWR, Chennai for studying the sea breeze during March–October 2003–2005

Item	Scan I	Scan II
Range (km)	250	120
Resolution (m)	500	500
Range sampling	2	2
Pulse	SP	SP
PRF 1	600	800
PRF 2	—	—
Antenna speed (^{o} */sec)*	9	9
Time sampling	66	88
Elevation angles	0.2, 1.0, 2.0, 3.0, 4.5, 6.0, 7.5, 9.0, 11.0, 13.0, 15.5, 18.0, 21.0	0.2, 1.0, 2.0, 3.0, 4.0, 5.5
Clutter to signal ratio (CSR)	10.0	7.0
Signal quality index (SQI)	0.25	0.25
doppler filter	9	6
Doppler filter notch width (m s^{-1}*)*	1.5	0.8

Note : LP : Long pulse (2μ sec) SP :Short Pulse (1μ sec)
PRF : Pulse repetition frequency (PRF1 and PRF2 refers to first PRF and second PRF respectively)
Scan I was repeated at every 30 minutes.
Scan II intended for rain rate estimation within 120 km radius is repeated at 15 minutes periodicity.

Acknowledgements

The author is thankful to the Director General of Meteorology, India Meteorological Department(IMD) for according permission to present the paper in the International Conference on Mesoscale Processes in Atmosphere, Ocean and Environmental Systems (IMPA-2006) held at the Indian Institute of Technology, New Delhi, India. Facilities extended by the Deputy Director General of Meteorology, Regional Meteorological Centre, IMD, Chennai are gratefully acknowledged.

References

Asnani, G.C., *Tropical Meteorology*, vol. I and II (Sindh Colony, Aundh, Pune 411007, India 1993).

Achtemeier, G.L. (1991), *The Use of Insects as Tracers for 'Clear–Air' Boundary-Layer Studies by Doppler Radar*, J. Atmos. Oceanic Technol. *8*, 746–765.

Arritt, R.W. (1993), *Effects of the large-scale flow on characteristic features of the sea breeze*, J. Appl. Meteor. *32*, 116–125.

Atkinson, B.W., *Meso-Scale Atmospheric Circulations* (Academic Press, London 1981).

Atlas, D., *Radar in Meteorology: Battan Memorial and 40th Anniversary Radar Meteoro. Conf* (ed. David Atlas) (Amer. Met. Soc., Boston 1990).

BHASKARA RAO, N.S., WILLIAMS, S.D., CHANDY, M., and Devi, U. (1984), *Study of sea breeze at Madras*, Mausam *35*, 537–538.

BHAT, G.S., SRINIVASAN, J., and GADGIL, S. (1996), *Tropical deep convection, convective available potential energy and sea surface temperature*, J. Meteor. Soc. Japan. *74*, 155–166.

BLANCHARD, D. and LOPEZ, R. (1985), *Spatial patterns of convection in south Florida*, Mon. Wea. Rev. *113*, 1282–1299.

BROWNING, K.A. and WEXLER, R. (1968), *The determination of kinematic properties of a wind field using doppler radar*, J. Appl. Meteor. *7*, 105–113.

DOVIAK, R.J. and ZRNIC, D.S., *Doppler radar and weather observations*, 2nd ed., (Academic Press, San Diego 1993).

PIELKE, R.A., *Mesoscale Meteorological Modeling* (Academic Press, Orlando 1984).

RAGHAVAN, S, *Radar Meteorology* (Kluwer Academic Publishers, Netherlands 2003).

RAJ, Y.E.A., SANKARAN, P.V., RAMAKRISHNAN, B., and PADMAKUMAR, P.L. (2002), *Climatological characteristics of sea breeze parameters at Chennai*, Mausam *52*, 31–44.

RHOME. J., NIYOGI. D., and RAMAN. S. (2003), *Assessing seasonal transport and deposition of agricultural emissions in eastern North Carolina, USA*, Pure Appl. Geophys. *160*, 117–141.

RINEHART, R.E., *Radar for Meteorologists* (Rinehart Publications, USA 1999).

SAUVAGEOT, H., *Radar Meteorology* (Artech House Publishers, Norwood, MA, 1992).

SAUVAGEOT, H. (1996), *The clear air coastal vespertine radar bands*, Bull. Am. Meteor. Soc. *77*, 673–681.

SIMPSON, J.E., *Sea Breeze and Local Wind* (Cambridge University Press, UK 1994).

SURESH, R. (2004), *On nowcasting wind shear induced turbulence over Chennai air field*, Mausam *55*, 103–118.

SURESH, R. RAVICHANDRAN, P. K., GUPTA, J. P., THAMPI, S.B., KALYANASUNADRAM, S., and RAJESH RAO, P. (2005), *On optimum rain rate estimation from a pulsed Doppler Weather Radar at Chennai*, Mausam *56*, 433–446.

SURESH, R., and BHATNAGAR, A.K. (2005), *Pre-convective environment of pre-monsoon thunderstorms around Chennai –a thermodynamical study*, Mausam *56*, 659–670.

WALDTEUFEL, P. and CORBIN, H. (1979), *On the analysis of single-doppler radar data*, J. Appl. Meteor. *18*, 532–542.

WILSON, J. W., WECKWERTH, T.M., VIVEKANANDAN, J., WAKIMOTO, R.M., and RUSSELL, R.W. (1994), *Boundary layer clear-air radar echoes: origin of echoes and accuracy of derived winds*, J. Atmos. Oceanic Technol. *11*, 1184–1206.

(Received March 20, 2006, accepted September 10, 2006)
Published Online First: June 19, 2007

Pure appl. geophys. 164 (2007) 1527–1545
0033–4553/07/091527–19
DOI 10.1007/s00024-007-0243-2

Pure and Applied Geophysics

Variability of Convective Activity over the North Indian Ocean and its Associations with Monsoon Rainfall over India

D. R. PATTANAIK

Abstract—The present study is an attempt to examine the variability of convective activity over the north Indian Ocean (Bay of Bengal and Arabian Sea) on interannual and longer time scale and its association with the rainfall activity over the four different homogeneous regions of India (viz., northeast India, northwest India, central India and south peninsular India) during the monsoon season from June to September (JJAS) for the 26 year period (1979 to 2004). The monthly mean Outgoing Long-wave Radiation (OLR) data obtained from National Oceanic and Atmospheric Administration (NOAA) polar orbiting spacecraft are used in this study and the 26-year period has been divided into two periods of 13 years each with period-i from 1979 to 1991 and period -ii from 1992 to 2004.It is ascertained that the convective activity increases over the Arabian Sea and the Bay of Bengal in the recent period (period -ii; 1992 to 2004) compared to that of the former period (period -i; 1979 to 1991) during JJAS and is associated with a significantly increasing trend (at 95% level) of convective activity over the north Bay of Bengal (NBAY). On a monthly scale, July and August also show increase in convective activity over the Arabian Sea and the Bay of Bengal during the recent period and this is associated with slight changes in the monsoon activity cycle over India. The increase in convective activity particularly over the Arabian Sea during the recent period of June is basically associated with about three days early onset of the monsoon over Delhi and relatively faster progress of the monsoon northward from the southern tip of India. Over the homogeneous regions of India the correlation coefficient (CC) of OLR anomalies over the south Arabian Sea (SARA) is highly significant with the rainfall over central India, south peninsular India and northwest India, and for the north Arabian Sea (NARA), it is significant with northwest India rainfall and south peninsular rainfall. Similarly, the OLR anomalies over the south Bay of Bengal (SBAY) have significant CC with northwest India and south peninsular rainfall, whereas the most active convective region of the NBAY is not significantly correlated with rainfall over India.It is also found that the region over northeastern parts of India and its surroundings has a negative correlation with the OLR anomalies over the NARA and is associated with an anomalous sinking (rising) motion over the northeastern parts of India during the years of increase (decrease) of convective activity over the NARA.

Key words: Outgoing long-wave radiation (OLR), convection, north Indian Ocean, homogeneous region, all India summer monsoon rainfall (AISMR).

1. Introduction

Meteorologically, one of the most intense heat source regions in the global tropics is the Bay of Bengal region located in the north Indian Ocean. The convective

India Meteorological Department, NHAC (NWP), Mausam Bhawan, Lodhi Road, New Delhi 110003, India. E-mail: pattanaik_dr@yahoo.co.in precipitation and associated release of latent heat of condensation in this region drive a large portion of the global tropics and midlatitudes (KRISHNAMURTI and SURGI, 1987; LAU and CHANG, 1987). PATTANAIK (2003) has made a comparative study of convective activity over the various convective regions during the southwest monsoon season from June to September (JJAS) over the globe and has shown that the head Bay of Bengal region is more convective compared to that of the western Pacific region, south Indian Ocean and the Arabian Sea. NITTA and YAMADA (1989) used Outgoing Longwave Radiation (OLR) as an index of tropical convective activity to investigate decadal variations of atmospheric circulations in the global tropical belt. With the arrival of the southwest monsoon the convective activity increases over the Arabian Sea and subsequently the convective area moves northward and covers the entire parts of the Indian subcontinent during the monsoon season. PATTANAIK *et al.* (2005) in their recent study have also found that the negative OLR (thus, convection) anomalies from January gradually strengthen in a west-northwest direction from the western Pacific region and become established over the southeast Asian region and adjoining eastern equatorial Indian Ocean by the month of May during excess monsoon years. However, during the deficient monsoon years the negative OLR anomalies established over the western Pacific region in January almost remain active over the same region until the month of May.

One of the parameters intimately linked with the convective activity is the Sea-Surface Temperature (SST). Again small changes in the SST may invoke a large planetary-scale climate response over the warm pool region of the western Pacific and also over the Indian Ocean region. Several studies (SPERBER and PALMER, 1996 and WEBSTER *et al.*, 1998) demonstrate the significance of the Indian Ocean SST and associated convective activity in influencing the monsoon. The role of the Indian Ocean gradients of SST within the oceans are important in determining the location of precipitation over the tropics, including the monsoon region (LINDZEN and NIGAM, 1987). It has been seen from many recent studies (WANG, 1995; RAJEEVAN *et al.*, 2000; PATTANAIK, 2005) that the SST over the Bay of Bengal region and equatorial Indian Ocean region is in the increasing trend. WANG (1995) had shown a rapid transition from the cold to warm state of the Indian Ocean in the late 1970s. RAJEEVAN *et al.*, (2000) have also shown that over the equatorial Indian Ocean, until the early 1980s, SSTs have shown a sharp increase, which continued steadily thereafter. Also shown by them was a simultaneous increase in low-cloud cover. With the important role of the Indian Ocean in the global climate system, it is desirable to know if there are any observable climate changes over recent years. In the scenario of increasing SST over the north Indian Ocean it is very useful to know the variability of convective activity over the Indian Ocean, which can influence the monsoon activity over different parts of India. Therefore the objective of the present study is to determine how the convective activity varies interannually and also on a longer time scale during the monsoon season over the north Indian Ocean (Bay of Bengal and Arabian Sea), particularly after the availability of satellite data (1979 onwards). The association of convective activity over the north Indian Ocean with the rainfall distribution over four homogeneous regions of India (viz., northeast India, northwest India, central India and south peninsular India) is also examined in the present study.

2. Data and Methodology

The OLR data measured from Advanced Very High Resolution Radiometers aboard National Oceanic and Atmospheric Administration (NOAA) polar orbiting spacecraft down loaded from the Web site at http://www.cdc.noaa.gov/ for a period of 26 year (1979–2004) during the monsoon season (JJAS) is used in this study. There are also missing grids, and missing values of OLR within grids basically due to satellite problems, archival problems or incomplete global coverage. These gaps are then filled with temporal and spatial interpolation. The details of the interpolation technique are avallable in LIEBMANN and SMITH (1996). The SST data for the same period have been obtained from NOAA extended SST data set available from the same website. In order to see the tendency of seasonal SST and OLR patterns over the north Indian Ocean, the same 26-year period considered for the study of convective activity has been divided into two periods of 13 years each with period-i (1979 to 1991) and period-ii (1992 to 2004). A comparison of convective activity is performed by taking the difference of OLR anomalies between these two periods during the monsoon season (JJAS). The rainfall over the four different homogeneous regions of India (viz., northeast India, northwest India, central India and south peninsular India). The four homogeneous regions of the country are identified by choosing regions in which the rainfall variation in each of the meteorological subdivisions comprising the region is positively and significantly correlated with the area-weighted rainfall variation over the region as a whole. It may be mentioned here that all the results of the present study are based on the OLR data used for a relatively shorter period of 26 years.

3. Results and Discussions

3.1. Seasonal Mean Convective Activity over the North Indian Ocean

The OLR (a proxy for deep convection) is used widely for the monsoon research (WEBSTER, 1995; KRISHNAN *et al.*, 2000; GADGIL *et al.*, 2004; PATTANAIK *et al.*, 2005). GADGIL *et al.* (2004) have suggested a possible link between the variation of deep convection over the equatorial Indian Ocean and monsoon rainfall by examining the OLR data over the Indian Ocean. Before the analysis of seasonal variations of OLR anomalies over the north Indian Ocean it is useful to know the tendency of SST patterns over the north Indian Ocean during the same period considered in the present

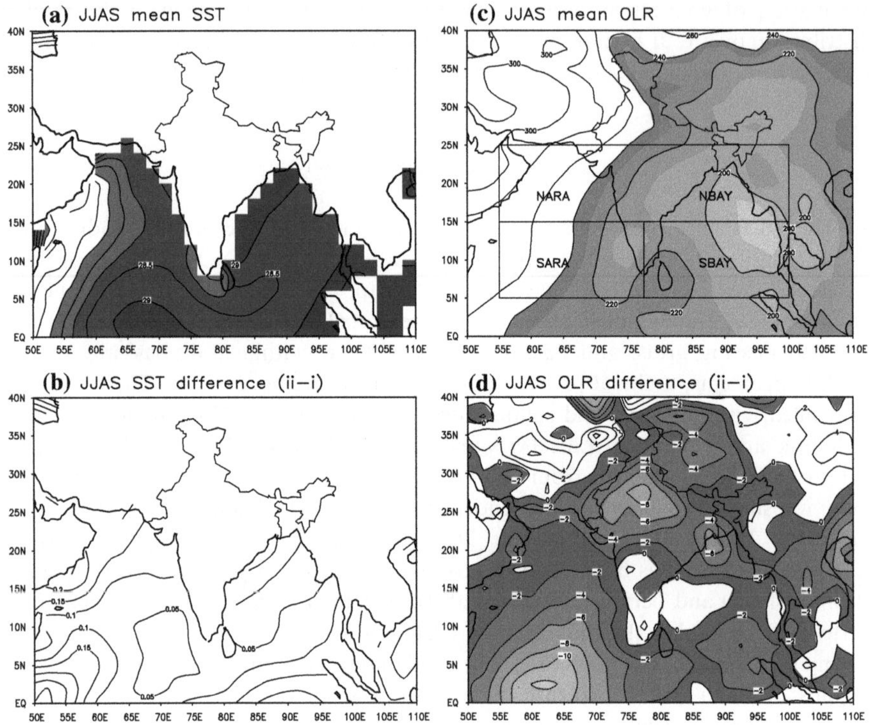

Figure 1
(a) JJAS mean SST with > 27°C is shaded, (b) difference of JJAS SST anomalies during period-ii (1992–2004) and period-i (1979–1991), (c) JJAS OLR mean with < 240 watts/m^2 is shaded and (d) same as 'b' but for OLR with shaded negative.

study. The mean SST and OLR during the monsoon season and the difference of its anomalies during the recent period (period-ii) to the former period (period-i) are shown in Figures 1a–1d. The OLR mean is calculated from data of 26 years (1979 to 2004) considered in the present study. It is seen from Figure 1a that in most of the Arabian Sea and the entire Bay of Bengal the mean SST is higher than 27°C, which can create favorable conditions for the occurrence of deep convection. Figure 1b also shows that the SST over the north Indian Ocean has increased during the recent period compared to the former period as indicated by a positive difference over the entire north Indian Ocean. PATTANAIK (2005) in his recent study has also shown significantly increasing trends (at 99.9% level) of SST over the Bay of Bengal region during the monsoon season by taking very long SST data series from 1891 to 2003.

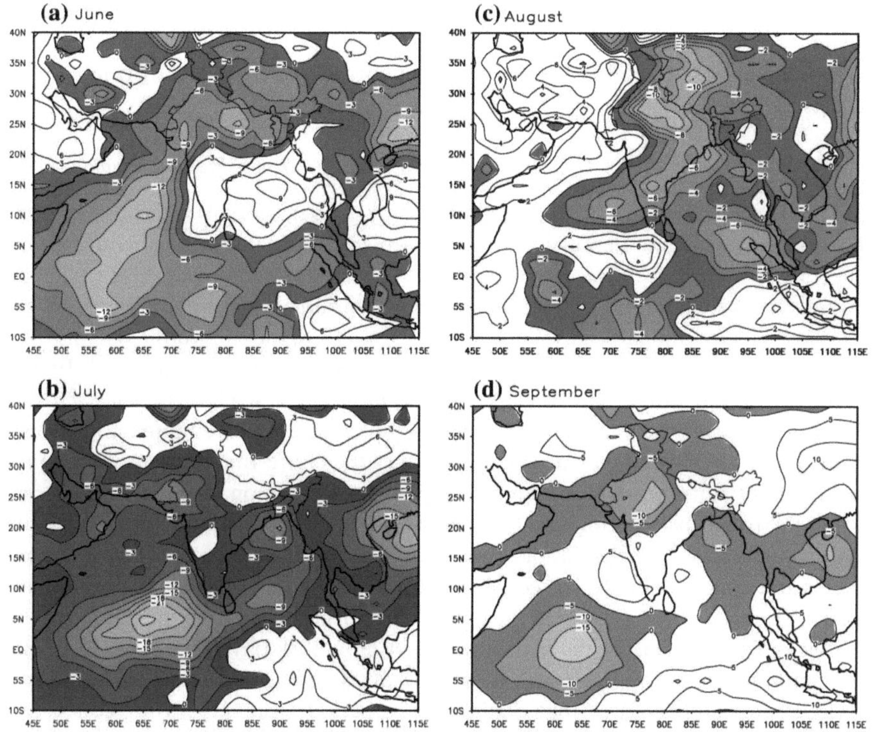

Figure 2
Difference of OLR anomalies during period-ii (1992–2004) and period-i (1979–1991) during (a) June, (b) July, (c) August and (d) September. The negative values are shaded.

The corresponding seasonal long-term mean OLR pattern (Fig. 1c) during the monsoon season shows that the low OLR values correspond to the area of deep convection (say the values less than 240 watts/m^2) which lies over most of the Bay of Bengal, the Andaman Sea and the eastern Arabian Sea in the north Indian Ocean. The higher value of mean OLR over the northwestern parts of the country is associated with reduced convective activity (Fig. 1c). Just like the increase in SST during the recent period, the difference of OLR anomalies (Fig. 1d) also indicates an increase in convective activity over the Arabian Sea and the Bay of Bengal, except some parts of southwest Bay of Bengal off the Andhra coast, where it is slightly positive. Figure 1d also indicates an increase of convective activity over the central and northern parts of India during the recent period compared to the former period. Again the magnitude of negative anomaly is very large over the southern parts of the Arabian Sea (of the order of -12 watts/m^2) compared to that over the Bay of Bengal during the monsoon season. Thus, the convective activity increases over the Arabian Sea and the Bay of

Bengal in the recent period compared to that of the former period during the monsoon season. A similar increase of convective activity over the equatorial Indian Ocean was also indicated by CHU and WANG (1997) during the monsoon season associated with decreasing trends of OLR.

3.2. Monthly Variability of Convective Activity over the North Indian Ocean

The Indian monsoon activity is maintained by propagation of convective systems of synoptic scale (lows, depressions, etc.) and planetary scale (tropical convergence zones) from the warm tropical oceans, onto the heated subcontinent (GADGIL, 2000 and references therein). As a result, the monsoon variability both on the subseasonal scales and on interannual scales is linked to a variation of the convective systems over the ocean. In this section the variability of convective activity over the north Indian Ocean during June to September (JJAS) on the monthly scale has been analyzed. On the monthly scale from June to September the difference of OLR anomalies between period-ii and period-i is shown in Figure 2 over the entire Indian monsoon region. The month of June (onset phase) shows very strong convection over the Arabian Sea during the recent period compared to that of the former period as indicated by the negative difference over the Arabian Sea (Fig. 2a), whereas, over the Bay of Bengal (almost to the north of 5°N) the convective activity decreases during the onset phase of June in recent time compared to that of the former period as indicated by the positive difference. During the active monsoon month of July (Fig. 2b) the oceanic convection increases over the north Indian Ocean including equatorial Indian Ocean in recent time compared to that of the former period with very strong convective activity over the south Arabian Sea in recent time. In the month of August (Fig. 2c) the Bay of Bengal is more active compared to the Arabian Sea in the recent period compared to the former period. During the withdrawal phase of September (Fig. 2d) the convective activity increases over the southern Arabian Sea and eastern parts of the Bay of Bengal to the east of about 87°E as indicated by the negative difference. Thus, the monthly OLR difference indicates an increase in convective activity during the recent period compared to the former period over the Arabian Sea during June and July, the central part of the Arabian Sea in August and mainly the southern part of it in September, whereas, over the Bay of Bengal region there is an increase in convective activity during the peak monsoon months of July and August in the recent period compared to the former period. Again the month of June shows reduced convective activity over the Bay of Bengal in recent time to the north of 5°N and the month of September shows reduced convective activity over the western part of the Bay of Bengal to the west of about 87°E. It is also seen from Figure 1d and Figure 2 that the large increase in convective activity over the southern parts of the Arabian Sea during JJAS as shown in Figure 1d is mainly dominated by the contribution from July as indicated by the large negative difference (of the order of -24 watts/m^2) over the region in the month of July (Fig. 2b).

As some earlier studies have indicated (DE *et al.*, 1998; GADGIL, 2000) during the peak monsoon months of July and August the rainfall activity over India fluctuates with the fluctuations of intensity of the continental tropical convergence zone (TCZ) over India and the TCZ over the equatorial Indian Ocean. Variation of the monsoon rainfall within the season (between active and weak spells) as well as of the seasonal rainfall from year to year (between the good monsoons and droughts) is associated with a variation in the intensity and location of the continental TCZ (GADGIL, 2000). Even after the continental TCZ is established in early July, the TCZ over the equatorial Indian Ocean occurs intermittently throughout the season. In order to see whether this increase in convective activity during recent time, particularly over the equatorial Indian Ocean during July, has influenced the monsoon seasonal cycle, the five days running mean of the daily composite anomalies of OLR over the core India region (73°E–82°E, 18°N–28°N) during the peak monsoon months of July and August during the two periods mentioned earlier is shown in Figure 3a. Similarly, the corresponding composite OLR anomalies over the south-equatorial Indian Ocean (70°E–85°E, 10°S–Eq) during the peak monsoon months of July and August during the two periods is also shown in Figure 3b. Though the active and break days are defined by different authors (DE *et al.*, 1998; GADGIL and JOSEPH, 2003, etc.) by considering different criteria, here the core India region considered was used earlier by KRISHNAN *et al.* (2000), who have used the daily OLR data for the same core India region mentioned above for defining break days. It is seen from Figure 3a that the monsoon activity cycle during the peak monsoon months of July and August (the active and break cycle) has been modified with respect to its intensity and phase during the two periods considered here. Figure 3a also shows that except after the end of the third week of July in the month of July and about 2 to 3 days around mid-August, the remaining periods show active convection (associated with large negative anomalies) during the recent period-ii (1992–2004) compared to the previous period (1979–1991). In other words, the present analysis indicates that the breaks are more frequent during about the last 10 days of July and a few days in mid-August during the recent period (1992–2004), which was more frequent during the remaining periods of July and August in the earlier period (1979–1991), and it is also interesting to see from Figure 3b that during these periods (last 10 days of July and few days in mid-August) the convective activity increases over the south-equatorial Indian Ocean in period–ii compared to period-i, which is similar to the strengthening of the continental TCZ and weakening of the oceanic TCZ associated with the active phase of the monsoon and weakening of the continental TCZ and strengthening of the oceanic TCZ associated with the break phase of the monsoon.

Again, to relate the changes in convective activity over the north Indian Ocean during the onset phase of June (as shown in Fig. 2a) with the onset and progress of the southwest monsoon over India, the onset dates over Kerala (southern tip of India) and Delhi during the period from 1979 to 2004 is considered. Normally it takes about 30 days of June for the monsoon to arrive at Delhi after arriving of the

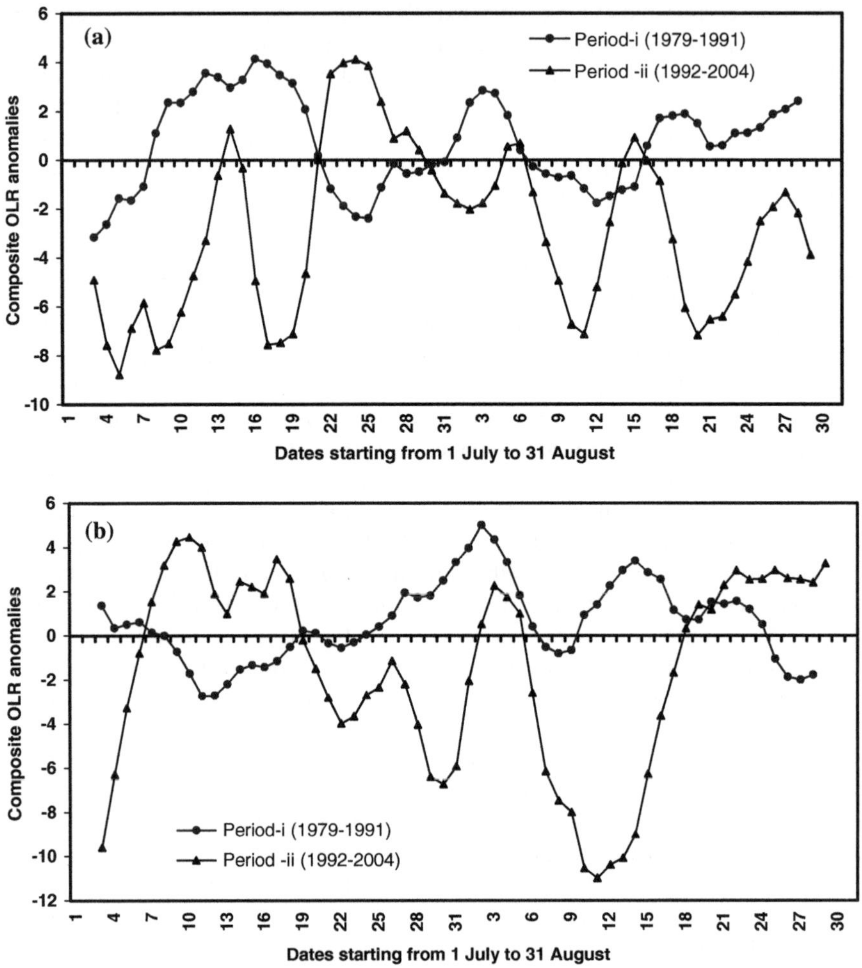

Figure 3
Pentad average of daily OLR composite anomalies in watts/m^2 during July and August during period-i (1979–1991) and period-ii (1992–2004). (a) For the core India region (73°E–82°E, 18°N–28°N) and (b) for equatorial south Indian Ocean (70°E–85°E, 10°S–Eq). Middle date is plotted in 'X' axis.

Kerala coast on 1st June. The onset dates of the southwest monsoon over Delhi (MOD) and that over Kerala (MOK) are shown in Figure 4 along with the mean onset lines for both the stations during period-i (1979–1991) and period-ii (1992–2004). It is seen from Fig. 4 that during the recent period (period-ii) the mean onset date over Delhi represented by the solid line at the top, is about three days earlier

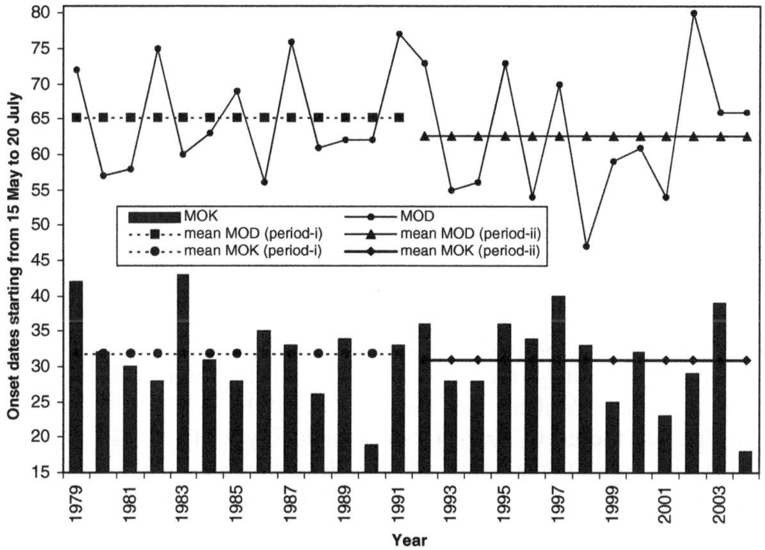

Figure 4
Onset date of southwest monsoon over Delhi (MOD) and over Kerala (MOK) with mean dates during the former period from 1979 to 1991 (dashed lines) and the recent period from 1992 to 2004 (solid lines).

than that of the mean onset date during the previous period (period-i) represented by the dotted line at the top in Figure 4. Whereas, the mean onset date over Kerala during period-i is 1 June, which is earlier by only one day (31 May) during period–ii. It is also observed from Figure 4 that the onset date over Delhi shows large variability (span of 34 days) with the earliest onset date of 16 June in 1998 and delayed onset date of 19 July in 2002, whereas, over Kerala the variability of the onset date is smaller compared to that over Delhi (span of 26 days) with the earliest onset date of May 18 during 2004 and the delayed onset date of 12 June in 1983. It is also observed from Figure 4 that except for the year 2002, the onset of the monsoon over Delhi for many years during period-ii (1992–2004) was earlier than the onset of the monsoon during period–i (1979 to 1991). It takes normally about 30 days of June for the monsoon to arrive at Delhi after the normal onset of the monsoon over the Kerala coast around 1 June (here the mean onset date over Kerala during period-i is 1 June and that during period-ii is 31 May, which is very close to 1 June). The duration of the monsoon to arrive at Delhi from Kerala during the period from 1979 to 2004 is also correlated significantly (at 99% level) with the Arabian Sea OLR averaged over the oceanic region bounded by 55°E–77.5°E, 5°N–25°N (identified as NARA and SARA in Fig. 1c) with a Correlation Coefficient (CC) of 0.53. Also the mean duration days of monsoon from Kerala to Delhi is about 2 days less during the recent period (period-ii) compared to the former period (period-i). Thus, the increase in convective activity, particularly over the Arabian Sea during the onset phase of June in the recent period, is basically associated with a slight early onset of the monsoon over Delhi and relatively faster progress of the monsoon northward from the southern tip of India compared to the earlier period.

3.3. Interannual Variability of Convective Activity over the North Indian Ocean

As it is seen in Figure 1d and Figures 2a to 2d the patterns of OLR difference between period-ii and period-i shows asymmetry in convective activity over the southern and northern part of the Arabian Sea and the Bay of Bengal during the monsoon season. In order to quantify the OLR difference in the both the southern and northern part separately, the north Indian Ocean has been divided into four blocks with the south Arabian Sea bounded by 55°E–77.5°E, 5°N–15°N (box 'SARA' in Fig. 1c), the north Arabian Sea bounded by 55°E–77.5°E, 15°N–25°N (box 'NARA' in Fig. 1c), the south Bay of Bengal bounded by 77.5°E–100°E, 5°N–15°N (box 'SBAY' in Fig. 1c) and the north Bay of Bengal bounded by 77.5°E–100°E, 15°N–25°N (box, 'NBAY' in Fig. 1c). While dividing these blocks the grid point over the Oceanic region is only considered excluding the land grid points. In order to see the interannual variability of convective activity, the OLR anomalies during JJAS averaged over the four regions viz., SARA, NARA, SBAY and NBAY, as identified in Figure 1c, are shown in Figure 5. The linear trend lines are also fitted over these regions. During the monsoon season the significantly decreasing trend of the OLR anomaly (thus an increase in convective activity) is noticed only over the NBAY region, the region where the monsoon lows and monsoon disturbances mainly form during the southwest monsoon season (Fig. 5c) and is the most active convective region as indicated by the lowest mean OLR value of 196.8 watts/m^2. The other three regions (NARA, SARA and SBAY), however show a decreasing trend of OLR (an increasing trend of convective activity) but this is not statistically significant. Also observed from the mean OLR values over the four oceanic regions in Figure 5 is the existence of a large difference of mean OLR over these four regions with active convection (lower OLR) over the Bay of Bengal than the Arabian Sea with most active convection over the NBAY. However, it is also observed from Figure 5 that there is only a slight difference in Standard Deviation (SD) of OLR over these regions and its values are very small (between 3 to 6 watts/m^2) compared to the mean. Thus, the OLR anomalies as shown in Figure 1d (maximum of the order of -12 watts/m^2) over the north Indian Ocean can imply a difference in convective activity as it is more than the departure of 2 SD value. It is also seen from Figures 5a to 5d that compared to the other three regions the positive and negative OLR anomalies over the SARA region are larger during many years associated with the highest SD value (Fig. 5c).

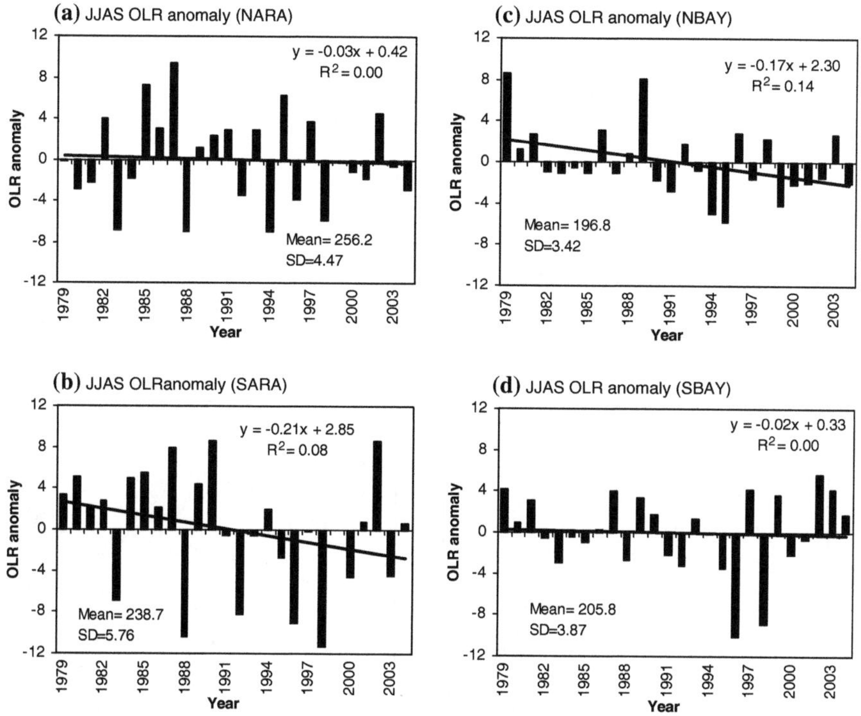

Figure 5
The interannual variability of OLR anomalies during JJAS over (a) NARA, (b) SARA, (c) NBAY and (d) SBAY over the north Indian Ocean as identified in Figure 1c.

It may be mentioned here that many recent studies (RAJEEVAN *et al.*, 2001; DASH *et al.*, 2004 and PATTANAIK, 2005 and references therein) have documented a significantly decreasing trend of frequency of monsoon disturbances (surface wind speed between 17 to 33 knots) and an increasing trend of the number of low-pressure areas (surface wind speed < 17 knots) formed over the north Indian Ocean in recent time. It is seen from these studies that in spite of an increasing trend of SST, the frequency of monsoon disturbances is decreasing over the north Indian Ocean region during the monsoon season. It has been shown by PATTANAIK (2005) that the decreasing trend of monsoon disturbances during the monsoon season is significant at 99% level, which is also true for the recent 26-year period from 1979 to 2004 considered in the present study (Fig. not shown). The increasing trend in convective activity over the NBAY during the monsoon season in recent time as shown in Figure 5c, in spite of the decreasing trend of monsoon disturbances, could be due to the increasing frequency of low pressure.

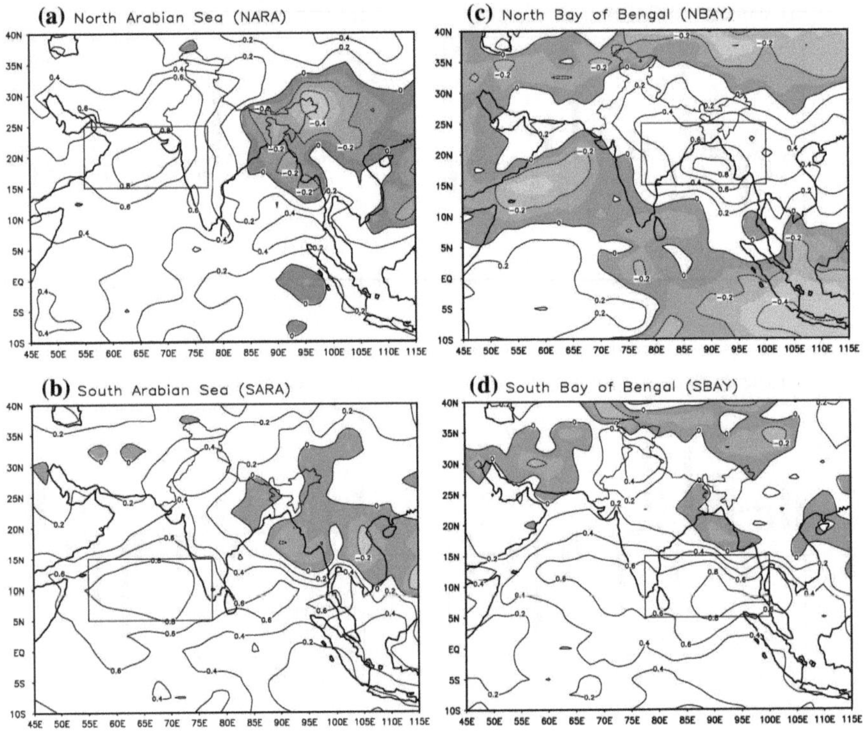

Figure 6
(a) The spatial maps of Correlation Coefficient (CC) between averaged oceanic OLR over NARA with the OLR at each grid point, (b) same as 'a' but for SARA region, (c) same as 'a' but for NBAY and (d) same as 'a' but for SBAY region. The negative CC is shaded.

3.4. Association of Convective Activity with the Rainfall Distribution over India

In order to examine the association of convective activity over the four regions of the north Indian Ocean with that of monsoon rainfall over India during the monsoon season, the phase relationship of convective activity over the four regions identified over north Indian Ocean (SARA, NARA, SBAY, NBAY) have been analyzed with that of surrounding regions. For this purpose, the OLR value averaged over the NARA and the SARA regions is correlated with that of OLR over each grid point and the Correlation Coefficients (CCs) maps are plotted in Figures 6a and 6b, respectively. Similarly the CCs maps of averaged OLR over the NBAY and the SBAY are also shown in Figures 6c and 6d, respectively. The negative CC region is shaded in Figure 6. It is seen from Figure 6a that the convective activity over the north Arabian Sea region has a negative correlation with that of the convective activity over northeast India and adjoining north Bay of Bengal and has a direct correlation over the rest of the country. Thus, the increase in convective activity over the NARA region is associated with the decrease in convective activity over northeast India and its surrounding. Similarly over the SARA region the convective activity has a direct relationship over the country with a weak inverse relationship over the northeastern parts of India (Fig. 6b). It is seen from Figure 6c that the region of very active convection over the NBAY has a positive correlation with that over the entire country, except the west coast of India, the southern tip of India and extreme northern parts of India, where the CC is slightly negative. Similarly the SBAY region (Fig. 6d) also has a positive correlation with most of India, excluding some eastern regions of the country. Thus, it is seen that the convective activity over the four parts of the north Indian Ocean correlates differently over the convective activity with the surrounding regions. To correlate the convective activity over the four oceanic regions of the north Indian Ocean with that of rainfall over India the four broad homogeneous regions of India are identified as shown in Figure 7. The mean and coefficient of variation (CV) of the rainfall during JJAS over the four regions are also given in the title of Figure 7, which indicate the highest variability over northwest India (NWI) and the lowest over northeast India (NEI). The CCs between OLR anomalies and rainfall departure over the four homogeneous regions of the country and with the all India summer monsoon rainfall (AISMR) for the period from 1979 to 2004 is given in Table 1. The superscript in Table 1 indicates the statistical significance level, and the negative CC between OLR and rainfall indicates the direct relation of convective activity with rainfall.

It is seen from Table 1 that the convective activity over SARA and NARA Sea is significantly correlated with AISMR. Whereas, the SBAY has a CC significant only at 90% level with AISMR. The most active convective region of NBAY, has a very weak correlation with AISMR, which can be interpreted from Figure 6c with a weak positive CC (of the order of 0.2) observed over the northwest parts of India, and a negative CC over the climatologically heavy precipitation region of the west-coast of India acts in an opposite manner. Lawrence and Webster (2001) have also shown earlier that the regression of seasonal mean OLR in the Indian Ocean basin onto the AISMR shows high anticorrelations over the Arabian Sea, whereas, the relationship between the AISMR and the seasonal mean OLR over the Bay of Bengal is minimal. Thus, the precipitation over India varies somewhat independently from precipitation over the north Bay of Bengal. Table 1 also shows that in the homogeneous regions of India the CC of OLR anomalies over NARA Sea is highly significant with the rainfall activity over Central India (CEI), south peninsular India (SPI) and NWI and for the SARA it is significant only with rainfall over NWI and SPI. It is also seen from Table 1 that NEI rainfall has a opposite relation with the convective activity over the oceanic regions, as indicated by positive CCs, except a slight negative CC with OLR anomalies over NBAY. The highest positive CC of 0.3 between OLR anomalies over NARA and rainfall over NEI as shown in Table 1 is also reflected in Figure 6a, with negative CCs over the northeastern parts of the country. This shows an anomalous

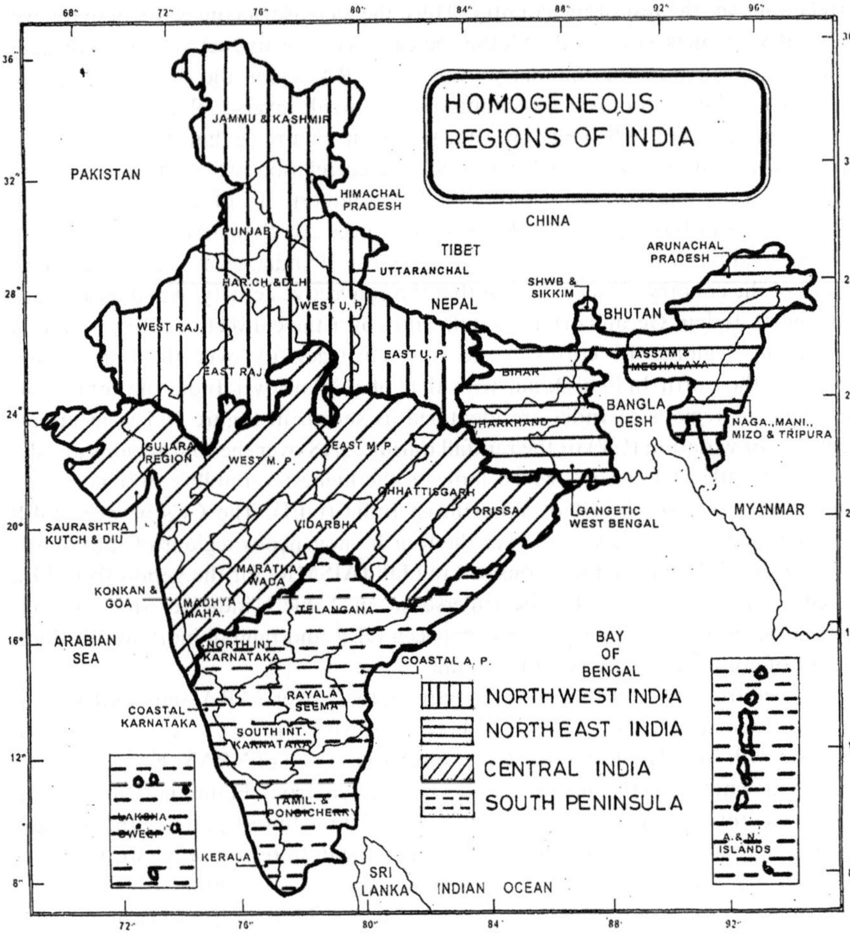

Figure 7

Four broad homogeneous regions of India with June to September (JJAS) mean rainfall (coefficient of variability) are: northwest India; 647 mm (18.4%), northeast India; 1367 mm (8.3%), central India; 994 mm (12.8%) and south peninsula; 722 mm (12.8%).

local east-west circulation with an anomalous rising motion over the NARA associated with increasing convective activity and anomalous sinking motion over the northeastern parts of India. In order to further quantify this, the standardized OLR anomalies over NARA and standardized Omega anomalies at 500 hPa over the northeastern parts of the country bounded by 90°E–100°E, 20°N–30°N are shown in Figure 8. The anomalous eastwest circulation is very clear from Figure 8 with the negative value of OLR anomalies over the NARA (increase of convective activity)

Table 1

Correlation Coefficients (CCs) of OLR anomalies over the four Oceanic regions identified in Figure 1c with the rainfall departures over the four homogeneous regions of India and AISMR. The superscript indicates the significance level

OLR anomalies over the Oceanic regions	Northeast India (NEI) Rainfall	Northwest India (NWI) Rainfall	Central India (CEI) Rainfall	South Peninsula (SPI) Rainfall	AISMR
South Arabian Sea (SARA)	0.09	$\mathbf{-0.48}^{98}$	−0.24	$\mathbf{-0.62}^{99.9}$	$\mathbf{-0.50}^{99}$
North Arabian Sea (NARA)	0.30	$\mathbf{-0.49}^{98}$	$\mathbf{-0.65}^{99.9}$	$\mathbf{-0.56}^{99}$	$\mathbf{-0.60}^{99}$
South Bay of Bengal (SBAY)	0.05	$\mathbf{-0.46}^{98}$	−0.05	$\mathbf{-0.48}^{98}$	$\mathbf{-0.37}^{90}$
North Bay of Bengal (NBAY)	−0.05	−0.20	−0.06	0.33	−0.07

mostly associated with positive values of Omega (sinking motion) over the northeastern parts of the country and *vice versa*. Thus, there exists an anomalous east-west circulation associated with the increase or decrease of convective activity over the north Arabian Sea through the local feed backs. To examine further, the composite velocity potential anomalies during the excess and deficient convective years over the NARA regions are plotted in Figure 9. The excess convective years (1983, 1988, 1994 and 1998) and deficient convective years (1985, 1987, 1995 and 2002) over the NARA regions are identified based on the departure of 1 standardized anomalies in Figure 8. The positive (negative) values of composite velocity potential indicate anomalous convergence (divergence). Figure 9a shows strong anomalous convergence over the Arabian Sea, which decreases as we go towards northeast India

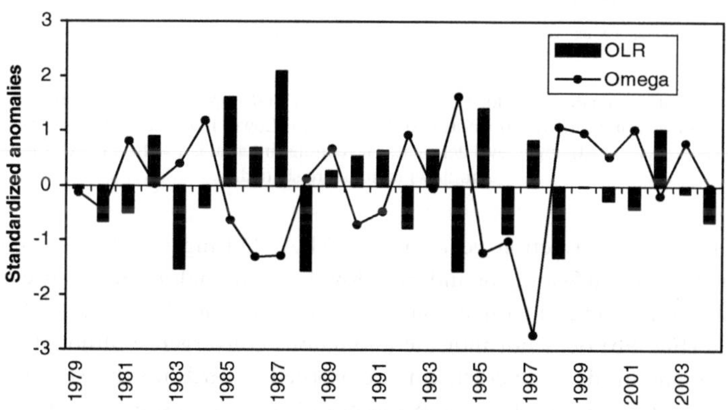

Figure 8

JJAS time series of oceanic OLR anomalies over NARA (55–77.5°E, 15–25°N) from 1979 to 2004 and corresponding time series of 500 hPa Omega (vertical velocity in Pa/sec) anomaly averaged over the northeastern parts of India and its surroundings (20–30°N, 90–100°E).

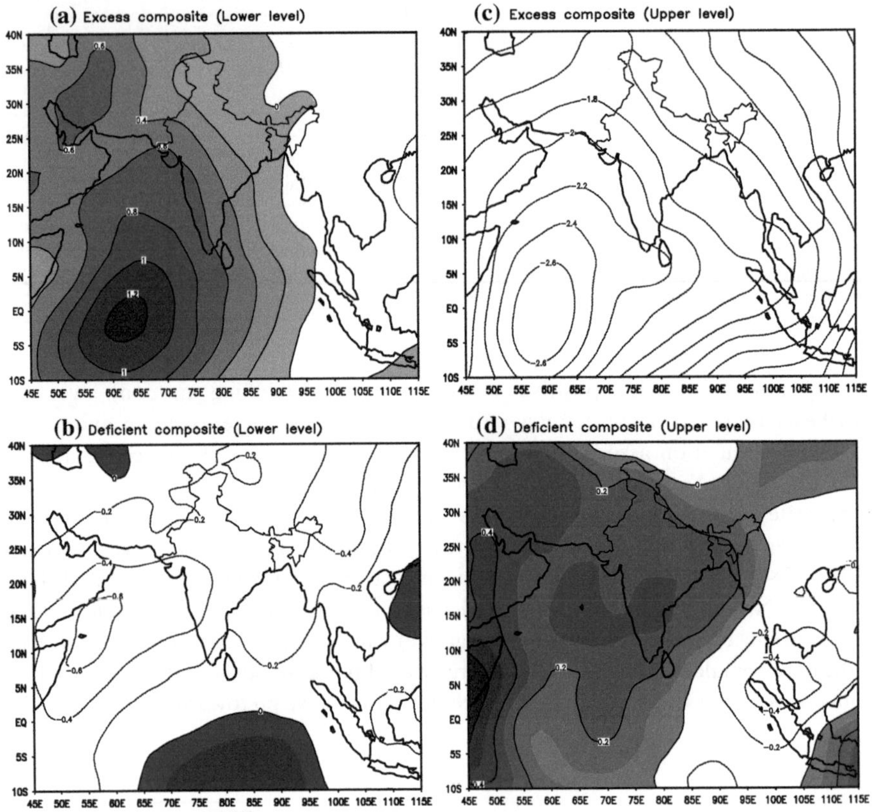

Figure 9

Seasonal composite anomalies of velocity potential at lower and upper troposphere levels during excess and deficient convection years over the north Arabian Sea. (a) Lower level excess, (b) lower level deficient, (c) upper level excess and (d) upper level deficient. The counter interval is 1×10^6 m^2 sec^{-1} and the positive (outflow) regions are shaded.

during the excess convection years over NARA. During the deficient composites (Fig. 9b) the Arabian Sea region indicates low-level anomalous divergence in place of anomalous convergence in Figure 9a. However, over northeast India the deficient composite (Fig. 9b) does not indicate anomalous convergence although it shows a decrease of magnitude of divergence as we proceed towards northeast India from the Arabian Sea region. Similarly the upper tropospheric composite anomalies of velocity potential during excess and deficient years indicate a large outflow over the Arabian Sea, which decreases as we go towards northeast India during the excess years (Fig. 9c) and is just the opposite during deficient years (Fig. 9d). Though the velocity potential anomalies shown in Figure 9 indicate the contrasting patterns at lower and upper levels during the excess and deficient convective years over the NARA, the tendency for an increase (decrease) of convective activity over the NARA region to be associated with sinking (rising) motion over northeast India is more prominent during many individual years shown in Figure 8 compared to that shown in Figure 9 as composite anomalies using the velocity potential. It could be due to the fact that the velocity potential indicates better representation of planetary scale motion compared to that of localized flow. Thus, the above analysis indicates that changes in Oceanic convection, particularly over the NARA, can change the circulation through local feedbacks in such a way that the increase (decrease) of convective activity over the NARA region is associated with an anomalous sinking (rising) motion over the northeastern parts of the country.

4. Conclusions

The following conclusions may be drawn from the present study:

During the monsoon season the SST over the north Indian Ocean evidences increasing tendency during the recent period and is associated with an increase in convective activity during the recent period (1992 to 2004) compared to that of the former period (1979–1991). It is determined that the convective activity increases over the Arabian Sea and the Bay of Bengal in the recent period compared to that of the former period during JJAS and is associated with a significantly increasing trend (at 95% level) of convective activity over the north Bay of Bengal (NBAY). On a monthly scale, July and August also show an increase in convective activity over the Arabian Sea and the Bay of Bengal during the recent period which is associated with slight changes in the magnitude and phase of the active and break cycle of the monsoon over India. During the onset phase of June the increase in convective activity, particularly over the Arabian Sea during the recent period, is basically associated with a slight early onset of the monsoon (about three days) over Delhi and relatively faster progress of the monsoon northward from the southern tip of India compared to the earlier period.

Over the homogeneous regions of India the CC of OLR anomalies over SARA is highly significant (above 98% level) with the rainfall over central India, south peninsular India and northwest India, and for the NARA it is significant only with northwest India rainfall and south peninsular rainfall. Over the SBAY region, significant CC is found with rainfall over northwest India and south peninsular India, whereas, the most active convective region of NBAY is not significantly correlated with rainfall over India.

It is also seen that the northeastern parts of India have an opposite correlation with the OLR anomalies over the NARA associated with anomalous sinking (rising) motion over the northeastern parts of India during the year of increase (decrease) of convective activity over the NARA.

Acknowledgements

The author is extremely grateful to the Director General of Meteorology (DGM) and Dr. H. R. Hatwar, ADGM (S), India Meteorological Department (IMD), New Delhi for their encouragement and for providing all the facilities to carry out this research work. Thanks are extended to the anonymous reviewers, whose suggestions were very useful for the quality improvement of the paper. Gratitude is expressed to DGM, IMD for allowing me to present this paper in the conference IMPA-2006 at IIT Delhi. Appreciation is also extended to NOAA for providing OLR and SST data used in this study.

References

Chu, P. S. and Wang, J. B. (1997), *Recent climate in the tropical western Pacific and Indian Ocean Regions as detected by Outgoing Longwave Radiation records*, J. Climate. *10*, 636–646.

Dash, S. K., Kumar, J. R., and Shekhar, M. S. (2004), *The decreasing frequency of monsoon depressions over the Indian region*, Current Science *86*, 1404–1411.

De, U. S., Lele, R. R., and Natu, J. C. (1998), *Breaks in southwest monsoon*, India Meteor. Dept. Report No. 1998/3.

Gadgil, S. (2000), Monsoon-ocean coupling, Current Science *33*, 309–323.

Gadgil, S. and Joseph, P. V. (2003), *On breaks of the Indian monsoon*, Proc. Indian. Acad. Sci. (Earth and Planet Sci), *112*, 529–558.

Gadgil, S., Vinayachandran, P. N., and Francis, P. A. (2004), *Droughts of the Indian summer monsoon: Role of clouds over the Indian Ocean*, Current Science *85*, 1714–1719.

Krishnamurti, T. N. and Surgi, N., *Observational Aspects of Summer Monsoon, Monsoon Meteorology* (ed. Chang, C. P. and Krishnamurti, T. N.) (Oxford University Press, 1987) pp. 3–25.

Krishnan, R., Zhang, C., and Sugi, M. (2000), *Dynamics of breaks in the Indian summer monsoon*, J. Atmos. Sci. 57, 1354–1372.

Lau, K.-M. and Chang, C. P., *Planetary Scale Aspects of the Winter Monsoon and Atmospheric Teleconnections Monsoon Meteorology* (ed. Chang, C. P. and Krishnamurti, T. N.) (Oxford University Press, 1987) pp. 161–202.

Lawrence, D. M. and Webster, P. J. (2001), *Interannual variations of the Intraseasonal Oscillation in the south Asian summer monsoon region*, J. Climate *14*, 2910–2922.

Liebmann, B. and Smith, C. A. (1996), *Description of a complete (interpolated) outgoing longwave radiation data set*, Bull. Am. Meteor. Soc. *77*, 1275–1277.

Lindzen, R. S. and Nigam, S. (1987), *On the role of sea-surface temperature gradients in forcing low-level winds and convergence in the tropics*, J. Atmos. Sci. *44*, 2418–2436.

Nitta, T. and Yamada, S. (1989), *Recent warming of tropical sea-surface temperature and its relationship to the Northern Hemisphere circulation*, J. Meteor. Soc. Japan *67*, 375–383.

Pattanaik, D. R. (2003), *Analysis of moist convective instability over Indian monsoon region and neighbourhood*, Mausam *54*, 659–670.

Pattanaik, D. R., Kalsi, S. R., and Hatwar, H. R. (2005), *Evolution of convection anomalies over the Indo-Pacific region in relation to indian monsoon rainfall*, Mausam *56*, 811–824.

Pattanaik, D. R. (2005), *Variability of oceanic and atmospheric conditions during active and inactive periods of storm over the Indian region*, Int. J. Climatology *25*, 1523–1530.

Rajeevan, M., De, U. S., and Prasad, R. K. (2000), *Decadal variation of sea-surface temperatures, cloudiness and monsoon depressions in the north Indian Ocean*, Current Science *79*, 283–285.

Rajeevan, M., Prasad, R. K., and De, U. S. (2001), *Cloud climatology of the Indian Ocean based on ship observations*, Mausam *52*, 527–540.

SPERBER, K. R. and PALMER, T. N. (1996), *Interannual tropical rainfall variability in general circulation model intercomparison project*, J. Climate *9*, 2727–2750.

WANG, B. (1995), *Interdecadal changes in El Niño onset in the last four decades*, J. Climate *8*, 267–285.

WEBSTER, P. J. (1995), *The annual cycle and the predictability of the tropical coupled Ocean-atmosphere system*, Meteor. Atmos. Phys. 56, 33–55.

WEBSTER, P. J., MAGANA, V. O., PALMER, T. N., SHUKLA, J., TOMAS, R. A., YANAI, M., and YASUNARI, T. (1998), *Monsoons, processes, predictability, and the prospects for prediction*, J. Geophys. Res. *103*, 14451–14510.

(Received March 28, 2006, accepted December 18, 2006)
Published Online First: June 29, 2007

Pure appl. geophys. 164 (2007) 1547–1576
0033–4553/07/091547–30
DOI 10.1007/s00024-007-0227-2

Pure and Applied Geophysics

Sensitivity of Mesoscale Surface Dynamics to Surface Soil and Vegetation Contrasts over the Carolina Sandhills

RYAN BOYLES, SETHU RAMAN, and AARON SIMS

Abstract—A region of contrasting soils exists over the Carolinas region of the southeastern United States. Previous research has shown an increase in mesoscale summertime precipitation over this region. Numerical simulations are analyzed to investigate the relationships between mesoscale surface dynamics and the transition from clay to sandy soils over this region. Numerical modeling experiments using four different soil and vegetation patterns suggest that the presence of the clay-to-sand transition zone produces a surface heat flux gradient and enhanced convergence. The soil contrasts appear to dominate over vegetation contrasts in affecting local surface heating and convergence in the model atmosphere.

Key words: Land surface variation, North Carolina, soils.

1. Introduction

In the southeastern United States, there exists a region of sharp change from dense clay soils to coarse sandy soils. This region is known as the Carolina Sandhills for its rolling sandy hills stretching from eastern North Carolina (NC) through South Carolina (SC) and Georgia (GA). The Sandhills region of the Carolinas lies along a zone generally parallel to the coastline and inland approximately 160 kilometers (100 miles) from the Atlantic Ocean. The transition from clay soils in the Piedmont region of the Carolinas to the Sandhills is believed to play a role in local precipitation dynamics.

The influence of soil type and land use variation on lower atmospheric energy exchange has been well documented (DEARDORFF, 1978; ANTHES, 1984; OOKOUCHI *et al.*, 1984; MAHFOUF *et al.*, 1987; SEGAL *et al.*, 1988; HONG *et al.*, 1995). KOCH and RAY (1997) documented a region of low pressure that forms during summers along the boundary of the Piedmont and Coastal Plain of North Carolina. This "Piedmont Trough" was found to be present 40% of the time during summer when convection occurred. KOCH and RAY (1997) found the "Piedmont Trough" to produce convection in the absence of other forcing boundaries, such as cold fronts or

State Climate Office of North Carolina, Department of Marine, Earth, and Atmospheric Sciences, North Carolina State University, Raleigh, NC 27695-7236, U.S.A.

sea-breeze fronts. KOCH and RAY (1997) determined that the "Piedmont Trough" enhanced convection in the presence of sea-breeze fronts, and that this region was second only to the sea-breeze as a source for summertime convective forcing.

More recently, RAMAN *et al.* (2005) used a summer climatology of National Weather Service Cooperative Observer gages and numerical model simulations to conclude that the contrasting soils and vegetation produce "significant horizontal gradients in the latent and sensible heat flux patterns in the Sandhills" and that these "contribute to the development of mesoscale circulations observed in this region".

The mesoscale dynamics associated with the Sandhills are further explored here using results from a numerical weather modeling study to identify the sensitivity of boundary-layer processes to the soil and vegetation variations of the Sandhills during a summertime event with weak synoptic flow.

2. *Numerical Modeling Experiments*

Observations of MPE on climatological scales suggest an increase in precipitation over the Sandhills region of the Carolinas (RAMAN *et al.*, 2005). A possible reason for this observed increase is enhanced convection due to the formation of a locally-forced thermal gradient and resulting vertical circulation, similar to a sea-breeze. A conceptual schematic of this circulation is provided in Figure 1. However, it is not known if the boundary layer and local circulations are more sensitive to the soil transition or the vegetation transition in this region. A series of numerical simulations are performed to investigate the sensitivity of the lower troposphere to soil and vegetation contrasts in the Sandhills. Using a numerical weather model, soil

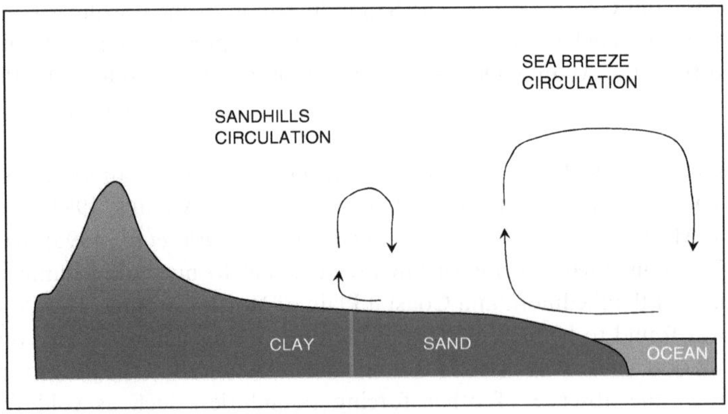

Figure 1
Diagram of the sea breeze and proposed Sandhills thermal circulations. Just as differential heating between land and water produces the classic sea breeze circulation, differential heating over different soil types may produce a circulation over the Sandhills region of the Carolinas.

and vegetation patterns are changed to identify the sensitivity of the boundary layer to the variation in land use and soil patterns that are typical of the Sandhills region. Specifically, the soil type and vegetation in the model are modified to remove the clay-to-sand transition zone and the vegetation contrasts and compare the results with a control simulation that includes standard soil and vegetation patterns.

2.1. Model Description

MM5 version 3.7 was used to simulate the atmosphere and surface dynamics for the period July 9–13, 2004. GRELL *et al.* (1995) provides details on the MM5 numerical weather modeling system. MM5 uses a sigma coordinate system that follows the terrain and a finite fixed resolution grid to solve the general atmospheric equations of motion, thermodynamics, and state.

For this study, a one-way single nest is used. The model domain is shown in Figure 2, with an outer domain grid spacing of 12 km and inner domain grid spacing of 4 km. 42 vertical levels were used, with 25 below 700 mb. Terrain, soil, and land-use data were obtained from the National Center for Atmospheric Research (NCAR)

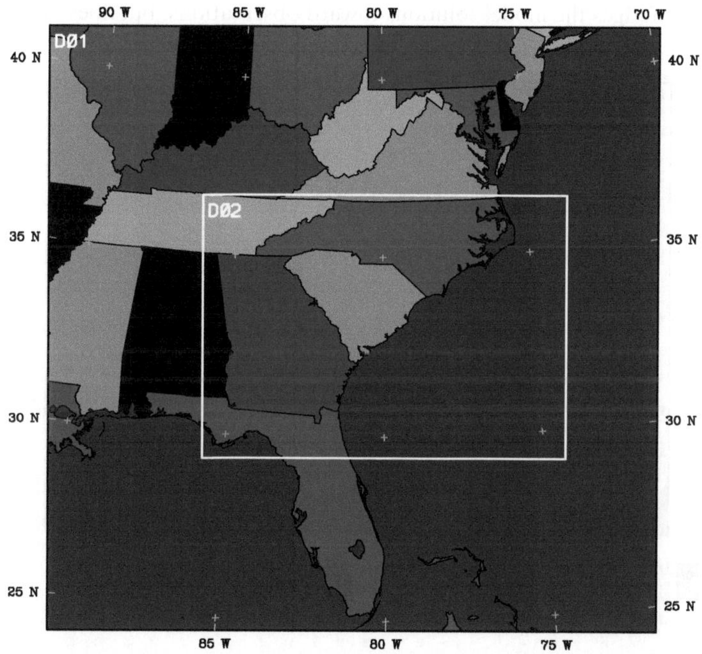

Figure 2
Outer and inner domains for MM5 simulations. The outer domain uses a 12-km grid spacing while the inner domain uses a 4-km grid spacing.

and use the 24-class vegetation/land use reference and 19-class dominant soils reference as developed by the US Geological Survey (USGS). The model vegetation pattern is shown in Figure 3 and dominant soil pattern is shown in Figure 4. The model characteristics associated with each land use/vegetation class are listed in Table 2, while model characteristics for each soil class are given in Table 1. The model physics options used for these simulations are given in Table 3. In all simulations, Reisner's mixed phase precipitation physics scheme, the Medium-Range Forecast planetary boundary layer (MRF PBL) scheme, the Rapid Radiative Transfer Model (RRTM) radiation scheme, and the NOAH land surface model were used. In the outer domain (155 × 155, 12-km grid spacing) the Kain-Fritsch 2 cumulus parameterization was used while the inner domain (202 × 277, 4-km resolution) used explicit cloud physics to resolve convection. Initial and lateral boundary conditions are derived from 3-hourly analyzed fields from the Eta Data Assimilation System (EDAS). Surface observations are assimilated into the model at three-hour time steps using available data from the in-house database at the State Climate Office of North Carolina. These observations, which include soil temperature and soil moisture data from the North Carolina Environment and Climate Observations Network (NC ECONet), are "nudged" into the model using the LITTLE_R module in MM5. Data assimilation nudges the model solutions toward observations, but does not shock the

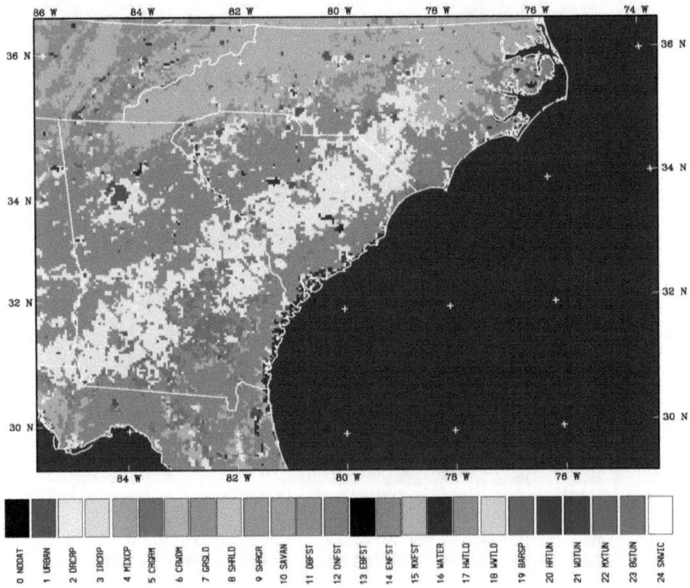

Figure 3
MM5 vegetation/land use classifications for the inner 4-km domain. Vegetation/land-use classes are described in Table 2.

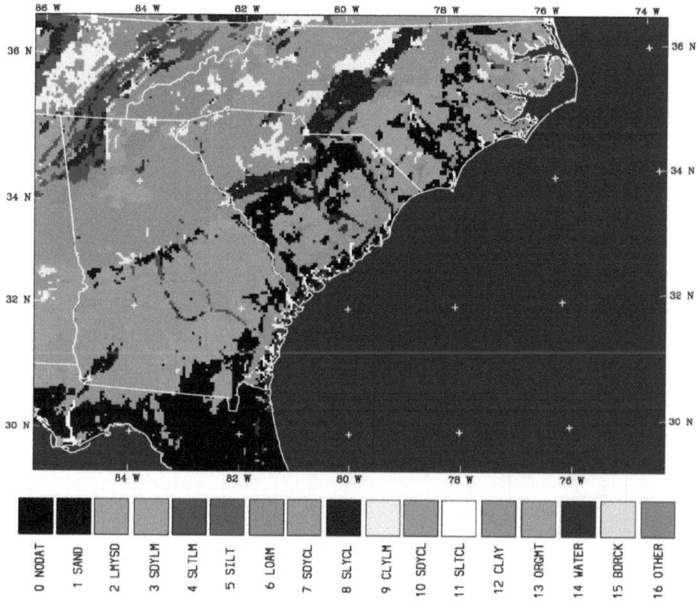

Figure 4
MM5 dominant soil classification for the inner domain. The soil classification properties are described in Table 1.

model with forced data, and therefore limits any spurious effects. A 24-hour model spin-up period is used prior to the period of analysis.

2.2. *Synoptic Conditions*

The period July 10–13, 2004 is chosen for simulations because there was weak synoptic forcing typical of summertime in the Carolinas and convection was observed in the satellite and radar imagery over the Sandhills region. Winds aloft are generally from the west to northwest over the 4-day period. Satellite imagery indicates generally clear conditions during the morning and early afternoon hours of July 10–13, 2004 over the Sandhills region (not shown). Each afternoon, convection forms over the Sandhills region, and then dissipates overnight. Convection is observed over the Sandhills and/or coastal plain on each day, but only July 12 is used for discussion of the modeling experiments.

2.3. *Experimental Design*

To investigate the sensitivities of the model atmosphere to soil and vegetation patterns, a series of four simulations are performed using MM5. The experimental

Table 1

Physical attributes associated with soil classes

USGS Soil Class	Soil Label	Slope of the Retention Curve	Wilting Point ($m^3\ m^{-3}$)	Soil Thermal Diffusivity / Conductivity Coefficient	Volumetric Water Content at Saturation ($m^3\ m^{-3}$)	Reference Field Capacity ($m^3\ m^{-3}$)	Saturated Soil Suction (m)	Hydraulic Conductivity at Saturation ($m\ s^{-1}$)	Soil Diffusivity Coefficient	Soil quartz content (%)
1	Sand	2.79	0.010	0.472	0.339	0.236	0.069	1.07E-06	6.08E-07	0.92
2	Loamy Sand	4.26	0.028	1.044	0.421	0.383	0.036	1.41E-05	5.14E-06	0.82
3	Sandy Loam	4.74	0.047	0.569	0.434	0.383	0.141	5.23E-06	8.05E-06	0.60
4	Silt Loam	5.33	0.084	0.162	0.476	0.360	0.759	2.81E-06	2.39E-05	0.25
5	Silt Loam	5.33	0.084	0.162	0.476	0.383	0.759	2.81E-06	2.39E-05	0.10
6	Loamy Sand	5.25	0.066	0.327	0.439	0.329	0.355	3.38E-06	1.43E-05	0.40
7	Sandy Clay Loam	6.66	0.067	1.491	0.404	0.314	0.135	4.45E-06	9.90E-06	0.60
8	Silty Clay Loam	8.72	0.120	1.118	0.464	0.387	0.617	2.04E-06	2.37E-05	0.10
9	Clay Loam	8.17	0.103	1.297	0.465	0.382	0.263	2.45E-06	1.13E-05	0.35
10	Sandy Clay Loam	10.73	0.100	3.209	0.406	0.338	0.098	7.22E-06	1.87E-05	0.52
11	Silty Clay Loam	10.39	0.126	1.916	0.468	0.404	0.324	1.34E-06	9.64E-06	0.10
12	Clay	11.55	0.138	2.138	0.468	0.412	0.468	9.74E-07	1.12E-05	0.25
13	Organic Material	5.25	0.066	0.327	0.439	0.329	0.355	3.38E-06	1.43E-05	0.05
14	Water	0.00	0.000	0.000	1.000	0.000	0.000	0.00E+00	0.00E+00	0.60
15	Bedrock	2.79	0.006	1.111	0.200	0.170	0.069	1.41E-04	1.36E-04	0.07
16	Other (Land-Ice)	4.26	0.028	1.044	0.421	0.283	0.036	1.41E-05	5.14E-06	0.25
17	Playa	11.55	0.030	10.472	0.468	0.454	0.468	9.74E-07	1.12E-05	0.60
18	Lava	2.79	0.006	0.472	0.200	0.170	0.069	1.41E-04	1.36E-04	0.52
19	White Sand	2.79	0.010	0.472	0.339	0.236	0.069	1.07E-06	6.08E-07	0.92

Table 2

Physical attributes associated with vegetation/land use classes

USGS Vegetation Class	Vegetation / Land Use Label	Albedo (%)	Soil Moisture Availability Fraction	Surface Emissivity Fraction	Surface Roughness Length (cm)	Surface Thermal Inertia (100×cal cm^{-2} K^{-1} $s^{-1/2}$)	Snow Cover Effect Fraction	Surface Heat Capacity (J m^{-3} K^{-1})
1	Urban and Built-Up Land	15	0.10	0.880	80	3	1.67	1.89E+06
2	Dryland Cropland and Pasture	17	0.30	0.985	15	4	2.71	2.50E+06
3	Irrigated Cropland and Pasture	18	0.50	0.985	10	4	2.2	2.50E+06
4	Mixed Dryland/Irrigated Cropland and Pasture	18	0.25	0.985	15	4	2.56	2.50E+06
5	Cropland/Grassland Mosaic	18	0.25	0.980	14	4	2.56	2.50E+06
6	Cropland/Woodland Mosaic	16	0.35	0.985	20	4	3.19	2.50E+06
7	Grassland	19	0.15	0.960	12	3	2.37	2.08E+06
8	Shrubland	22	0.10	0.930	5	3	1.56	2.08E+06
9	Mixed Shrubland/Grassland	20	0.15	0.950	6	3	2.14	2.08E+06
10	Savanna	20	0.15	0.920	15	3	2	2.50E+06
11	Deciduous Broadleaf Forest	16	0.30	0.930	50	4	2.63	2.50E+06
12	Deciduous Needleleaf Forest	14	0.30	0.940	50	4	2.86	2.50E+06
13	Evergreen Broadleaf Forest	12	0.50	0.950	50	5	1.67	2.92E+06
14	Evergreen Needleleaf Forest	12	0.30	0.950	50	4	3.33	2.92E+06
15	Mixed Forest	13	0.30	0.970	50	4	2.11	4.18E+06
16	Water Bodies	8	1.00	0.980	0.01	6	0	9.00E+25
17	Herbaceous Wetland	14	0.60	0.950	20	6	1.5	2.92E+06
18	Wooded Wetland	14	0.35	0.950	40	5	1.14	4.18E+06
19	Barren or Sparsely Vegetated	25	0.02	0.900	1	2	0.81	1.20E+06
20	Herbaceous Tundra	15	0.50	0.920	10	5	2.87	9.00E+25
21	Wooded Tundra	15	0.50	0.930	30	5	2.67	9.00E+25
22	Mixed Tundra	15	0.50	0.920	15	5	2.67	9.00E+25
23	Bare Ground Tundra	25	0.02	0.900	10	2	1.6	1.20E+06
24	Snow or Ice	55	0.95	0.950	5	5	0	9.00E+25

Table 3

Model physics schemes used in domains

Outer Domain (155 × 155–12 km resolution)	
Cumulus parameterization	Kain-Fritsch 2 (w/shallow convection) (KAIN and FRITSCH, 1993)
Precipitation microphysics	Reisner's mixed-phase (REISNER *et al.*, 1998)
Planetary boundary layer processes	MRF (HONG and PAN, 1996)
Surface layer processes	NOAH land-surface model (CHEN and DUDHIA, 2001)
Atmospheric radiation	RRTM long-wave (MLAWER *et al.*, 1997)
Inner Domain (202 × 277–4 km resolution)	
Cumulus parameterization	Explicit cloud physics (SCHULTZ, 1995)
Precipitation microphysics	Reisner's mixed-phase (REISNER *et al.*, 1998)
Planetary boundary layer processes	MRF (HONG and PAN, 1996)
Surface layer processes	NOAH land-surface model (CHEN and DUDHIA, 2001)
Atmospheric radiation	RRTM long-wave (MLAWER *et al.*, 1997)

design for the four sensitivity simulations, including the differences in model soil and vegetation/land use for each simulation, is given in Table 4. In the CONTROL simulation, the reference USGS vegetation and soil spatial patterns are used. In CASE1, the USGS reference vegetation classes are used and all soils in the inner and outer domains are assigned as sand. Sand is used since it is typical of eastern NC (Sandhills area and east). Therefore, CASE1 has a uniform soil pattern but maintains a variation in vegetation. In CASE2, vegetation over land is prescribed to be only mixed forest and soils to be all sand. A mixed forest vegetation class is used since it is typical of vegetation in central and western NC, and is appropriate to represent typical vegetation dynamics across the southeastern US. In CASE2, uniform soil (sand) and uniform vegetation (mixed forest) are used in both the inner and outer domain of the model. In CASE3, the USGS standard soils are used, but the vegetation surface is changed to mixed forest, creating uniform vegetation but variations in soil. Each of the four model simulations is analyzed to investigate differences in surface convergence, wind fields, and heat fluxes, and regional precipitation amounts. By adjusting only the soil and vegetation patterns

Table 4

Experimental design for simulations

SIMULATION	SOIL	VEGETATION/LANDUSE
Reference (CONTROL) Case	USGS Standard	USGS Standard
CASE1	Sand	USGS Standard
CASE2	Sand	Mixed Forest
CASE3	USGS Standard	Mixed Forest

in each simulation, the resulting changes in atmospheric dynamics can be isolated as contributions from soil variation, vegetation variation, or both.

3. Discussion of Model Results

3.1. Control Simulation and Validation

A control simulation of the atmosphere is performed using the standard USGS soil and vegetation layers. Over the five-day study period, July 12 is used for analysis and discussion. To evaluate the model performance on this day, atmospheric profiles and surface measurements from the control simulation are compared with observations.

Analysis of model profiles compared with radiosonde observations (not shown) for Charleston, SC (CHS) and Greensboro, NC (GSO) show the model overall does a good job of simulating the wind and thermodynamic at heights above 1000 meters in the morning (08LT), though the northerly (-V) component of the wind is too strong through the entire troposphere. The model overestimates potential temperature and underestimates the mixing ratio near the surface. More importantly, the model does not properly capture the boundary layer inversion seen in the observed soundings.

A time history of the u and v components of the wind for the CONTROL simulation compared with observations for the Jackson Springs ECONet station (JACK) on July 12 is shown in Figure 5. Jackson Springs is located at latitude 35.18782° and longitude −79.68437° - in the middle of the Sandhills region of NC. Model simulated winds on July 12 overall have a northerly component that is too strong, but generally captures the weak U component over this day. Model simulated temperature and dew point for the entire four day simulation are shown in Figure 6. The model slightly overestimates the air temperature during the day and largely overestimates the temperature during nocturnal hours.

The model simulated 24-hour accumulated precipitation for July 12, 2004 is shown in Figure 7. Figure 8 shows the Multi-sensor Precipitation Estimated (MPE) total precipitation for July 12 (see LIN and MITCHELL, 2005 for more on MPE). The model captures the location of the daily precipitation in southeastern Georgia, northeast SC, and western NC, but missed the precipitation seen in MPE over central parts of NC and SC, and in northern Georgia. The CONTROL simulation produces precipitation over eastern NC that is not seen in MPE. The model also tends to produce much higher amounts in locations where rainfall is simulated–values often exceed one inch (25.4 mm) over areas where precipitation is simulated. In contrast, MPE shows a much smaller geographic area with amounts in excess of one inch (25.4 mm). However, it is rare that a model accurately represents both timing, location, and the amount of rainfall (especially convective

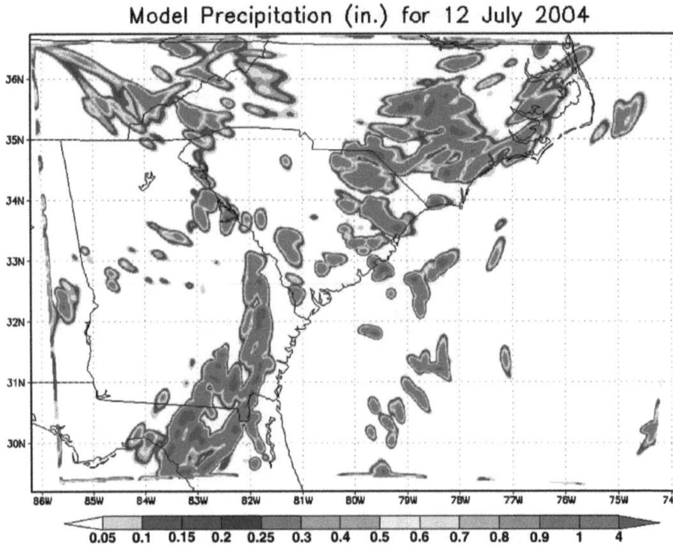

Figure 7
24-hour accumulation of model-simulated precipitation (in inches) using the CONTROL simulations.

eastern parts. Much lower values (< 200 W/m^2) are observed in both simulations over regions where the model simulates precipitation.

A contrasting pattern in sensible heat flux (SHF) is observed during the study period. During overnight hours, the sensible heat flux pattern is uniformly zero except for the higher SHF observed over the urban areas (less than 25 W/m^2 – figure not shown). As daytime heating begins, the sandy soil tends to heat slower than the clay-based soils. SHF patterns are shown for 14Z and 16Z on July 12, 2004 in Figures 11 and 12, respectively. In the CONTROL case, a SHF difference of 150 W/m^2 is observed between the clay soils and the Sandhills. In the Sand/Std. Vegetation simulation, a smaller difference of ~50 W/m^2 is observed. At 16Z (12LT), the difference is more pronounced (see Fig. 12). Areas with clay soils in the CONTROL simulation have SHF values of ~450 W/m^2, while in the sensitivity experiment with uniform sandy soils the SHF values over this region are generally less than 200 W/m^2.

Based on the latent and sensible heat flux patterns, a heat flux gradient associated with the clay-to-sand transition zone exists in the CONTROL simulation, but not simulated in the Sand/Std Vegetation sensitivity case. The heat flux gradient in the CONTROL simulation should be associated with surface convergence, while little or no convergence should be along the Sandhills in CASE1 since the heat flux gradient is weak.

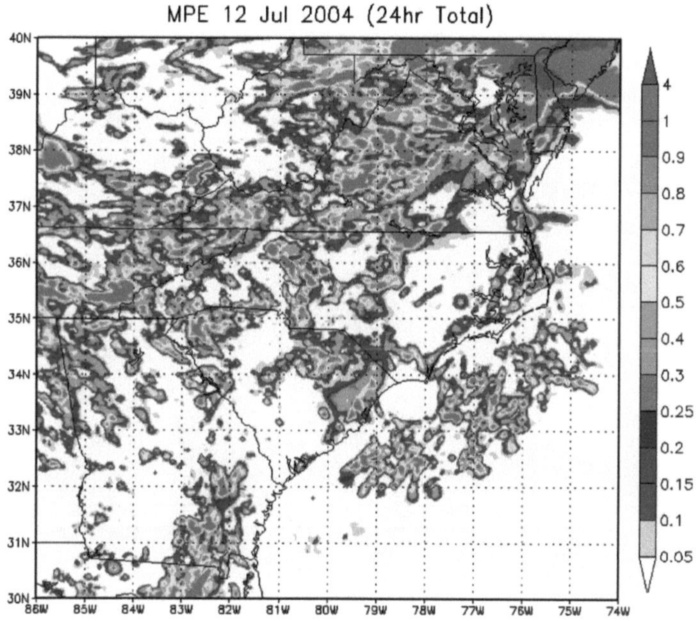

Figure 8
24-hour accumulated Multi-sensor Precipitation Estimates (MPE) (in inches).

Simulated wind vectors and convergence at 10 meters for the CONTROL and CASE1 simulations at 14Z and 16Z are shown in Figures 13 and 14, respectively. In both of these figures, strong convergence is observed over coastal areas and is associated with a sea breeze circulation. This circulation is not associated with dynamics along the Sandhills region. At 14Z (10 LT), wind vectors suggest convergence in NC and SC along the clay-to-sand transition. Wind vectors also appear to change in the sensitivity case, but the directional shift is less pronounced. However, at 16Z (12 LT), the differences between the two cases are obvious (see Fig. 14). At this time, a line of weak convergence and divergence is observed along the clay-to-sand transition zone in the CONTROL case with convergence values near 0.0004 s^{-1}. Convergence is generally on the west side of the transition, suggesting rising motion over the clay soils and sinking motion further east. This line of convergence is noticeably absent in the Sand/Std Vegetation sensitivity simulation.

Based on an analysis of the model surface heat fluxes, it is apparent that the model atmosphere is sensitive to soil variations and produces a surface heat flux gradient in the CONTROL simulation that is much weaker in the sensitively simulation. This gradient appears to be of sufficient strength to produce local convergence and a weak vertical circulation over the clay-to-sand transition zone in

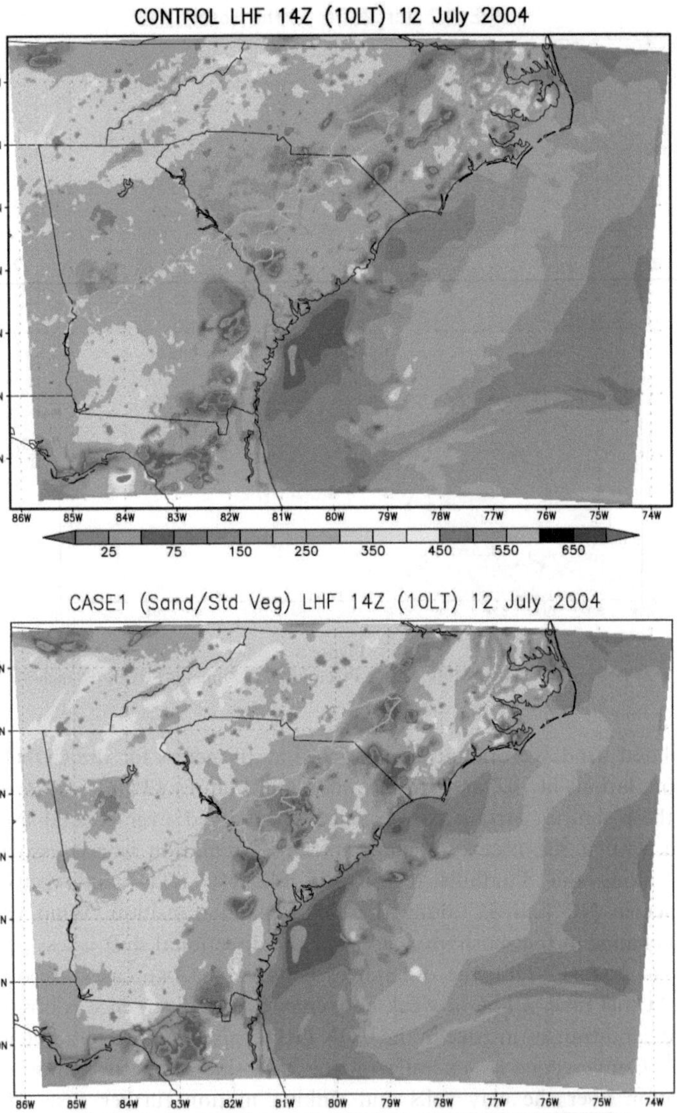

Figure 9

Surface latent heat flux (W/m^2) at 14Z (10LT) on July 12, 2004 for the CONTROL (top) and Sand/Std Vegetation (bottom) simulations. Differences in LHF are observed over the western parts of NC/SC/GA, where the clay soils exist in the CONTROL simulation but not in Case 2.

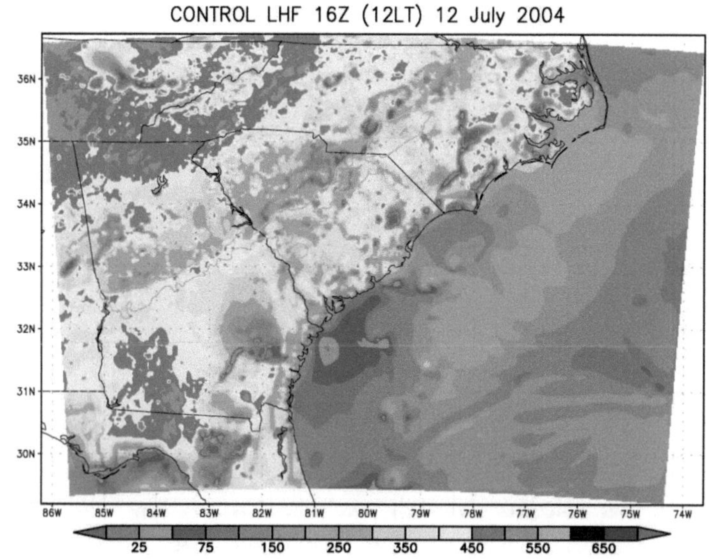

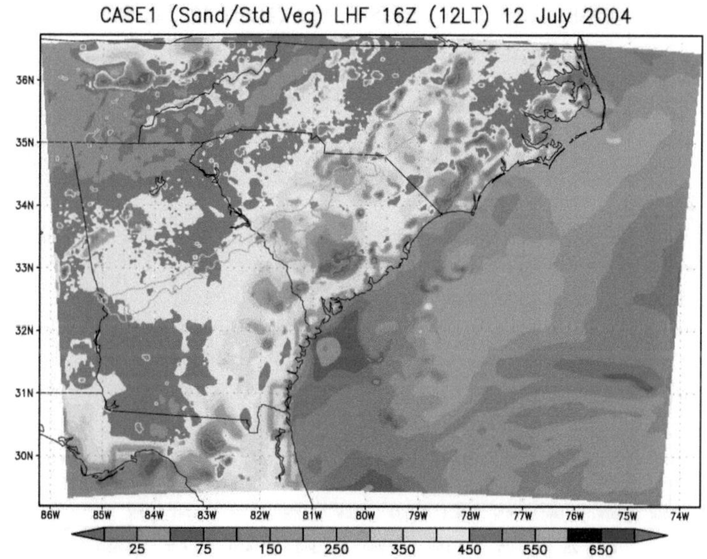

Figure 10

Surface latent heat flux (W/m^2) at 16Z (12LT) on July 12, 2004 for the CONTROL (top) and Sand/Std Vegetation (bottom) simulations. The heat-flux gradient that is observed in the CONTROL case is not observed in CASE1 west of the Sandhills.

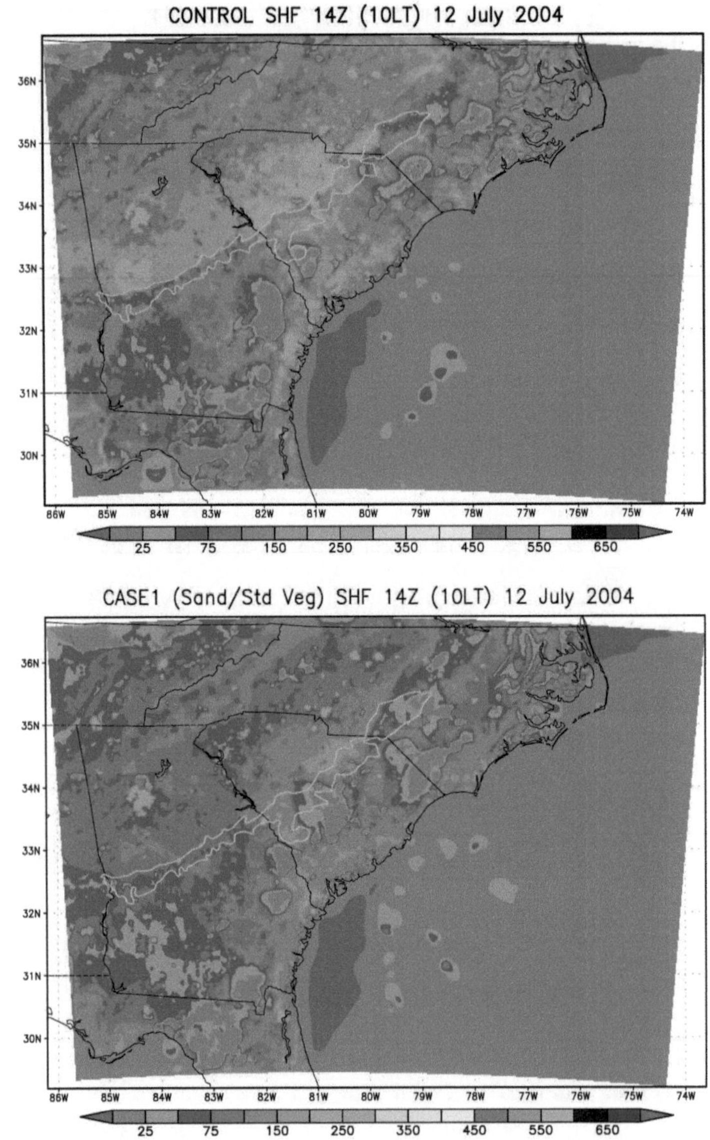

Figure 11

Surface sensible heat flux (W/m^2) at 14Z (10LT) on July 12, 2004 for the CONTROL (top) and Sand/Std Vegetation (bottom) simulations. Higher sensible heat fluxes are observed over the clay soils in the CONTROL simulation.

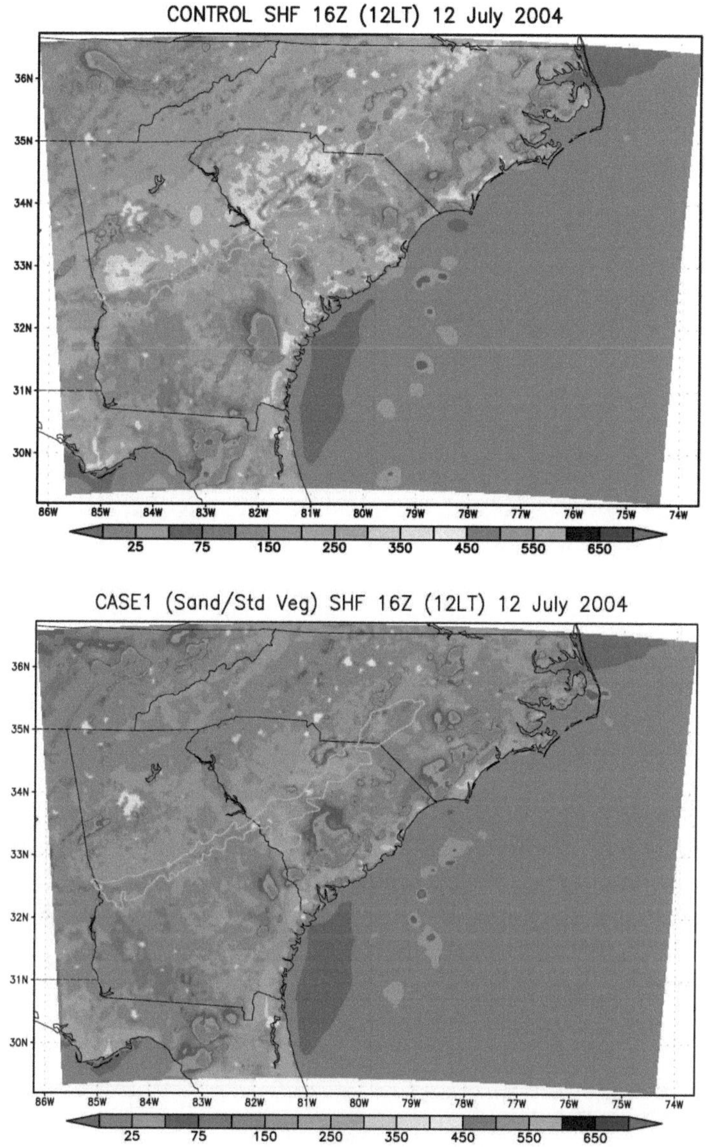

Figure 12

Surface sensible heat flux (W/m^2) at 16Z (12LT) on July 12, 2004 for the CONTROL (top) and Sand/Std Vegetation (bottom) simulations. The SHF gradient in the sensitivity simulation is not as large as in the CONTROL case.

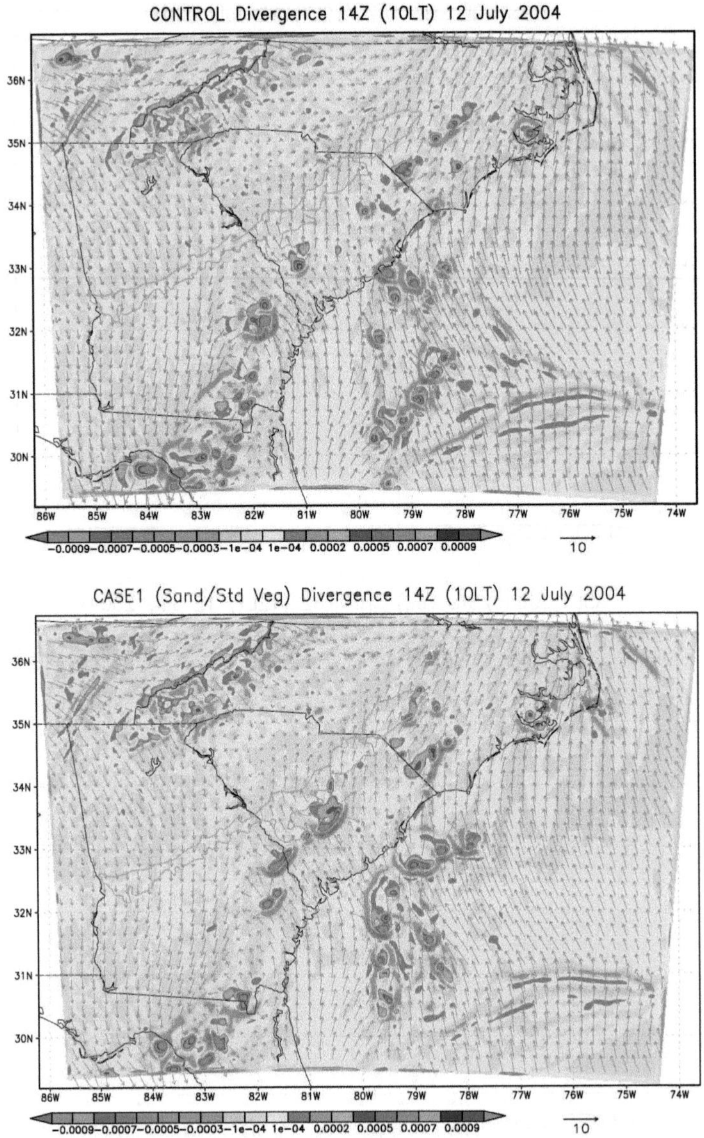

Figure 13

Surface winds and divergence for the CONTROL and CASE1 simulations at 14Z (10 LT) on July 12, 2004. Convergence is given as negative values (cool colors). Coastal convergence is associated with a sea breeze circulation.

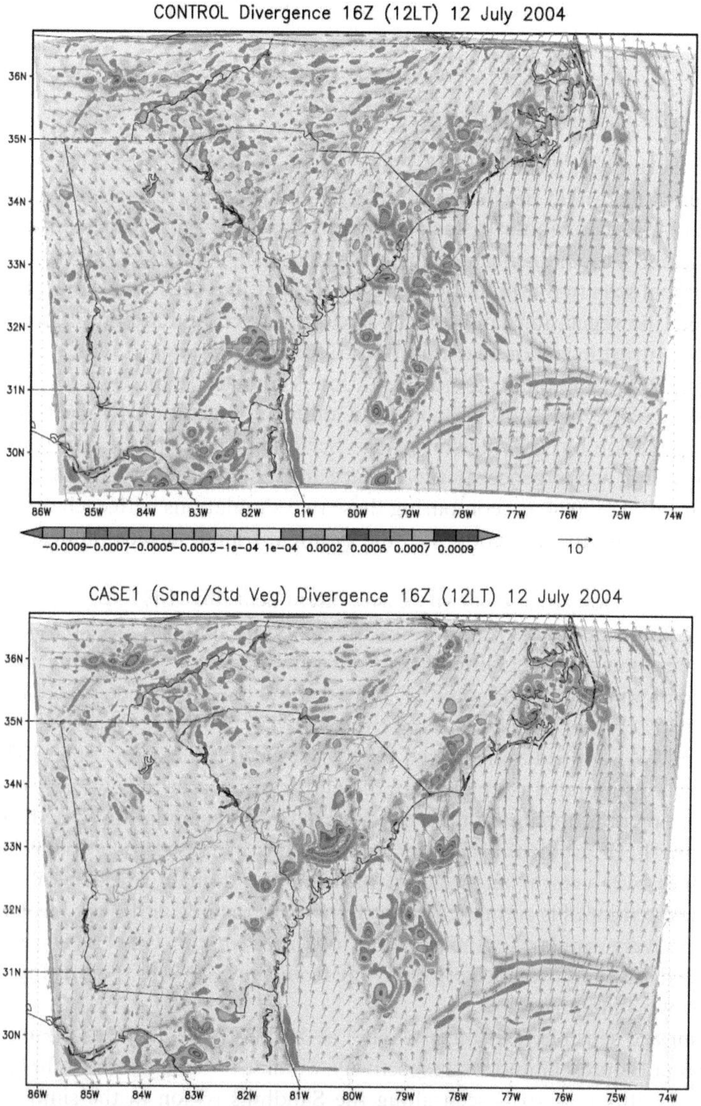

Figure 14

Surface winds and divergence for the CONTROL and CASE1 simulations at 16Z (12 LT) on July 12, 2004. A line of convergence through central NC/SC/GA is observed in the CONTROL case that is absent in the sensitivity case.

the CONTROL simulation. This feature is absent in the Sand/Std Vegetation sensitivity case.

3.3. Sensitivity to Vegetation/Land Use

While the analysis discussed in Section 3.2 offers strong evidence for the influence of the soils transition in the model simulation, contrasts in vegetation types over the region may also contribute to local heat flux gradients and local convergence. In order to identify the influence of vegetation contrasts compared with soil contrasts, two additional sensitivity simulations are performed. In CASE2, all soils are assigned as sand and the vegetation/land use classifications over land are prescribed as mixed forest (no contrast in soil or vegetation). In CASE3, the standard soils are used (same as in CONTROL) and the land use/vegetation is changed to mixed forest everywhere over land. CASE2 has uniform soils and uniform vegetation, while CASE 3 has uniform vegetation but varying soils. As in Section 4.2, latent heat flux, sensible heat flux, winds and convergence at the surface are analyzed at 14Z (10LT) and 16Z (12LT) on July 12, 2004. By comparing these two simulations with each other and the simulations discussed in Section 4.2, the dominant land surface influences may be evident.

As in the previous two sensitivity cases, latent heat flux (LHF) patterns in CASE2 and CASE3 during the overnight hours are generally uniform (25–50 W/m^2) across the domain (not shown). Differences between these two simulations are evident during the daytime hours. Figures 15 and 16 show the simulated surface latent heat fluxes at 14Z (10LT) and 16Z (12LT) on July 12, 2004 for CASE2 (Sand/Mixed Forest) and CASE3 (Std Soils/Mixed Forest), respectively. At this time, LHF of 50–100 W/m^2 are simulated over the clay-to-sand transition in SC and GA in CASE3, while no gradient is apparent in the Sand/Mixed Forest simulation (CASE2). At 16Z (12 LT), the LHF gradient along the Sandhills has increased to $\sim$ 200 W/m^2, while LHF values over central and eastern parts of the domain are generally uniform (except where rainfall is simulated). The LHF patterns in CASE3 (Std Soils / Mixed Forest) are very similar to the patterns observed in the CONTROL simulation (Figs. 9 and 10), and are not very different from the CONTROL as compared with the other two simulations.

The surface sensible heat fluxes (SHF) at 14Z and 16Z for CASE2 and CASE3 are shown in Figures 17 and 18, respectively. Similar to the latent heat flux pattern, a sensible heat flux pattern is seen along the Sandhills region in the simulation with contrasting soils (CASE3) at both 14Z and 16Z, ranging from 150–200 W/m^2. As with the latent heat flux, the sensible heat flux differences along the Sandhills in CASE3 are much closer in magnitude to the CONTROL simulation than the other sensitivity cases. Based on the surface heat flux patterns, it appears that the effect of the contrasting soils dominates over the vegetation contrasts to produce the heat flux gradient in the CONTROL simulation. Indeed, the SHF difference along the

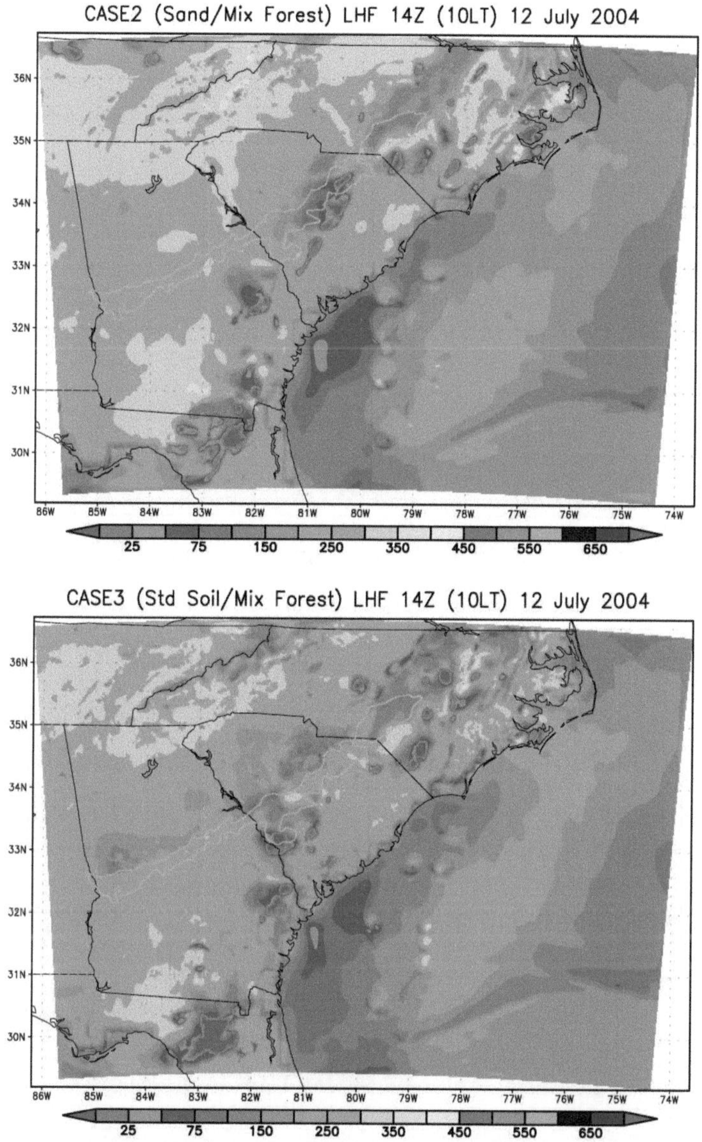

Figure 15

Surface latent heat flux (W/m^2) at 14Z (10LT) on July 12, 2004 for CASE2 (Sand/Mixed Forest) and CASE3 (Std Soils/Mixed Forest). LHF differences of 50–100 W/m^2 are observed over the clay-to-sand transition in SC and GA in CASE3, while no gradient is apparent in the Sand/Mixed Forest simulation. Values less than 100 W/m^2 are associated with recent rainfall over those areas.

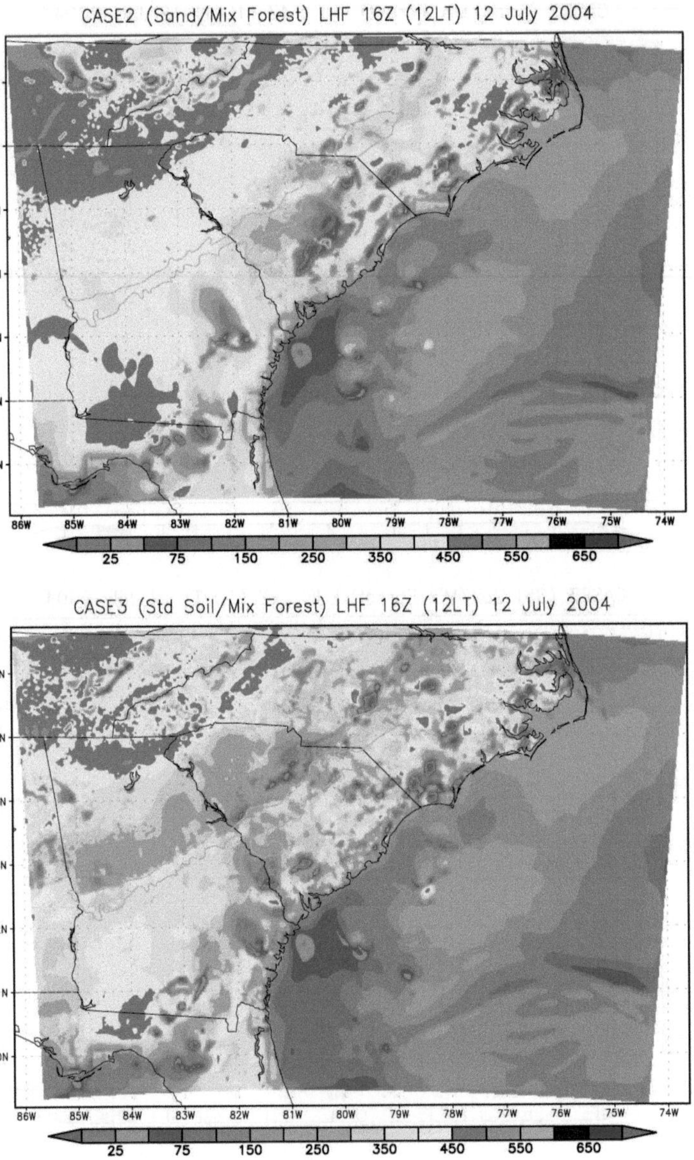

Figure 16

Surface latent heat flux (W/m^2) at 16Z (12LT) on July 12, 2004 for CASE2 (Sand/Mixed Forest) and CASE3 (Std Soils/Mixed Forest). LHF difference of 200 W/m^2 are observed over the clay-to-sand transition in SC and GA in CASE3. Values less than 100 W/m^2 are associated with recent rainfall over those areas.

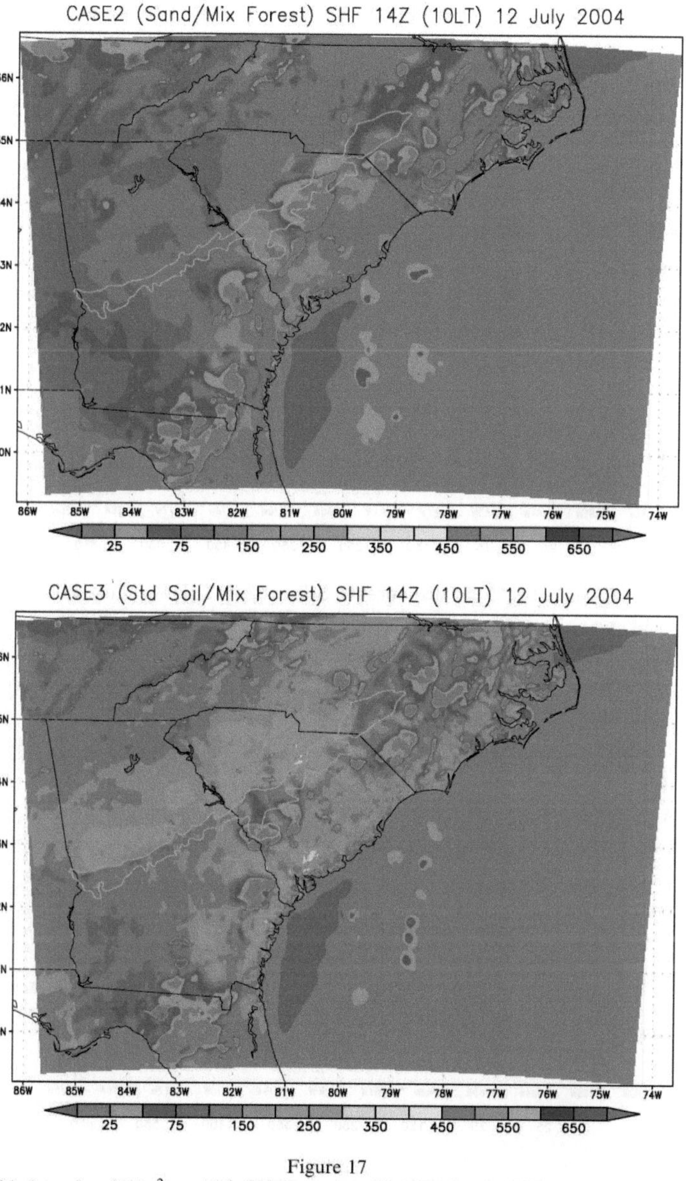

Figure 17

Surface sensible heat flux (W/m^2) at 14Z (10LT) on July 12, 2004 for CASE2 (Sand/Mixed Forest) and CASE3 (Std Soils/Mixed Forest). A gradient exists along the Sandhills in CASE3 that is not observed in CASE2.

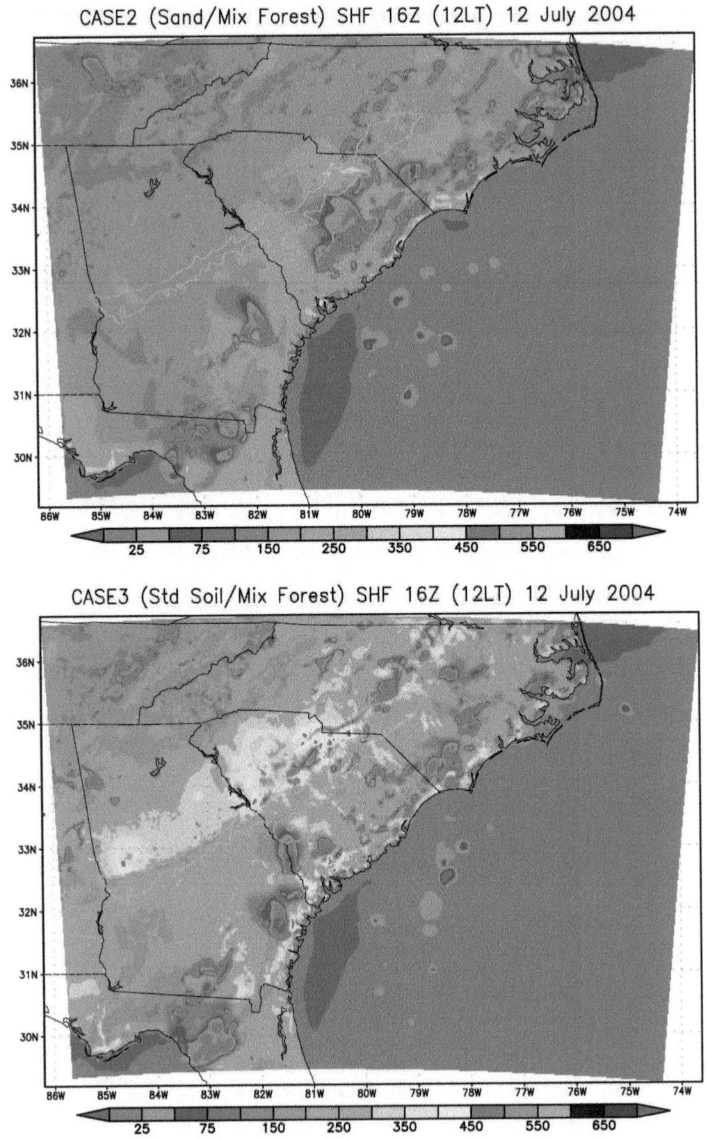

Figure 18

Surface sensible heat flux (W/m^2) at 16Z (12LT) on July 12, 2004 for CASE2 (Sand/Mixed Forest) and CASE3 (Std Soils/Mixed Forest). SHF gradients are similar to those observed in the CONTROL simulation.

Sandhills associated with the soil contrast is nearly 300% of the difference associated with the vegetation contrasts observed in CASE1 (Figs. 11 and 12).

Further analysis suggests that the heat flux gradients associated with the soils contrast along the Sandhills is sufficient to produce local convergence without the vegetation contrast. Figures 19 and 20 show the surface wind vectors and convergence for CASE2 and CASE3 on July 12, 2004 at 14Z and 16Z, respectively. Again, the patterns observed in CASE3 are similar to the CONTROL simulation. At 16Z (12LT) a local convergence zone along the Sandhills is observed just as in the CONTROL simulation. However, since CASE3 has a uniform vegetation surface, the convergence zone observed in CASE3 and the CONTROL simulation are likely forced by the soils contrast.

The differences in surface heat fluxes can largely be accounted for by differences in soil heat capacity. The surface energy budget for a layer has a relationship of the form:

$$R_N = H_S + H_L + H_G + \Delta H_S,$$

where R_N is net radiation, H_S is sensible heat flux, H_L is latent heat flux, H_G is ground heat flux, and ΔH_S is the change in the energy storage. The rate of change of energy storage of a soil layer is given as

$$\Delta H_S = \int \frac{\partial}{\partial t}(CT)\, dz,$$

where T is the absolute temperature of the soil layer and C is the heat capacity, which is a product of the mass density and specific heat of the soil. ARYA (2001) lists the heat capacity in $\mathrm{J\,m^{-3}\,K^{-1}} \times 10^6$ for dry (saturated) sand as 1.28 (2.96) and clay as 1.42 (3.10). The heat capacity for clay is approximately $0.14\,\mathrm{J\,m^{-3}\,K^{-1}} \times 10^6$ higher at both dry and saturated states. Thus, the rate of heating for a clay soil layer is between 5% and 11% higher than a sandy soil layer (depending on moisture content).

As described by CHEN and DUDHIA (2001), the model ground heat flux is based on a similar diffusion equation for soil temperature (T):

$$C(\Theta)\frac{\partial T}{\partial t} = \frac{\partial}{\partial z}\left(K_t(\Theta)\frac{\partial T}{\partial z}\right),$$

where C is the volumetric heat capacity and K_t is the thermal conductivity—both described as a function of Θ—the fraction of volumetric soil occupied by water. The soil thermal conductivity coefficient used in the model simulation for clay is 2.138, while the thermal conductivity coefficient for sand is 0.472 (see Table 4.1). The ground heat flux is therefore higher over clay soils than sandy soils, and energy is transferred to the atmospheric boundary layer at higher rates over clay soils as compared with sand. Sandy soil has many more pores and tends to be filled with air (with significantly less heat capacity as the soil dries) as compared to clay. This difference in the soil heat capacity and thermal conductivity coefficient largely

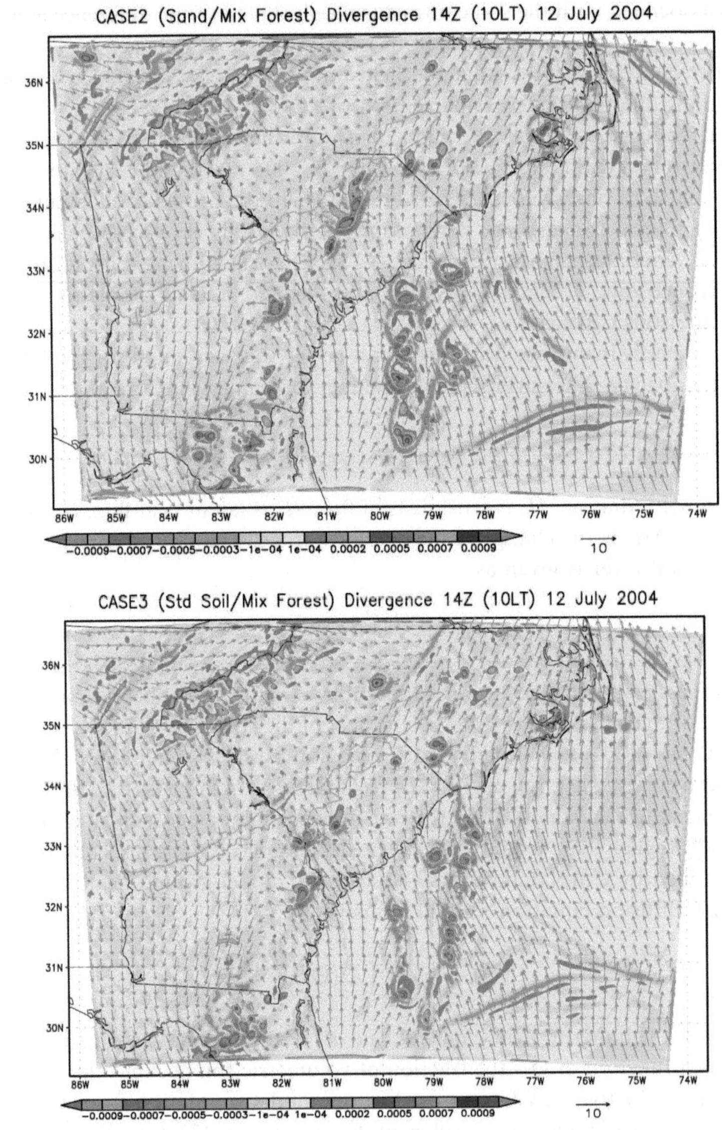

Figure 19

Surface winds and divergence for CASE2 (top) and CASE3 (bottom) simulations at 14Z (10 LT) on July 12, 2004.

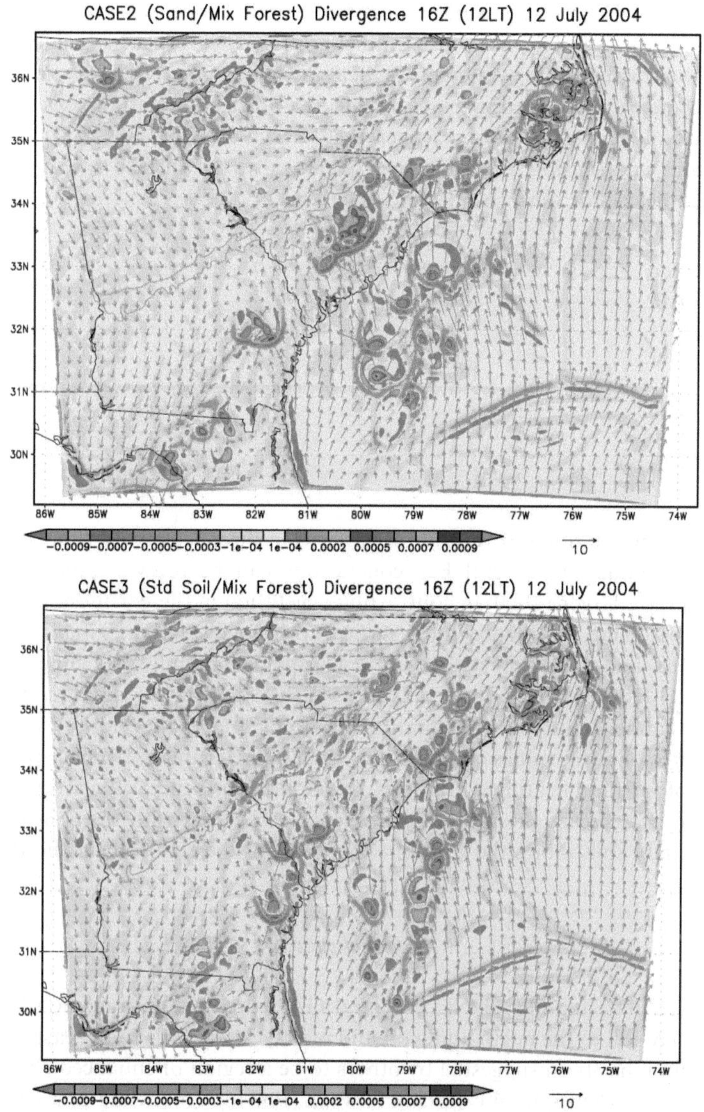

Figure 20
Surface winds and divergence for CASE2 (top) and CASE3 (bottom) simulations at 16Z (12 LT) on July 12, 2004.

accounts for the differences in surface heat fluxes in the model sensitivity simulations. In particular, the sensible heat flux gradients drive the vertical circulation in this region of contrasting soils by increasing local instability over clay soils similar to the role that land surface heating plays in the sea-breeze circulation. Over clay, the ground flux and change in storage is increased, which directly leads to higher ground temperatures and sensible heat fluxes. Schematically, the surface energy balance is represented below for sand (black) and clay (grey):

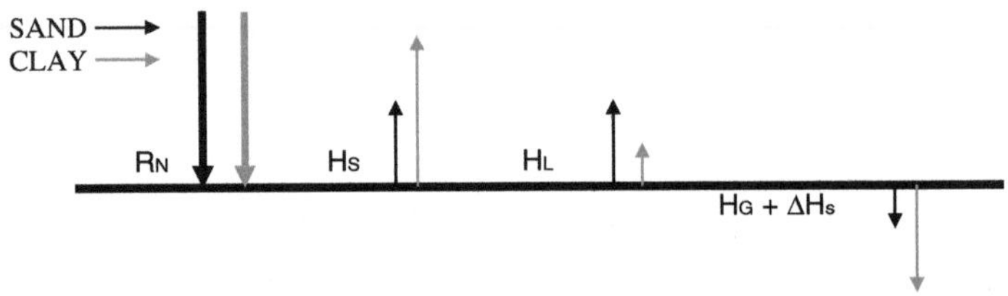

By adjusting the soil and vegetation patterns in three sensitivity simulations with the control simulation, several key conclusions can be drawn. Contrasting soils appear to play a larger role than contrasting vegetation in the partitioning of surface energy fluxes. There is evidence in the surface convergence patterns that the presence of the clay to sand transition zone over the Carolinas Sandhills may enhance surface convergence. However, it is important to note that the control simulation does not accurately handle observed precipitation locations and timing, although the areal averaged precipitation values compare well. Still, there is ample evidence to show the existence of surface heating variations due to the presence of contrasting soils, resulting in an increase in surface convergence associated with the transition zone from the sandy soils along the Carolina Sandhills to the clay soils in the Piedmont.

4. Conclusions

The Carolina Sandhills of NC, SC, and GA represent the transition between a primarily clay soil to the north and west and mostly sandy soils to the east and south. This transition has been suggested by others to be a region of enhanced precipitation during summer when synoptic forcing is weak. Previous modeling and climatology research have suggested that the variation in soil and vegetation patterns over this area may serve as a "trigger" for local convection.

Numerical weather simulations are analyzed to identify the sensitivity of the model atmosphere to variations in soil and vegetation patterns over the Carolina Sandhills. By comparing simulations with and without soil and vegetation contrasts with a control simulation of July 11–12, 2004, the atmospheric response to soil and vegetation variations are isolated. Simulations reveal that the presence of the Sandhills and the transition from clay to sand in this region could support the dynamics needed for locally enhanced precipitation over this region. Local variations in soils are associated with heat flux gradients, which in turn affect local temperature and pressure gradients in a manner similar to a sea breeze circulation. The variation in pressure over the Sandhills leads to momentum and mass convergence that is observed in model wind and convergence fields. Simulations with contrasting soils showed surface heat flux gradients and increased surface convergence while simulations with uniform soils lacked these features.

REFERENCES

ANTHES, R. (1984), *Enhancement of convective precipitation by mesoscale variations in vegetative covering in semiarid regions,* J.Climate Appl. Meteor. *23*, 541–554.

ARYA, S.P., *Introduction to Micrometeorology*, Second Edition (Academic Press, San Diego 2001), 415 pages.

CHEN, F. and DUDHIA, J. (2001), *Coupling an advanced land-surface/hydrology model with the Penn State/ NCAR MM5 modelling system. Part I: Model Implementation and sensitivity,* Mon. Wea. Rev. *129*, 569–585.

DEARDORFF, J.W. (1978), *Efficient prediction of ground surface temperature and moisture with inclusion of a layer of vegetation,* J. Geophys. Res. *20*, 1889–1903.

GRELL, G., DUDHIA, J., and STAUFFER, D. (1995), *A description of the the fifth-generation Penn State / NCAR mesoscale model (MM5),* NCAR Technical Note NCAR/TN-398+STR, 122 pp.

HONG, S.Y. and PAN, H.L. (1996), *Nonlocal boundary layer vertical diffusion in a medium-range forecast model,* Mon. Wea. Rev. *124*, 2322–2339.

HONG, X., LEACH, M.J., and RAMAN, S. (1995), *Role of vegetation in generation of mesoscale circulation,* Atmos. Environ. *29*, 2163–2176.

KAIN, J. S. and FRITSCH, J. M., *Convective parameterization for mesoscale models: The Kain-Fritsch scheme. The representation of cumulus convection in numerical models* (eds. K. A. Emanuel and D. J. Raymond) (Am. Meteor. Soc. 1993), 246 pp.

KOCH, S.E. and RAY, K. A. (1997), *Mesoanalysis of summertime convective zones in central and eastern North Carolina,* Wea. Forecasting *12*, 56–77.

LIN, Y. and MITCHELL, K.E. (2005), *The NCEP Stage II/IV hourly precipitation analyses: Development and applications.* 19 Conf. *Hydrology*, San Diego, CA, Am. Meteor. Soc. 1.2.

MAHFOUF, J.-F., RICHARD, E., and MASCART, P. (1987), *The influence of soil and vegetation on the development of mesoscale circulations,* J. Climate Appl. Meteor. *26*, 1483–1495.

MLAWER, E. J., TAUBMAN, S. J., BROWN, P.D., IACONO M. J. (1997), *Radiative transfer for inhomogeneous atmospheres: RRTM, a validated correlated-k model for the longwave,* J. Geophys. Res. *102*, 16663–16682.

OOKOUCHI, Y., SEGAL, M., KESSLER, R.C., and PIELKE, R.A. (1984), *Evaluation of soil moisture effects on the generation and modification of mesoscale circulations,* Mon. Wea. Rev. *112*, 2281–2292.

RAMAN, S., SIMS, A., ELLIS, R., and BOYLES, R. (2005), *Numerical simulation of mesoscale circulations in a region of contrasting soil types,* Pure Appl. Geophys. *162*, 1689–1714.

REISNER, J., RASMUSSEN, R. J., and BRUINTJES, R. T. (1998), *Explicit forecasting of supercooled liquid water in winter storms using the MM5 mesoscale model,* Quart. J. Roy. Meteor. Soc. *124B*, 1071–1107.

SEGAL, M., AVISSAR, R., MCCUMBER, M.C., and PIELKE, R.A. (1988), *Evaluation of vegetation effects on the generation and modification of mesoscale circulations,* J. Atmos. Sci. *45*, 2268–2293.

SCHULTZ, P. (1995), *An explicit cloud physics parameterization for operational numerical weather prediction,* Mon. Wea. Rev. *123*, 3331–3343.

SIMS, A. (2001), *Effect of mesoscale processes on boundary layer structure and precipitation patterns : A diagnostic evaluation and validation of MM5 with North Carolina ECONet observations.* Masters Thesis, Department of Marine, Earth, and Atmospheric Sciences, North Carolina State University.

(Received April 1, 2006, accepted November 14, 2006)
Published Online First: July 19, 2007

Pure appl. geophys. 164 (2007) 1577–1591
0033–4553/07/091577–15
DOI 10.1007/s00024-007-0246-z

Pure and Applied Geophysics

The Assimilation of GPS Radio Occultation Data and its Impact on Rainfall Prediction along the West Coast of India during Monsoon 2002

Ching-Yuang Huang,[1] Ying-Hwa Kuo,[2] Shu-Ya Chen,[1] Anisetty S. K. A. V. Prasad Rao,[1] and Chien-Ju Wang[1]

Abstract—In this study, the Weather Research and Forecasting (WRF-2.0.3.1) model with three-dimensional variational data assimilation (3DVAR) was utilized to study a heavy rainfall event along the west coast of India with and without the assimilation of GPS occultation refractivity soundings in the monsoon period of 2002. The WRF model is a next-generation mesoscale numerical weather prediction system designed to serve both operational forecasting and atmospheric research communities. The Global Positioning System (GPS) radio occultation (RO) refractivity data, processed by UCAR, were obtained from the CHAMP and SAC-C missions. This study investigates the impact of thirteen GPS occultation refractivity soundings only, as assimilated into the WRF model with 3DVAR, on the rainfall prediction over the western coastal mountain of India. The model simulation, with the finest resolution of 10 km, was in good agreement with rainfall observations, up to 72-h forecast. There are some subtle but important differences in predicted rainfalls between the control run CN (without the assimilation of refractivity soundings) and G13 (with the assimilation of thirteen GPS RO soundings). In general, the assimilation run G13 gives a better prediction in terms of both rainfall locations and amounts at later times. The moisture increments were analyzed at the initial and forecast times to assess the impact of GPS RO data assimilation. The results indicate that remote soundings in the forcing region could have significant impacts on distant downstream regions. It is anticipated, based on this study, that considerably occultation soundings available from the six-satellite constellation of FORMOSAT-3/COSMIC would have even more significant impacts on weather prediction in this region.

Key words: WRF, 3DVAR, GPS RO data, FORMOSAT-3/COSMIC.

1. Introduction

The monsoon system is characterized by a reversal of prevailing wind directions and by alternating wet and dry seasons. In India, the circulation of the wet season, called the southwest monsoon, occurs from about June to September, when winds from the Indian Ocean carry moisture-laden air across the subcontinent, causing heavy rainfall and often considerable flooding. Most of India receives 70–90% of its

[1] Department of Atmospheric Sciences, National Central University, Jhongli, Taiwan.
E-mail: hcy@atm.ncu.edu.tw
[2] University Corporation for Atmospheric Research, Boulder, Colorado, USA.

annual rainfall during the summer monsoon period. The southwest monsoon generally follows two distinct branches, one initially flowing eastward from the Arabian Sea and the other northward from the Bay of Bengal. The former begins by lashing the west coast of peninsular India and rising over the adjacent Western Ghats where mesoscale mountains are laid on. When crossing these mountains, the air cools (thus losing its moisture-bearing capacity) and deposits rain copiously on the windward side of the highland barrier. The west coast of India receives heavy to very heavy rainfall during the southwest monsoon, and usually more than half of the seasonal total rainfall during June and July. This paper presents a preliminary study of a heavy rainfall event along the west coast of India during the monsoon season of 2002 using the WRF (Weather Research and Forecasting) model with and without the assimilation of GPS radio occultation (RO) refractivity soundings at the time of the Arabian Sea Monsoon Experiment (ARMEX-1). The ARMEX-1 experiment was held from mid-June to mid-August 2002. The main objective of the ARMEX-1 experiment was to study the Arabian Sea convection associated with intense rainfall along the west coast of India.

The GPS RO soundings are particularly useful for providing moisture information, which may help to improve the prediction of rainfall resulting from the intense southwest monsoon flow impinging upon mesoscale topography. Currently, the GPS system consists of 28 satellites, which continuously broadcast radio signals in two frequencies. Receivers onboard low-Earth-orbit (LEO) satellites measure the phase and amplitudes of these radio signals as they pass through the Earth's atmosphere. With precise knowledge of the positions and velocities of both GPS and LEO satellites, the delayed phase of each ray, due to wave propagation in the atmospheric medium, can be determined and then used to retrieve some atmospheric properties. For a detailed description of GPS RO limb sounding technique, the readers are referred to Kursinski *et al.* (2000). The first proof-of-concept GPS radio occultation mission, known as GPS/MET, was led by UCAR (Ware *et al.*, 1996), and launched in 1995. This was then followed by the SAC-C (Satélite de Aplicaciones Científicas-C) and CHAMP (CHAllenging Minisatellite Payload) missions. The two satellites, CHAMP and SAC-C, may collectively receive about 350 soundings daily. The upcoming FORMOSAT-3/COSMIC (Constellation Observing Systems for Meteorology, Ionosphere and Climate), a joint Taiwan-USA project, was launched on 15 April, 2006 and will produce ~ 2500 RO soundings per day for real-time applications. The details of COSMIC and the comparisons of GPS RO observations with conventional radiosonde soundings were given by Rocken *et al.* (1997, 2000). Kuo *et al.* (2005) showed that the retrieved temperature from the GPS RO soundings is of accuracy compatible or higher than radiosondes. Data assimilation studies have shown that GPS RO data can have significant positive impacts on regional as well as global weather predictions (Kuo *et al.*, 1997; Zou *et al.*, 1999, 2000; Liu and Zou, 2003; Huang *et al.*, 2005; Healy *et al.*, 2005; Healy and Thepaut, 2006).

Depending on the level of processing, GPS RO observations in various forms can be used for assimilation, ranging from raw excess phases to retrieved moisture and/or temperature profiles (see KUO *et al.*, 2000). GPS RO refractivity is an intermediate product after taking an Abel transform of bending angles defined for each incident ray tangent to its perigee point. HUANG *et al.* (2005) assimilated several GPS RO refractivity soundings into the MM5 model with a three-dimensional variational data assimilation system (3DVAR), and showed that both track and accumulated rainfall predictions were improved in simulations of two typhoons which made landfall on Taiwan, namely Nari in 2001 and Nakri in 2002.

The southwest Indian monsoon is usually quite moist over oceanic regions and may bring copious rainfalls along the west coast of India. Since GPS RO soundings contain valuable moisture and temperature information, it would be interesting to see whether such additional data will improve model performances in simulating the Indian monsoon rainfall. This study will use the newly developed regional model (WRF) to simulate the Indian monsoon rainfall case and utilize the WRF 3DVAR system to assimilate the GPS RO refractivity soundings. We will briefly introduce the methodology, including model setups, in section 2. The model results and performance assessments for the monsoon rainfall case will be presented and discussed in section 3. Finally, we give our conclusions in section 4.

2. *The Methodology*

The WRF model is a next-generation mesoscale numerical weather prediction system designed to meet both operational forecasting and atmospheric research needs. The model has been developed by a team involving many scientists from several institutes and organizations in the U.S. The details of the WRF model can be found on the web site (http://www.wrf-model.org) as well as in SKAMAROCK *et al.* (2005). The model is compressible and nonhydrostatic. It features multiple dynamical cores with high-order numerics. A 3DVAR system has recently been developed for WRF to allow incorporation of remote sensing data into the model (BARKER *et al.*, 2003, 2004).

The WRF 3DVAR was developed based on the MM5 3DVAR, and they share the same observation operators (for details of the WRF/MM5 3DVAR, see BARKER *et al.*, 2003). By minimizing a cost function, which calculates the total magnitude of differences between the analysis and first guess (as the background), and between the analysis and observations, the ingested observations can be used to adjust the initial analysis through proper weighting, depending on observation and background error statistics. A detailed introduction of the WRF 3DVAR minimization can be found in BARKER *et al.* (2003). KUO *et al.* (2004) showed that the fractional observation error of RO refractivity usually reaches a maximum near the surface over the tropics and does not exceed 6%. The forward observation operator of local refractivity is related to pressure, temperature and moisture in the atmosphere and one common form is given by

$$N = (n-1) \times 10^6 = 77.6\frac{P}{T} + 3.73 \times 10^5 \frac{P_w}{T^2}, \tag{1}$$

where n is the index of refractivity, P atmospheric pressure (hPa), and P_w the water vapor pressure (hPa) related to the specific humidity of the moisture. This forward operator simply relates refractivity (N) to temperature (T) and moisture (P_w). The first term in (1) accounting for dry atmospheric effects is considerably larger than the second term which accounts for moisture effects. Since relative temperature changes in the troposphere normally are much smaller than relative moisture changes, the second term actually contributes to most of the refractivity increments. If ancillary data (e.g., pressure or temperature) are available, then (1) can be used to retrieve the moisture. Equation (1) is simple and its associated linear tangent and adjoint operators are much easier to implement as compared to ray-tracing operators for bending angle assimilation (e.g., ZOU *et al.*, 1999). Details on the implementation of the forward operator shown in (1) and its tangent and adjoint codes into the MM5 3DVAR can be found in CUCURULL *et al.* (2006). The GPS RO refractivity soundings are obtained from the COSMIC Data Analysis and Archive Center (CDAAC) at UCAR that routinely processes RO data observed by CHAMP and SAC-C satellites.

The background errors and the observation errors are briefly described in HUANG *et al.* (2005). The radio occultation observational covariance matrix used in the experiments is diagonal and thus all vertical correlations between the retrieved refractivity measurements at different heights are ignored. The diagonal elements (variances) are prescribed as a profile exponentially decreasing from 3 N-units at 100 hPa to 10 N-units at 1000 hPa. The 10 N-units observational error value near the surface is consistent with the 3% refractivity difference between CHAMP radio-occultations and ECMWF analysis found at 1000 hPa as reported by KUO *et al.* (2004).

A detailed description of calculating the background errors in the cost function can be found on the web site (http://www.mmm.ucar.edu/wrf/users/docs/arw_v2.pdf) or BARKER *et al.* (2003). The background error covariance matrix is prescribed as monthly mean forecast error variances derived from the NCAR MM5 forecast. The effect of spatial error correlations, existing in the error covariance matrix, on the analysis is modeled by a recursive filter (LORENC, 1992). The background field can be given by either the previous forecast or first guesses.

In India, monsoon rainfalls occur quite often, in particular, over the west coast. This can be partly attributed to the orographic effects by the mesoscale mountains along the west coast. Two intensive rainfall periods were observed during 14–16 and 20–22 June, 2002. Two other heavy rainfall events were 26–28 June and 7–10 August,

2002. These four cases of heavy rainfall events were observed along the west coast of India during the ARMEX-1 campaign. In this study, the 14–16 June case was selected since the associated rainfall activity was strong. The daily rainfalls observed along the west coast of India from 14–16 June, 2002 at 0300 UTC are shown in Figure 1. The heavy rainfall was observed at stations between Mangalore and Cochin on 14 June (MADAN *et al.*, 2005). This system then moved northward and rainfall became even more intense between Ratnagiri and Goa on 15 June. On 16 June the region of heavy rainfall remained stationary, but its intensity was decreased. The surface chart (not shown) from India Meteorological Department indicates that an offshore trough was present, extending from Konkan-Goa to the Kerala coast.

Impact of the GPS RO refractivity soundings assimilated through WRF 3DVAR on the heavy rainfall prediction will be explored in this study. Thus, we included only the GPS observation data in the assimilation. The model simulations were carried out to 72 hours by WRF version 2.0.3.1 with three domains (90-km, 30-km and 10-km resolution, respectively), as shown in Figure 2, using the cloud microphysics scheme of LIN *et al.* in all the domains, and cumulus parameterization of Kain-Fritsch scheme in the two outer domains. The first guess was taken from the NCEP

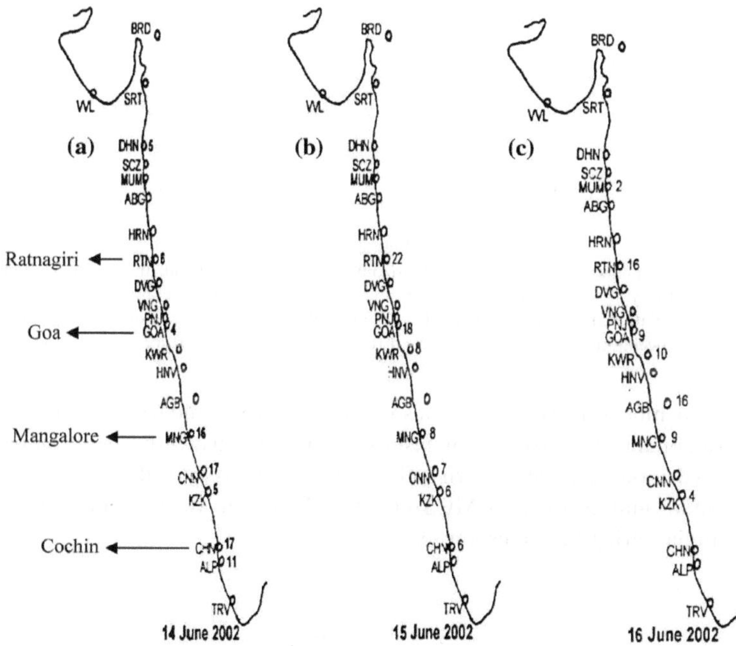

Figure 1

Observed 24-h accumulated rainfall (in cm) along the west coast of India at 0300 UTC. (a) 14 June, 2002, (b) 15 June, 2002, and (c) 16, June 2002. (From MADAN *et al.*, 2005).

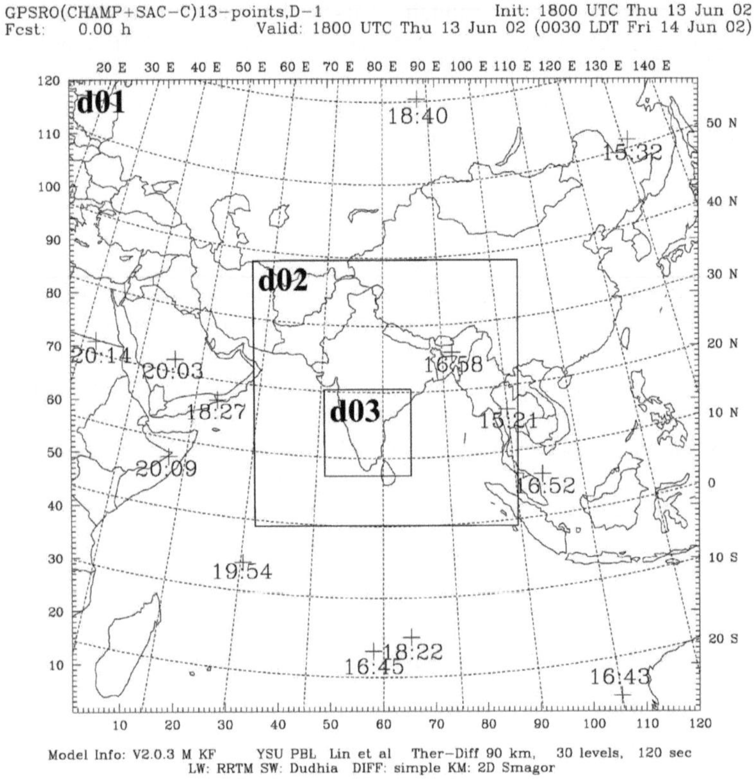

Figure 2
The three nested domains used for simulation. Horizontal resolutions are 90, 30, and 10 km for domains 1, 2, and 3, respectively. Thirteen GPS Occultation points in the outermost domain are available in a 6-h window for assimilation initialized at 1800 UTC 13 June, 2002. The plus signs and numbers indicate the occultation positions and their corresponding UTC times, respectively.

AVN global model analysis which also provided the boundary condition for the outermost domain. The assimilation results will be presented with a focus on regional rainfall prediction over the western Indian coast. This study will help shed light on the potential impact of FORMOSAT-3/COSMIC on weather prediction over western India during the monsoon season.

3. *The Results*

The experiments of control run (CN, without GPS refractivity soundings), and G13 (with ten GPS refractivity soundings from CHAMP and three from SAC-C) were conducted for comparisons using identical model and physical schemes mentioned above. The GPS occultation soundings are rather scattered in the outermost model domain as shown in Figure 2. The "+" sign indicates the GPS occultation locations and the number designates the sounding observation time (in UTC) at that point. Totally, there are thirteen occultation points occurring within six hours centered at the model initial time. The three SAC-C soundings are indicated in blue color by 20:14 UTC, 20:09 UTC and 18:22 UTC in Figure 2. In the monsoon region, there are five soundings (two near the Arabian and east-African coasts and the other three south of India) which produce larger negative increments on moisture analyses at lower levels as seen in Figure 3. There is also a region of notable analysis increments induced by three soundings to the east and southeast of India, extending from Bengal to Indonesia.

The differences between the first guess and the observations are generally small and are less than 5% near the surface and 1 N-unit (about 1%) about 10 km. However, at some point (e.g., the RO point at 19:54 UTC), the difference could reach 12% (about 24 N-units) near the surface. To illustrate the patterns of retrieved moisture increment (the difference between the analysis and the first guess), we plot in Figure 4 the vertical cross section through the four occultation points as indicated in Figure 3. For the two RO points far south of India, moisture increments are negative (positive) below (above) about 3 km height (Figs. 4b,d). Maximum positive moisture increments are produced at the height of about 5 km. Clearly, the assimilation of GPS RO soundings produces drying in the boundary layer and moistening up to the depth of the monsoon. For the other two points near the coast, the moisture increments are entirely negative throughout the depth of the monsoon flow (Figs. 4a,c). Maximum reduction on water vapor is about 2.7 g kg^{-1} at the height of 2 km (Fig. 4a). In general, these vertical east-west cross sections show that the soundings produce larger negative moisture increments at lower levels. Thus, to the south and west of India, the assimilation of GPS RO soundings produces a drier analysis in the marine boundary layer, with a maximum reduction in moisture greater than 2 g kg^{-1}. Induced wind changes (not shown) are very small in 3DVAR analysis, which are similar to the results of MM5 3DVAR analysis for the typhoon cases (HUANG *et al.*, 2005).

The initial analysis and prediction at 24 h show that the west coast of India is impinged by the recurved strong southwesterly monsoon flow as seen in Figure 5 for the control run CN. The southwest monsoon extends up to the mid-troposphere (about 400 mb), above which easterly winds prevail over the India subcontinent. At the lower levels, the southeasterly flow that crosses the Equator turns northeastward toward west India, as a Somali jet just off the coast of east Africa. The jet intensity is nearly constant through the end of the simulation period (72 h), reaching 21 ms^{-1} near the surface (figures not shown). East of India, the flow is a downstream extension of the southwesterly monsoon, but appears more cyclonic in the Bay of Bengal. For both cases (CN and G13), the simulated synoptic monsoon circulations are similar (figures not shown).

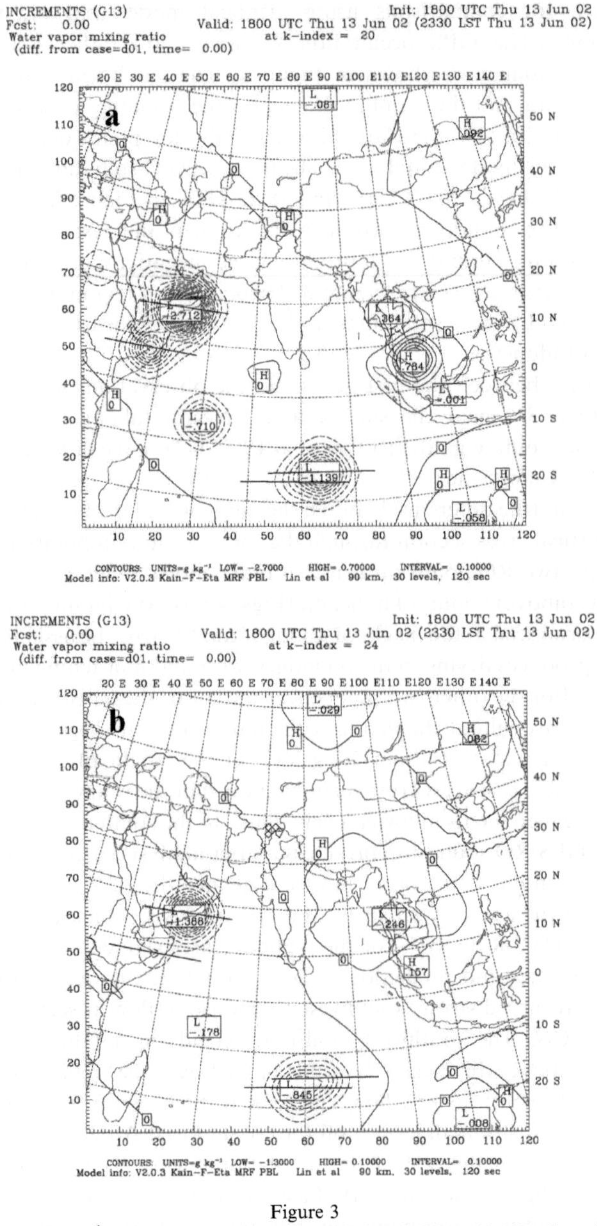

Figure 3

Moisture differences (g kg^{-1}) at the initial time between G13 and CN at (a) 807 mb and (b) 910 mb. CN: the control run (with no GPS sounding), G13 (with 13 GPS soundings). The locations of the 13 GPS soundings are indicated in Figure 2. Bold lines indicate the plotted cross sections in Figure 4.

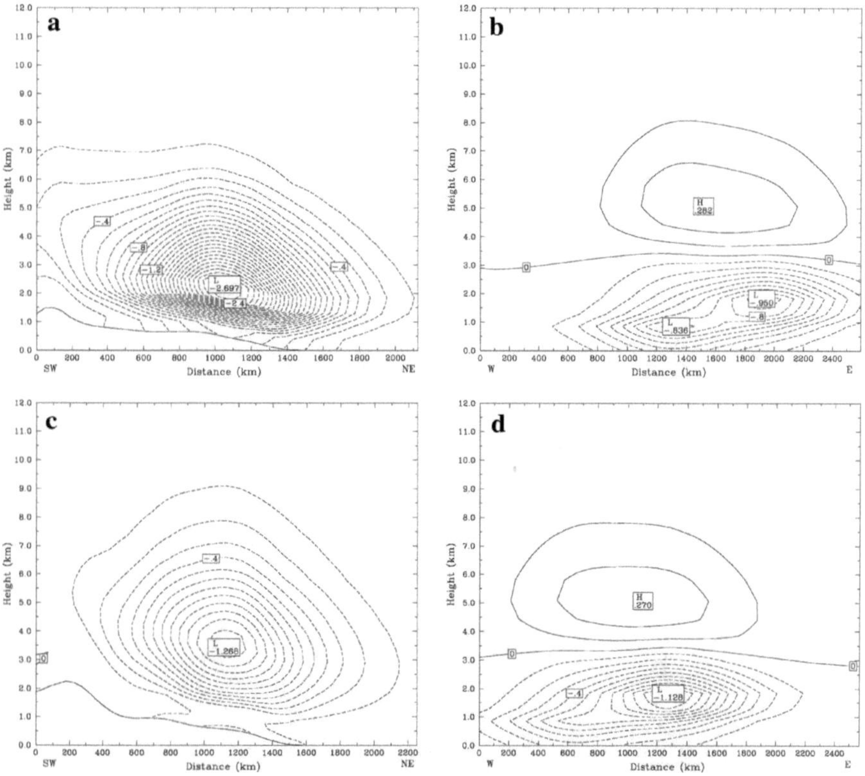

Figure 4
The initial moisture increments for Run G13 at the vertical east-west cross sections through the occultation points occurred at time (a) 18:27 UTC, (b) 16:45 UTC, (c) 20:09 UTC, and (d) 18:22 UTC. Contour intervals are 0.1 g kg^{-1} in all panels.

With the quasi-stationary southwest monsoon flow crossing east Africa and the Indian Peninsula, intense large-scale rainfalls (up to 150 mm daily) occur over the three regions of strong low-level convergence, namely south of the Arabian coast, east of Africa, and southeast of India (figures not shown). In particular, orographic lifting results in copious rainfalls on the west coast of India when the intense monsoon flow passes over the Ghats Mountains. The observed daily rainfalls on 14 June, 2002, as denoted in Figure 1, are 170 mm at Cochin and 160 mm at Mangalore. As can be found in Figure 6, the simulated 24-h accumulated rainfall for the control run is about 265 mm near Cochin and 179 mm near Mangalore, both being larger than the observed. At a later time, the observed rainfalls on 15 June, 2002 are 180 mm at Goa and 220 mm at Ratnagiri, and the corresponding 24-h accumulated rainfalls (from 24 h to 48 h) for the control run are underpredicted, e.g.,

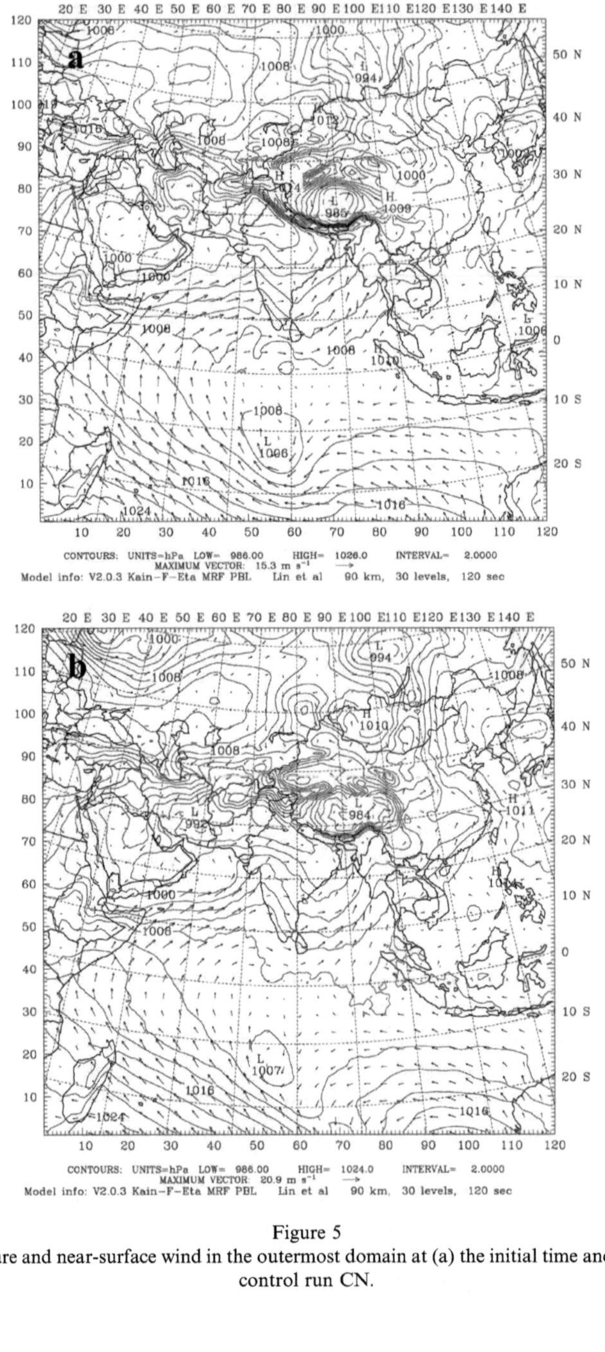

Figure 5
Sea-level pressure and near-surface wind in the outermost domain at (a) the initial time and (b) 24 h for the control run CN.

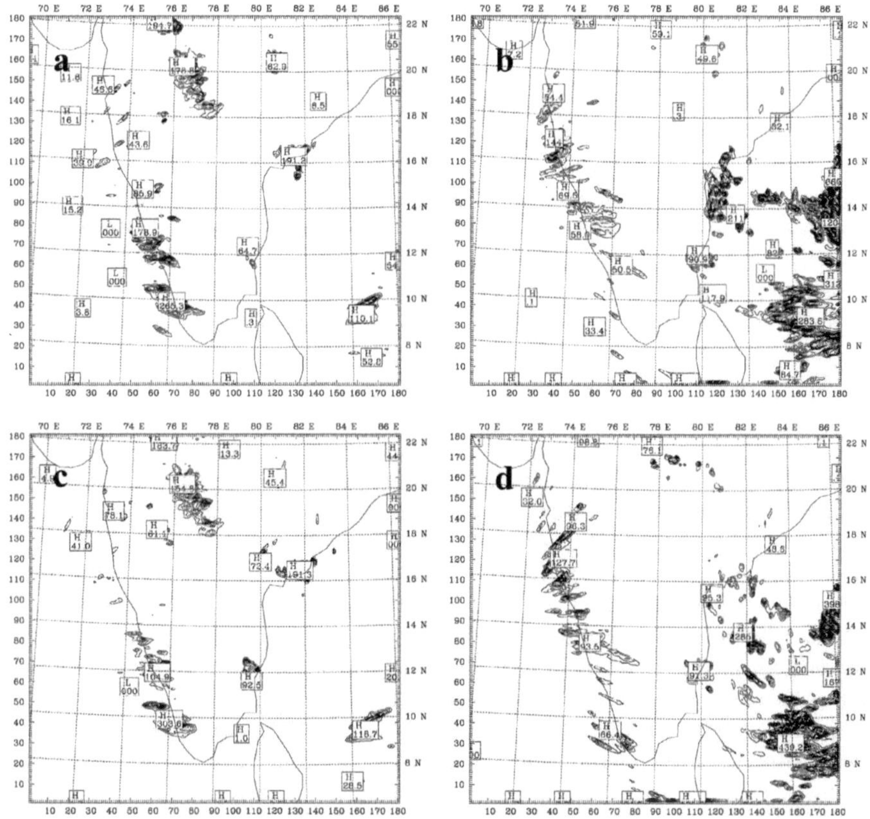

Figure 6
The simulated 24-h accumulated rainfalls (mm) for (a) 0–4 h and (b) 24–48 h for Run CN, (c) 0–24 h and (d) 24–48 h for Run G13. The contour interval is 10 mm in all panels.

giving 69 mm near Goa and 144 mm at Ratnagiri. The 24-h accumulated rainfalls for G13 are similar to those from CN in spatial distribution, however the first-day rainfall at the mid-western India coast is considerably less and is more consistent with the observations. For the second day, the rainfall around 14°N for G13 is about 93 mm near Mangalore, which is comparable with the observed value of 80 mm and gives a lessoned underprediction than that (58 mm) for CN. In particular, the observed rainfall (about 60 mm) near the southern tip of the coast on the second day is well simulated by G13 (giving 66 mm), is completely unpredicted by CN. The simulated 48–72 h rainfall of G13 (not shown) is also in agreement with the observations (Fig. 1). In summary, the accumulated maximum rainfalls for the assimilation experiment are slightly overpredicted (underpredicted) at (north of)

Cochin on the first day but are increased on the second day in better agreement with the observations.

The rainfall prediction is affected by many factors, such as model precipitation physics as well as assimilated observations. To understand how the differences in the accumulated rainfalls are induced, we may examine the moisture differences at different forecast times. The correlation between the evolving moisture increments and the ensuing rainfall could be complicated, although we herein aim to illustrate the processes of how the observation information is spread with time. To clarify physical processes responsible for producing differences over the west coast of India, the increments at the outermost domain are further analyzed. We analyze the increment of wet refractivity (the second term of (1)) that results from the contribution of moisture in local refractivity. We plot the wet refractivity since it can reveal the amount relative to the total refractivity (which is observable) and is also representative of moisture information. Based on (1), both temperature and moisture contribute to local refractivity and the dry refractivity (the first term on the right-hand side of (1)) is usually two to three times larger in magnitude than the wet refractivity. However, most of the increments are generated by the wet refractivity, not the dry refractivity, with a maximum magnitude larger than 10 units. As shown in Figure 7, the differences in analysis increment of wet refractivity between G13 and CN are being advected by the monsoon flow. At 12 h, the initial increments at lower heights, south of India, have already spread outward (figures not shown). Large negative increment differences near the southern Arabian coast have been transported toward India by 18 h as seen in Figure 7a, which resulted from the initial large negative modifications at the RO points as seen in Figure 3. These increment differences generally move to higher latitudes with time, due to the monsoon flow. By 24 h, these increments southwest of India remain unchanged, and vertically coherent below 7 km, and are associated with strong prevailing wind blowing toward India during the simulation period. By 36 h, these initial increments have almost spread over the entire Indian Ocean (Fig. 7b). In general, most of the induced increment differences are negative at lower levels just west and southwest of India. There is a zone with positive differences just offshore of the mid-western coast of India, supporting a possibility of more rainfalls over the west coast of India on the second day as seen in the G13 run.

4. Conclusions

This study uses the WRF model to investigate a heavy rainfall event that took place along the west coast of India during the monsoon period of 2002. In this period, the Indian scientists performed an experiment, known as Arabian Sea Monsoon EXperiment (ARMEX-2002), to study the off-shore trough and associated convective rainfall. GPS occultation refractivity soundings, ten from CHAMP and three from SAC-C, were assimilated into WRF-3DVAR, and the WRF model with

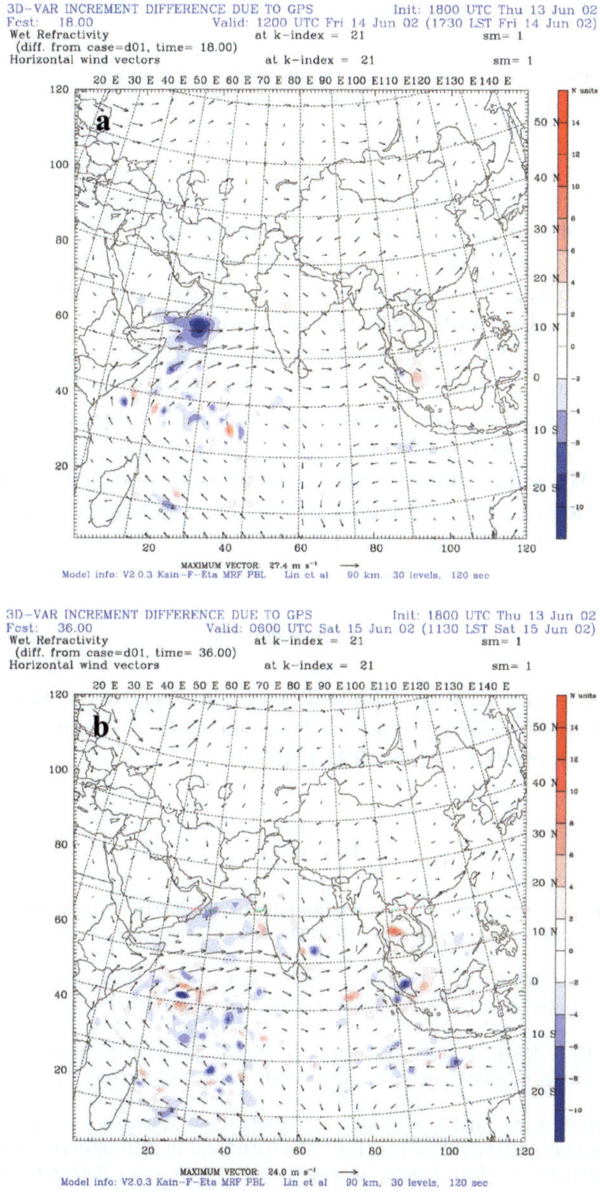

Figure 7

Differences in wet refractivity between G13 and CN at (a) 18 h and (b) 36 h at the height of $\sigma = 0.81$ (about 835 mb). The horizontal wind vectors at the plotted level are also overlapped.

three nested domains was integrated for 72 hours. A total of thirteen refractivity soundings was assimilated to improve the moisture analysis. The assimilation of GPS refractivity soundings was shown to have a positive impact on rainfall prediction over the western Indian coast. The WRF model without the assimilation of GPS RO data predicted the observed rainfall reasonably well, although overpredicting (underpredicting) the maximum rainfall in the first (second) day. With the assimilation of thirteen GPS RO soundings only, underprediction of rainfall over the west coast of India was improved in the second day.

We found that the impacts on the rainfall prediction over the west coast of India are mainly attributed to the assimilated soundings over the swath of the southwest monsoon. When these GPS RO soundings are further removed, the simulated rainfalls over the west coast of India are much closer to those in the control run (without assimilation). With the limited GPS RO soundings, the impacts on rainfall prediction, though modest, are positive and promising for this 2002 Indian monsoon rainfall case. This study aims at illustrating the potential impact of the GPS observations. We will continue to explore other cases associated with heavy rainfall during the ARMEX-2002 and conduct assimilation experiments with other available observations, in particular, using the FORMOSAT-3/COSMIC GPS RO data.

Acknowledgments

This study is supported by National Science Council in Taiwan under Grants NSC 93-2119-M-008-008-AP1 and by National Space Program Organization in Taiwan under Grant 93-NSPO(B)-RS3-FA07-01.

References

Barker, D., Huang, W., Guo, Y.-R., and Bourgeois, A. (2003), *A three-dimensional variational (3DVAR) data assimilation system for use with MM5*. NCAR Technical Note NCAR/TN-453 + STR, 68 pp.

Barker, D., Huang, W., Guo, Y.-R., Bourgeois, A. J., and Xiao, Q. N. (2004), *A three-dimensional variational data assimilation system for MM5: Implementation and initial results,* Mon. Wea. Rev. *132*, 897–914.

Cucurull, L., Kuo, Y.-H., Barker, D., Rizvi, S. R. H. (2006), *Assessing the impact of simulated COSMIC GPS radio occultation data on weather analysis over the Antarctic: A case study*, Mon. Wea. Rev., in press.

Healy, S. B., Jupp, A. M., and Marquardt, C. (2005), *Forecast impact experiment with GPS radio occultation measurements*, Geophys. Res. Lett. *32*, L03804, doi:10.1029/2004GL020806.

Healy, S. B. and Thepaut, J.-N. (2006), *Assimilation experiments with CHAMP GPS radio occultation measurements,* Quart. J. Roy. Meteor. Soc. *132*, 605–623.

Huang, C.-Y., Kuo, Y.-H., Chen, S.-H., and Vandenberghe, F. (2005), *Improvements on typhoon forecast with assimilated GPS occultation refractivity,* Wea. Forecasting *20*, 931–953.

KUO, Y.-H., ZOU, X., and HUANG, W. (1997), *The impact of GPS data on the prediction of an extratropical cyclone: An observing system simulation experiment,* J. Dyn. Atmos. Ocean *27*, 413–439.

KUO, Y.-H., SOKOLOVSKIY, S. V., ANTHES, R. A., and VANDENBERGHE, F. (2000), *Assimilation of GPS radio occultation data for numerical weather prediction,* Terr. Atmos. Oceanic Sci. *11*, 157–186.

KUO, Y.-H., WEE, T.-K., SOKOLOVSKIY, S., ROCKEN, C., SCHREINER, W., HUNT, D., and ANTHES, R. A. (2004), *Inversion and error estimation of GPS radio occultation data,* J. Meteor. Soc. Japan *82*, 507–531.

KUO, Y.-H., SCHREINER, W. S., WANG, J., ROSSITER, D. L., and ZHANG, Y. (2005), *Comparison of GPS radio occultation soundings with radiosondes,* Geophys. Res. Lett. *32*, L05817, doi:10.1029/2004GL021443.

KURSINSKI, E. R., HAJJ, G. A., LEROY, S. S., and HERMAN, B. (2000), *The GPS radio occultation technique,* Terr. Atmos. Oceanic Sci. *11*, 53–114.

LIU, H. and ZOU, X. (2003), *Improvements to GPS radio occultation ray-tracing model and their impacts on assimilation of bending angle,* J. Geophys. Res. *108*(D17), 4548, doi: 10.1029/2002JD003160.

LORENC, A. C. (1992), *Iterative analysis using covariance functions and filters,* Quart. J. Roy. Meteor. Soc. *118*, 569–591.

MADAN, O.P., MOHANTY, U.C., IYENGER, GOPLA, SHIVHARE, R.P., PRASAD RAO, A.S.K.A.V., SAM, N.V., and BHATLA, R. (2005), *Offshore trough and very heavy rainfall events along the west Coast of India during ARMEX-2002,* Mausam *56*, 1, 37–48.

ROCKEN, C., ANTHES, R., EXNER, M., HUNT, D., SOKOLOVSKIY, S., WARE, R., GORBUNOV, M., SCHREINER, W., FENG, D., HERMAN, B., KUO, Y., and ZOU, X. (1997), *Analysis and validation of GPS/MET data in the neutral atmosphere,* J. Geophys. Res. *102*, 29, 849–29, 866.

ROCKEN, C., KUO, Y.-H., SCHREINER, W. S., HUNT, D., SOKOLOVSKIY, S., and MCCORMICK, C. (2000), *COSMIC system description,* Terr. Atmos. Oceanic Sci. *11*, 21–52.

SKAMAROCK, W. C., KLEMP, J. B., DUDHIA, J., GILL, D. O., BARKER, D. M., WANG, W., and POWERS, J. G. (2005), *A Description of the Advanced Research WRF Version 2,* NCAR Technical Note NCAR/TN-468 + STR, 100 pp.

WARE, R., EXNER, M., FENG, D., GORBUNOV, M., HARDY, K., HERMAN, B., KUO, Y., MEEHAN, T., MELBOURNE, W., ROCKEN, C., SCHREINER, W., SOKOLOVSKIY, S., SOLHEIM, F., ZOU, X., ANTHES, R., BUSINGER, S., and TRENBERTH, K. (1996), *GPS soundings of the atmosphere from low earth orbit: Preliminary results,* Bull. Am. Meteor. Soc. *77*, 19–40.

ZOU, X., WANG, B., LIU, H., ANTHES, R. A., MATSUMURA, T., and ZHU, Y.-J. (2000), *Use of GPS/MET refraction angles in 3D variational analysis,* Quart. J. Roy. Meteor. Soc. *126*, 3013–3040.

ZOU, X., VANDENBERGHE, F., WANG, B., GORBUNOV, M. E., KUO, Y.-H., SOKOLOVSKIY, S., CHANG, J. C., SELA, J. G., and ANTHES, R. (1999), *A raytracing operator and its adjoint for the use of GPS/MET refraction angle measurements,* J. Geophys. Res. Atmos. *104*, 22, 301–322, 318.

(Received April 13, 2006, accepted September 11, 2006)
Published Online First: July 14, 2007

Pure appl. geophys. 164 (2007) 1593–1615
0033–4553/07/091593–23
DOI 10.1007/s00024-007-0244-1

Pure and Applied Geophysics

An Experiment Using the High Resolution Eta and WRF Models to Forecast Heavy Precipitation over India

Y. V. Rama Rao,[1] H. R. Hatwar,[1] Ahmad Kamal Salah,[2] and Y. Sudhakar[3]

Abstract—In the present study using the Weather Research and Forecasting (WRF) and Eta models, recent heavy rainfall events that occurred (i) over parts of Maharastra during 26 to 27 July, 2005, (ii) over coastal Tamilnadu and south coastal Andhra Pradesh during 24 to 28 October, 2005, and (iii) the tropical cyclone of 30 September to 3 October, 2004/Monsoon Depression of 2 to 5 October 2004, that developed during the withdrawal phase of the southwest monsoon season of 2004 have been investigated. Also sensitivity experiments have been conducted with the WRF model to test the impact of microphysical and cumulus parameterization schemes in capturing the extreme weather events. The results show that the WRF model with the microphysical process and cumulus parameterization schemes of Ferrier *et al.* and Betts-Miller-Janjic was able to capture the heavy rainfall events better than the other schemes. It is also observed that the WRF model was able to predict mesoscale rainfall more realistically in comparison to the Eta model of the same resolution.

Key words: Heavy rainfall events, WRF, Eta model, mesoscale rainfall prediction.

1. Introduction

Prediction of heavy rainfall is one of the many challenging problems in meteorology, but very important for issuing timely warnings for the agencies engaged in disaster preparedness and mitigation. With the advances in observation technology, especially weather satellites and buoys, there is considerable improvement in the quality of observational data over the ocean around a tropical system such as cyclones/depressions. The availability of high computing power, improved models with better parameterization schemes and also the availability of real-time initial and boundary conditions from global models, has enabled numerous forecasting centers to run high resolution mesoscale models for operational use and to utilize these models in order to improve the forecasting ability of meteorologists over short time scales.

---

[1] India Meteorological Department, New Delhi, 110003, India. E-mail: ramarao@imdmail.gov.in
[2] Egyptian Meteorological Authority, Cairo, Egypt.
[3] Hinditron Infosystems, New Delhi, India.

Recently the India Meteorological Department (IMD) implemented the Advanced Research Weather Research and Forecasting (ARW, Version 2) and Eta models in research mode. Using these models some case studies were conducted to examine the usefulness of these high resolution advanced mesoscale models for the Indian region. In the present study we have selected three cases viz. (i) Tropical Cyclone 30 September to 3 October, 2004/Monsoon Depression of 2 to 5 October, 2004, the multiple weather systems that occurred during the withdrawal phase of the southwest monsoon season of 2004 and also recent heavy rainfall events that occurred (ii) over parts of Maharastra during 26 to 27 July, 2005 and (iii) coastal Tamilnadu and south coastal Andhra Pradesh during 24 to 28 October, 2005. Also sensitivity experiments were conducted with the WRF model to test the impact of various microphysical and cumulus parameterization schemes in capturing the extreme weather events. Using the WRF and Eta models, the forecasts up to 72 hours were produced and the results are discussed in this paper.

2. Forecast Model

2.1. WRF Model

The WRF model is a next-generation mesoscale modeling system (Skamarock and Klemp, 1992; Wang *et al.*, 2004; Skamarock *et al.*, 2005) developed as a collaborative effort among the NCAR Mesoscale and Microscale Meteorology (MMM) Division, the National Oceanic and Atmospheric Administration's (NOAA) National Centers for Environmental Prediction (NCEP), Forecast System Laboratory (FSL), the Department of Defense's Air Force Weather Agency (AFWA), Naval Rcscarch Laboratory (NRL), the Center for Analysis and Prediction of Storms (CAPS) at the University of Oklahoma, and the Federal Aviation Administration (FAA), along with the participation of a number of university scientists. The model is designed to place new and existing research and operational models in the USA under a common software architecture. In the WRF system, it is possible to mix and match the dynamic cores and physics packages of different models to optimize performance, since each model has strengths and weaknesses in different areas. This feature is particularly advantageous for inter-model comparison and sensitivity studies. In the present study the NCAR Version-2 of Advanced Research WRF (ARW) has been used. The model domain consists of 10°S to 45°N and 50°E to 115°E and has 20-km horizontal resolution with 31 vertical sigma levels.

The ARW equations are formulated using a terrain-following hydrostatic-pressure vertical coordinate denoted by η and defined as

$$\eta = (ph - pht)/\mu \text{ where } \mu = phs - pht, \tag{2.1}$$

where ph is the hydrostatic component of the pressure, and phs and pht refer to values along the surface and top boundaries, respectively and μ is the hydrostatic pressure difference between the surface and the top of the model. The coordinate definition (2.1), proposed by LAPRISE (1992), is the traditional σ coordinate used in many hydrostatic atmospheric models, named the η–coordinate system by the WRF developers. The value of η varies from 1 at the surface to 0 at the upper boundary of the model domain. This vertical coordinate is also called a mass vertical coordinate. Since $\mu(x, y)$ represents the mass per unit area within the column in the model domain at (x, y), the appropriate flux form variables are

$$V = \mu v = (U, V, W), \quad \Omega = \mu\dot{\eta}, \quad \Theta = \mu\theta. \tag{2.2}$$

$v = (u, v, w)$ are the covariant velocities in the two horizontal and vertical directions, respectively, while $\omega = \eta$ is the contravariant 'vertical' velocity. θ is the potential temperature, $\varphi = gz$. The prognostic variables are: $(U, V, W, \Theta, \phi', \mu', Q_m)$.

2.1.1. Initial conditions

The initial conditions for the real-data cases are preprocessed through a separate package called the Standard Initialization (SI). The output from the SI is passed to the preprocessor in the ARW, which generates initial and lateral boundary conditions. The SI is a set of programs that takes terrestrial and meteorological data (typically in GriB format) and transforms them for input to the ARW preprocessor program for real-data cases. In the present experiments NCEP Global Final Analyses (FNL) at $1° \times 1°$ lat./long. resolution are used for the initial and lateral boundary conditions.

2.1.2. Physical process

The simulations from numerical models are known to be sensitive to the representation of the physical processes. In order to obtain realistic results it is necessary to incorporate appropriate physics into the model. WRF physics options fall into several categories, each containing several options. The physics categories are (1) microphysics, (2) cumulus parameterization, (3) planetary boundary layer (PBL), (4) land-surface model, and (5) radiation. Diffusion may also be considered part of the physics. In the present experiments, sensitivity experiments have been conducted with two sets of microphysical (LIN *et al.*, 1983 and FERRIER *et al.*, 2002) and cumulus (KAIN-FRITSCH, 1993) and Betts-Miller-Janjic (Janjic, 1994, 2000) schemes. Here we set forth a brief description of the microphysics and cumulus parameterization schemes used in the experiment.

2.1.2.1. MICROPHYSICS

Microphysics includes explicitly resolved water vapor, cloud, and precipitation processes. The model is general enough to accommodate any number of mass mixing-ratio variables, and other moments such as number concentrations. ARW, microphysics is carried out at the end of the time-step as an adjustment process, and does not provide tendencies. The rationale for this is that condensation adjustment should be at the end of the time-step to guarantee that the final saturation balance is accurate for the updated temperature and moisture.

(i) **Lin scheme.** In this scheme six classes of hydrometeors are included: Water vapor, cloud water, rain, cloud ice, snow, and graupel. All parameterization production terms are based on LIN *et al.* (1983) with some modifications, including saturation adjustment and ice sedimentation. This is a relatively sophisticated microphysics scheme in WRF, and it is more suitable for use in research studies. The scheme is taken from the Purdue cloud model, and the details can be found in CHEN and SUN (2002).

(ii) **Ferrier scheme**. This is also known as the Eta-Ferrier scheme. The scheme predicts changes in water vapor and condensate in the forms of cloud water, rain, cloud ice, and precipitation ice (snow/graupel/sleet). The individual hydrometeor fields are combined into total condensate, and it is the water vapor and total condensate that are advected in the model. Further description of the scheme can be found in Sec. 3.1 of the November 2001 Technical Procedures Bulletin (TPB) at http://www.emc.ncep.noaa.gov/mmb/mmbpll/ eta12tpb/.

2.1.2.2. Cumulus Parameterization

The effects of convective clouds on the large-scale environment are parameterized in terms of the grid scale variables in different ways. In the present experiment, a mass flux scheme after Kain-Fritsch (KF) (1990, 1993) and Betts-Miller-Janjic (BMJ) (JANJIC, 1994, 2000) is used. The KF scheme uses a Lagrangian Parcel method along with vertical momentum dynamics to estimate the properties of cumulus convection, It incorporates a trigger function, a mass flux formulation and closure assumption. The trigger function identifies the potential updraft source layers associated with convection, whereas the mass flux formulation calculates the updraft, down draft and environmental mass flux associated with that. The scheme assumes conservation of mass, thermal energy, total moisture and momentum, and since it represents different processes associated with convection it is chosen for the present study. The BMJ scheme is similar to that used in the Eta model (described Section 2.2.1), however, recently attempts have been made to refine the scheme for higher horizontal resolutions, primarily through modifications of the triggering mechanism.

The model land surface, radiation and PBL schemes used are given in Appendix 1. The detailed description of model formulation, horizontal and vertical discretization and time integration scheme used in the experiment is given in SKAMAROCK *et al.* (2005).

2.2. *Eta Model*

The National Centers for Environmental Prediction (NCEP), USA developed a step-mountain eta (η) coordinate model generally known as the Eta model and was made operational in 1993 (BLACK, 1994). The workstation version 3.1 (March 2002) of the same model has been recently installed at IMD. It is an advanced numerical weather prediction model with complete physical processes. The Eta model is a hydrostatic mesoscale weather forecast model with an accurate treatment of complex topography using eta (η) vertical coordinate system and step-like mountains (MESINGER, 1984), which eliminates errors in computation of pressure gradient force over steeply sloped terrain present in sigma (σ) coordinate system (MESINGER and BLACK, 1992). The model topography is represented as discrete steps whose tops coincide exactly with one of the model's 45 vertical layer interfaces (BLACK, 1994). In determining their elevations, each horizontal grid box is first divided into 16 sub-boxes. Mean elevations for each of these 16 sub-boxes are calculated from the United States Geological Survey (USGS) topographical data. The model version used here follows that described by BLACK (1994). The present version is a hydrostatic model with a horizontal grid spacing of approximately 22 km and 45 vertical levels, with layer depths that range from 20 m in the planetary boundary layer to 2 km at 50 mb. The model considered in this study consists of the domain 4°S to 42°N; 52°E to 110°E. The initial and lateral boundary conditions are obtained by direct interpolation from the NCEP 6 hourly global model forecasts fields.

2.2.1. *Physical processes*

The model uses the convective precipitation, based on the Betts-Miller cumulus parameterization (BETTS 1986, BETTS and MILLER, 1986) convective precipitation with some of the modifications based on work by JANJIC (1994). Convective precipitation is computed every eight adjustment time steps, or about every ten minutes. The model calculates large-scale precipitation by using an explicit cloud water parameterization scheme (ZHAO *et al.*, 1991). The mixing ratio of cloud water and ice has recently been added to the temperature, specific humidity, wind components, surface pressure, and turbulent kinetic energy as a fundamental prognostic variable. This explicit cloud water parameterization takes into account the physical processes of evaporation, condensation, melting, freezing, sublimation, and deposition that occur in the atmosphere. There is currently no direct interaction between the cloud water scheme and the Betts-Miller convection parameterization, thus the convective scheme produces no cloud water. The radiation package used in the model is one developed at GFDL, USA. The planetary boundary layer (PBL) scheme uses a modified Mellor-Yamada Level 2.5 scheme (BLACK 1994). In this approach, turbulent kinetic energy (TKE) is a fully prognostic variable that is recalculated every eight advective time steps. The details of the above physical schemes used in the model are described in JANJIC (1994) and some of the major features of the model are given in Appendix 1.

3. Forecast Experiment

3.1. Case 1: Tropical Cyclone of 30 September to 3 October 2004/Monsoon Depression of 2–5 October, 2004

3.1.1. Synoptic situation

Towards the end of the south-west monsoon season (June-September) a low pressure system formed over east central Arabian Sea on 29 September, 2004. Moving in a north-northwesterly direction, the system intensified into a depression and further into a cyclonic storm on 1 October. The system subsequently moved in a northerly direction and lay centered near latitude 21.0° N/longitude 66.5° E at 0300 UTC of 2 October. It subsequently recurved north-eastwards and further intensified into a severe cyclonic storm at 0900 UTC on 2 October. The system weakened rapidly into a cyclonic storm on the 3rd and subsequently into a depression/low over the same region over the Sea area close to the Gujarat coast, the northwest coast of India. At the same time another low pressure area formed over southeast Bay of Bengal on 30 September, 2004. Moving in a northwesterly direction it concentrated into a depression on 2 October. Subsequently moving in a northwest direction it crossed the south coastal Andhra Pradesh on 4 October and weakened into a low pressure system. The initial 850 hPa wind fields and observed tracks of these systems are given in Figures 1(a, b).

Under the influence of the Arabian Sea system, significant rainfall occurred over the northern parts of India on 3 October and in association with the Bay of Bengal system, heavy to very heavy rainfall occurred over Andhra and Orissa coast on 4 October and widespread rainfall over south coastal Orissa (on 5 October) of peninsular India.

3.1.2. Model forecast

In the present experiment, using the initial conditions of 2 October, 2004, 72-hour forecasts are generated with the WRF and Eta models. The WRF model control run with microphysical process due to LIN *et al.* and cumulus parameterization by the Kain-Fritsch scheme and experiment run with microphysical process of Ferrier *et al.* and cumulus parameterization scheme of Betts-Miller-Janjic, were selected. As outlined in the preceding paragraphs, a cyclonic storm was located over the Arabian Sea and another depression over the Bay of Bengal. The 48 and 72-hours forecast 850 hPa wind flow (speed in knots) and rainfall (mm) forecast valid for 00 UTC of 4 and 5 October, 2004 based on 2 October initial fields are given in Figures 1 and 2. The initial analysis of 2 October

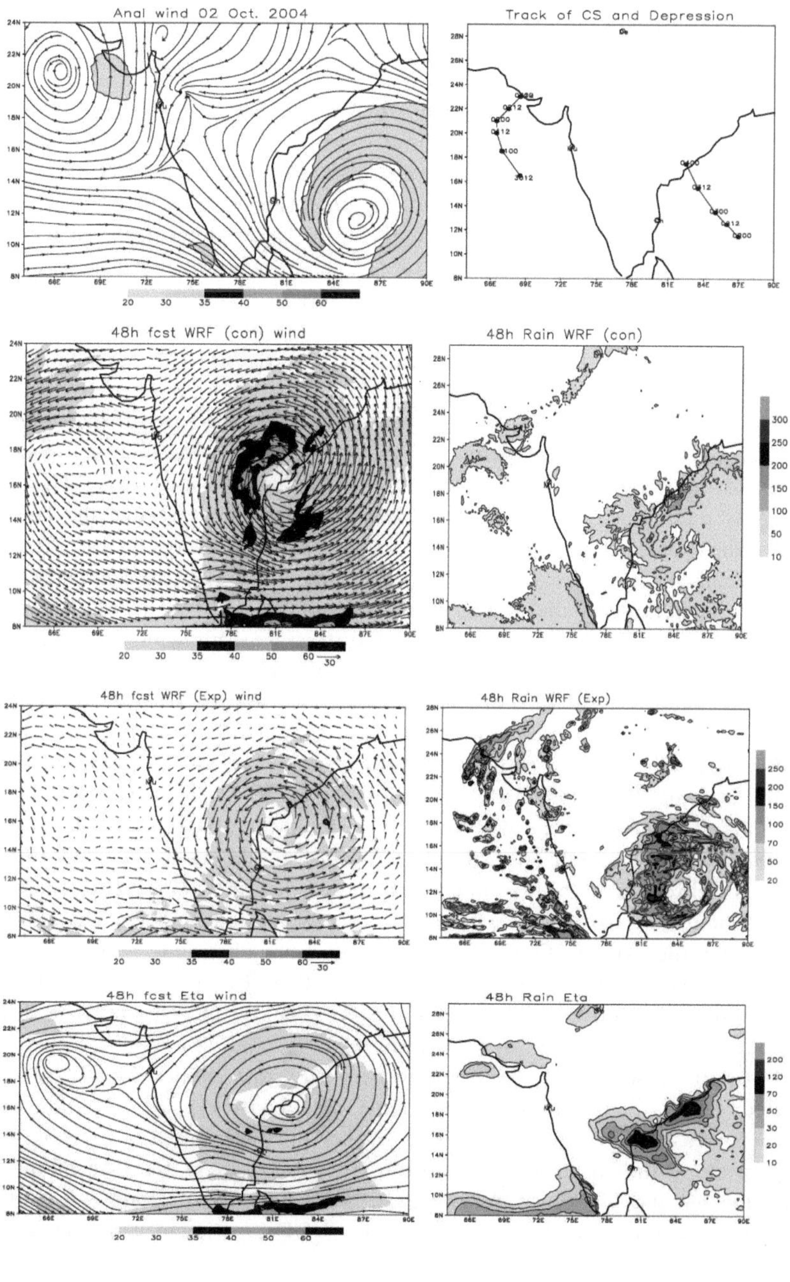
Anal wind 02 Oct. 2004
20 30 35 40 50 60
Track of CS and Depression
48h fcst WRF (con) wind
20 30 35 40 50 60
30
48h Rain WRF (con)
300
250
200
150
100
50
10
48h fcst WRF (Exp) wind
20 30 35 40 50 60
30
48h Rain WRF (Exp)
250
200
150
100
70
50
20
48h fcst Eta wind
20 30 35 40 50 60
48h Rain Eta
200
120
70
50
30
20
10

◄ Figure 1
(a, b) Initial 850 hPa wind flow (speed in knots) based on 00 UTC of 02 October, 2004 and observed track and 48-hour forecast 850 hPa wind fields and rainfall (mm) valid for 00 UTC of 4 October, 2004 based on WRF model control run (c,d); WRF experiment run (e,f) and Eta model (g,h).

shows a cyclonic circulation over the north Arabian Sea and another circulation over the east Bay of Bengal. In the 24-hour forecast, wind fields valid for 00 UTC of 3 October based on the WRF model control, experiment and Eta model forecast, the Arabian Sea system shown, the intensification of the system into a cyclonic storm with sustained wind strength of 30–35 kts along northern sector of the system (figures not shown) and the 48 and 72-hour forecasts showed a weakening of the system over the sea itself as observed (Fig 1, 2). In the case of the Bay of Bengal system, the day-one to day-three forecast by the WRF (control and experiment run) and the Eta model showed that the system moved northwestwards and crossed the south Andhra coast on 4 October as observed. However, the WRF model control run overestimated the intensity of the system as the forecasts showed a strengthening of the system into a cyclonic storm in the 24 hour forecast with an intensity of 35–40 kts (figure not shown). The same intensity is maintained in the 48 and 72-hour forecasts even after the landfall, whereas the observed system weakened into a low pressure system after the landfall. In the case of the WRF experiment run, the 48-hour forecast wind field showed a weakening of the system after landfall on the 4 with wind speed of 20–30 kts and the 72-hour forecast showed further weakening of the system and moved northwestwards. In case of the Eta model forecast, the 48-hour forecast showed asymmetric circulation oriented northeast to southwest close to the north Andhra coast with wind speed of 20–30 kts and in the 72-hour forecast showed further weakening of the system and moved northwestwards. In association with both the systems, the rainfall predictions by the models were more realistic. The heavy rainfall on 3 October over northern parts of India (figure not shown) and moderate to heavy rainfall on 4 and 5 October over the east coast of peninsular India was well captured by the models. In the case of the WRF control run, the model 48-hour forecast was able to capture heavy rainfall with maxima of 10 to 15 cm along the Andhra-Orissa coast on the 4 as observed, however, in the 72-hour forecast the model over-predicted rainfall of 10 to 15 cm along interior parts of north Andhra-Orrisa coast on the 5th (Figs. 1, 2). This may be due to overprediction of the strength of the system in the 48 and 72-hour forecasts by the model. In the case of the WRF experiment run, the model 48-hour forecast predicted 10 to 15 cm along the Andhra-Orissa coast on the 4th and in the 72-hour forecast, the rainfall belt moved further northwest with heavy rainfall of 15–20 cm along the coast which is close to the observed. In the case of the Eta model, the 48 and 72-hour forecasts showed heavy rainfall along the Andhra-Orrisa coast on the 4th and 5th, similar to the WRF experiment run, however the

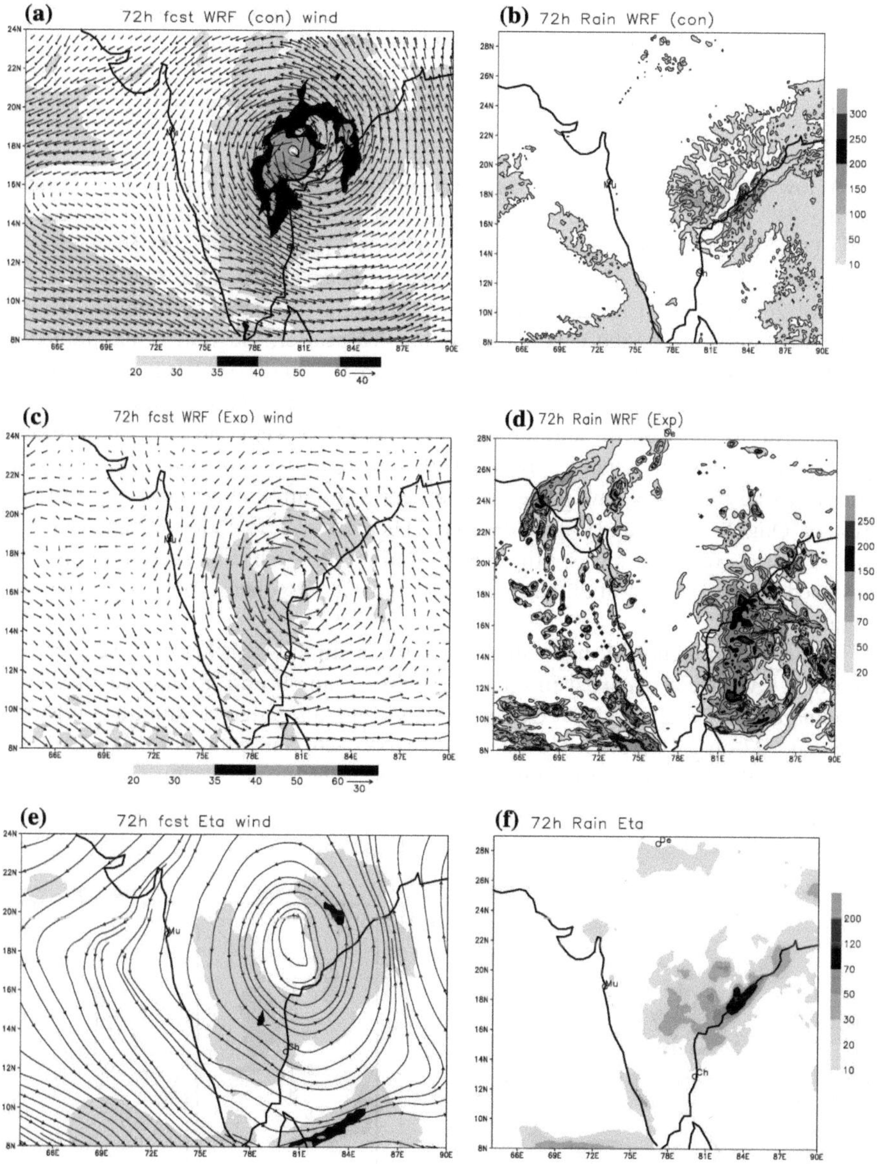

Figure 2

Forecast 72 hours 850 hPa wind fields and rainfall (mm) valid for 00 UTC of 5 October, 2004 based on WRF model control run (a,b); WRF experiment run (c,d) and Eta model (e,f).

maximum rainfall amounts of 7 to 12 cm predicted by the Eta model are less than the WRF experiment run and also the observed rainfall.

3.2. Case 2: Deep Depression over Bay of Bengal (October 26–29, 2005)

3.2.1. Synoptic situation

A low pressure area formed over the southwest and adjoining Bay of Bengal on the 25th and concentrated into a depression over southwest Bay of Bengal at 0300 UTC of 26 October near latitude 12.0°N and longitude 84.5°E. Moving in a northwesterly direction, the system was located near 12.5 N and 84.0 E at 1200 UTC of 26 October. It remained practically stationary and further intensified into a Deep Depression at 1800 UTC of same day. Thereafter, the system maintained a steady northwesterly track and lay centered at 15.5°N and 80.0°E, close to Ongole on the Andhra coast at 0300 UTC of 28 October. The system became somewhat sluggish in its movement and crossed the coast, as suggested by the surface observations, close to Ongole around 0800 UTC of the same day. Immediately after crossing the coast it weakened into a depression and lay centered close to the west of Ongole (over land) at 1200 UTC. Thereafter, it further weakened into a low pressure area over the south coastal Andhra Pradesh and surrounding. The observed track and rainfall (cm) for 27 and 28 October, 2005 are given in Figure 3.

Under the influence of this system widespread rainfall with heavy to very heavy rain at a few places occurred in coastal Andhra Pradesh and at isolated places in Tamilnadu on 27, 28 and 29 October. Significant amounts of rainfall (cm) are: Chennai—27 on 27 October, Kavali—35, Nellore—26, Sompeta—14, Mandasi—11 of Andhra Pradesh and Tiruvallur—27, Redhills—24, Ponneri—20, Chennai—16,

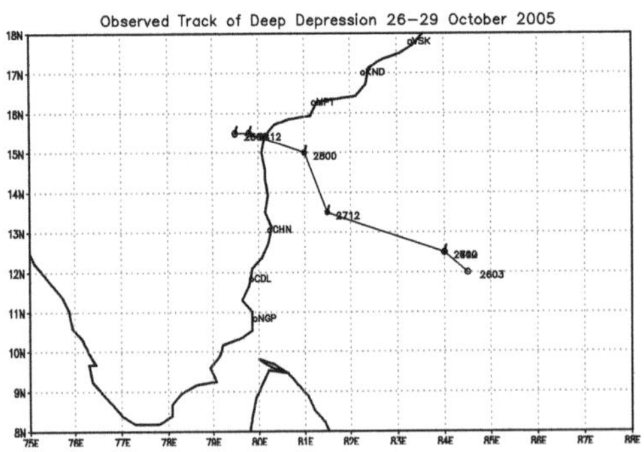

Figure 3
Observed track of deep depression 26–29 October, 2005 and observed rainfall (cm) for 27 and 28 October, 2005.

Tambaram—13 of Tamilnadu on 28 October and Kavali—8, Visakhapatnam—7 of Andhra Pradesh and Kalwakurthi—27, Porumammala and Cumbum—13 each of Tamilnadu on 29 October, 2005. Due to heavy rainfall along the coast of Andhra Pradesh, most of the paddy and property damage occurred in Nellore and Prakasam districts, and railway tracks were submerged at many places in these districts.

3.2.2. Model forecast

In this case, using the initial conditions of 00 UTC of 25 October, 2005, 72-hours forecasts are generated with the WRF (control run and experiment run) and Eta models. The initial 850 hPa wind flow (speed in knots) and mean sea level pressure (hPa) analysis for 00 UTC of 25 October, 2005 are given in Figures 4(a,b). The analysis shows a circulation in the southeast Bay of Bengal with wind speed of 20–30 kts over the south and also over the north and northeast sectors of the system with the lowest sea-level pressure of 1005 hPa.

3.2.2.1. WRF Model Forecasts

The 24 and 48-hour forecasts based on WRF experiment run 850 hPa wind flow (speed in knots) and rainfall (mm) valid for 00 UTC of 26 and 27 October, 2005 based on 25 October, 2005 initial fields are given in Figures 4(c-f), and the 72-hour forecast 850 hPa wind fields and rainfall based on WRF control and experiment run are given in Figures 5(a-d). In the case of the control run, the 24-hour forecast wind fields showed the symmetric structure of the system with 30–35 kts winds in the southeast sector, in 48 hours the system moved northwest with a strengthening of wind speed 40–50 kts to the southeast sector (figures not shown). In the 48 to 72-hour forecasts, the system recurved north-northeast and continued the same intensity (Fig. 5a). Also in this case the WRF control run has shown an overestimation of strength of the system as predicted in the previous case of the monsoon depression over the Bay of Bengal. Because of the north-northeast movement of the system, the 72-hour forecast rainfall maxima was shown in the southwest to northeast sector distant to the east coast of India (Fig. 5b). In the case of the WRF experiment run, the 24 and 48-hour forecasts wind fields show a well organized asymmetric structure of the system with strong winds 30–40 kts over the southeast and northeast/northern sectors of the system (Figs. 4c,e). The 72-h forecast shows the system close to Andhra Pradesh coast (Fig. 5c). The forecast shows a weakening of the system with a decrease of wind speed in the southeast and northwest sector and the system becoming more symmetric. The movement of the system in the day-one and day-two forecasts is northwest which is nearly 100 km west of the observed position and in the day-three forecast, the predicted movement is further northwest which is 50 km south of the observed position. The model day-one forecast 24-hour rainfall (mm) up to 00 UTC of 26 October shows a heavy rainfall area over the northeast sector of the system with isolated heavy rainfall of 20–25 cm to the north of the system and another maxima of 10 to 15 cm over the Tamilnadu coast (Fig. 4d). In day-two forecast, the 24-hour rainfall (mm) to up 00 UTC of 27 October shows the maximum rainfall belt shifted close to the east of Chennai (Fig. 4f) and in day-three forecast, the 24-hour rainfall (mm) up to 00 UTC of 28 October, the rainfall belt further shifted to the north along the coastal Andhra Pradesh where heavy to very heavy rainfall was reported on 28 October (Fig. 5d). It is also interesting to note that the model was able to capture the mesoscale rainfall features with isolated maxima over the northeast sector of the system in association with strong winds and moisture incursion from the Bay of Bengal in this sector.

3.2.2.2. Eta Model Forecast

The 72-h forecast 850 hPa wind flow (speed in knots) and rainfall (mm) valid for 00 UTC of 28 October, 2005 based on 25 October, 2005 initial fields are given in Figures 5(e,f). The wind field shows a similar structure to that of the WRF model forecast, however, the Eta model showed a slow northeast movement. In the 24-h forecast, the system is 50 km south and in the 48-h forecast it is 50 km west of the observed position. However, in the 72-h forecast, due to slow forecast movement, the system is far east of the observed position. The day-one forecast 24-h rainfall valid for 00 UTC of 26 October showed a broad area of 7–13 cm rainfall belt to the southwest of the system over the southwest Bay of Bengal. In the day-two forecast, the 24-h rainfall valid for 00 UTC of 27 October showed the rainfall area of 7–13 cm concentrated over the coast of Tamilnadu (figures not shown) and the in day-three forecast, the 72-hour rainfall valid for 00 UTC of 28 October shows that the rainfall belt shifted to northeast along the coast of Andhra Pradesh (Fig. 5f). Due to the slow forecast movement of the system, the model predicted rainfall in day-three is remotely east of the observed heavy rainfall over the Andhra Pradesh coast. It is also seen that although the model was able to capture a broad area of heavy rainfall belt, the model was not able to capture the isolated heavy rainfall which was well captured by WRF model experiment run.

3.3. Case 3: Heavy Rainfall over Mumbai, West Coast of India

Mumbai (Lat. 18.50°N, Long. 72.52°E) situated on the West Coast of India is the commercial and financial capital of India. On 26 July different stations of Mumbai and its suburbs have reported exceptionally heavy rainfall (Fig. 6a). Some of these are: Mumbai Airport (Santa Cruz) 94.4 cm, also some lake stations northeast of it within 15-km distance recorded 104.4 cm while Bhandup reported 81.5 cm and Tulsi 60.1 cm. Even stations which are further to the northeast of Santa Cruz, for example

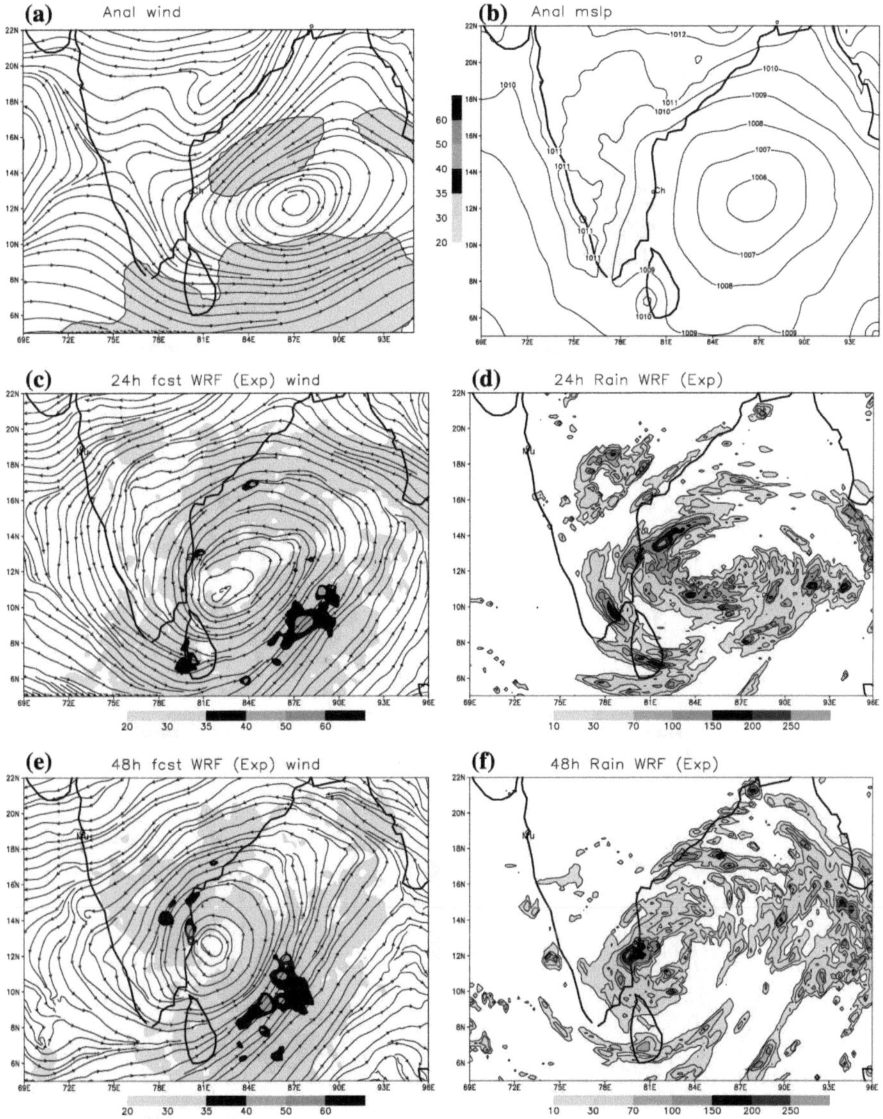

Figure 4

(a,b) Initial 850 hPa wind flow (speed in knots) and mslp (hPa) analysis for 00 UTC of 25 October and WRF model experiment run (b,c) 24-hour forecast, (e,f) 48-hour forecast wind (kt) and rainfall (mm) valid for 00 UTC of 26 and 27 October, 2005.

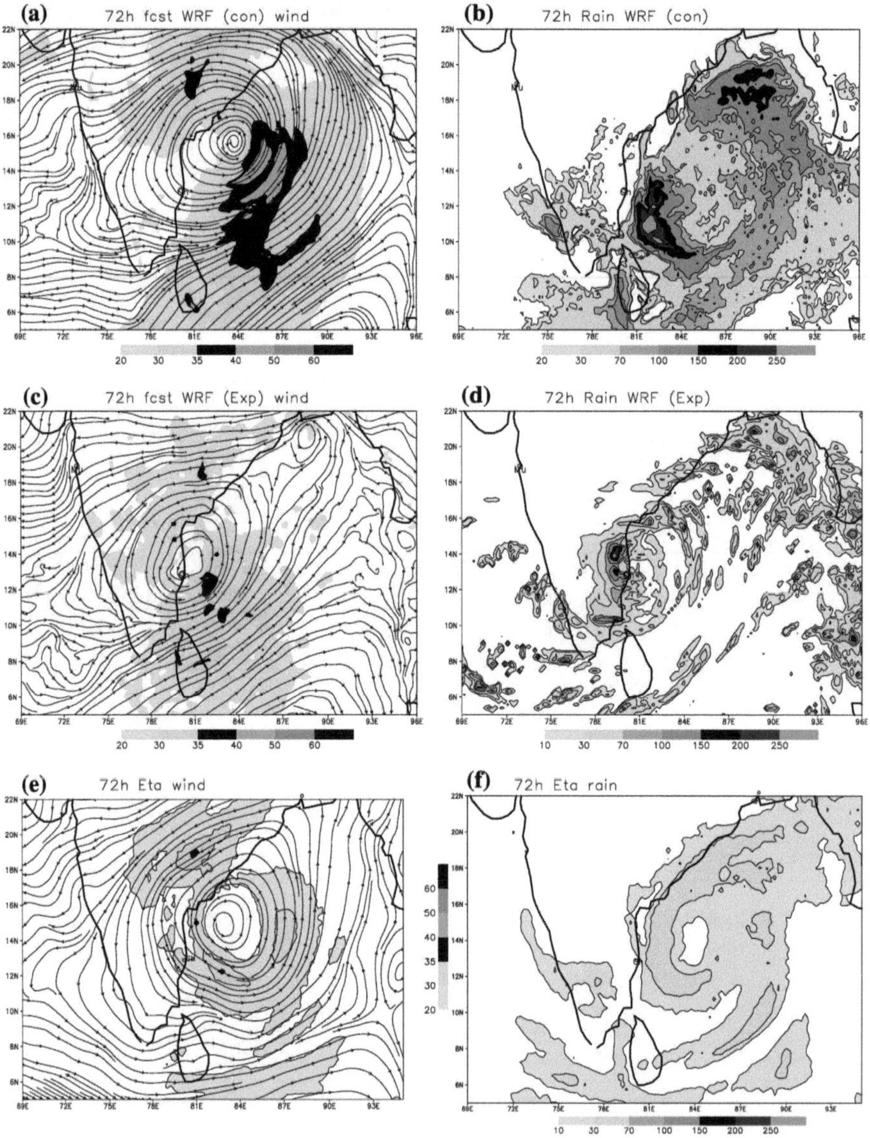

Figure 5
(a,b) WRF (con), (c,d) WRF (exp) and (e,f) Eta model 72-h forecast 850 hPa wind flow (speed in knots) and rainfall (mm) valid for 00 UTC of 28 October, 2005 based on 25 October, 2005 initial conditions.

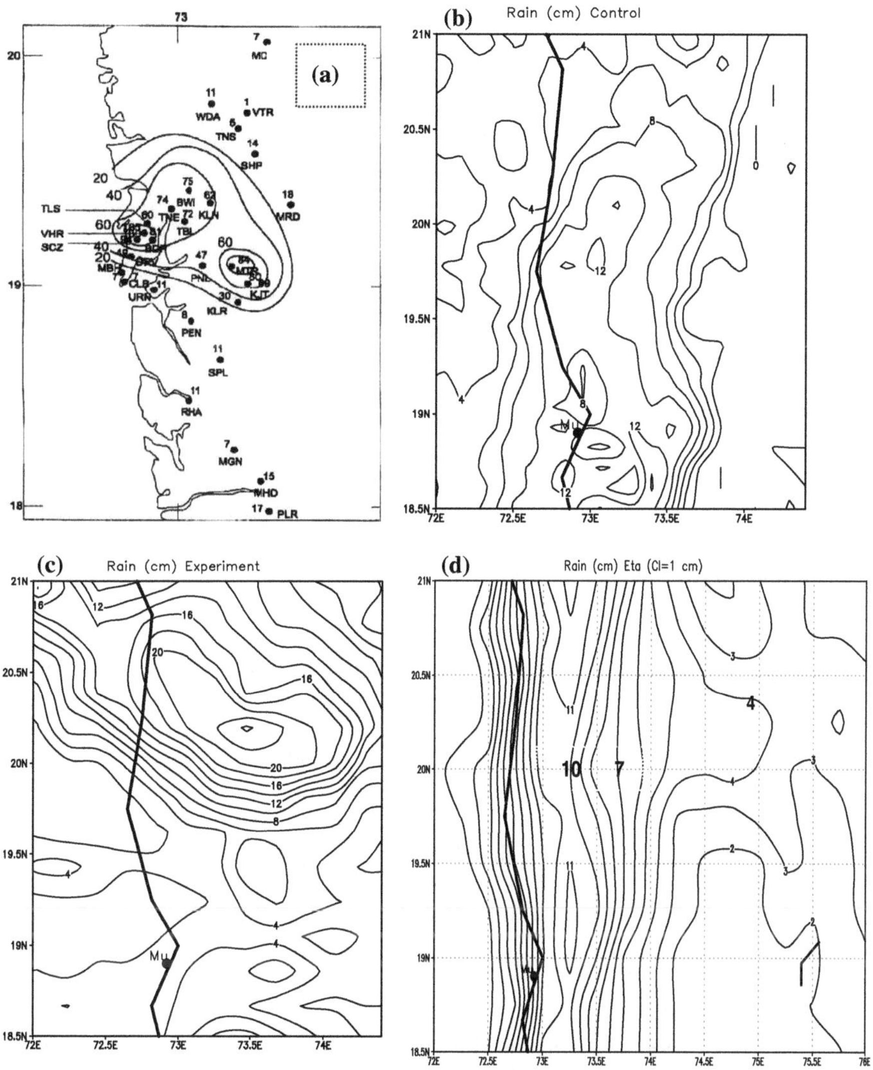

Figure 6
(a) 24-hours observed rainfall (cm) recorded at suburban stations up to 03 UTC of 27 July, 2005, (b) WRF control run, (c) WRF experiment, (d) Eta forecast valid for 00 UTC of 27 July, 2005.

Bhivandi, Kalynana and Thane reported 62, 75 and 74 cm, respectively. However, some lake stations, which are beyond 50 km and lie far to the northeast (*e.g.*, Tansa and Vaiterna), reported 5 and 1 cm of rainfall; a substantial reduction of rainfall from Santa Cruz. South of Santa Cruz, Colaba about 15–20 km away recorded only 7.4 cm rainfall. Isohytal contour analysis shows that rainfall contour of 60 cm with a spatial scale nearly 35 × 35 km, which lies just to the northeast of Santa Cruz with the highest contour of 90 cm around 10 × 10 km just to the northeast of it. It is also very interesting to note the occurrence of another very high rainfall belt nearly 45–50 km away distant of Santa Cruz over MTR and Karjat which recorded 84 cm and 59 cm, respectively.

3.3.1. Synoptic situation

A low pressure area formed over north Bay of Bengal off Gangetic west Bengal-Orissa coast on 23 July. This system persisted over the coastal area for next 2 days and became well marked on the 25th. It moved inland and lay over Orissa on the 26th and over East Madhya Pradesh on the 27th, and moving westwards it became less marked on the 29th over southeast Rajasthan. Another low-pressure area formed over northwest Bay of Bengal off Orissa coast on 28 July and moving westwards it intensified into a depression on the 29th close to Chandbali of Orissa and became a deep depression on the 30th inland over Orissa. It moved further westwards to Chattisgarh on the 1 August and to Madhya Pradesh on the 2nd and weakened into a low over northwest Madhya Pradesh on the 3rd.

The 850 hPa wind field analysis (speed in knots) for 00 UTC of 26, 27, 28 and 29 July, 2005 are given in Figure 7. As described earlier, the increase in rainfall from 24 July to the 27th over the west coast, including record rainfall in Mumbai on 26–27 July, was due to the strengthening of low level winds over the west coast and the westward movement of the monsoon disturbances from the Bay of Bengal. In addition to this period an off-shore trough was present at sea level from Konkan coast to Karnataka coast. As per rainfall reports there was a temporary reduction in rainfall activity over the west coast including Mumbai for two days after 27 July when the first low pressure system moved far westwards and weakened. There was again a second spell of very heavy rainfall over the state including Mumbai from 31 July to 2 August. Santa Cruz recorded 20.8 cm on 1 August.

3.3.2. Model forecast

In this case, using the initial conditions of 00 UTC of 26 July 2005, 72-hour forecasts are generated with the WRF (control run and experiment run) and Eta models. In the case of the control run, the 24-hour rainfall forecast up to 00 UTC of 27 July (Fig. 6b) indicated a broad rainfall belt with a northeast to south orientation and a maximum of 10–12 cm along the west coast of India. The Eta model (Fig. 6d) also showed similar with a maximum of 10–12 cm along the coast oriented north to south. However, in the experiment run, the model was able to capture the localized heavy rainfall belts with maxima of 20–25 cm to the north of the observed location of Mumbai (Fig. 6c). To understand the mechanism of the mesoscale heavy rainfall

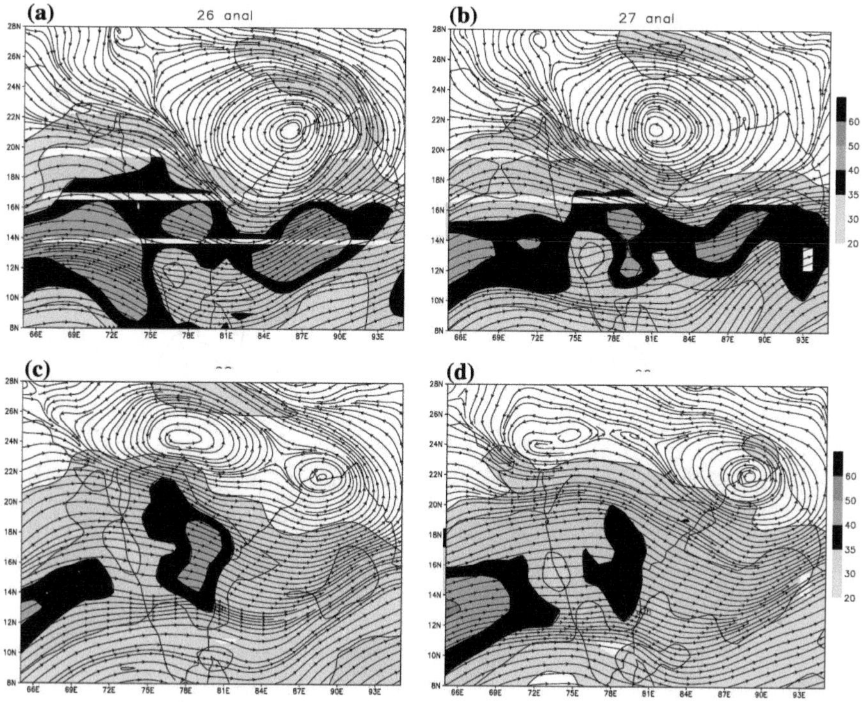

Figure 7
Initial 850 hPa wind fields analysis (speed in knots) for 00 UTC of 26, 27, 28 and 29 July, 2005.

episode, the six-hourly forecast rainfall, wind fields, vorticity and divergence fields are investigated. The model predicted 6, 12, 18 and 24-hour rainfall is given in Figure 8. The initial 6-hour model predicted 5–10 cm rainfall over the north of Mumbai, and subsequent 12 and 18 hours the model predicted rainfall intensity increased up to 20–25 cm at the same location and from 18 to 24 hour, there is no significant increase in rainfall intensity at the same location, however the rainfall belt moved slightly northeast. The initial 850 hPa wind fields for 00 UTC of 26 July, 2005 and 06, 12, 18 and 24-hour forecasts 850 hPa wind fields are given in Figures 7a and 9(a-d). The initial fields show a circulation along the Orrisa coast with strong westerly current of 30–40 kts along the west coast, including peninsular India. Over western India there is wind convergence between westerlies from the northwest and easterlies from the northeast in association with a low pressure system over Orrisa. The forecasts based on these initial conditions showed a strengthening of wind convergence in the 06- and 12-hour forecast over western India and formation of

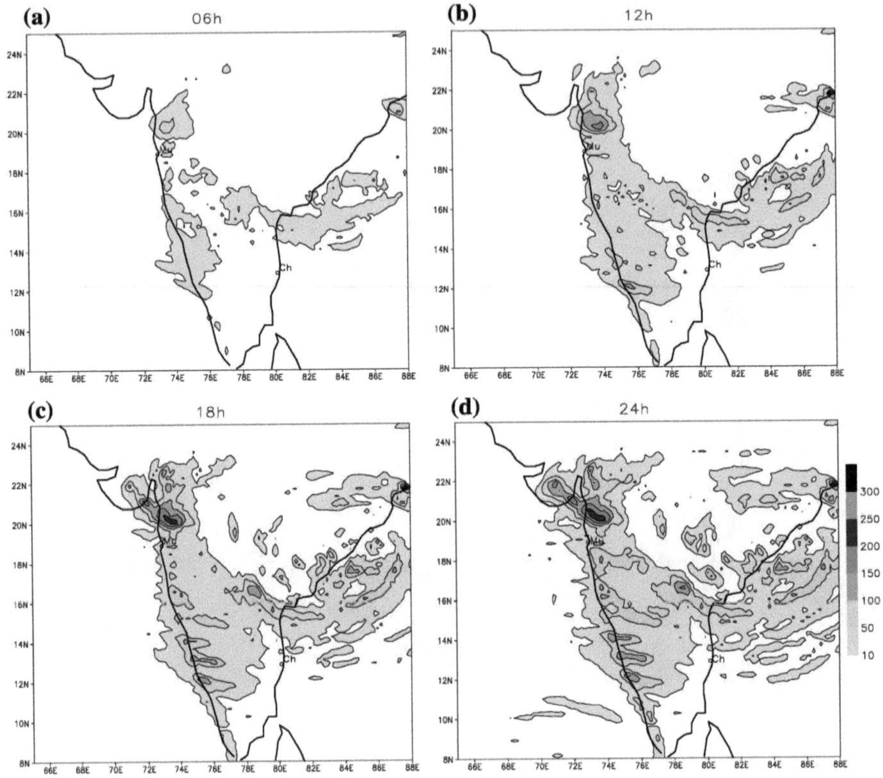

Figure 8
WRF model (exp) 06, 12, 18 and 24-hour forecast rainfall (mm) valid for 06, 12, 18 UTC of 26 and 00 UTC of 27 July, 2005.

circulation over the area near north Maharastra which is about 200 kms northeast of Mumbai. In the 18 and 24-hour forecasts, the circulation shows gradually weakening with westward movement of the low pressure system. For further investigation we have considered the 6-hourly 850 hPa forecast vorticity and divergence fields. The initial vorticity ($*10^{-5}$/sec) shows a broad area of 4 to 8 along western India (figure not shown). The subsequent 06-hour forecast shows the strengthening of vorticity to more than 30, and in the 12-hour forecast the maximum voticity area further strengthened with extension to the north and in subsequent 18 and 24-hour forecasts it moved northwest (figures not shown). The divergence field 6-hour forecast shows strengthening of negative divergence ($*10^{-5}$/sec) (which is closely associated with vertical velocity) from -2 to -15 over the northeast of Mumbai, in the 12-hour forecast, it moved slightly westward and in subsequent 18 and 24 hours it moved

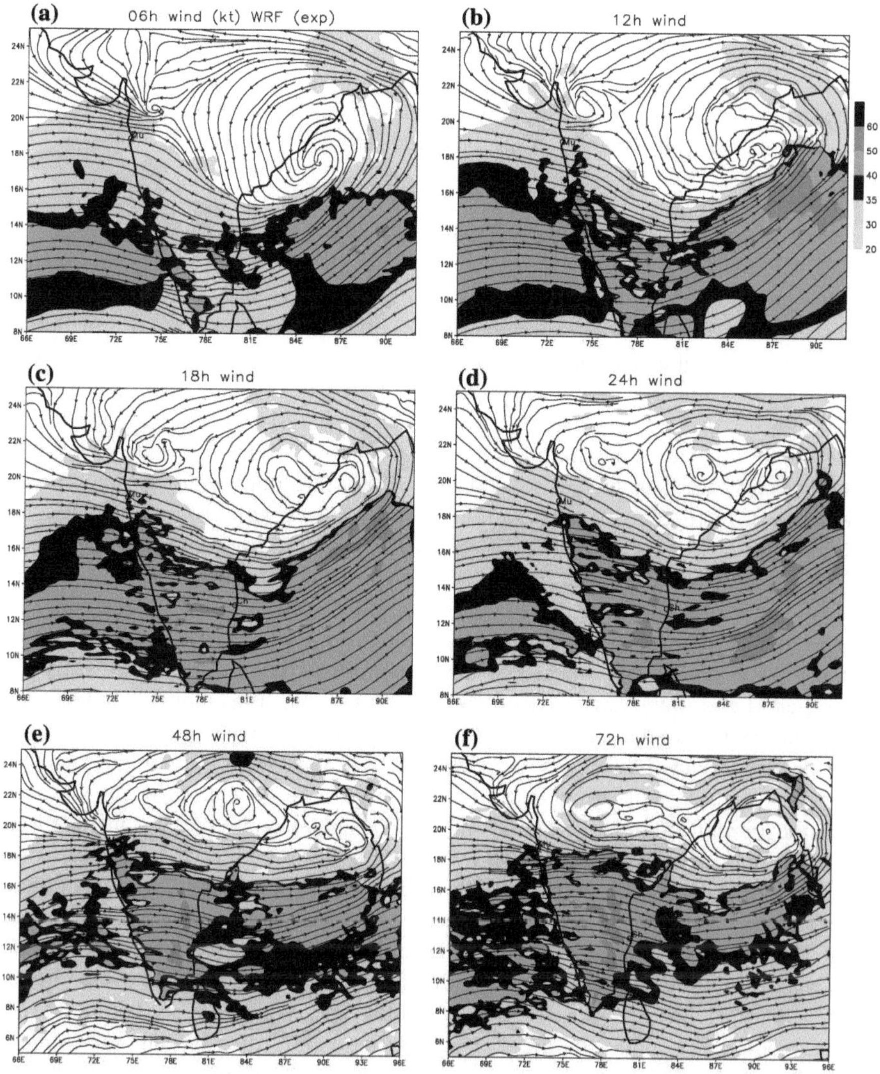

Figure 9
WRF model (exp) 850 hPa wind fields (speed in knots): (a-c) 06, 12, 18-hour forecasts valid for 06, 12, 18 UTC of 26 July and (d-f) 24, 48 and 72-hours forecasts valid for 00 UTC of 27, 28 and 29 July, 2005, based on 26 July, 2005 initial conditions.

further northwest over to the Gujarat region (figure not shown). It is also noteworthy to see (Figs. 9(d-f)) that with the movement of low pressure westwards in day-one forecast, the model day-two forecast showed the formation of another low pressure system over the Bay of Bengal and subsequent organization of the system in day-three forecast valid for 00 UTC of 29 July, 2006 as observed (Fig. 7).

4. Concluding Remarks

The preliminary study of comparisons of the WRF and Eta model results shows that the WRF model experiment run with Ferrier microphysics and Betts-Miller-Janjic cumulus parameterization scheme showed better prediction of the movement of tropical systems and rainfall 72 hours in advance compared to the WRF control run and Eta model. The Eta model has shown underprediction of rainfall amounts with a maximum of 13 cm over a broad area, whereas, the WRF model was able to predict the isolated heavy rainfall amounts more than 25 cm, which are close to the observed rainfall in association with tropical depression/cyclones. In the first case, the WRF control run with Lin microphysics and Kain-Fritsch cumulus scheme overestimated the intensity of the Bay of Bengal system. However, in experiment run the model was able to show the weakening of the system after landfall as observed. In two other cases also considered it was shown that the WRF model experiment run forecast products had better agreement with the observations than the control run. In case one, Eta model showed the movement of the system similar to the WRF experiment run, however, in case-two, the Eta model showed a slow northeast bias and the WRF control run has shown a northeast recurvature with strengthening of the system. Also, the 24-hour maximum rainfall amount predicted by the Eta model is 5 to 12 cm in all three cases, whereas, the WRF model showed isolated heavy rainfall amounts more than 25 cm, which is close to the observed. In case-three, the 24-hour rainfall forecast with the WRF model experiment run was able to capture the heavy rainfall event over Mumbai with a location error of 50 km north of the observed. However, the WRF control run and Eta model forecasts had shown a broad rainfall belt along the west coast of India. It is also observed that the WRF model was able to predict better mesoscale distribution of rainfall compared to the Eta model of the same resolution. This appears to be due to the better combination of microphysical and cumulus parameterization schemes in the WRF model. Although the present study demonstrates that the combination of the Ferrier microphysics scheme coupled with a Betts-Miller-Janjic cumulus parameterization is able to simulate the heavy rainfall events as well as the track of the systems, more cases are to be examined to test the success of the model. These aspects will be expected to be tested in the future program of model development.

Appendix 1
Description of the WRF Model

Non-hydrostatic version, single domain, horizontal resolution: 20 km,

Vertical resolution: 31 sigma levels (1.000, 0.993, 0.980, 0.966, 0.950, 0.933,0.913, 0.892, 0.869, 0.844, 0.816, 0.786,0.753, 0.718, 0.680, 0.639, 0.596, 0.550,0.501, 0.451, 0.398, 0.345, 0.290, 0.236,0.188, 0.145, 0.108, 0.075, 0.046, 0.021, 0.000).

Domain: 10°S–45°N; 48°E–116°E; Time step:72 sec

Input Data Types

Real-data using Standard Initialization (SI) conversion from Grib files, 6-hourly boundary conditions NCEP Global Final Analyses (FNL) at 1° × 1° lat./long. resolution.

Terrestrial data: 30 sec topography, land use (USGS Version 2.0), soil type (data from 30 sec STATSGO (US) and 5 min UN/FAO (global)), 10 min 12 monthly vegetation fraction data, 1 degree deep soil temperature data.

Post processing: A simple conversion program to convert NetCDF data to GrADS

Physics

Planetary boundary layer (Yonsei University (S. Korea))
Surface layer (similarity theory)
Land-surface models (NOAH)
Slab soil model (5-layer thermal diffusion)
Longwave radiation (RRTM)
Shortwave radiation (simple MM5 scheme / Goddard)
Sub-grid turbulence (Smagorinsky)
Control run: Microphysics (LIN *et al.*); cumulus parameterization (Kain-Fritsch scheme)

Experiment run: Microphysics (FERRIER *et al.*); cumulus parameterization (Betts-Miller-Janjic scheme)

Description of Mesoscale Eta Model

(Workstation version 3.1i, March 2002, NCEP)

Grid: Semi-staggered Arakawa E-grid (ARAKAWA and LAMB, 1977), horizontal resolution: 22 km, vertical levels: 45

Model Domain: 52°–111° E and 4°S–42°N (variable); model time step (Δt): 60 seconds

Topography: Silhouette step topography

Model dynamics: (BLACK, 1994)
Fundamental prognostic variables: T, u, v, q, Ps, TKE, cloud water/ice
Inertial gravity wave adjustment: forward-backward scheme ($\Delta t = 60\,\mathrm{s}$)
Vertical advection: Euler-backward scheme centred in space piecewise linear for q
Horizontal advection: Modified Euler-backward scheme, Janjic advection in space

Conservative, (nearly) shape-preserving scheme for H_2O, upstream advection near boundaries

Model physics: (JANJIC, 1994)
Betts-Miller-Janjic convection
Mellor-Yamada level 2.5 turbulent exchange
GFDL Radiation
Explicit cloud water/ice prediction; 4 layer soil scheme with vegetation; Δ^2 horizontal diffusion

Initial and Lateral boundary conditions: GFS, NCEP global analysis and 6-hourly global model forecasts

Time-dependent surface fields: SST, snowdepth, surface analysis from NCEP global model

REFERENCE

ARAKAWA, A. and LAMB, V.R. (1977), *Computational design of the basic dynamical processes of the UCLA general circulation model, Methods Computer*, Phys. *17*, 173–265.

BETTS, A.K. (1986), *A new convective adjustment scheme. Part I: Observational and theoretical basis*, Quart. J. Roy. Meteor. Soc. *112*, 1306–1335.

BETTS, A.K. and Miller, M. J. (1986), *A new convective adjustment scheme. Part II: Single column tests using GATE wave, BOMEX, and arctic air-mass data sets*, Quart. J. Roy. Meteor. Soc. *112*, 693–709.

BLACK, T.L. (1994), *The new NMC mesoscale Eta model: Description and forecast examples*, Wea. Forecasting *9*, 265–278.

CHEN, S.-H. and SUN, W. Y. (2002), *A one-dimensional time-dependent cloud model*, J. Meteor. Soc. Japan *80*, 99–118.

FERRIER, B. S., JIN, Y., LIN, Y., BLACK, T., ROGERS, E., and DIMEGO, G. (2002), *Implementation of a new grid-scale cloud and precipitation scheme in the NCEP Eta model*, Preprints, 15th Conf. On Numerical Weather Prediction, San Antonio, TX, Amer. Meteor. Soc., 280–283.

JANJIC, Z.I. (1994), *The step-mountain Eta coordinate model: Further developments of the convection, viscous sublayer, and turbulence closure schemes*, Mon. Wea. Rev. *122*, 927–945.

JANJIC, Z.I. (2000), *Comments on "Development and Evaluation of a Convection Scheme for Use in Climate Models"*, J. Atmos. Sci. *57*, 3686.

KAIN, J.S. and FRITSCH, J.M. (1990), *A one-dimensional entraining/ detraining plume model and its application in convective parameterization*, J. Atmos. Sci. *47*, 2784–2802.

KAIN, J.S. and FRITSCH, J.M. (1993), *Convective parameterization for mesoscale models: The Kain-Fritcsh scheme, the representation of cumulus convection in numerical models* (K.A. Emanuel and D.J. Raymond, eds.) (Amer. Meteor. Soc. 1993) 246 pp.

LAPRISE, R. (1992), *The Euler equations of motion with hydrostatic pressure as independent variable*, Mon. Wea. Rev. *120*, 197–207.

LIN, Y.-L., FARLEY, R.D., and ORVILLE, H.D. (1983), *Bulk parameterization of the snow field in a cloud model*, J. Climate Appl. Meteor. *22*, 1065–1092.

MESINGER, F. (1984), *A blocking technique for representation of mountains in atmospheric models*. Rev. Meteor. Aeronaut. *44*, 195–202.

MESINGER, F. and BLACK, T.L. (1992), *On the impact of forecast accuracy of the step-mountain (Eta) vs. sigma coordinate*, *Meteorol. Atmos. Phys. 50*, 47–60.

SKAMAROCK, W.C. and KLEMP, J.B. (1992), *The stability of time-split numerical methods for the hydrostatic and the nonhydrostatic elastic equations*, Mon. Wea. Rev. *120*, 2109–2127.

SKAMAROCK, W.C., KLEMP, J.B., DUDHIA, J., GILL, D.O., BARKER, D.M., WANG, W., and POWERS, J.G. (2005), *A Description of the Advanced Research WRF Version 2*, NCAR Technical Note, 468.

WANG, W., BARKER, D., BRUY`ERE, C., DUDHIA, J., GILL, D., and MICHALAKES, J. (2004), *WRF Version 2 modeling system user's guide*, http://www.mmm.ucar.edu/wrf/users/docs/user guide/.

ZHAO, Q., CARR, F.H. and LESINS, G.B. (1991), *Improvement of precipitation forecasts by including cloud water in NMC's Eta Model*, Preprints, Ninth Conf. on *Numerical Weather Prediction*, Denver, CO, Amer. Meteor. Soc.

(Received May 25, 2006, accepted December 18, 2006)
Published Online First: July 4, 2007

Pure appl. geophys. 164 (2007) 1617–1637
0033–4553/07/091617–21
DOI 10.1007/s00024-007-0245-0

Pure and Applied Geophysics

Sensitivity of Mesoscale Model Forecast During a Satellite Launch to Different Cumulus Parameterization Schemes in MM5

V. Rakesh, R. Singh, P. K. Pal, and P. C. Joshi

Abstract—The identification of the model discrepancy and skill is crucial when a forecast is issued. The characterization of the model errors for different cumulus parameterization schemes (CPSs) provides more confidence on the model outputs and qualifies which CPSs are to be used for better forecasts. Cases of good/bad skill scores can be isolated and clustered into weather systems to identify the atmospheric structures that cause difficulties to the forecasts. The objective of this work is to study the sensitivity of weather forecast, produced using the PSU-NCAR Mesoscale Model version 5 (MM5) during the launch of an Indian satellite on 5th May, 2005, to the way in which convective processes are parameterized in the model. The real-time MM5 simulations were made for providing the weather conditions near the launch station Sriharikota (SHAR). A total of 10 simulations (each of 48 h) for the period 25th April to 04th May, 2005 over the Indian region and surrounding oceans were made using different CPSs. The 24 h and 48 h model predicted wind, temperature and moisture fields for different CPSs, namely the Kuo, Grell, Kain-Fritsch and Betts-Miller, are statistically evaluated by calculating parameters such as mean bias, root-mean-squares error (RMSE), and correlation coefficients by comparison with radiosonde observation. The performance of the different CPSs, in simulating the area of rainfall is evaluated by calculating bias scores (BSs) and equitable threat scores (ETSs). In order to compute BSs and ETSs the model predicted rainfall is compared with Tropical Rainfall Measuring Mission (TRMM) observed rainfall. It was observed that model simulated wind and temperature fields by all the CPSs are in reasonable agreement with that of radiosonde observation. The RMSE of wind speed, temperature and relative humidity do not show significant differences among the four CPSs. Temperature and relative humidity were overestimated by all the CPSs, while wind speed is underestimated, except in the upper levels. The model predicted moisture fields by all CPSs show substantial disagreement when compared with observation. Grell scheme outperforms the other CPSs in simulating wind speed, temperature and relative humidity, particularly in the upper levels, which implies that representing entrainment/detrainment in the cloud column may not necessarily be a beneficial assumption in tropical atmospheres. It is observed that MM5 overestimates the area of light precipitation, while the area of heavy precipitation is underestimated. The least predictive skill shown by Kuo for light and moderate precipitation asserts that this scheme is more suitable for larger grid scale (> 30 km). In the predictive skill for the area of light precipitation the Betts-Miller scheme has a clear edge over the other CPSs. The evaluation of the MM5 model for different CPSs conducted during this study is only for a particular synoptic situation. More detailed studies however, are required to assess the forecast skill of the CPSs for different synoptic situations.

Key words: Mesoscale model, CPSs, entrainment/detrainment, skill scores, TRMM, synoptic validation.

Atmospheric Sciences Division, Meteorology and Oceanography Group, Space Applications Centre (ISRO), Ahmedabad-380015, India. E-mail: rakeshv82@gmail.com

1. Introduction

Weather plays a crucial role in the planning and execution of spacecraft launch operations. Each launch vehicle has a specific tolerance for wind shear, cloud cover, temperature, whereas lightening constraints are common for all types of vehicles. A recent satellite launch (PSLV-C6) was made by Indian Space Research Organization (ISRO) in May 2005. The launch site (SHAR) is on the east coast of India (Sriharikota, 13.72°N, 80.2°E) and falls in the vicinity of the tropical cyclone (SINGH *et al.*, 2005) landfall zone. Consequently weather forecast is very important during satellite launch. The presence of convective storm activity in the vicinity of the spacecraft launch site can be significant due to the lightening strikes and electrostatic discharges, which can damage or destroy the launch vehicle and/or its payload. Cumulus convection is one of the processes that plays an important role in weather by influencing the dynamic and thermodynamic state of the atmosphere. Since the convective elements are normally of the order of 1 km in size, fine resolution is required in the numerical model for an explicit treatment of the convective process. On the other hand parameterization schemes allow us to use coarser grid in numerical simulation where the horizontal scale of the phenomena is of the order of several thousands of kilometers. Therefore the selection of a proper cumulus physics option is extremely important in real-time applications of numerical weather prediction models.

The objective of this study is to observe the sensitivity of real-time forecasts produced during the previously mentioned satellite launch using the fifth-generation Pennsylvania State University-National Center for Atmospheric Research (NCAR) Mesoscale Model version 5 (MM5; GRELL *et al.*, 1994), to different cumulus parameterization schemes in the model. One benefit to the mesoscale modeling community from such real-time evaluation studies is that in such cases the model in general is subjected to a particular synoptic condition and not tuned to a specific case to obtain better results. Also due to the large number of forecasts generated during such experiments, statistical evaluation works as an appropriate tool in identifying model forecast errors and its sensitivity to various physics options in the model. Ultimately this information may become helpful not only for selecting proper physics options but also for sorting out the sources of model forecast errors, which in turn is useful for the efforts to improve the assumptions made in the formulation of different physics options.

Many cumulus parameterization schemes (CPSs) have been developed and implemented in numerical weather prediction models that implicitly account for the associated subgrid exchanges of mass, heat, and moisture. The basic feature that differentiates CPSs is the closure assumptions upon which they are based. The schemes vary in complexity from simple moist convective adjustment to sophisticated mass flux-type schemes (DAS *et al.*, 2002). It is recognized that convection in nature often develops mesoscale organization and these resolvable mesoscale structures develop from initially unresolvable cumulonimbus clouds that in itself provides a major challenge to mesoscale modelers. The representation of convection in the models is strongly scale-dependent (MOLINARY and DUDEK, 1992). The adequate representation of convective processes is particularly important in numerical models, but there is no universally accepted framework for representing such processes with grid scales that prohibit fully explicit representation (GOCHIS *et al.*, 2002). CPSs are developed for specific synoptic conditions, and are evaluated in a limited number of cases (KUO *et al.*, 1996; WANG and SEAMAN, 1997; PENG and TSUBOKI, 1997; YANG *et al.*, 2000; GOCHIS *et al.*, 2002; YANG and TUNG, 2003; DAS *et al.*, 2002). YANG and TUNG (2003) performed an evaluative study to discern the performance of four CPSs, the Anthes-Kuo (ANTHES, 1977), Betts-Miller (BETTS and MILLER, 1986), Grell (GRELL, 1993), and Kain-Fritsch scheme (KAIN and FRITSCH, 1993) in the rainfall forecasts over Taiwan using MM5. They showed that none of these CPSs consistently outperforms the others in all measurements of forecast skill. The skill of the MM5 model in simulating the Indian summer monsoon is studied by DAS GUPTA *et al.* (2005) and the results show that the model performs reasonably well except for heavy rainfall events. In general, for simulations with fine horizontal resolutions (~15 km) MM5 tends to overpredict the area of light rainfall, and underpredict the area of heavy rainfall (YANG and TUNG, 2003; GALLUS, 1999).

During the initial days of forecast, the Bay of Bengal was quite calm without any synoptic systems. Most parts of the Bay of Bengal and the region around SHAR are associated with very low (2–4 m/s) wind speed. A well-marked trough over south central India was also observed during the period. A region of high wind speed (9-10 m/s) was seen on the western coast of India. Significant rainfall was observed over south east Indian coasts during 29 April to 1 May. Heavy rainfall with thunderstorm activity was observed over the launch site SHAR on 1 May. There were light rainfall events over the Indian region throughout the study period. Even in the case of light rainfall events CPSs can become active and can have significant impact on the circulation. The light rainfall at one place in the domain can change the wind field at another place.

Parameterized moist downdrafts are crucial for reproducing many of the observed mesoscale characteristics, as well as correct large-scale temperature fields. Therefore verification of these mesoscale features is an important element of evaluating CPS performance. Conclusions of these evaluative studies of CPSs are often speculative because of the complexity of the parent model and case dependency of the results. We will not attempt here to make an absolute statement about which parameterization is best as a whole, but will try to evaluate their performance in this particular synoptic condition and the statistical effect they have on the model forecast error. The model predicted meteorological fields such as wind speed, temperature and relative humidity by different CPSs are statistically evaluated by comparison with the radiosonde observations from 25 April to 05 May, 2005. The first 24 h model predicted rainfall by different CPSs, during the simulation period, is evaluated by comparing with the TRMM observed rainfall and calculating statistical skill scores.

2. Model Description

In the present study we use a well-studied mesoscale model, namely MM5. MM5 is the latest version of the mesoscale model originally developed by ANTHES and WARNER (1978) known as Fifth Generation NCAR/Penn State Mesoscale Model. This non-hydrostatic version employs reference pressure as the basis for terrain following vertical coordinate and with a fully compressible system of equations. In combination with multiple-nest capability, a four-dimensional data assimilation technique and a variety of physics options make the model capable of simulations on finer scales, limited only by data resolution, quality and computer resources. The model was run with 25 km horizontal resolutions with a single domain. Twenty-seven unevenly spaced full-sigma levels were used in the vertical, with the maximum resolution in the boundary layer. Ten-minute averaged terrain/landuse data were interpolated to the 25-km model grids. Figure 1 shows the model domain used for the simulation.

For the present study, the following physics options are employed:

(i) *Planetary Boundary Layer (PBL)*—the PBL technique is the Medium Range Forecast (MRF) of the National Centre for Environmental Prediction (NCEP) by HONG and PAN (1996). In MM5 version 3.7 MRF PBL was made compatible with polar physics, particularly sea-ice fraction. Another modification in MRF PBL is to remove the effect of convective velocity on the surface momentum stress during the day time, and which should help to alleviate a day-time low wind speed bias (DUDHIA, 2005).

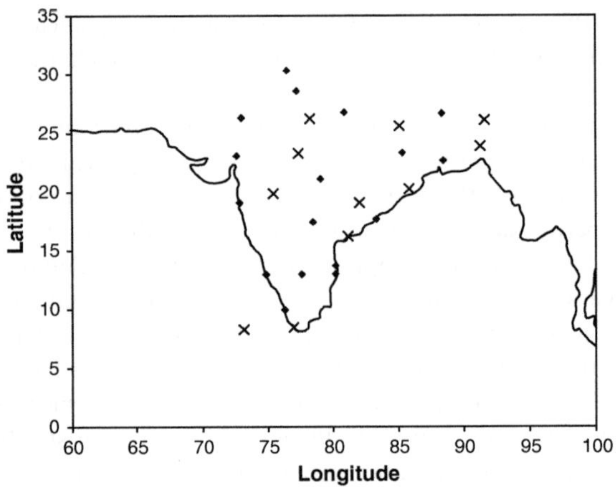

Figure 1
Model domain and radiosonde locations (×, ♦) where local data is used to enhance model initial condition. Locations shown by symbol "♦" are selected for point validation.

(ii) *Cloud microphysics*—A simple treatment of cloud microphysics (DUDHIA, 1989) in which both ice and liquid phases are permitted for cloud and precipitation.
(iii) *Cumulus parameterization*—Four CPSs have been selected on the basis of their widespread use in numerical models and the representation of different closure assumptions and scale considerations. Since numerous references are available on the conceptual formulation of these CPSs, here we are not describing them in detail. The CPSs selected for the present study are as follows

a. Anthes- Kuo scheme (Kuo)

The original KUO (1974) scheme determines the amount of rainfall, convective heating, and moisture convergence. The Kuo scheme used here is a modified version described in ANTHES *et al.*, (1987) and GRELL *et al.*, (1994), in which the moistening is vertically distributed based on relative humidity.

b. Grell Scheme (Grell)

The Grell scheme (described in GRELL *et al.* 1994) is essentially a simplified Arakawa-Schubert scheme. Its closure is based on the quasi-equilibrium assumption following the ARAKAWA-SCHUBERT (1974) scheme.

c. Kain-Fritsch scheme (KF)

The KAIN and FRITSCH (1993) scheme is an improved version of the FRITSCH and CHAPPELL (1980) parameterization. In the cloud model (KAIN and FRITSCH 1990) that is used to determine vertical profiles of heating and drying, entrainment and detrainment rates depend on the buoyancy of mixtures of clear and cloudy air.

d. Betts-Miller scheme (BM)

The deep convection parameterization developed by BETTS (1986) adjusts the grid-column sounding toward a reference sounding that is designed to resemble the quasi-equilibrium thermodynamics sounding observed in association with deep convection. In MM5, this deep convection algorithm is used in combination with the shallow convection scheme of JANJIC (1994). More details about this scheme are available in BETTS and MILLER (1993).

(iv) *Cloud-radiation scheme*—The radiation parameterization used is NCAR Community Climate Model (CCM2). It has multiple spectral bands in shortwave and longwave, but the radiative properties of clouds are based on the grid point relative humidity values from the model (HACK *et al.*, 1993). It also provides radiative fluxes at the surface.
(v) *Surface scheme*—The land surface model (Noah-LSM) is capable of predicting soil moisture and temperature in four layers, as well as canopy moisture and water equivalent snow depth. It also outputs surface and underground run-off accumulation. The LSM makes use of vegetation and soil type in handling evapotranspiration, and effects such as soil conductivity and gravitational flux of moisture. The Noah-LSM can also optionally use satellite derived climatological albedo, instead of relating albedo to landuse type.

3. Data Used

3.1. Initial Conditions for MM5

Initial data were generated for a 25-km resolution by interpolating the NCEP/AVN global analysis at 75-km resolutions to the model grid. The same model forecasts for every 6 hours were linearly interpolated in time to provide the lateral boundary conditions. Daily 00 GMT analysis as well as forecast validation up to 48-h forecast were downloaded from the NCEP site (ftp://ftpprd.ncep.noaa.gov/pub/data/nccf/com/gfs) during 25 April to 04 May, 2005.

3.2. Radiosonde/in situ

Since global field becomes interpolated from coarse resolution to mesoscale, the interpolated fields may not represent the actual mesoscale feature. Therefore, enhancement of global fields by local observations has to be done to derive realistic mesoscale simulations. The asynoptic data (wind, moisture and temperature profile) from radiosonde and microwave remote sensing is most useful for this purpose. In this study the local radiosonde data available at different stations in the domain is used to improve the initial fields. Figure 1 shows the stations from where the local data are incorporated in the model initial conditions. The local data were injected using Cressman objective analysis schemes of MM5. Daily 00 GMT upper air data were downloaded during 25 April to 04 May, 2005 from Wyoming University Website (http://weather.uwyo.edu/upperair/sounding.html). The 00 GMT SHAR local GPS data of wind profile over SHAR, provided in real time by SHAR met facility from 28 April onwards, was also included in the model initial conditions using an objective analysis technique. These data were of very high vertical resolution and taken only during satellite launch.

3.3. TRMM - Microwave - Infrared Merged Precipitation Data - 3B42

The Tropical Rainfall Measuring Missions (TRMM) was launched in November 1997 in a 350-km altitude 35° inclination orbit for the quantitative measurements of precipitation over the entire tropics on a continuous basis. The low inclination (35°) of the orbit permits documentation of the important diurnal rainfall cycle (KISHTAWAL and KRISHNAMURTI, 1998). In this study TRMM Microwave Imager (TMI) derived rainfall product 3B42RT is used for the period 25 April to 3 May, 2005. The algorithm 3B-42 produces TRMM-adjusted microwave merged - infrared (IR) precipitation (ADLER *et al.*, 2000). The precipitation product 3B42 uses the high quality precipitation estimates from TRMM as the calibrating mechanism for estimates from other geostationary satellite data. This scheme allows estimation of surface rain at finer time and space scales than is available from TRMM microwave imager. The gridded rainfall estimates are on a 3-hour temporal resolution and

$0.25°\times0.25°$ spatial resolution in a global belt extending from 50° South to 50° North latitude. The daily accumulated rainfall is obtained by combining eight, 3-hourly observations from TRMM.

4. *Forecast and Evaluation Method*

4.1. *Forecast Method*

The standard method of MM5 initialization is used, which is discussed in MM5 user's manual. GRIB formatted data on pressure levels are read and interpolated to MM5 grid. The next step is to assimilate the local radiosonde observations using Cressman objective analysis scheme of MM5. Then vertical interpolation is carried out to generate model initial, lateral and lower boundary conditions. Daily, from 25 April to 04 May 2005, the model was initialized with 00GMT for 48-h period integration using four different CPSs.

4.2. *Evaluation Method*

4.2.1. *Data comparison method*

We have selected 17 radiosonde locations (stations indicated by the symbol "♦" in Fig. 1) for point validation on the basis of the availability of data at additional vertical levels. Model forecasted temperature, wind and relative humidity were verified with these radiosonde observations. After the forecast calculation is completed, bilinear interpolation is conducted to obtain model forecast data at the radiosonde locations.

4.2.2. *Statistical parameters*

For calculating statistical parameters we selected the same radiosonde locations mentioned above. Mean bias (MB), root-mean-squares error (RMSE), and correlation coefficient were used to assess temperature, wind speed, and relative humidity forecast. The MB associated with model tendency to overforecast or underforecast the observed quantity (WILKS, 1995) is given by

$$MB = \frac{1}{N}\sum_{i=1}^{N}(o_i - f_i),$$

where N is number of forecast/observation pairs in the sample, f_i is the forecast, and o_i is the observation.

The precipitation forecasts by the four different CPSs are compared to the observations and are evaluated quantitatively calculating statistical skill scores (such as the threat and bias scores; ANTHES, 1983; ANTHES *et al.*, 1989) for different rainfall thresholds (0.1, 1, and 10 mm). BSs and ETSs are calculated from Table 1, which is

Table 1

Rain contingency table used for verification. Each element of the matrix (A, B, C, and D) holds the number of occurrence in which the observations and/or the model forecasts reach a precipitation threshold amount for a given forecast period

Forecasted	Observed	
	Rain	No rain
Rain	A	B
No rain	C	D

equivalent to a 2 × 2 matrix, where A, B, C, and D holds the number of occurrences, in which the observation and model did, or did not reach certain threshold amounts of precipitation for a given period of the forecast (24-h accumulated in this study). A detailed description of rain contingency tables and formulas used for calculating skill scores is available in YANG and TUNG (2003) and COLLE *et al.* (1999).

5. *Results*

5.1. *Point Validation of Wind, Temperature, and Moisture Field*

In this section the results of point validation of wind speed, temperature and relative humidity over some selected stations are presented. The 24-h and 48-h predicted wind, temperature and relative humidity profiles using different CPSs are compared with radiosonde over 17 selected stations (stations with "♦" in Fig. 1). For brevity, we are presenting only the result obtained for the 24-h forecast by different CPSs over SHAR valid at 00 GMT of 29 April (Figs. 2–4). From Figure 2a it is seen that model predicted westerly jet location by all the CPSs is slightly shifted downward (~10 km) compared to the radiosonde observed jet location (~11 km). The vertical trend in wind speed is well predicted by all the CPSs. The differences between model predicted and observed wind varies from −6 to 10 m/s. Among the CPSs Betts-Miller scheme differs more from observation in reproducing wind speed. Figure 3a indicates that model predicted temperature profile is well matching with the radiosonde profile. The differences between model predicted and radiosonde observed temperature ranges from −3 to 2°C (Fig. 3b). Compared to other CPSs the predicted temperature by Betts-Miller scheme differs more from the observation, particularly at lower levels (~1 km) and upper levels (~13 km). The predicted relative humidity (Fig. 4a) is always higher than the observed humidity in the upper levels and differences as high as 70% are observed.

We have taken 17 stations for the computation of statistical parameters (stations with "♦" in Fig. 1). The 24-h and 48-h predicted fields by different CPSs were compared with radiosonde observation and the statistical parameters such as correlation coefficient, RMSE and bias were calculated. The statistical parameters

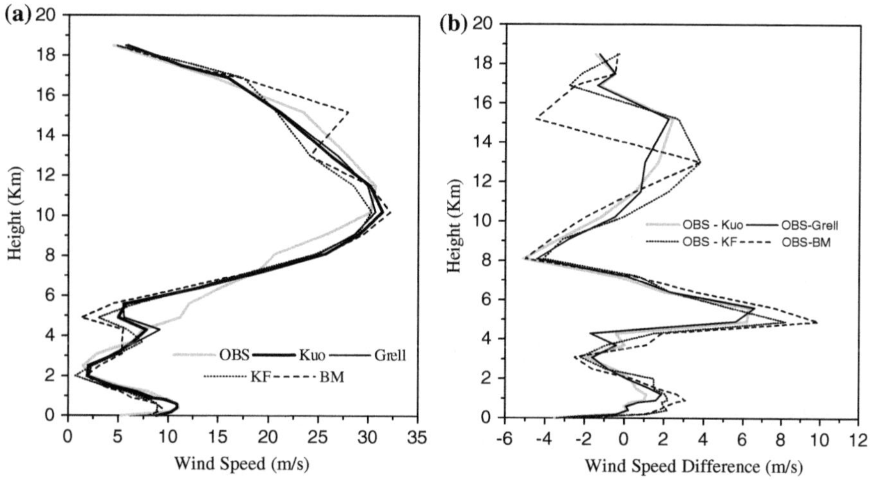

Figure 2
(a) Comparison of 24-h model predicted wind speed (m/s) with observation over SHAR valid at 00 GMT of 29 April, (b) difference between the observed and model predicted wind speed (m/s).

were computed for individual station as well as for individual height. These results are for all simulations carried out during 25 April to 04 May, 2005. All the simulations were of 48-h duration, except the simulation from 04 May, 2005, which was of 30-h duration. Thus for each height, there are at least 10 samples available in computing statistical parameters for a single variable at a particular station.

The vertical variation of RMSE and biases in the 24-h and 48-h predicted wind speed, temperature and relative humidity by different CPSs is depicted in Figs. 5–10. Wind speed (Figs. 5, 6) is underestimated by all CPSs except in the upper levels. There is a sharp overestimation in the 24-h (Fig. 7) and 48-h (Fig. 8) predicted temperature by all the CPSs at heights 14–16 km. The forecast by all the CPSs overestimates humidity (Figs. 9, 10) except at lower levels. The 24-h and 48-h forecast by the Grell scheme shows less rms error and bias particularly in the upper levels. Compared to other CPSs, error in the predicted humidity is less for the Kain-Fritsch scheme in the lower troposphere. Among the CPSs the predicted wind, temperature and humidity fields by the Betts-Miller scheme differ more from observation.

From the overall statistics obtained for all vertical levels from 24-h and 48-h forecast by different CPSs based on the 17 selected stations, it is clear that the RMSE of wind speed, temperature and relative humidity show no significant differences among the four CPSs. Compared to other CPSs, the Grell scheme shows slightly less RMSE for wind speed, temperature and relative humidity in both 24-h and 48-h

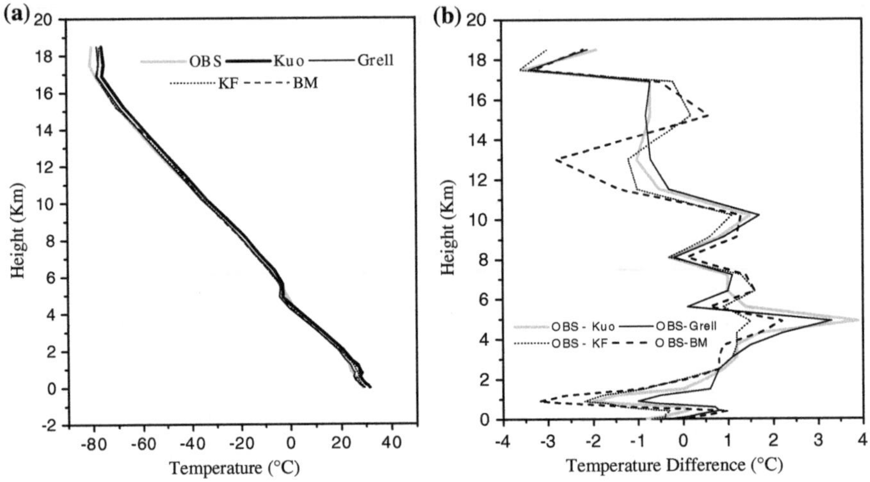

Figure 3
As in Figure 2, but for temperature (°C).

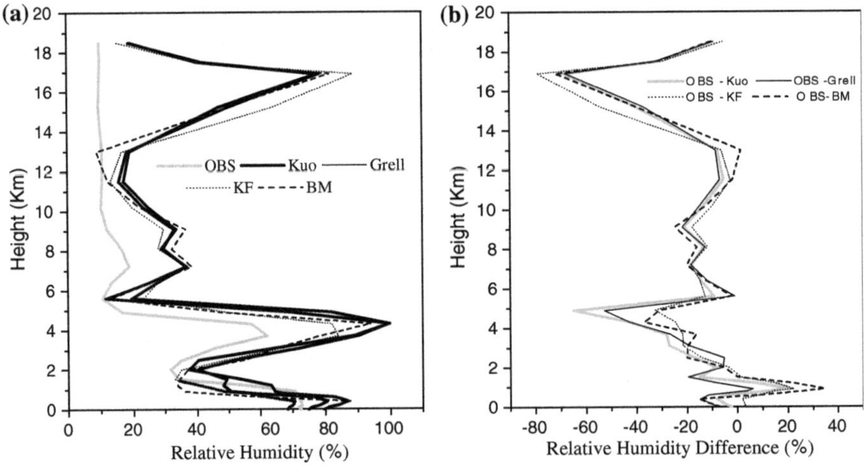

Figure 4
As in Figure 2, but for relative humidity (%).

prediction. From the overall statistics it is observed that all the CPSs overestimate temperature and relative humidity in the 24-h and 48-h prediction, while wind speed is underestimated.

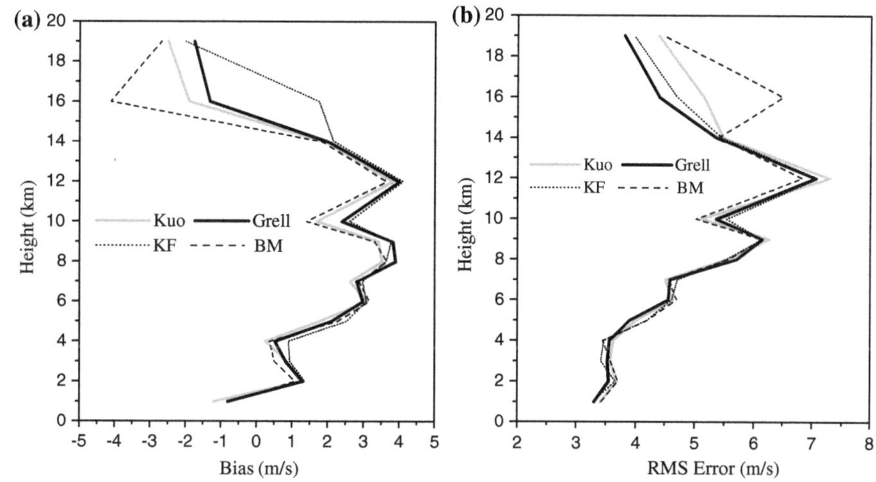

Figure 5
Vertical profile of, (a) bias, (b) RMS error, in the model predicted wind speed (m/s) obtained by all 24-h forecasts over selected stations during the period of study.

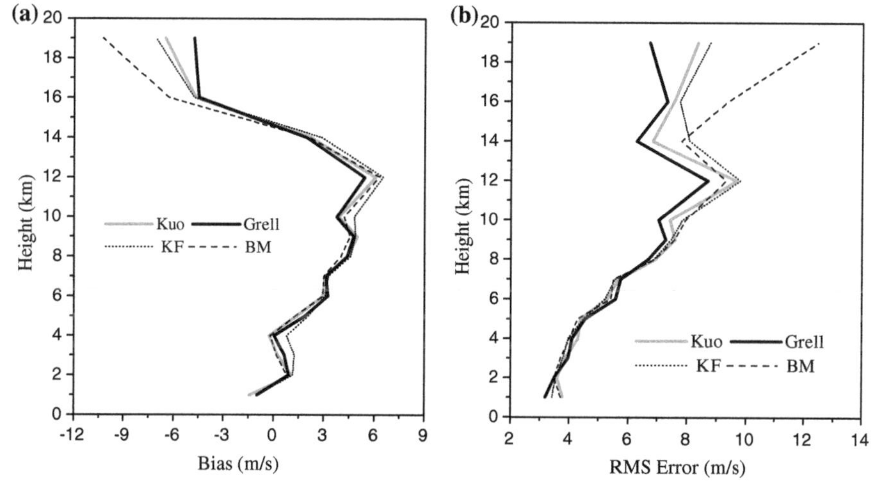

Figure 6
As in Figure 5, but for 48-h forecast.

5.2. Statistical Validation of Rainfall

Twenty four hours predicted rainfall from each simulation is compared with the TRMM observed rainfall during the study period. For evaluating the predictive skill

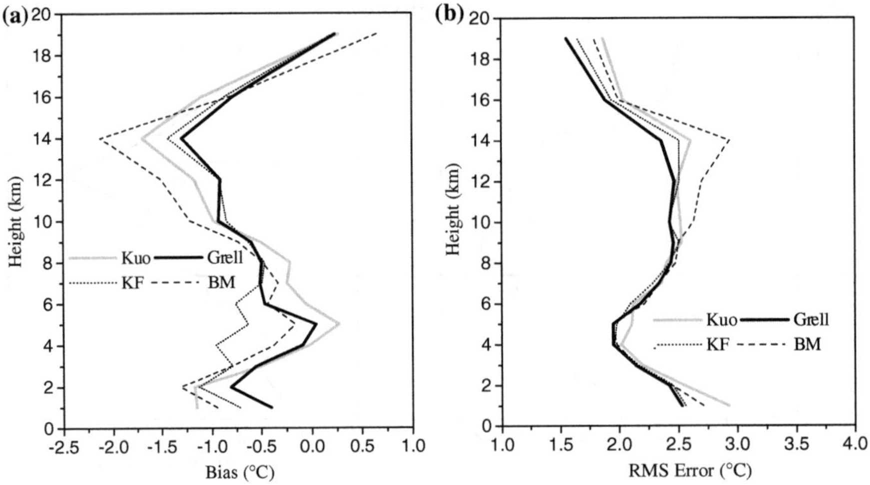

Figure 7
Vertical profile of, (a) bias, (b) RMS error, in the model predicted temperature (°C) obtained by all 24-h forecasts over 17 selected stations during the period of study.

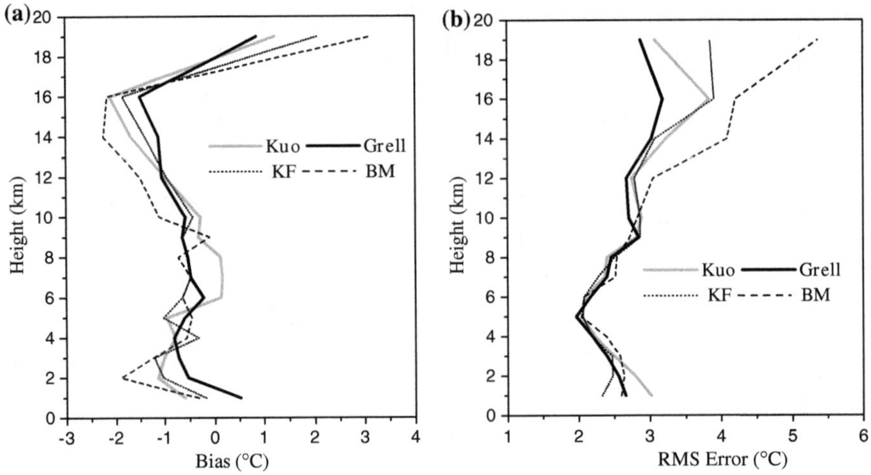

Figure 8
As in Figure 7, but for 48-h forecasts.

of different CPSs in simulating an area of rainfall, Bias scores (BSs) and Equitable threat scores (ETSs) are computed for different rainfall thresholds (0.1,1 and 10 mm). The number of grid points observed in each of the rainfall thresholds is

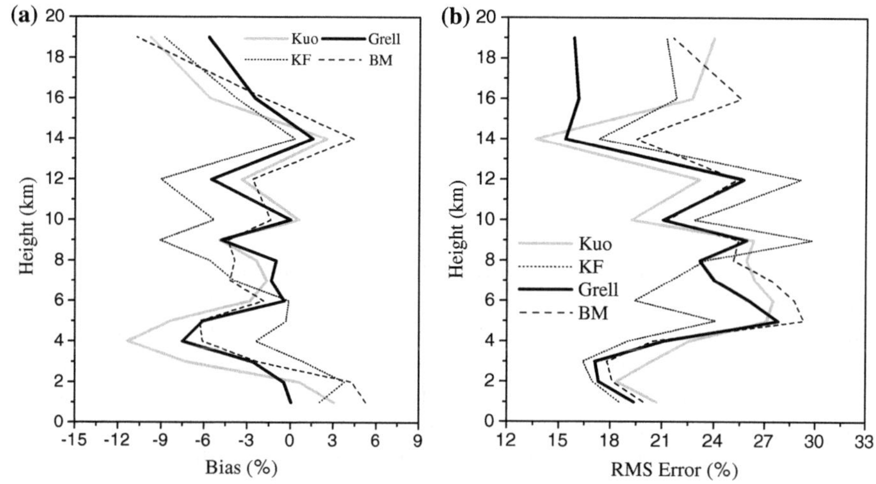

Figure 9
Vertical profile of, (a) bias, (b) RMS error, in the model predicted relative humidity (%) obtained by all 24-h forecasts over 17 selected stations during the period of study.

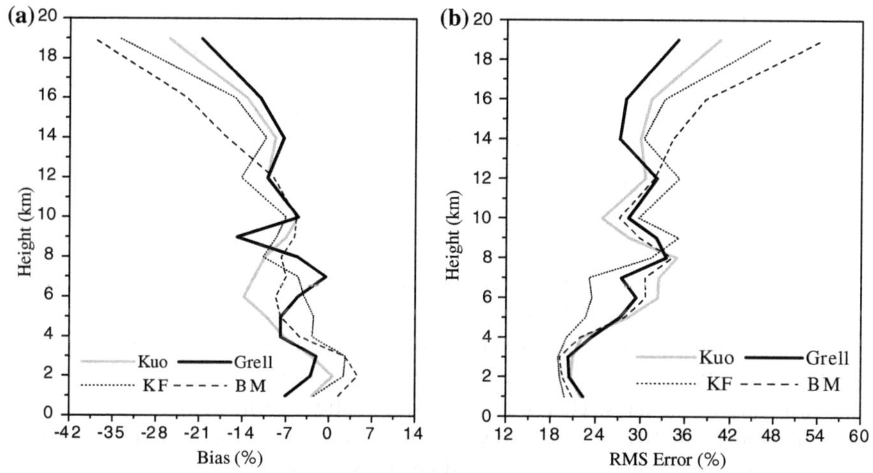

Figure 10
As in Figure 9, but for 48-h forecasts.

shown in Table 2. Figure 11 shows the BSs obtained for 24-h prediction by different CPSs at different thresholds. It is clear that the forecast given by all the CPSs estimate/underestimate the rainfall area for light/heavy precipitation. The forecast

Table 2

The number of grid points in each of the rainfall thresholds considered for computing the skill scores

Cases	1	2	3	4	5	6	7	8	9
> = 0.1<1	1295	1250	1170	1288	1636	1095	1310	1312	1299
> = 1<10	1551	1431	1749	1909	2234	1580	2119	1872	1908
>10	0922	0498	1021	0716	0546	0697	1396	1224	1325

using Betts Miller scheme always lies closer to the observation. The ETSs obtained at different rainfall thresholds are shown in Figure 12. The low averaged ETSs by all the CPSs for the area distribution of precipitation demonstrate that considerably more improvement is needed both in the model resolutions and the assumptions made in the formulation of different CPSs for better forecast skill. In the predictive skill for the area of light precipitation (0.1 mm/day) the Betts-Miller scheme has a clear edge over the other CPSs (Fig. 12a). This may be due to the incorporation of shallow convection with the normal deep convection in this scheme where parameterized shallow convective elements rise without precipitation formation through the top of the boundary layer into the free atmosphere. In case of moderate rainfall (1 mm/day), Grell scheme demonstrates better forecast skill in the majority of the cases (Fig. 12b) and for heavy rainfall (10 mm/day) CPSs give mixed results (Fig. 12c). The least predictive skill shown by Kuo for light and moderate precipitation asserts that this scheme is more suitable for larger grid scale (> 30 km). The performance of the CPSs in predicting rainfall area varies with the rainfall thresholds for which they have used. The predictive skill of CPSs in simulating area of rainfall is more (less) for low (high) thresholds, which is consistent with the earlier studies (WANG and SEAMAN, 1997; YANG and TUNG, 2003).

5.3. Synoptic Validation of Rainfall

The simulated rainfall features reproduced by different CPSs during the forecast period are validated against TRMM observed rainfall throughout the study period. There were light rainfall events over the Indian region throughout the study period. Significant rainfall events were observed over the southeast Indian coast on 29 April and two subsequent days. Heavy rainfall with thunderstorm activity was observed near the launch site SHAR on 1 May. The first 24-h forecasted rainfall by different CPSs is compared with the corresponding TRMM observation for all the days during the time period. For brevity we are presenting only the rainfall features obtained for 1 May. The observed and model simulated accumulated rainfall using different CPSs for 1 May is presented in Figures 13 and 14. It is clear, as seen in the previous section, that except in the case of Betts-Miller scheme, MM5 simulations using the CPSs overestimate low rainfall while heavy rainfall is underestimated. Among the CPSs Kuo produces the worst simulation, which is similar to the results

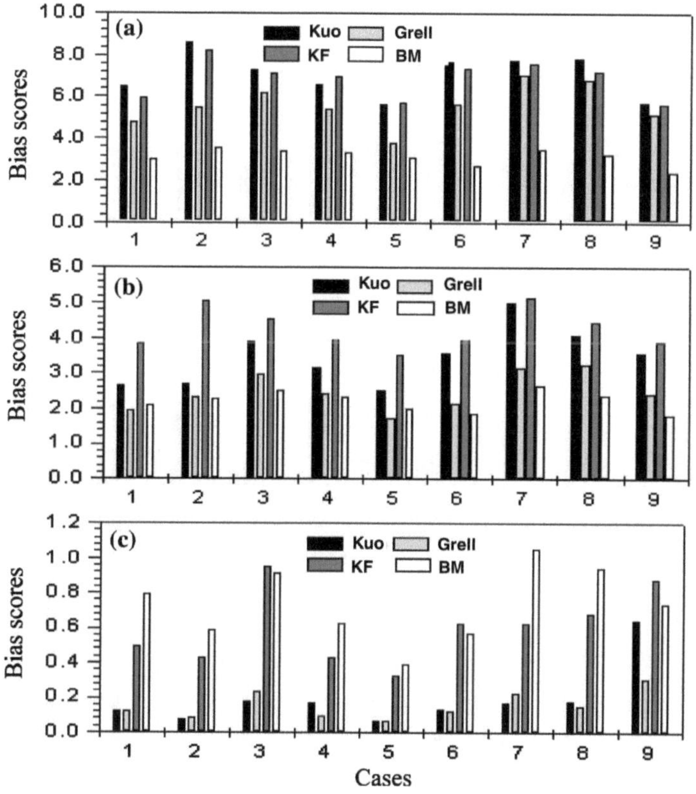

Figure 11
Bias scores (BSs) at different thresholds (a) 0.1, (b) 1 and (c) 10 obtained for the first 24-h prediction of accumulated rainfall from 9, 48-h forecasts (from 25 April to 3 May).

obtained in 5.2. This indicates that this scheme is minimally useful in model simulation with high resolution. The criteria used in this scheme that large-scale moisture convergence associated with a high degree of conditional instability is required for the activation of CPS, may be one of the reasons for the worst simulation. The underestimation of rainfall shown by Betts-Miller over the Indian land mass may be due to the basic assumption used in this scheme such that adjusting the grid-column sounding towards a reference sounding in order to resemble the quasi-equilibrium thermodynamic state associated with deep convection over the region. Even though none of the CPSs simulated rainfall features correctly, the simulation by the Kain–Fritsch scheme lies closer to the observed one.

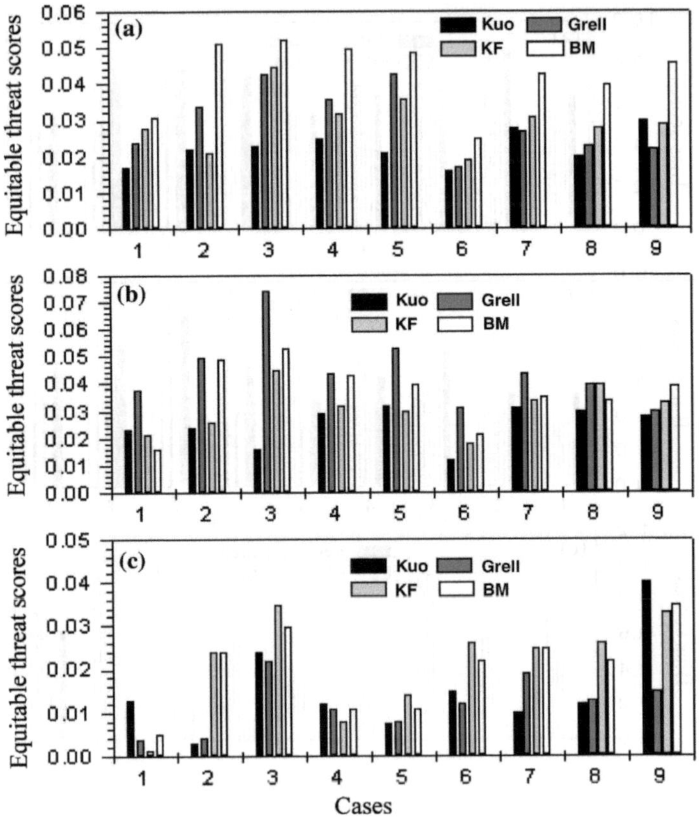

Figure 12
As in Figure 11, but for equitable threat scores (ETSs).

6. Discussion and Conclusions

The mesoscale model MM5 has been integrated during pre-launch and up to PSLV-C6 launch from SHAR. The model initial condition is enhanced by available radiosonde profiles from nearby land stations. The model predicted wind, temperature and moisture for different cumulus schemes were compared with local radiosonde observations over selected stations for 24-h and 48-h predictions. The statistical parameters (RMS error, bias and correlation coefficient) were calculated for different CPSs to evaluate the model forecast. The analysis indicates that all the CPSs overestimate temperature and relative humidity but underestimate the wind speed. It was found that model simulated wind speed and temperature by all the CPSs are in reasonable agreement with that of radiosonde observation. The downward shifting of the position of westerly jet in the simulated wind field can be due to the model's failure in capturing the proper terrain features. The model simulation displays differences in the predicted circulation features among the CPSs throughout the model domain for 00GMT 29 April, 2005. During that period rainfall activity prevailed only over the southeast Indian coast (Figure not shown). This demonstrates that representation of convection in mesoscale models has a marked influence not only on the simulated precipitation over that region, but also on the simulated circulation pattern and moisture fields over nearby regions. That is, a light rainfall at one place in the domain can change the wind and moisture fields at another place. Also the light rainfall observed in all the cases over some locations in the domain can result from CPS activation. These results are consistent with the previous studies (STENSRUD, 1996; GOCHIS *et al.*, 2002).

Among the CPSs, the Grell scheme gives slightly better forecasts of wind, temperature and moisture fields particularly in the upper levels which can be due to the assumption that no direct mixing is allowed between updraft and downdraft and with the surrounding atmosphere, and thereby not disturbing conserved mass flux in a vertical column of cloud. Which suggests that representing cloud entrainment/detrainment may not necessarily be a beneficial assumption in tropical regions. The low error shown by the Kain–Fritsch scheme in the predicted relative humidity, particularly at lower levels can be attributed to the explicit treatment of convective downdrafts by this scheme. The high error shown by Betts-Miller scheme in reproducing temperature and moisture field may be ascribed to the absence of explicit parameterization of subgrid-scale cloud and mesoscale processes in this scheme. The forecasts given by all the CPSs overpredicted (underpredicted) the area of low (high) precipitation. A similar result is also obtained for GULLUS (1999) and

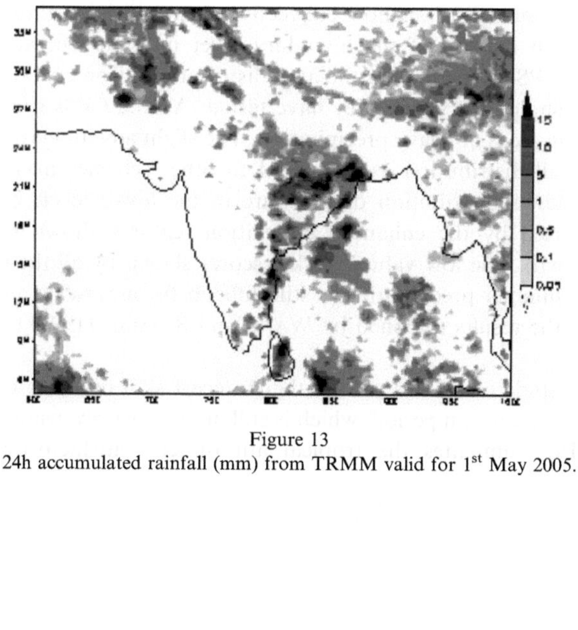

Figure 13
24h accumulated rainfall (mm) from TRMM valid for 1st May 2005.

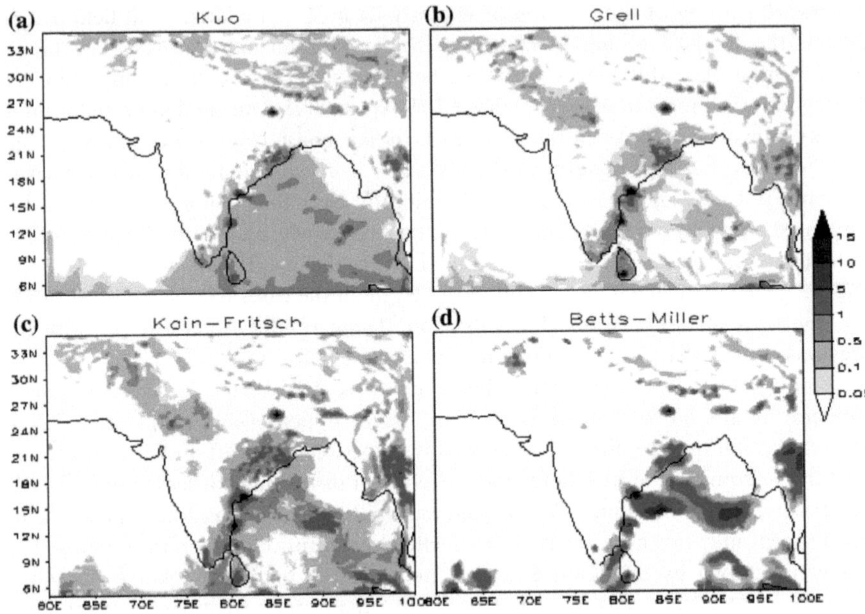

Figure 14
24-h forecasts of accumulated rainfall (mm) for different CPSs valid at 00 GMT of 1 May, 2005.

YANG and TUNG (2003). In predictive skill for the area of light precipitation, Betts-Miller scheme has a clear edge over other CPSs, which can be due to the incorporation of shallow convection with the normal deep convection in this scheme. In such cases parameterized shallow convective elements rise without precipitation formation through the top of the boundary layer into the free atmosphere. The performance of CPSs in predicting the area distribution of precipitation varies with the rainfall thresholds for which they have tested. All the CPSs show less forecast skill in simulating areas of heavy precipitation. The slight advantage of KF scheme in simulating rainfall distribution as compared to other schemes may be due to the comparatively better distribution of moisture in the lower levels by this scheme, which is supported by the enhanced circulation features shown by this scheme (Figure not shown). The low values of skill scores shown by all the CPSs indicates that predicting summer precipitation is still difficult for mesoscale models, which is consistent with the results obtained by WANG and SEAMAN (1997) over the United states.

The present study elucidates the weather forecast skill of MM5 over the Indian region for the pre-monsoon period, which is still not extensively evaluated. The study shows that MM5 simulates the tropical atmosphere satisfactorily even though predicted humidity differs highly with the observation. The study substantiates the sensitivity of simulated meteorological features by the model to the way in which the convective processes in the atmosphere are parameterized in the model. Although there are some common features, the comparative performance of the different schemes differs throughout the simulation period. Even though these differences can be attributed to the different type of assumptions used in these schemes, their sensitivity to different cases (different synoptic situation and conditions) makes their application in mesoscale models highly complicated. The characterization of the model errors for different CPSs provides more confidence on the model outputs and qualifies which CPSs are to be used for better forecasts. Cases of good/bad skill scores can be isolated and clustered into weather systems to identify the atmospheric structures that cause difficulties to the forecasts. The evaluation of MM5 model for different CPSs conducted during this study is only for a particular synoptic situation, more detailed study however is required to assess the forecast skill of the CPSs for different synoptic conditions.

Acknowledgements

The globally analyzed data provided by NCEP/AVN in real time is acknowledged with thanks. The radiosonde data provided by the University of Wyoming are also acknowledged. The authors also thank Director SAC, for his keen interest in this study. The discussions with Dr. G. V. Rama and Sri. Appa Rao from the meteorological facility of SDSC, SHAR were very useful in conducting the present studies. The authors are grateful to the anonymous reviewers for their constructive criticism, which enhanced the presentation of this study. The first author (RV) acknowledges the award of research fellowship by UGC.

References

Adler, R. F., Bolun, D. T., Curtis, S., and Nelkin, E. J. (2000), *Tropical rainfall distributions determined using TRMM combined with other satellite and rain-gauge information*, J. Applied Meteor. *39*, 2007–2023.

Anthes, R.A. (1977), *A cumulus parameterization scheme utilizing a one- dimensional cloud model*, Mon. Wea. Rev. *105*, 270–286.

Anthes, R.A (1983), *Regional models of the atmosphere in middle latitudes*, Mon. Wea. Rev. *111*, 1306–1335.

Anthes, R.A. and Warner, T. T. (1978), *Development of hydrodynamical models suitable for air pollution and other mesometeorological studies*, Mon. Wea. Rev. *106*, 1045–1078.

Anthes, R.A., His, E.-Y., and Kuo, Y.H. (1987), *Description of Pen State/NCAR Mesoscale model version 4(MM4)*. NCAR/TN-282 + STR, National Center for Atmospheric Research.

Anthes, R.A, Kuo, Y.-H., Hsie, E.-Y., Low-Nam, S., and Bettge, T.W. (1989), *Estimation of skill and uncertainty in regional numerical models*, Quart. J. Roy. Meteor. Soc. *115*, 763–806.

ARAKAWA, A. and SCHUBERT, W.H. (1974), *Interaction of a cumulus cloud ensemble with the large-scale environment. Part I,* J. Atmos. Sci. *31,* 674–701.

BETTS, A.K. and MILLER, M.J. (1986), *A new convective adjustment scheme;Part II: single column tests using GATE wave, BOMEX,ATEX, and arctic air-mass data* set, J. Roy. Meteor. Soc. *112*, 693–709.

BETTS, A.K. and MILLER, M.J. (1993), *The Betts-Miller scheme. The representation of cumulus convection in numerical models* (K.A. Emanuel and D. J. Raymond, eds.), (Am. Met. Soc. 1993) 246 pp.

COLLE, B.A., WESTRICK, K.J., and MASS, C.F. (1999), *Evaluation of MM5 and Eta-10 precipitation forecasts over the Pacific northwest during the cool season,* Wea. Forecasting *14*, 137–154.

DAS, S., MITHRA, A.K., IYENGAR, G.R., and SINGH, J. (2002), *Skill of medium-range Forecasts over the Indian region using different parameterizations of deep cumulus convection,* Wea. Forecasting *17,* 1194–1210.

DAS GUPTA, M., ASHRIT, R., GEORGE, J. P., DAS, SOMESWAR, and BOHRA, A.K. (2005), *Validation of MM5 model over Indian region during Monsoon 2004, WRF/MM5* User's Workshop, June 2005.

DUDHIA, J. (1989), *Numerical study of convection observed during the winter monsoon experiment using mesoscale two-dimensional model,* J. Atmos. Sci. *46*, 3077–3107.

DUDHIA, J. (1993), *Nonhydrostatic version of Penn State –NCAR mesoscale model validation tests and simulation of an Atlantic cyclone and cold front,* Mon. Wea. Rev. *121*, 1493–1513.

DUDHIA, J. (2005), *MM5 Version 3.7 (The Final Version),* WRF/MM5 User's Workshop. IIT Delhi, India.

FRITSCH, J.M. and CHAPPEL, C.F. (1980), *Numerical prediction of convectively driven mesoscale pressure system. Part I: Convective parameterization,* J. Atmos. Sci. *37*, 1722–1732.

GALLUS, W. A., Jr. (1999), *Eta simulations of three extreme precipitation events: Sensitivity to resolution and convective parameterization,* Wea. Forecasting *14*, 405–426.

GOCHIS, D.J., SHUTTLEWORTH, W.J., and YANG, Z.L. (2002), *Sensitivity of the modeled North American monsoon regional climate to convective parameterization,* Mon. Wea. Rev. *130*, 1282–1298.

GRELL, G.A (1993), *Prognostic evaluation of assumptions used by cumulus parameterizations,* Mon. Wea. Rev. *121*, 764–787.

GRELL, G.A., DHUHIA, J., and STAUFFER, D.R. (1994), *A description of the fifth-generation Penn State/ NCAR mesoscale model (MM5),* NCAR/ Tn-398 + STR, 117 pp.

HACK, J.J, BOVILLE, B.A., BRIEGLEB, B.P., KIEHL, J.T., RASCH, P.J., and WILLIAMSON, D.L. (1993), *Description of the NCAR community climate model (CCM2),* NCAR Tech. Note NCAR/TN-382 + STR, 108 pp.

HONG, S. Y., and Pan, H. L. (1996), *Nonlocal boundary layer vertical diffusion in a medium range forecast model,* Mon. Wea. Rev. *124*, 2322–2339.

JANJIC, Z.I. (1994), *The step-mountain eta coordinate model: Further developments of the convection, viscous sublayer, and turbulence closure schemes,* Mon. Wea. Rev. *122*, 927–945.

KAIN, J. S. and FRITSCH, J.M. (1990), *A one-dimensional entraining /detraining plume model and its application in convective parameterization,* J. Atmos. Sci. *47*, 2784–2802.

KAIN, J. S. and FRITSCH, J.M., *Convective parameterization for mesoscale models.* The Kain-Fritsch scheme. *The representation of cumulus convection in numerical models,* (K. A. Emanuel and D. J. Raymond, eds.), (Amer. Meteor. Soc. 1993) 246 pp.

KISHTAWAL, C.M. and KRISHNAMURTI, T.N. (1998), *TRMM: a user's manual for modelers,* Available at Florida State University. 31 pp.

KUO, H.L. (1965), *On the formation and intensification of tropical cyclone through latent heat release by cumulus convection,* J. Atoms. Sci. *22*, 40–63.

KUO, H.L. (1974), *Further studies of the parameterization of the influence of cumulus convection on large-scale flow,* J. Atoms. Sci. *31*, 1232–1240.

KUO, H.L., REED, R.J., and LIU, Y.-B. (1996), The ERICA IOP 5 storm. Part III: *Mesoscale cyclogenesis and precipitation parameterization,* Mon. Wea. Rev. *124*, 1409–1434.

MOLINARI, J. and DUDEK, M. (1992), *A critical review,* Mon. Wea. Rev. *120*, 326–344.

PENG, X. and TSUBOKI, K. (1997), *Impact of convective parameterizations on mesoscale precipitation associated with the Baiu front,* J. Meteor. Soc. Japan *75*, 1141–1154.

SINGH, R., PAL, P.K., KISHTAWAL, C.M., and JOSHI, P.C. (2005), *Impact of bogus vortex for track and intensity prediction of tropical cyclone,* J. Earth System Sci. *114*, 427–436.

STENSURD, D. J. (1996), *Effects of persistent, mid-latitude mesoscale regions of convection on large-scale environment during the warm season,* J. Atmos. Sci, *53*, 3503–3527.

WANG, W. and SEAMAN, N.L. (1997), *A comparison study of convective parameterization schemes in a mesoscale model,* Mon. Wea. Rev. *125*, 252–278.

WILKS, D.S., *Statistical Methods in the Atmospheric Sciences (Academic Press 1995)*, 467 pp.

YANG, M.-J., CHIEN, F.-C., and CHENG, M.-D. (2000), *precipitation parameterizations in a simulated Mei-Yu front,* Terr. Atmos. and Oceanic Sci. *11*, 393–422.

YANG, M.-J., and TUNG, Q.C. (2003), *Evaluation of rainfall forecasts over Taiwan by four cumulus parameterization schemes,* J. Meteor. Soc. Japan, *81*, 1163–1183.

(Received March 27, 2006, accepted October 5, 2006)
Published Online First: June 23, 2007

C. Seasonal Monsoon

Pure appl. geophys. 164 (2007) 1641–1665
0033–4553/07/091641–25
DOI 10.1007/s00024-007-0242-3

Pure and Applied Geophysics

Experimental Seasonal Forecast of Monsoon 2005 Using T170L42 AGCM on PARAM Padma

J. Venkata Ratnam,[1] D. R. Sikka,[2] Akshara Kaginalkar,[1] Amit Kesarkar,[1] N. Jyothi,[1] and Sudipta Banerjee[1]

Abstract—As a part of the Experimental Extended Range Monsoon Prediction Experiment, ensemble mode seasonal runs for the monsoon season of 2005 were made using the National Centre for Environmental Prediction (NCEP), T170L42 AGCM. The seasonal runs were made using six initial atmospheric conditions based on the NCEP operational analysis and with forecast monthly sea-surface temperature (SST) of the NCEP Coupled forecast system (CFS). These simulations were carried out on the PARAM Padma supercomputer of Centre for Development of Advanced Computing (C-DAC), India. The model climatology was prepared by integrating the model for ten years using climatological SST as the lower boundary. The climatology of the model compares well with the observed, in terms of the spatial distribution of rainfall over the Indian land mass.The model-simulated rainfall compares well with the Tropical Rainfall Measuring Mission (TRMM) estimates for the 2005 monsoon season. Compared to the model climatology (7.81 mm/day), the model had simulated a normal rainfall (7.75 mm/day) for the year 2005 which is in agreement with the observations (99% of long-term mean). However, the model could not capture the observed increase in September rainfall from that of a low value in August 2005. The circulation patterns simulated by the model are also comparable to the observed patterns. The ensemble mean onset is found to be nearer to the observed onset date within one pentad.

Key words: Monsoon seasonal forecast, PARAM Padma, T170L42, CFS SST.

1. Introduction

The long-range forecasting of the Indian Summer Monsoon, using dynamical models, is a challenge faced by the meteorological community as research has shown that the simulation of the monsoon rainfall of India is a very tough problem. The dynamical seasonal forecasting, if successful, would provide a wonderful tool for researchers and operational meteorologists to forecast the Indian Summer Monsoon Rainfall (ISMR) which is of value to the applications for community and national agencies. There have been many notable studies undertaken to understand the

[1] Centre for Development of Advanced Computing, Pune university campus, Ganeshkhind, Pune, 411 007, India. E-mail: jvratham@cdac.in
[2] 240 Mausam Vihar, New Delhi, India.

dynamical processes of the Indian Monsoon (RAMAGE, 1971; RAO, 1976; SIKKA and GADGIL, 1980; YASUNARI, 1980; CHANG and KRISHNAMURTI, 1987) and to improve the forecast skill of the dynamical models (KRISHNAMURTI *et al.*, 2000a, 2000b, 2001; KRISHNAMURTI, 2005 and others). The relatively poor skills of the numerical models can be attributed to the limitations of the physical parameterization schemes in the models and also to the low resolution of the models used in several earlier attempts at the monsoon simulation. SPERBER *et al.* (1994) found that an increase in resolution had a positive impact on the simulation of the Indian Monsoon. JHA *et al.* (2000), using Florida State University Atmospheric General Circulation Model (AGCM) at resolutions of T42 and T170 for the month of July, found that the higher resolution model simulated the monsoon features more realistically. The improvement in the monsoon precipitation in the high-resolution models is attributed to the improvement of the regional rainfall resulting from better resolution of the topographical features of the Indian region.

This paper presents an ensemble forecast of ISMR 2005, carried out experimentally by the Centre for Development of Advanced Computing (C-DAC) under the Extended Range Monsoon Prediction (ERMP) initiative of the Indian Climate Research Program (ICRP) of the Department of Science and Technology (DST), India. The main objectives of this program are to evaluate and to improve upon the AGCMs being used by the research community in India for the long-range monsoon forecasting. The simulations reported here were carried out on C-DACs supercomputer PARAM Padma using the NCEP T170L42 AGCM. PARAM Padma is a distributed memory supercomputer with 256 IBM processors connected with an indigenously developed PARAMNet switch. The computer has a peak performance of one Teraflop. More details about the machine are available at http://www.cdac.in/html/ctsf/ctsfidx.asp. The primary aim of this attempt was to test the use of a high resolution AGCM for foreshadowing the monsoon season (June-September) rainfall over India in advance through SST forcing and to evaluate its performance at the end of the season. Therefore, it necessitated using initial conditions at the beginning of May with the forecast SST provided by a coupled ocean-atmosphere climate model. The option used in this study for the forecast SST for May, June, July, August and September 2005 was the data provided by the NCEP Coupled Forecast System (CFS) (SAHA *et al.*, 2006) and the initial conditions used were those provided by the operational NCEP analysis for the six member ensemble.

The observed monsoon of 2005 showed some interesting features with regard to its evolution on subseasonal scale. The onset of the monsoon over the southwestern coast of India (Kerala State) took place on 5–6 June, 2005. Its progress along the West Coast was arrested between 7–17 June, followed by rapid progress to 23°N within the next week. It advanced over the Gangetic Plain rather rapidly between 20–27 June such that the entire country was in its sweep by 30 June. The performance of the monsoon on the monthly scale in terms of the observed rainfall for India as a whole was below normal for June (−12%) and August (−28%), but for July and September it was above normal by 14% and 17%, respectively. For the season as a whole the observed rainfall of the country was 99% of the long-period normal. On the subseasonal scale the All India rainfall for 2005 was above normal between 17 June to 6 August and it was mostly below normal for the month of August 2005 (LAL *et al.*, 2006). The mid-season deficit in rainfall between 6 August to 3 September was made up by the revival of the monsoon, resulting in excess rainfall between 5–23 September. These were the most conspicuous features of the observed performance of the monsoon 2005 in terms of rainfall. In all, one cyclonic storm, five monsoon depressions, and six low pressure areas formed in the season, which approaches normal for the season. Of these, three depressions and three low pressures areas formed in June and July 2005 and one cyclonic storm, two depressions, and two low pressure areas formed during 7 to 24 September in an overlapping manner. Hence June, July and September witnessed good cyclogenic activity, while in August the situation remained rather quiet and weak monsoon conditions prevailed.

In section 2 the model description in brief and the data used for the model initialization are presented. A discussion of the model precipitation climatology, forecast of monsoon 2005 and the simulation of the large-scale episodes of the monsoon are presented in section 3. Summary and concluding remarks are given in section 4.

2. *Model Description and Initialization*

The NCEP T170L42 AGCM was used for making the seasonal forecasts of monsoon for the year 2005. The model has 512×256 horizontal Gaussian grid points and 42 vertical sigma levels. The physics used in the model was the Simplified Arakawa Schubert scheme (PAN and WU, 1995) for convection, Rapid Radiation Transfer Model (RRTM) (MLAWER *et al.*, 1997) for the longwave radiation, and the parameterization of HOU *et al.* (2002) for the short-wave radiation. Detailed model documentation can be found at http://www.emc.ncep.noaa.gov/gmb/moorthi/gam.html.

An ensemble of the model seasonal runs was made using six initial conditions of the 1, 2, 3, 5, 6 and the 7 May, 2005. The model was integrated from the above six initial conditions up to 30 September. The analysis files of the NCEP Global Forecast System (of T254L64 resolution converted to T170L42) were used for the initial data. The monthly forecast sea-surface temperature (SST) of the NCEP Coupled Forecast System (CFS) was used as the lower boundary. We followed the methodology suggested by SAHA (2004) for removing the biases, if any, in preparing the ensemble mean CFS SST data from the member ensembles of 1 May, 2005. Our purpose in this study is not to determine the impact of SST bias on the monsoon rainfall simulation *vis-à-vis* the simulations under the climatological SST. We use the

Table 1

Monthly and seasonal ensemble mean rainfall (mm/day) as simulated for 2005 monsoon with CFS SST, climate SST and observed SST

	June	July	August	September	Seasonal
CFS SST	8.71	10.60	7.10	4.76	7.81
Climate SST	6.68	9.47	7.73	5.18	7.29
Observed SST	7.51	9.58	7.59	4.32	7.27

CFS forecast SST merely to provide viable values at the lower boundary prior to the season (with May initial conditions) to be utilized for operational long-range forecasting needs. Therefore, we believe that biases, if any, in the evolution of the monthly SST in the CFS forecasts would not be a serious handicap in our results. However, we ran the same ensemble with the observed SST, which were available after the hindcast and also with the climatological SST to understand the role played by the SST on the ensemble mean rain simulation for the 2005 monsoon season. Table 1 gives the comparison of the rainfall simulation for the monthly and seasonal rainfall with these three sets of SSTs. It is observed that there are differences in the monthly average rainfall such that the rainfall for climate SST is the lowest of the three in June and July. It is highest of the three in August and September. The rainfall simulated by the CFS SST showed the reverse behavior in June and July. We shall revert back to it in section 3. Hence the SST bias, if any, in the CFS SST apparently did not play a major role in the rainfall simulation over India. We show in Figure 1 the difference between the Reynolds (REYNOLDS *et al.*, 1994) observed SST and the CFS forecast SST for the months of May–Sep. 2005. The figure shows that the CFS SST is cooler over most parts of the equatorial ocean throughout the season. However, the CFS forecasts underestimated the SST by 1–2°K over parts of the Indian Ocean at isolated places. Comparison of the CFS SST with the climatological SST showed that the CFS SST was warmer by 0.5°K over most parts of the Indian Ocean.

The climatology of the model was generated by integrating the model for ten years using the climatological SST. Work is planned to obtain climatology of the model with observed SST. The discussion on ensemble forecast of monsoon 2005, precipitation climatology of the model and the inter-ensemble variability is based on simulated five-day averaged model outputs. The model simulated monthly and seasonal precipitation for the year 2005 is also compared with the Tropical Rainfall Measuring Mission (TRMM) (SIMPSON *et al.*, 1988) 3-hourly rainfall (3B42 V6) estimates averaged for the whole monsoon season. The climatological rainfall is also compared with the monthly long-term mean XIE-ARKIN (1997) (XA) precipitation. The simulated winds are compared with the NCEP-NCAR Reanalysis by interpolating the model output to the reanalysis grid.

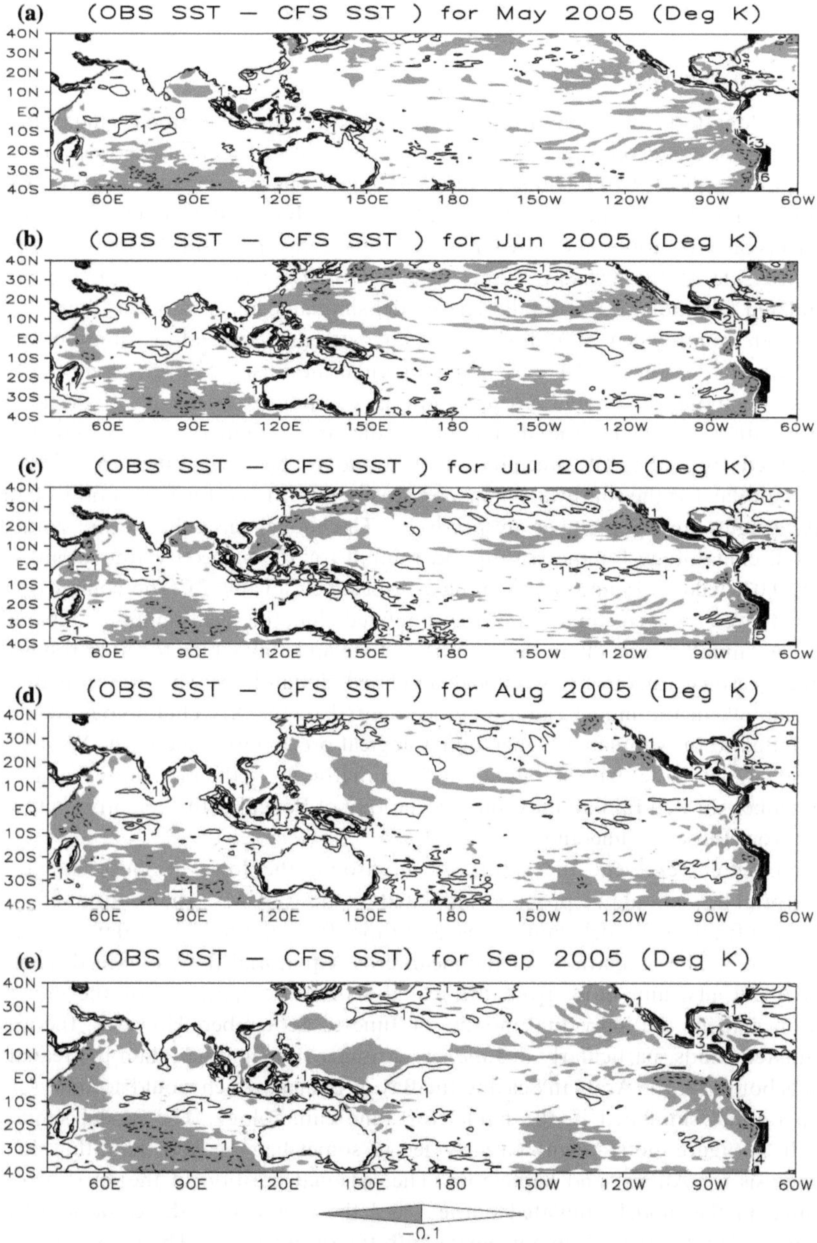

Figure 1
Differences between the observed SST and the CFS forecast SST. Negative values are shaded. Contour interval 1°K.

3. Results and Discussion

a) Model Climatology

The long-term mean monthly XA precipitation and the model climatological precipitation based on a ten year average are shown in Figure 2. The seasonal XA precipitation shows two maxima in the rainfall distribution, one over the West Coast of India and the other near northeast India and the Bay of Bengal. Qualitatively the model is able to capture these two maxima realistically. The rain shadow zone on the leeward side of the Western Ghats of India is also well captured in the model simulated climatological rainfall. The model-simulated mean monthly rainfall and the mean monthly XA rainfall show similar patterns of spatial distribution for the four months season (JJAS) as well as individual months. However, along the West Coast of India the model precipitation is comparatively higher than the XA precipitation. Over the Indian landmass, the area-weighted rainfall climatology computed from the XA precipitation is found to be 6.07, 5.04, 7.62, 6.93 and 4.69 mm/day for the southwest monsoon season (JJAS) and for the months of June, July, August and September, respectively. The model ensemble mean simulated rainfall climatology for the season and rainfall for the respective four months are 7.75, 9.02, 9.00, 7.91 and 5.06 mm/day, respectively (Table 2). The comparison of both showed an overestimation of the simulated climatological rainfall compared to the XA rainfall, especially in the months of June and July which together resulted in higher seasonal rainfall by the model simulated climatology. The overestimation in the rainfall, in the model climatology compared to the XA climatology, could be attributed to the simulated higher climatological precipitation along the West Coast of India. Also, note that the rainfall in the climatology of the model for August and September showed rainfall maximum in the Bay of Bengal with a slightly southward shift from those of June and July.

To better represent the southwest monsoon, the lower tropospheric cross-equatorial flow should be properly simulated by the AGCMs. The winds at 850 hPa, obtained from the model simulations, and those from the reanalysis data are shown in Figure 3. The magnitudes of the mean cross-equatorial flow, obtained from the model outputs, and those from the reanalysis, are comparable for the monsoon season and for the individual months of June to September. However, the model simulated winds, particularly in June, are stronger in magnitude than the reanalysis winds both over the Arabian Sea and the Bay of Bengal, which would account for the high rainfall simulated in the model for June climatology. The model simulated 200 hPa (figure not shown) easterly jet is somewhat weaker than that in the reanalysis for August and September. The seasonal position of the jet is properly located in the model simulated outputs and the flow across the equator between 80–100 E is from the NE in agreement with the observations. The monsoon trough location is fairly well simulated in July, both in position and strength, whereas it

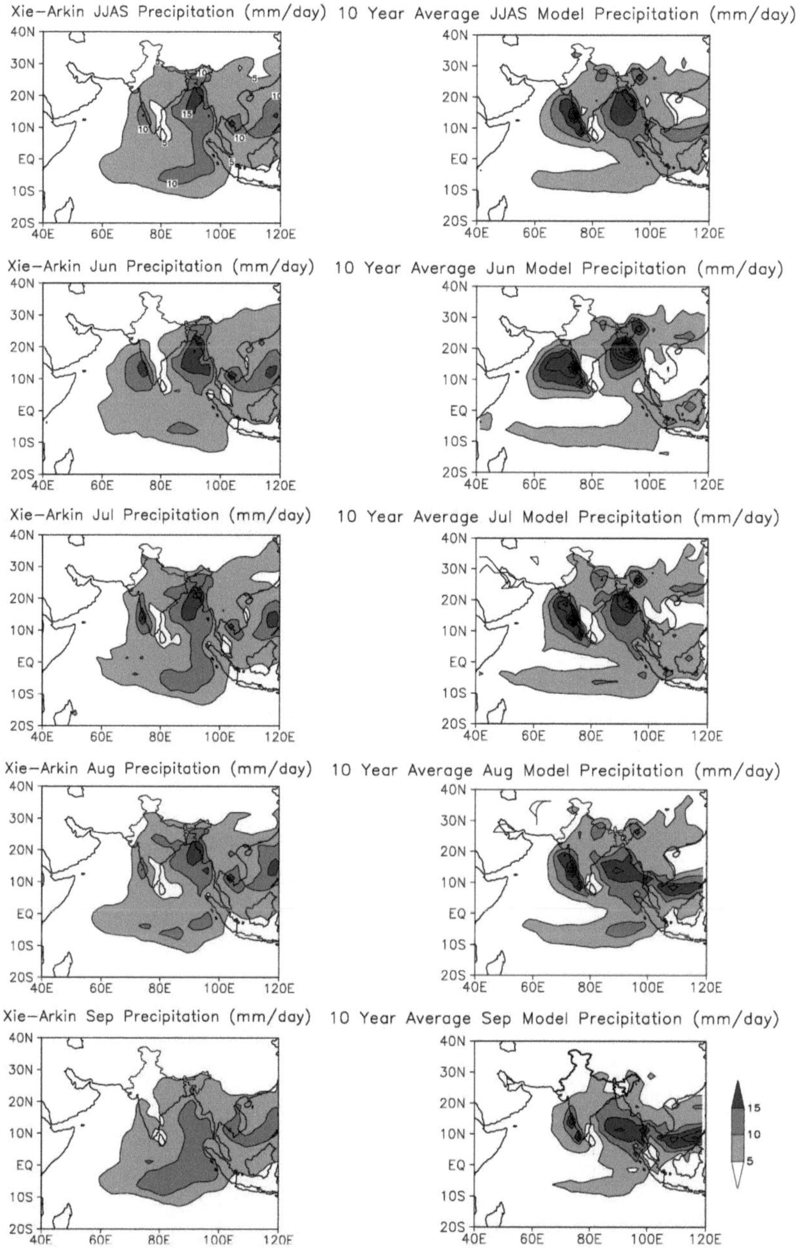

Figure 2
Long-term mean XA precipitation and the ten-year average model simulated climatological precipitation (mm/day). Contour interval 5 mm/day.

Table 2

Area weighted all India summer monsoon rainfall (mm/day). Departure of area weighted summer monsoon rainfall from model climatological precipitation given in bracket

Months /Season	Initial Conditions							Observed 2005 Season		Seasonal rainfall Climatology	
	1 May	2 May	3 May	5 May	6 May	7 May	Ensemble Mean	IMD	TRMM	Model	Obs.
June	7.12 (−21.1)	8.68 (−3.8)	10.42 (15.5)	9.12 (1.1)	7.01 (−22.3)	9.93 (10.1)	8.71 (−3.4)	**4.55** (−12)	**4.63**	**9.02**	**5.18**
July	9.3 (3.3)	12.16 (35.1)	9.81 (9.0)	10.82 (20.3)	12.08 (34.3)	9.45 (5.0)	10.60 (17.9)	**10.82 (+14)**	**10.22**	**9.0**	**9.49**
August	7.01 (−11.3)	6.7 (−15.3)	7.38 (−6.7)	8.14 (3.0)	6.86 (−13.3)	6.53 (−17.4)	7.10 (−10.2)	**6.02** (−28)	**5.90**	**7.91**	**8.35**
Sept.	3.86 (−23.6)	5.03 (−0.5)	4.17 (−17.5)	5.9 (16.8)	4.97 (−1.6)	4.6 (−9.06)	4.76 (−5.9)	**6.7 (+14)**	**6.79**	**5.06**	**5.73**
Season	6.84 (−11.7)	8.16 (5.3)	7.96 (2.6)	7.96 (2.6)	7.76 (0.0)	7.63 (−1.5)	**7.81 (0.7)**	**7.15** (−0.9)	**6.90**	**7.75**	**7.22**

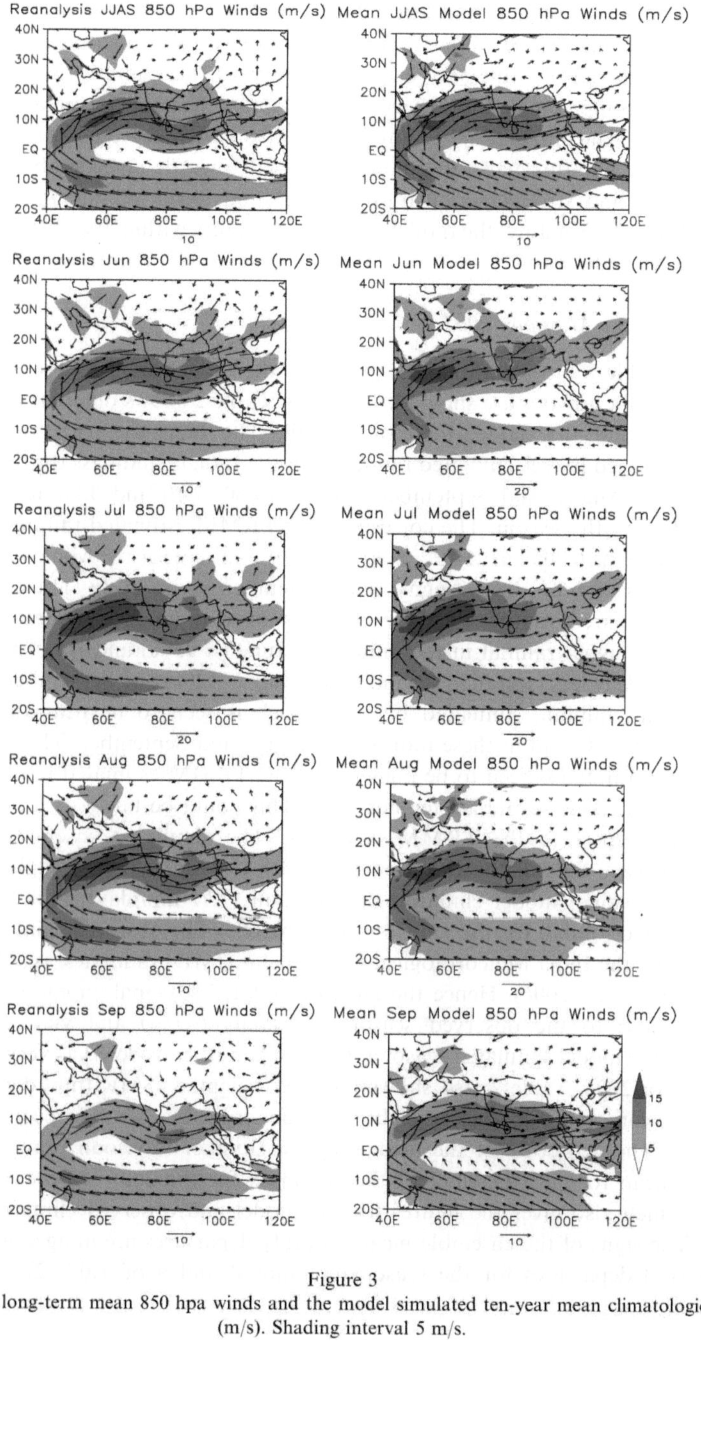

Figure 3
Reanalysis long-term mean 850 hpa winds and the model simulated ten-year mean climatological winds (m/s). Shading interval 5 m/s.

shows more intensity in June and less intensity in August and September compared to the observations. These differences would support the higher intensity of the model simulated rain in June and slight weakening of the rainfall from August to September as compared to the observations. Overall the model-simulated rainfall and the large-scale circulation features are satisfactory on the seasonal basis. However, on a monthly basis the model is biased towards considerably (+75%) higher rainfall and stronger circulation in June and lower rainfall in September (−12%) (Table 2). For July, the model climatology agrees within 5% of the observed rainfall on All India basis.

b) Forecast of the Monsoon 2005

The ensemble monsoon rainfall forecast by the model and as observed in the TRMM rainfall estimates are shown in Figure 4. Table 2 shows different features of the observed and the model forecast rainfall. The monthly averaged and the ensemble averaged model simulated rainfall over the Indian landmass for the months of June, July, August and September is 8.71, 10.60, 7.10 and 4.76 mm/day and 7.81 mm/day for the season. The corresponding TRMM estimated rainfall over the Indian landmass is 6.90, 4.63, 10.22, 5.90 and 6.79 mm/day for the whole season and for the months of June, July, August and September, respectively. Thus, it can be seen that compared to the TRMM estimates over India, the model simulated comparatively higher rainfall in the months of June (+95%) and August (+34%) and comparatively less rainfall in the months of July (−12%) and September (−25%). The maximum rainfall, simulated by the model, is seen to be near the eastern equatorial Indian Ocean in these months of August and September. The equatorial Indian Ocean rainfall is seen to be higher than the TRMM estimated rainfall in the months of July, August and September 2005. This could be taken as a bias of the model with respect to the TRMM estimates. The simulated ensemble average seasonal forecast rainfall for the monsoon 2005 (7.81 mm/day) is slightly above (0.8%) the seasonal model climatological rainfall (7.75 mm/day), indicating very close to normal rainfall for the monsoon season 2005. The observed rainfall for the season, as per the India Meteorological Department (IMD) estimates, is 99% of the normal (LAL *et al.*, 2006). Hence the model simulated seasonal anomaly (+0.8%) was very close to the observed seasonal anomaly (−1%) and viewed in that perspective the model simulated seasonal forecast can be considered as very good for the season of 2005. Table 2 also shows the area weighted ISMR (mm/day) for the ensemble mean and for each member of the ensemble with respective initial conditions. The departure of the area weighted summer monsoon rainfall for the ensemble mean for each member of the ensemble for the 2005 season is given in Table 2, which also gives the figures for the model and observed climatologies of rainfall. The signs of the ensemble mean monthly departures are in agreement with the observed departures for the season (columns 8 and 9 of Table 2) 2005. The

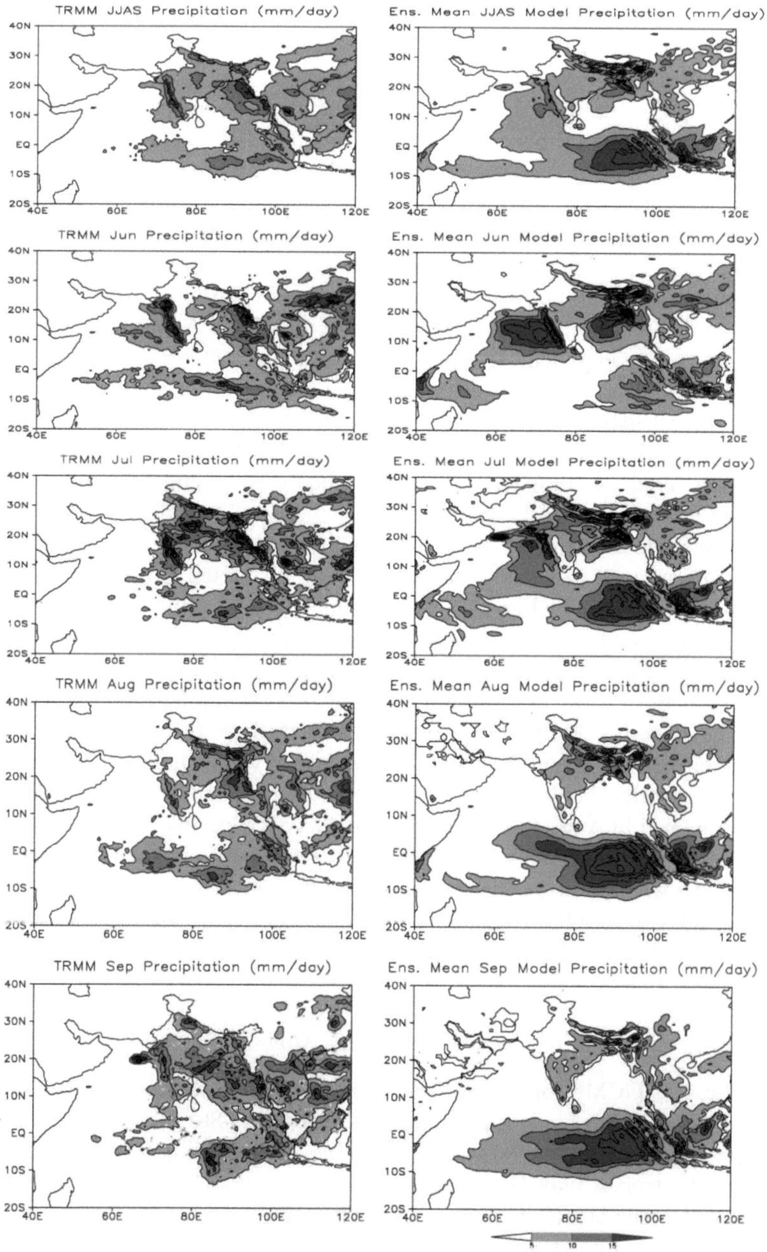

Figure 4

TRMM precipitation and the ensemble mean model simulated precipitation (mm/day) for 2005. Contour interval 5 mm/day.

magnitudes of the ensemble mean rainfall for different months are within one standard deviation of the observed values. These features show that the ensemble mean gave a satisfactory measure of model rainfall on the monthly scale.

The 850 hpa winds (Fig. 5) for the year 2005, as seen in the reanalysis and as simulated by the model, show that the model simulated wind speeds are somewhat smaller than that in the reanalysis. The model winds are reversed (easterly flow) over the central Bay of Bengal in the months of August and September. This reversal is due to high amounts of rainfall in the equatorial Indian Ocean, as seen in Figure 4, which resulted in the building up of subsidence and consequent ridge over the central Bay of Bengal extending up to the South China Sea. The simulated 200 hPa winds also showed that the easterly jet weakened in the months of August and September. Also there was a development of a southeasterly flow at 200 hPa across the equator in the model simulations instead of the normally observed northeasterly flow. Thus the major model bias, with respect to the observations for 2005 monsoon, is the excess rain over the near-equatorial belt along 80°–120°E and the correspondingly less rainfall over the central Bay of Bengal which even extended up to the South China Sea.

To understand the intra-seasonal variability in the ISMR 2005 within the ensemble forecasts we calculated the pentad rainfall for the entire season for the region 8°–32°N, 70°–90°E (Fig. 6) for all the individual members of the ensemble. The figure displays a large variation in the simulated rainfall within the members of the ensembles however, all the members show an increase of rainfall in the month of July and reduction of rainfall in the months of August and September. The intra-ensemble differences in the pentad rainfall are expected but the consensus regarding the reduction of rainfall in the individual members would indicate a bias in the model, as the model simulations show an increase in rainfall along 80°–120°E near the equatorial belt in August and September with a decrease over the Bay of Bengal. It is difficult to explain this bias with regard to the increase in rainfall in the near-equatorial belt which leads to the corresponding drying up of the monsoon over the Bay of Bengal, due to subsidence on the flanks of a region of highly organized near-equatorial convection. We suspect that the model convection scheme may be responsible for this behavior.

c) Simulated Features of the Main Phases of the Monsoon 2005

The use of AGCMs for simulating the interannual variability of the seasonal climate in the tropics rests on the premise (CHARNEY and SHUKLA, 1981 and several others since then) that while the atmospheric predictability of the large-scale weather in the tropics may be restricted to 10–15 days, due to dynamical reasons there may be some signal in the seasonal climate prediction in the tropics in a statistical sense on a long-enough time basis (say a month or a season) and on a large basis (such as the size of India). This signal is considered to result from the

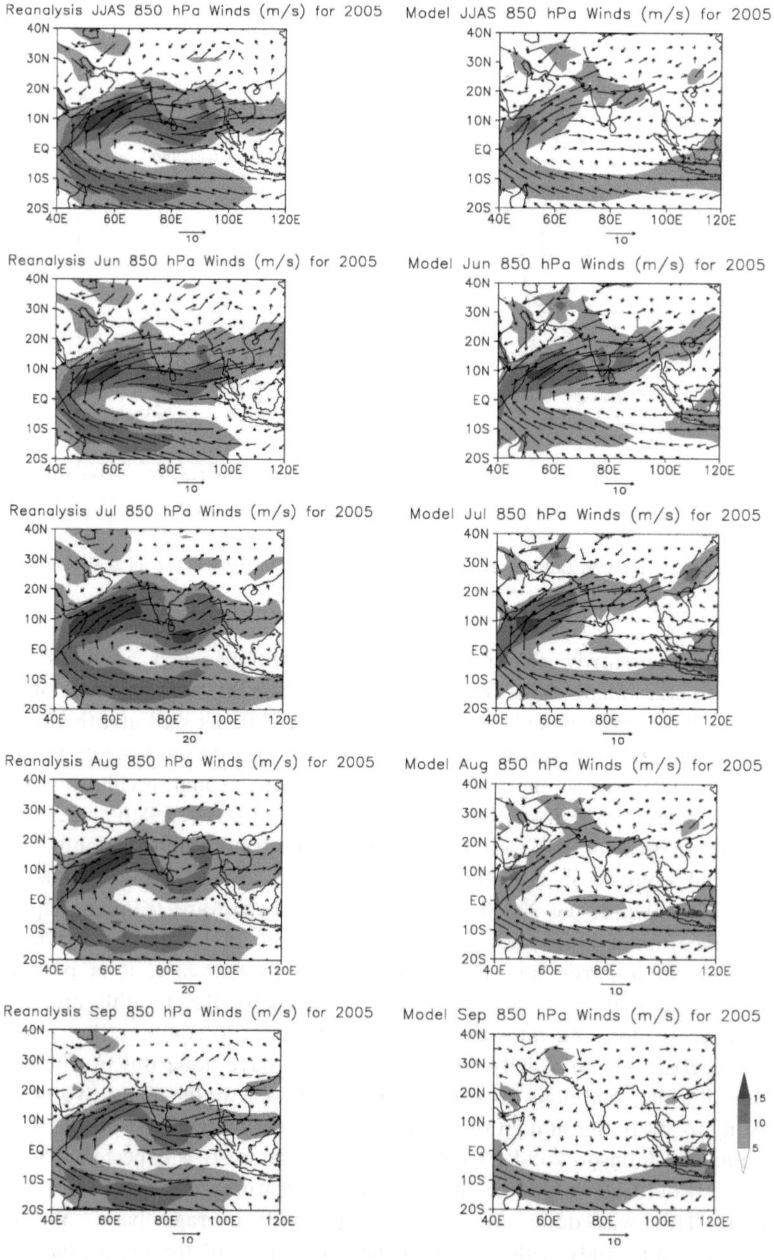

Figure 5
Reanalysis 850 hpa winds and the model simulated ensemble mean winds (m/s) for 2005. Shading interval 5 m/s.

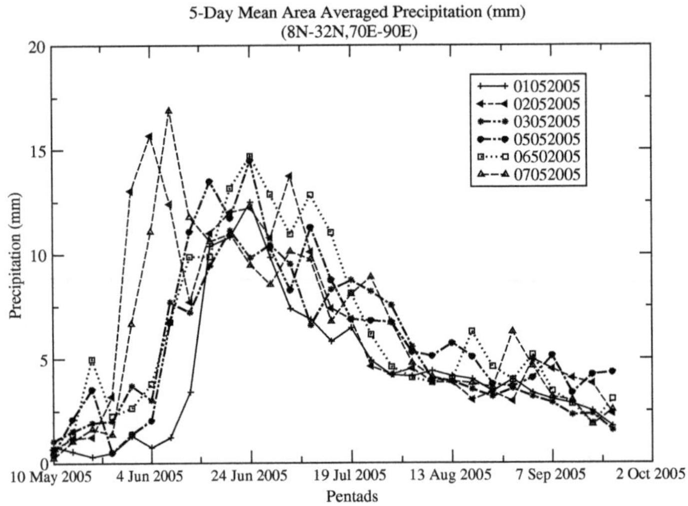

Figure 6
5-day averaged precipitation (mm) over the region 8°N–28°N, 70°E–90°E for all the members of the ensemble.

influence of the slowly changing boundary conditions such as SSTs. Several attempts so far made with regard to using AGCM for seasonal monsoon predictions have given mixed results (KANG *et al.*, 2002 as well as others). This is perhaps due to the complexities introduced by the coupled land-ocean-atmosphere monsoon system as the monsoon has strong intraseasonal oscillations on two prominent scales: 10–20 days (KRISHNAMURTI and ARDUNEY, 1980) and 30–50 days (SIKKA and GADGIL, 1980; YASUNARI, 1980), which result due to a combination of internal and coupled ocean–atmosphere dynamics. While the 10–20-day mode propagates westward, the 30–50-day mode moves northward from the equator to 30°N in the Indian longitudes. Hence, in the study of simulating the monsoon performance in a particular year, such as 2005 in our case, it is pertinent to compare the simulations as a result of SST forcings for the monthly and seasonal performance of monsoon rains and circulation and not to compare the simulations against observations on a daily basis in detail. We have done this on the monthly scale for all India rainfall as discussed above. However, we felt that it would also be interesting to study the simulations with respect to the main phases in the evolution of the regional monsoon and its subseasonal variability. It is expected that a good simulation would show all the major observed features in the evolution of the monsoon. This was done on the 5-day (pentad) average basis. We discuss the results of our study with regard to the simulations of the main phases of the monsoon in the following paragraphs.

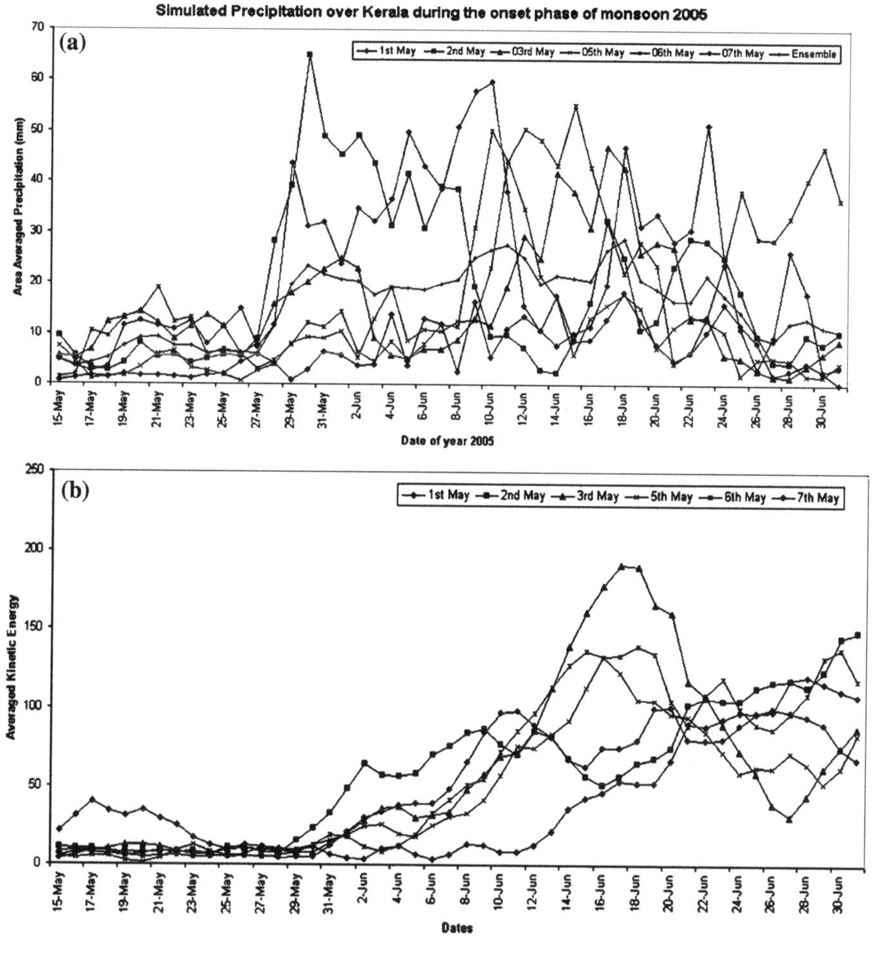

Figure 7
(a) Model simulated precipitation over Kerala (8°–12°N and 75°–77°E) during the onset phase of the monsoon 2005 by various members of the ensemble. (b) model Similated 850hpa area averaged Kinetic Energy (Joule) over the area 50°–70° E, equator –10° N by individual members of the ensemble.

(i) The Monsoon Onset

The onset of the monsoon in the observations is characterized by a sudden spell of increased precipitation over Kerala and its persistence for at least a few days or so. To determine the monsoon onset in the model simulation, an area-weighted average of the model forecast precipitation over Kerala (8°–12°N and 75°–77°E) was computed for each initial condition and for the ensemble mean (Fig. 7a). We note that before 25 May simulated precipitation amounts, based on all the initial conditions and their ensemble mean, were below 1 cm. The sudden rise in the simulated rainfall, based on the 2nd and the 7th of May initial conditions, is observed on 27 May and 28 May, respectively. With the 2nd May initial conditions the simulated rainfall was about 5 cm and with that of the 7th May initial conditions, it was 4 cm. The 3rd May initial condition showed a gradual rise in the amount of precipitation to about 2 cm from 27 May to 2 June. The simulated rainfall, based on 1, 5 and the 6 May initial conditions, showed less precipitation activity during the last week of May and the first week of June. The sudden rise in the simulated precipitation amounts was observed on 8, 9 and 16 June for the initial conditions of 1, 5 and the 6 May, respectively. Thus it can be seen that the initial conditions of 2, 3 and the 7 May, showed an early onset whereas the other three initial conditions showed a delayed onset as compared to the date of onset of the 5th of June declared by the IMD. The ensemble mean predicted rainfall of all the initial conditions showed a sudden rise in rainfall on 30 May and its persistence up to 25 June. From this it could be concluded that the dispersion of date of onset, simulated by the initial conditions of 2, 3, 5, 6, and 7 May, 2005 and that of the ensemble mean is in the range of one pentad with respect to the observed onset date of IMD. However, the initial conditions of 1 May showed a large dispersion of about 11 days from the observed date of onset. It is *a priori* difficult to diagnose the probable cause for the delayed onset simulated with the initial condition of 1 May, 2005. The initial conditions of a particular date may not be found favorable for a skillful forecast. Since there is a scatter in the monsoon onset date simulations among different initial conditions, an ensemble mean would provide a better guidance for this purpose.

Some investigators have used other methods for determining the date of monsoon onset. For example VERNEKAR and JI (1999) calculated the onset of the monsoon by computing an area-weighted average rainfall over the region 8°N–28°N, 65°E–85°E (figure not shown). Judged in this perspective, the calculated area-weighted average for the ensemble mean over this region in our study showed that the model simulated onset date was 28 May, 2005. This is one week ahead of the onset declared by the IMD but only two days earlier from our ensemble mean date of 30 May, 2005. However, the application of the VERNEKAR and JI (1999) methodology on the individual ensemble member simulations gave the onset dates as 11 June, 26 May, 6 June, 2 June and 6 June for the members with initial conditions of 1, 2 , 5, 6 and 7 May, 2005, respectively. This shows large intra-member dispersion compared to one pentad as per the criterion applied by us for the onset date to be fixed with respect to rainfall increase over the Kerala grid. The initial conditions with the 3rd May did not show any sudden increase in rainfall. Based on this limited study, it is suggested that it would be good to stick to the criterion of sudden increase in pentad rainfall over Kerala for determining the pentad in which the monsoon would set over the Kerala region. The analysis of the pentad precipitation (Fig. 8) showed an increase in precipitation by a large amount and its persistence thereafter over the Kerala region

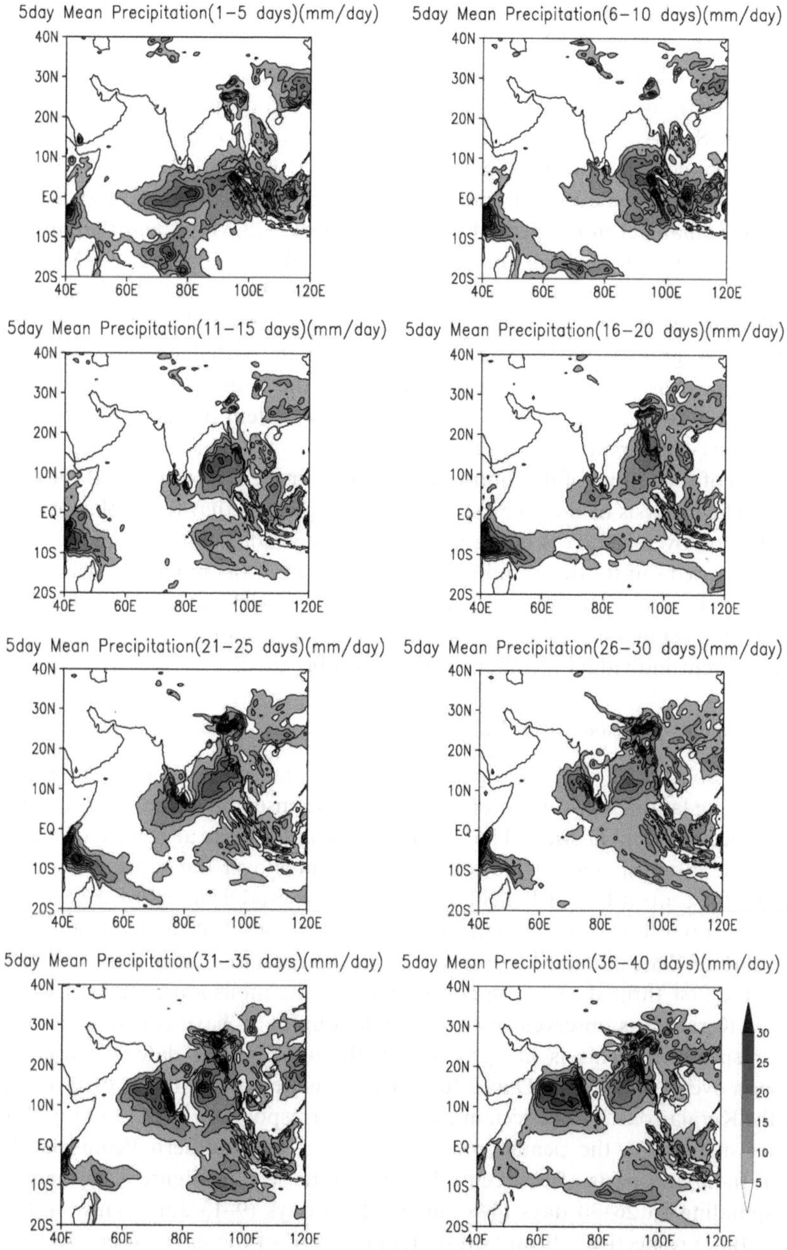

Figure 8
5-day averaged model simulated ensemble mean precipitation (mm/day). Day 1 corresponds to 8 May, 2005. Contour interval 5 mm/day.

from the pentad corresponding to 28 May to 2 June. The first day of our pentad corresponds to 8 May, 2005.

It is also well known that the onset of the monsoon is characterized by an increase in the wind speed and changes in its direction in the lower troposphere by at least 120° over the Arabian Sea (ANANTHAKRISHANAN *et al.*, 1967) and hence the Kinetic Energy (KE) over the oceanic region near the equator (KRISHNAMURTI and RAMANATHAN, 1982). The analysis of the pentad mean wind at 850 hPa in our simulations showed the gradual setting of Somali-Jet in the pentad of 11–15 days (19–23 May), and the monsoonal winds sweeping over the Arabian Sea in the pentad of 16–20 days (24–28 May).

The model simulated area-average (50°–70°E and 0°–10°N) KE per unit mass at 850 hPa for all the initial conditions and for the ensemble mean is shown in Figure 7b. A gradual increase in the KE for all the initial conditions and their ensemble mean is seen from 28 May. Also from the figure it can be seen that the simulated KE in the first week of June showed a persistent increase of more than 25 Joule, which corresponds to winds of more than 5 m/s. This can be treated as an indication of the setting of the monsoonal winds over the Arabian Sea off the coast of Kerala. The analysis of the increase in KE also indicated that the probable onset of the monsoon over Kerala is in the first week of June 2005. Thus, from the different analyses carried out for determining the date of onset of the monsoon, using AGCM simulations in our study, we can conclude that the model was capable of indicating the date of onset in a satisfactory way with a dispersion of only one pentad in five out of six different individual members of the ensemble used.

(ii) Advance of Monsoon over India in 2005

The establishment of the monsoonal flow over the southeast Bay of Bengal and the adjoining landmass could be observed from the ensemble mean winds at 850 hPa to have occurred on 27 May. The averaged ensemble winds at 850 hPa on 29 May showed the establishment of the southwesterly winds over the entire Andaman Sea and over the central Bay of Bengal on 31 May. The establishment of the Somali Jet, cross-equatorial flow over the African coast and the onset of the monsoon over the entire Bay of Bengal could be observed from the ensemble mean winds at 850 hPa during the first three days of June. The onset of the monsoon over Sri Lanka and parts of Kerala was observed from the model output to have occurred on 5 and 6 June, respectively. This is in agreement with the observed date of onset of the monsoon over Kerala by IMD as 5 June. The simulated precipitation of 10 mm / day over the Kerala coast can be seen in the pentad corresponding to 21–25 May (Fig. 8). Further increase in the pentad precipitation over the southern Peninsular India, central India and parts of east central India can be seen in Figure 8, in the pentads corresponding to 26–30 days (4–8 June), 31–35 days (9–13 June) and 36–40 days (10–14 June) respectively from 8 May. Examination of the daily ensemble averaged winds shows that progress of the monsoon over the southern parts of South

Peninsular India occurred by 8 June; over entire peninsular India by 22 June, central India and parts of east central India by 25 June and entire India by 30 June. This progress of the monsoon is in good agreement with the observed progress of the monsoon provided by the IMD. The establishment of the monsoon trough could also be seen in the ensemble mean winds at 850 hPa. Thus the model could simulate the advance of the monsoon over different parts of India in a satisfactory manner.

(iii) Mid-Season Active-Break Spell and Revival of the Monsoon 2005

The simulated monsoon, after its full establishment by the end of June, showed good activity during the whole of July with a small south-north fluctuation in the monsoon trough and formation of synoptic scale disturbances which looked like monsoon lows/depressions. Also we noticed the pulsatory strengthening of the low-level southwesterly flow over the Arabian Sea and the Bay of Bengal. Tropical easterly jet at 200 hPa also showed fluctuation in its strength and position. All these features, simulated by the model, resemble the transient fluctuations in the monsoon. Hence the model has simulated the transient behavior of the monsoon rather realistically.

The weakening of the monsoon is characterized by the migration of the monsoon trough from its normal position to the foothills of the Himalayas. This behavior of the monsoon trough is also seen from the model simulated outputs. During the last week of July model 5-day averaged outputs of 850 hPa winds showed the migration of the monsoon trough from its normal position to the foothills of the Himalayas. From the end of July until the end of August, the simulated location of the monsoon trough was seen near the foothills of the Himalayas, which is by and large close to the observed behavior in August 2005. However the model was unable to simulate the revival of the monsoon which was observed in the beginning of September 2005 from the near-equatorial region in the Bay of Bengal. Associated with this development in early September 2005, the rainfall over the Peninsular and central India remained high up to the 4th week of September over the Bay of Bengal. In the observations this had happened due to the overlapping formation of four low pressure systems (one cyclone, 1 depression and two low pressure areas) which moved over the region as a result of the northward moving intra-seasonal oscillation (ISO) of the monsoon and formation of a depression in the northeast Arabian Sea. However, the model simulations did not show these observed features. We noticed that the model monsoon tended to revive from the formation of a low in the monsoon trough *in situ* by activation of the monsoon trough over north Bay of Bengal and adjoining land rather than from the influence of the northward propagating mode of the ISO from the near-equatorial zone to the monsoon trough zone. It is hard to know whether such a feature is a general characteristic of the model or if it occurred only for September 2005. This would need examination of the details of individual years of the twenty-year model climatology being made with the observed SST as the lower boundary. However, the monthly mean 850 hPa and 200 hPa circulations did not suggest any specific biases with regard to the increase in the near-equatorial rains and the decrease in the Bay of Bengal rains with the mean SST climatology as shown in Figure 2. This deficiency in the model may be due to it being stand-alone AGCM and without any explicit feedback from the ocean and also may be due to the deficiencies in the cumulus parameterization of the model. For the northward propagating modes to be properly simulated the coupling of the atmosphere with ocean is essential. The model convective parameterization is perhaps oversensitive to the prescribed SSTs as the climatological SST pattern is quite smooth and slightly cooler than CFS SST and the observed SST.

(iv) Withdrawal of the Monsoon

The start of the weakening of the southwesterlies over central and north India and the setting of the northeasterly winds over northeast India is observed in the ensemble pentad averaged winds from early September onward. The withdrawal of the monsoon from central India and setting of northeast monsoon is seen from the model outputs as early as 15 September. However, in actual observations withdrawal was delayed due to the prevalence of circulation patterns associated with the overlapping formation of the low pressure systems discussed above, as a result of strong ISOs, which caused rainfall over central India and even over northwest India until the last week of September. The model was unable to simulate this feature. The monsoon had weakened in the model in mid-August 2005. This is brought out in a striking manner by the build-up of an anticyclonic circulation along 10°–15°N from 70°–110°E from mid-August. This was responsible for the persistence of the easterly low-level flow along this belt, in the model simulations from mid-August. We also noticed the development of a strong convergence zone between the lower tropospheric monsoonal westerlies between 70°–80°E and easterlies between 80°–120°E in the near-equatorial belt. This obviously enhanced the rainfall in this belt during August and September in the model simulations and the corresponding sinking motion on the northern flank in the central Bay of Bengal, which dried up the monsoon rains in the Bay of Bengal and up to even the South China Sea. Figure 9 shows a comparison of the pentad averaged relative vorticity along 88–92°E over the Bay of Bengal from early August to the end of September 2005 as simulated by the model and as in the reanalysis. There is an episode of the northward migration as seen in the observation which began in early September and mid-September near 3–6°N and continued up to the end of the September. Positive relative vorticity persisted in the observations under the influence of the northward moving ISOs. These features are missing in the model simulations. Figure 9b shows a belt of positive vorticity in mid-September prevailing in the belt 5–10°N and occasional appearance of positive vorticity near 18–20°N, but no regional characteristic northward propagating ISO as seen in the observations (Fig. 9a). This explains the lack of agreement in the revival of the monsoon between the observations and the simulations.

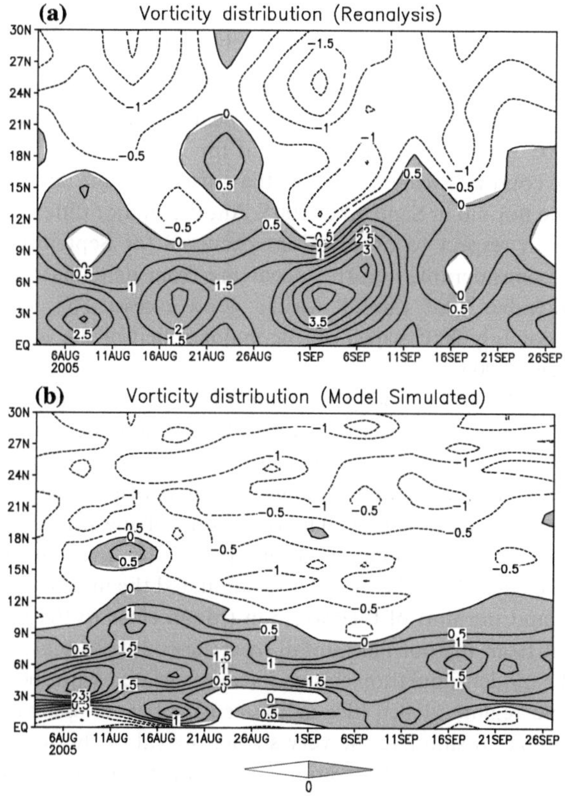

Figure 9
Vorticity ($\times 10^{-5} s^{-1}$) variations along 90° E from 3 August to 30 September, 2005. (a) Reanalysis, (b) model simulated ensemble mean. Positive Vorticity is shaded.

4. Summary and Concluding Remarks

In this study we simulated the Indian monsoon using the high resolution T170L42 NCEP global model. As the study was done in May 2005 to foreshadow the long-range seasonal performance of the monsoon 2005, as such CFS SST forecasts for individual months of May to September were used with the belief that the biases, if any, in the CFS forecast SST had been corrected by the methodology followed in adopting CFS SST after SAHA (2004). The climatology of the model showed that the model was able to capture the main features of the monsoon rainfall quite well. However, it is seen that the model has a positive bias in simulating heavy rainfall off the Western Ghats in June. The analysis of the large-scale circulation patterns in the lower and the upper troposphere showed that the model was able to simulate the

Somali Jet in the lower troposphere and the tropical easterly Jet in the upper troposphere realistically well. Some broad conclusions are given below.

i) Compared with the model's rainfall climatology, the model simulated ensemble mean rains, for India as a whole, were near-normal for June, August and September 2005 while it was above normal for July. In actual observations the rainfall on the country's scale was below the normal in June and August 2005 and quite above the normal in September 2005. Another major difference of the model simulations with respect to observations occurred for September 2005 rainfall, which showed considerable negative departure from its climatology whereas the observed September 2005 rainfall was in good excess. This together with the excess rainfall of July 2005 wiped off the major deficit of the observed rainfall in June and August 2005.

ii) The model has shown a good performance in foreshadowing the onset of the monsoon and its advance over the country. It also simulated higher than normal rainfall in July for the ensemble mean as well as for the individual members of the ensemble. The model could well simulate the onset of the monsoon through the build-up of cross-equatorial flow. The dispersion of the predicted date of onset from the observed one is found to be one pentad for five out of six individual members of the ensemble. The model also simulated the mid-season weakening of the monsoon and the shift of the monsoon trough close to the Himalayas which led to a deficit from the normal rainfall over the country's scale in the month of August. In the model simulation this was also associated with the development of an anticyclonic regime at 850 hPa in the central Bay of Bengal. However, the model was unable to simulate the very strong revival of the southwest monsoon which was witnessed in the observations. Some revival in the rainfall activity did occur in the model simulation toward the end of August to early September 2005 and up to mid-September 2005, resulting from the activation of the monsoon trough *in situ* over north Bay of Bengal and the adjoining landmass of India. Conversely, in actual observations the strong revival had taken place due to the activation of a northward moving convective organization mode of the ISO from the near-equatorial central and the eastern Indian Ocean. During this phase five overlapping monsoon low-pressure systems (one cyclonic storm, 2 depressions and one low-pressure area) had formed between 7–24 September, 2005, some of which moved over Peninsular India towards northwest India. The simulation of such a northward moving convective organization is not observed in this AGCM in 2005 as they result in observations from the coupling of the ocean—atmospheric processes on the intraseasonal scale for whose simulations a good coupled ocean-atmosphere model is needed. Even several of the presently available coupled models are not able to capture ISOs of the monsoon realistically both in intensity and phase propagation (INNESS and SLINGO, 2003). However, some investigators like FU *et al.* (2002) and others have shown that coupled processes help in simulating the correct phase and structure of monsoon ISOs. Besides coupled processes, physical parameterization and other physical processes are also important in the models. Our opinion which agrees with the opinion of several others is that long-range monsoon forecasting could be only partially successful through the application of AGCMs and the hope of better success is only through the use of coupled models, provided they could realistically simulate both the intensity and the northward phase propagation of the ISO.

iii) Our results for the monsoon season of 2005 showed that the model did forecast normal seasonal rainfall compared to the model climatology (+0.7% anomaly). Compared to the TRMM estimates, the model had simulated a higher seasonal rainfall (23% of the TRMM) than that observed which was as a result of the very strong monsoon simulated by the model for June with high rainfall particularly over the Western Coast of India. This may be mainly due to the bias in the model of simulating heavy rainfall on the windward side of the Western Ghats.

iv) Even though the model seasonal rainfall agreed rather well with the observed performance of the seasonal rainfall on the All-India basis for 2005, nonrtheless on the monthly scale it occurred due to wrong reasons. The model simulated rainfall for June 2005 was much higher than that observed in a quantitative basis but it was lower than that observed for September 2005. The model had simulated the good performance of the monsoon from June to mid-July and the weakening of the monsoon from mid-August to the end of September. The model simulated a weakening of the monsoon and suppression of rainfall over the Bay of Bengal from mid-August itself. This was due to the development of a large-scale convergence in the model simulations in the lower tropospheric winds along 80°–120°E near the equator (6°S–6°N) and the build-up of a ridge at 850 hPa with an easterly flow over most of the Bay of Bengal. The persistence of the subsidence over the central Bay of Bengal resulted from the near-equatorial convective organization. This in turn was responsible for the development of the anticyclonic flow in the Bay of Bengal at 850 hPa. It is difficult to pinpoint the specific causes for this. We suspect that it could result in the model's convective parameterization being sensitive to prescribed SST.

Several years of monsoon simulations in multi-member ensemble mode with different initial forecasts and observed SST conditions, are needed to determine biases, if any, in the model simulations. Such a study is underway and the results are to be presented in another paper. One aspect which is clear in the model simulations as well as in observations for the season 2005 is that, the seasonal rainfall may become normal as a result of compensation taking place on the subseasonal scale. Hence, a good monsoon forecast should also have a good skill for the monthly performance of the rains over India. This appears to be a challenging problem and only the future holds the answer to the capability of an AGCM for such a purpose.

Acknowledgments

The authors are thankful to NCEP, USA for providing the global model and the CFS SST. The authors are also grateful to NOAA-CIRES Climate Diagnostics Centre, Boulder, Colorado, USA, for providing the reanalysis data, the Reynolds sea-surface temperature and Xie-Arkin precipitation data through their web site http://www.cdc.noaa.gov/. The authors wish to thank the Centre for Development of Advanced computing, India, for providing the computing facilities for carrying out the simulations. The authors also express appreciation to the two anonymous reviewers whose suggestions considerably improved the paper.

References

Ananthakrishanan, R., Acharya, U. R., and Ramakrishnan, A. R., *On the criteria for declaring the onset of southwest monsoon over Kerala in India*, Meteorological Department Forecasting Manual (FMU Rep. IV-18.1, India Meteorological Department, 1967).

Chang, C.-P. and Krishnamurti, T. N., *Monsoon Meteorology* (Oxford University Press. New York 1987).

Charney, J.G. and Shukla, J., *Predictability of Monsoons in Monsoon Dynamics* (eds. James Lighthill and R.P. Pearce) (Cambridge University Press, 1981) pp. 99–109.

Fu, X., Wang, B., and Li, T. (2002), *Impacts of air-sea coupling on the simulation of the mean Asian summer monsoon in the ECHAM-4 model*, Mon. Wea. Rev. *130*(12), 2889–2904.

Hou, Y.-T., Moorthi, S., and Campana, K.A., *Parameterization of solar radiation transfer in the NCEP models* (NCEP Office Note 441, 2002).

Inness, P.M. and Slingo, J.M. (2003), *Simulation of the Madden-Julian oscillation in a coupled general circulation model I: Comparisons with observations and an atmosphere-only GCM*, J. Climate *16*, 345–364.

Jha, B., Krishnamurti, T. N., and Christides, Z. (2000), *A note on horizontal resolution dependence for monsoon rainfall simulations,* Meteor. Atmos. Phys. *74*, 11–17.

Kang, I.-S., Jin, K., Wang, B., Lau, K.-M., Shukla, J., Krishnamurthy, V., Schubert, S.D., Waliser, D.E., Stern, W.F., Kitoh, A., Meehl, G.A., Kanamitsu, M., Galin, V.Y., Satyan, V., Park, C.K., and Liu, Q. (2002), *Intercomparison of atmospheric GCM simulated anomalies associated with the 1997–98 El Niño,* J. Climate *15*, 2791–2805.

Krishnamurti, T. N. and Ardunay, P. (1980), *The 10 to 20 day westward propagating mode and "breaks in the monsoons,"* Tellus *32*, 15–26.

Krishnamurthi, T. N., Ramanathan, Y. (1982), *Sensitivity of the monsoon onset of differential heating*, J. Atmos. Sci. *39*, 1290–1306.

Krishnamurthi, T. N. and Kishtawal, C.M., LaRow, T., Bachiochi, D., Zhang, Z., Williford, C.E., Gadgil, S., and Surendran, S. (2000a), *Multimodel ensemble forecast for weather and seasonal climate*, J. Climate *13*, 4196–4216.

Krishnamurthi, T. N., Kishtawal, C. M., Shin, D. W., and Williford, C. E. (2000b), *Improving tropical precipitation forecasts from a multianalysis superensemble*, J. Climate *13*, 4217–4227.

Krishnamurthi, T. N., Surendran, S., Shin, D.W., Correa-Torres, R.J., Vijaya Kumar, T.S.V., Williford, C.E., Kummerow, C., Adler, R.F., Simpson, J., Kakar, R., Olson, W.S., and Turk, F.J. (2001), *Real time multianalysis – multimodal superensemble forecast of precipitation using TRMM and SSM/I products*, Mon. Wea. Rev. *129*, 2861–2883.

Krishnamurthi, T. N. (2005), *Weather and seasonal climate prediction of Asian summer monsoon*, Review Topic B1a: Numerical Modeling – Forecast, 1–34. (www.weather.nps.navy.mil/˜cpchang/IWM-III/R4-B1a-NumericalModeling-forecast.pdf).

PAN, H.-L. and WU., W.-S., *Implementing a Mass Flux Convection Parameterization Package for the NMC Medium-Range Forecast Model* (NMC Office Note, No. 409, 1995), (Available from NCEP, 5200 Auth Road, Washington, DC 20233).

LAL, B., JAYANTHI, N., PRASAD, T., RAJEEVAN, M., SUNITHA DEVI, SRIVASTAVA, A.K., and PAI, D.S., *Monsoon 2005-A report.* (IMD Met. Monograph, Synoptic Met.-3/2006, 2006).

MLAWER, E.J., TAUBMAN, S.J., BROWN, P.D., IACONO, M.J., and CLOUGH, S.A. (1997), *Radiative transfer for inhomogeneous atmospheres: RRTM, a validated correlated-k model for the longwave*, J. Geophys. Res. *102*, 16663–16682.

RAMAGE, C. S., *Monsoon Meteorology* (Academic Press, New York and London, 1971).

RAO, Y. P., *Southwest Monsoon,* In Meteor. Monograph, Synoptic Meteor. 1/1976 (India Meteor. Dept., 1976).

REYNOLDS, R. W. and SMITH, T. M. (1994), *Improved global sea surface temperature analysis using optimum interpolation*, J. Climate *7*, 929–948.

SAHA, S. (2004), *Documentation of operational NCEP CFS data files.* (Can be obtained from http://www.emc.ncep.noaa.gov/gmb/ssaha/cfs_data/cfs_data.pdf).

SAHA, S., NADIGA, S., THIAW, C., WANG, J., WANG, W., ZHANG, Q., VAN DEN DOOL, H.M., PAN, H.-L., MOORTHI, S., BEHRINGER, D., STOKES, D., PENA, M., LORD, S., WHITE, G., EBISUZAKI, W., PENG, P., and XIE, P. (2006), *The NCEP Climate Forecast System,* In press, J. Climate (Can be obtained from http://cfs.ncep.noaa.gov/).

SPERBER, K. R., HAMEED, S., POTTER, G.L., and BOYLE, J.S. (1994), *Simulation of northern summer monsoon in the ECMWF model: Sensitivity to horizontal resolution,* Mon. Wea. Rev. *122*, 2461–2481.

SIKKA, D. R. and GADGIL, S. (1980), *On the maximum cloud zone and the ITCZ over India longitude during the southwest monsoon*, Mon. Weather Rev. *108*, 1840–1853.

SIMPSON, J., ADLER, R.F., and NORTH, G.R. (1988), *A proposed Tropical Rainfall Measuring Mission (TRMM) satellite*, Bull. Am. Meteor. Soc. *69*, 278–295.

VERNEKAR, A.D. and JI, Y. (1999), *Simulation of the onset and intraseasonal variability of two contrasting summer monsoons*, J. Climate *12*, 1707–1725.

XIE, P. and ARKIN, P. (1997), *Global precipitation: a 17 year monthly precipitation based on observations, satellite estimates, and numerical model outputs*, Bull. Am. Meteor. Soc. *78*, 2539–2558.

YASUNARI, T. (1980), *A quasi-stationary appearance of 30–40 day period in the cloudiness fluctuation during the summer monsoon over India*, J. Meteor. Soc. Japan *58*, 225–229.

(Received March 30, 2006, accepted September 10, 2006)
Published Online First: June 29, 2007

Pure appl. geophys. 164 (2007) 1667–1681
0033–4553/07/091667–15
DOI 10.1007/s00024-007-0247-y

Pure and Applied Geophysics

Sensitivity of Indian Monsoon to Entrainment and Detrainment in Mass Flux Schemes

S. K. Deb,[1,2] H. C. Upadhyaya,[1] O. P. Sharma,[1] and J. Y. Grandpeix[3]

Abstract—A new analytical formulation of entrainment and detrainment in the Tiedtke's mass flux cumulus parameterization is presented here in which cloud height is one of the key parameters. The proposed analytical profiles of entrainment and detrainment are tested in GCM for long-term simulation and are evaluated in the light of the results from the original Tiedtke's scheme and another mass flux scheme due to Emanuel. The variations of Indian monsoon rainfall have been examined with these schemes in a general circulation model. Evaluation of the simulated rainfall against observations is done by empirical orthogonal function (EOF) analysis for the Indian Monsoon region. It is noted that the spatial and temporal variations of the all-India monsoon rainfall are sensitive to the formulation of entrainment and detrainment in a mass flux scheme, and that the new formulation can effectively represent the increased dilution with height in deep clouds.

Key words: Entrainment, cumulus convection, EOF analysis, Indian Monsoons.

1. Introduction

The treatment of entrainment of environmental air into cloud columns is one of the challenging problems of cumulus parameterization. The entrainment of environmental air into clouds represents one of the major uncertainties in the formulation of cumulus parameterization, which is not yet properly resolved (Arakawa, 2004). Due to their finite life spans, the entrainment rate is a highly variable quantity (Ooyama, 1969) in clouds. Generally, entrainment inflow is due to turbulent lateral mixing and its representation in a mass flux scheme is done via plume theory which establishes that entrainment is inversely proportional to plume radius (Turner, 1962). But some observations (Warner, 1970; Pruppacher and Klett, 1997) have shown that dilution in clouds indeed increases with height, which is contrary to the inverse relationship of entrainment to cloud radius. Other studies,

[1] Centre for Atmospheric Sciences, Indian Institute of Technology Delhi, New Delhi, India.
E-mail: hcdhyaya@cas.iitd.ernet.in
[2] Atmospheric Science Division, Meteorology and Oceanography Group, Space Applications Centre, ISRO, Ahmedabad, 380015, India.
[3] Laboratoire de Météorologie Dynamique, 4, Place Jussieu, Paris, France.

based on large-eddy simulation of shallow cumuli, have shown that with suitable entrainment and detrainment rates, a mass-flux decomposition can represent quite well the simulated vertical fluxes of conserved thermodynamic variables (SIEBESMA and CUIJPERS, 1995). For an ensemble of deep clouds, LIN (1999) demonstrated, in mass flux cumulus parameterizations, wide variations in entrainment rates simulated by a cloud-resolving model (CRM). GRANT and BROWN (1999) suggested that entrainment rate could be estimated from turbulent kinetic energy generation rate by buoyancy, depth of the cloud layer and cloud base mass flux. GREGORY (2001) links entrainment rate to buoyant production of vertical kinetic energy but argues that his formulation implies a decrease in entrainment with height analogous to that found by Lin. In the study of shallow cumulus convection that includes a feedback between entrainment rate and vertical velocity, the distributions of thermodynamic properties at the cloud base were sufficient to explain the variability in cumulus clouds (NEGGERS *et al.*, 2002). The above studies suggest that the inverse relationship of entrainment to cloud radius may not have serious impact for shallow clouds, although it may not be the case for deep convective clouds in the tropics. Since cloud types are easily distinguishable from their heights, a re-look at a formulation of turbulent entrainment should consider cloud height as an important parameter. To this effect, entrainment is expressed analytically in terms of cloud radius and height in such a manner that its value changes for deep clouds without any major modification in its value for shallow clouds. Such a formulation of the turbulent entrainment is incorporated in Tiedtke's parameterization, which is referred to in the text as Tiedtke's scheme with analytical formulation of entrainment (Tiedtke_AFE).

The sensitivity of Indian monsoon to different types of cumulus parameterization has been studied by several researchers. SLINGO *et al.* (1988) studied the sensitivity of monsoon onset using the operational European Centre for Medium Range Weather Forecast (ECMWF) model with Kuo's scheme (KUO, 1974) and found a strong dependence of the simulated Indian summer monsoon (ISM) onset to cumulus convection. SUD *et al.* (1992) found a more active hydrological cycle from the use of ARAKAWA and SCHUBERT (1974) scheme over the monsoon area, in comparison to simulation without cumulus convection, in the Goddard Laboratory for Atmosphere (GLA) GCM. ZHANG (1994) investigated the sensitivity of the ISM to cumulus parameterization in the Canadian Climate Centre global model by comparing the results from a mass flux cumulus scheme to those from simulations carried out with and without a convective scheme. He found that mass flux scheme performed better especially for the simulation of the Somali jet over the Arabian Sea and precipitation over India. GADGIL and SAJANI (1998) have analyzed tropical precipitation in AMIP simulations with more emphasis on the ISM and found that Geophysical Fluid Dynamics Laboratory (GFDL) model using a moist convective adjustment scheme (MANABE *et al.*, 1965) was able to outperform the others. Later, EITZEN and RANDALL (1999) also found realistic simulations of ISM by relating cumulus mass flux to the cumulus kinetic energy. SCINOCCA and MCFARLANE (2004) studied the temporal variability of modeled tropical precipitation with Canadian Centre for Climate Modelling and Analysis (CCCma) third generation atmospheric general circulation model (AGCM3) using different cumulus schemes and found that large-scale precipitation, not cumulus convection, is responsible for the resolved tropical Kelvin and mixed Rossby-gravity waves in the GCM. The above studies suggest that further experimentation with a GCM is well in order to examine the sensitivity of ISM to cumulus convection with different formulations of entrainment/detrainment rates. For this purpose, simulations have been performed with Laboratoire de Météorologie Dynamique (LMDZ4) general circulation model (HOURDIN *et al.*, 2006) with prescribed initial and boundary conditions.

Next, evaluation and intercomparison of these simulations have been undertaken by using the empirical orthogonal functions (EOFs). The EOF analysis is one particular method to identify and compare patterns by decomposing observed and simulated data in modes. It provides both spatial and temporal patterns and helps to identify dominant oscillations and non-essential noise in the data. The analysis of the simulated rainfall and its variability has been performed here in terms of the EOFs of GPCP precipitation (XIE *et al.*, 2003). Earlier studies (BEDI and BINDRA, 1980; RASMUSSON and CARPENTER, 1983) show that the first four principal components (PCs) are sufficient to explain a significant part of the total variance of rainfall over India. If a model is able to simulate the observed circulation and its variability, then patterns of principal components of simulated and observed parameters should resemble. With this view in mind, simulated rainfall has been projected on the EOFs of the observed rainfall. The differences in the simulated and observed rainfall are then visible in the time series of EOFs of GPCP and that of the projections of simulated rainfall. The projection method also permits us to identify the differences in the annual cycle of Indian monsoon rainfall as simulated in the LMDZ4 general circulation model with different mass flux cumulus parameterizations. The schemes employed in the LMDZ4 model are TIEDTKE's (1989), EMANUEL's (1999) and the Tiedtke_AFE.

2. *An Analytic Formulation of Entrainment/Detrainment*

The turbulent contribution for entrainment/detrainment parameterization is given by Tiedtke as,

$$E_u^1 = \varepsilon_u M_u; \quad D_u^1 = \delta_u M_u,$$

where the fractional entrainment (ε_u) and detrainment (δ_u) rates are written as

$$\varepsilon_u = \frac{\lambda_u}{R_u}; \quad \delta_u = \frac{\lambda_u}{R_u}.$$

Here M_u is the upward convective mass flux. TIEDTKE (1989) used the fixed values of ε_u and δ_u, however we have used the analytical expressions for the parameters in Tiedtke_AFE. For that a non-dimensional variable, $x = \Delta p_{\mathrm{cld}}/\Delta p_{\mathrm{max}}$, is defined. Here Δp_{cld} represents the cloud depth calculated from the undiluted adiabatic ascent and Δp_{max} is the maximum possible cloud height set as 800 hPa. The analytical profile of λ_u and R_u are assumed as

$$\lambda_u = \frac{\lambda_0}{(1+\alpha^2 x^2)}; \quad R_u = \frac{R_0}{(1+\alpha_1^2 x^2)}, \qquad 0 < x \leq 1,$$

where λ_0 denotes the maximum value of entrainment coefficient λ_u which is set to 0.35. The shape of the profile of λ_u and R_u depends on the values of the constant $\alpha(\geq 0)$ and $\alpha_1(\geq 0)$; R_0, maximum cloud radius, is equal to 2000 m if $\Delta p_{\mathrm{cld}} < 200$ hPa and R_0 is 600 m if $\Delta p_{\mathrm{cld}} \geq 200$ hPa. Thus the profile of entrainment rate can thus be expressed as (TURNER 1962),

$$\frac{1}{M}\frac{dM}{dz} = \frac{\lambda_0}{R_0} f(x); \quad f(x) = \frac{(1+\alpha_1^2 x^2)}{(1+\alpha^2 x^2)}; \quad \alpha_1 > \alpha, \quad 0 < x \leq 1.$$

The values of α and α_1 are chosen as 2.2 and 3.5 after a series of single column model experiments using GATE Phase-III data sets. The same parameterization is applied to downdraft (though the parameters α, and α_1 could be assigned different values), therefore one writes

$$\varepsilon_d = \frac{\lambda_d}{R_d}; \quad \delta_d = \frac{\lambda_d}{R_d}; \quad \lambda_d = \frac{\lambda_0}{\left(1+\alpha^2(1-x)^2\right)}; \quad R_d = \frac{R_0}{\left(1+\alpha_1^2(1-x)^2\right)}; \quad 0 < x \leq 1.$$

The quantities λ_0, R_0, α, α_1 and x have already been defined above.

3. Numerical Simulations

The LMDZ4 general circulation model has been utilized here to perform numerical simulations with different cumulus parameterization schemes. The boundary conditions remain identical in all simulations analyzed here. The basic equations and their horizontal and vertical discretizations have been described by SADOURNY and LAVAL (1984). It is a grid point GCM which uses Arakawa's C-type grid for spatial discretization. For this study, the grid points are regularly spaced along longitude and latitude. The horizontal discretization is achieved by setting 96 points along longitude as against 72 points along the latitude. It uses a hybrid coordinate in the vertical with 19 unequally spaced layers. The time integration scheme is a combination of two explicit schemes: The basic time step of 30 minutes is split into one Euler-backward (Matsuno) step followed by four leap-frog steps. The physics of the model is fairly complete. An important omission here is the specification of vegetation in these simulations. Three seven-year simulations have been performed with LMDZ4 using TIEDTKE (1989), EMANUEL and ROTHMAN (1999) and Tiedtke_AFE schemes in the model.

The main aim of this study is to investigate how well the above analytical profiles of entrainment and detrainment are able to simulate the variability of the Indian monsoon in the long-term simulations with LMDZ4. All simulations in this study start from 1 January, 1993 and end on 31 December, 1999. For preparing the initial conditions, atmospheric parameters (such as winds, moisture, zonal, meridional and vertical velocity, temperature, geopotential) have been extracted from the ECMWF datasets. ECMWF analysed SSTs have been used as surface boundary conditions. The surface albedo, soil moisture and sea-ice are prescribed from climatology. For validation of the model results, the $2.5° \times 2.5°$ GPCP monthly rainfall data are utilized. Apart from the analysis of the direct model output like large-scale circulation and rainfall, the simulated monthly rainfall has been subjected to EOF analysis. The simulated rainfall is projected on the EOFs of GPCP rainfall over the Indian subcontinent. This technique, based on a sound mathematical foundation, appears to be most suitable for evaluating and comparing many numerical simulations of longer durations with greater ease and reliability (MOLTENI *et al.*, 2003).

4. Results and Discussion

In this section, the difference fields of simulated and observed large-scale circulation and of the corresponding rainfall patterns over the Indian subcontinent are discussed. The simulated rainfall from each cumulus schemes has been evaluated by projecting it on the EOFs of the GPCP rainfall. The time series of each mode is then utilized for the sake of comparison and evaluation of simulated rainfall with observations.

4.1. 850 hPa Winds

For each seven-year (1993–1999) model run, mean June, July and August (JJA) differences of simulated and observed (ECMWF analysis) 850 hPa wind fields are shown in Figure 1. The cross-equatorial flow, the Somali jet and the dominant features of the monsoon circulation over the Arabian Sea have been reproduced satisfactorily in the simulations, though there are differences. In the simulation with Tiedtke scheme, underestimation in the strength of the Somali Jet by 3 m/s around 5°N (Fig. 1a) may be noted. The easterlies are weaker in equatorial Indian Ocean and stronger in the northeastern region extending to Japan as compared to the ECMWF analysis. The model simulates the realistic flow configuration of the Somali Jet with Emanuel scheme (Fig. 1b), but it is a relatively stronger jet with onshore flow

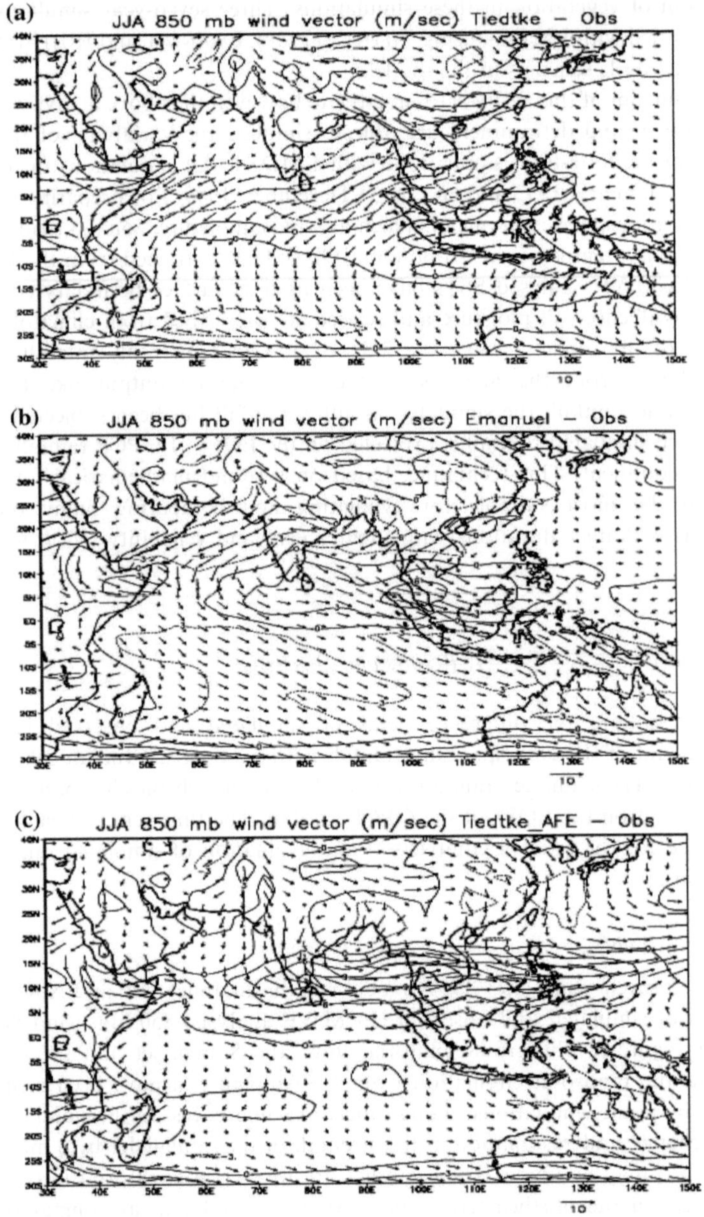

Figure 1
June-July-August (JJA) mean difference plots of simulated and observed (ECMWF analysis) winds (m/s) at 850 hPa for the period 1993–1999: (a) Tiedtke, (b) Emanuel and (c) Tiedtke_AFE.

over the west coast of India and Southeast Asia from 0°N to 15°N. The meridional component of the wind is also simulated satisfactorily. In the simulation with Tiedtke_AFE, the position and strength of the Somali jet are also realistically reproduced although it is stronger in comparison to the ECMWF analysis in Southeast Asia and the western Pacific regions. The strong westerly flow covers the west Pacific region, which is an unrealistic feature in this simulation. Though the flow over the South Indian Ocean is produced very well in this simulation, nonetheless an unrealistic westerly flow is simulated over the Bay of Bengal. The position of ITCZ is reproduced very well in all simulations.

Figure 2 shows December, January and February (DJF) mean differences of simulated and ECMWF analysis 850 hPa wind vectors from the three simulations. Each scheme has produced easterlies over the south and westerlies over the north of the Indian monsoon region with comparable magnitude as in observations. The strength of the simulated easterlies over the South Indian Ocean is relatively stronger than the observed when Tiedtke_AFE is employed (Fig. 2c). However, the weak easterlies along 5°S are simulated, such as in the observations, with Tiedtke and Emanuel schemes but not with Tiedtke_AFE. The simulated flow over Central China by all schemes is unrealistic. Nevertheless, the position of ITCZ and the simulated northeasterly flow over India compare well with the 7-year climatology.

4.2. Mean Rainfall

The JJA mean differences of simulated and observed (GPCP) rainfall (1993–1999) from simulations are shown in Figure 3. Stronger JJA mean rainfall is simulated with Tiedtke scheme (Fig. 3a) in comparison to observations over most parts of India and the equatorial Indian Ocean. But over northern India and the Bay of Bengal, simulated rainfall amounts are inferior to GPCP values. However, JJA mean rainfall patterns, simulated with Emanuel scheme (Fig. 3b), show remarkable similarities with observations, both over land and ocean, though a slight overestimation of rainfall over south India and southern Bay is noticeable. The peak rainfall over the west coast is somewhat higher (4 mm/day). The centre of maximum rainfall is located in the central Bay and the simulation is weaker in northern and northeastern India. The overall pattern of rainfall simulated by Tiedtke_AFE scheme (Fig. 3c) agrees with the observed distribution in the central and eastern part of India and a significant improvement may be noted in the rainfall over Indian Ocean that resembles GPCP over the original Tiedtke scheme. However this scheme has simulated strong unrealistic rainfall in the region starting from Western Ghats extending up to the western Pacific. This is one of the major deficiencies noted in the Tiedtke_AFE where more research is necessary to eliminate it. Like the other two schemes, Tiedtke_AFE also simulates dry bias in the northern part of India and over the Himalayas. The DJF mean differences of simulated and observed rainfall (Fig. 4) from three cumulus schemes also compare favourably with observations. In the

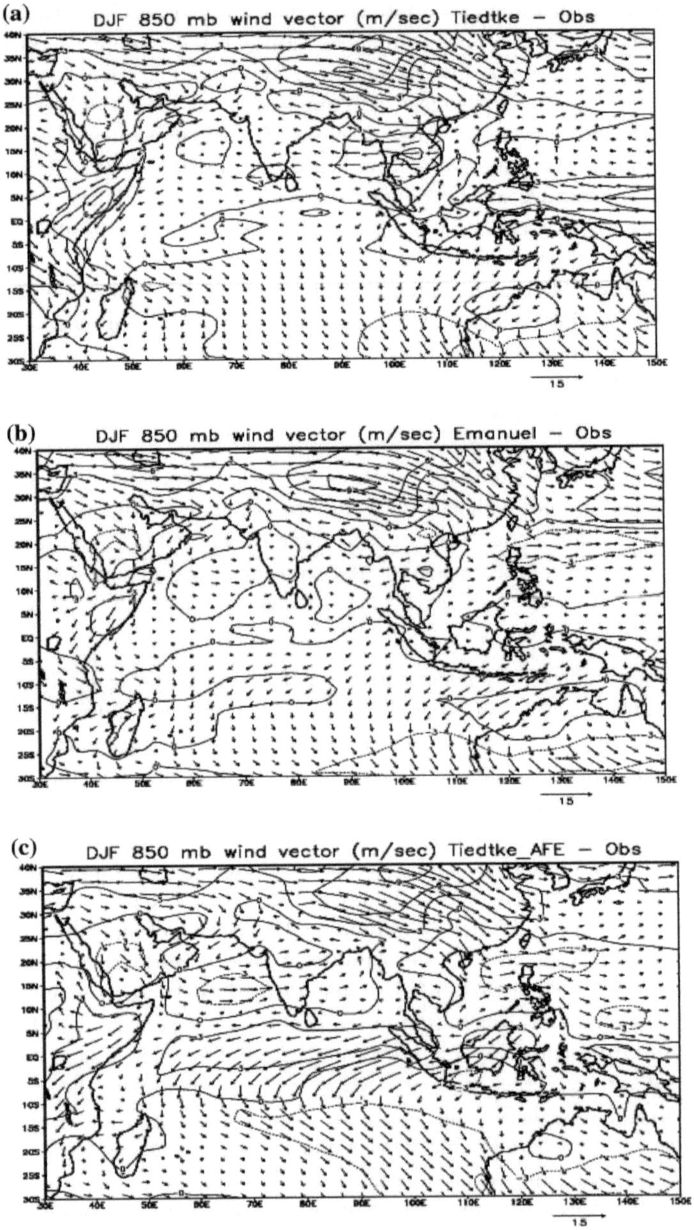

Figure 2
December-January-February (DJF) mean difference plots of simulated and observed (ECMWF analysis) winds (m/s) at 850 hPa for the period 1993–1999: (a) Tiedtke, (b) Emanuel and (c) Tiedtke_AFE.

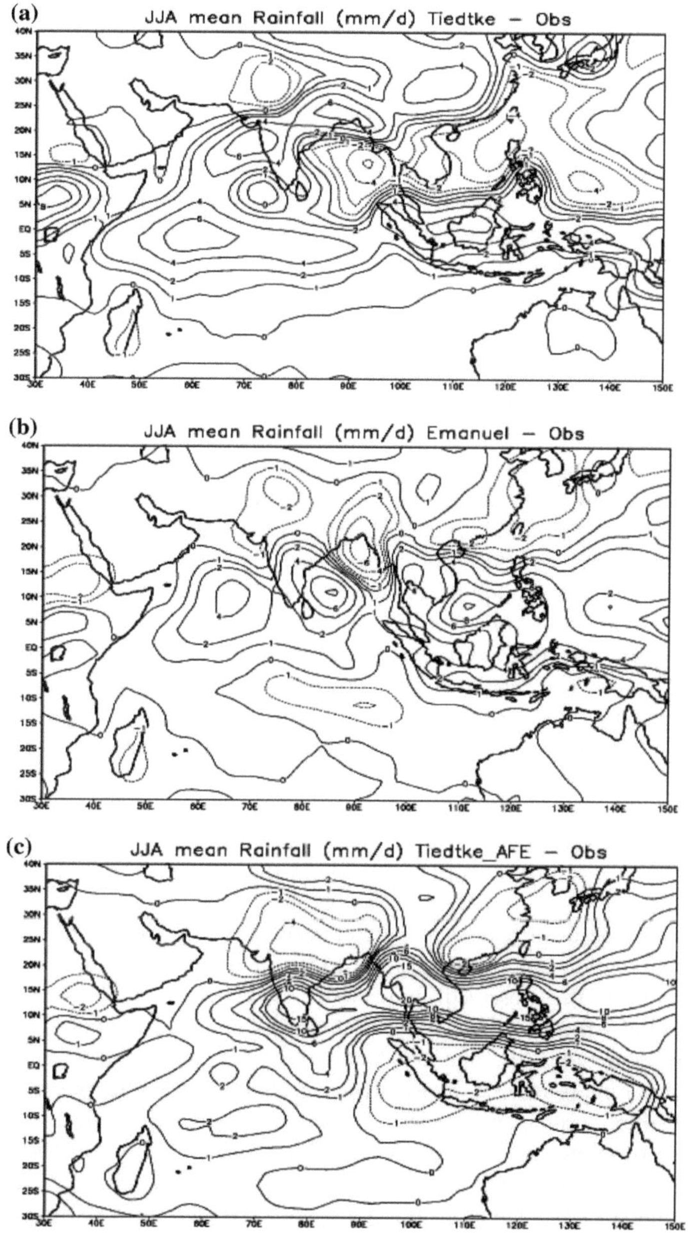

Figure 3

June-July-August (JJA) [1993–1999] mean difference plots of simulated and observed (GPCP) rainfall (mm/day): (a) Tiedtke, (b) Emanuel and (c) Tiedtke_AFE.

simulations using Tiedtke and Emanuel schemes, the center of minimum rainfall is located around 5°S, which is near 5°N for Tiedtke_AFE. The time series of simulated rainfall with Emanuel scheme (Figure not shown) in January closely agrees with observations, although in some years the July rainfall is underestimated by about 3 to 4 mm/day. However, some temporal variability in the Tiedtke_AFE (Figure. not shown) has been observed but not so significant as to disturb the annual cycle of rainfall. When it concerns simulation of the annual cycle of rainfall, Tiedtke scheme does better than the other two schemes, nevertheless it is desirable to use a unified convection scheme suggested by Krishnamurti and Sanjay (2003) in a GCM for improved estimates of rainfall (both spatially and temporally).

4.3. Principal Component Analysis of Observed and Simulated Rainfall

The EOFs of the GPCP monthly rainfall series with 84 time levels are calculated over India, which comprises 56 model grid points. The first four EOFs of the monthly rainfall represent most of the intra-seasonal and inter-annual variability of the IMR that arises from its oscillations or is due to convective systems that are advected into this region. The first EOF (Figure not shown) of observed rainfall explains a significant part (81%) of its total variance. The second, third and fourth EOFs (Figures not shown) jointly explain 12% of the total variance. Thus, the first four EOFs explain 93% of the total variance of the rainfall. The EOF1 has a unimodal spatial structure with the maximum in correlations (+ 0.9) occurring over central India. The region of positive correlations covers entire India and its homogeneous pattern is associated with high or low precipitation. The second, third and fourth EOFs also present some interesting results even though they explain only 12% of total variance. EOF2 divides Indian monsoon region into two parts: positive correlations (+ 0.2) over Western Ghats and head Bay of Bengal and negative correlations (–0.3) over the northwestern part of India. This mode portrays an east-west/north-south oscillation in the rainfall. The EOF3 depicts the north-south oscillations in rainfall and finally the EOF4 describes the contributions from weather-scale disturbances to rainfall variability. The simulated precipitation from each simulation is projected on the EOFs of the GPCP rainfall as discussed above. Therefore the observed and model simulated rainfall anomalies have common spatial structures defined by the EOFs of observed rainfall. Consequently the differences between the simulated and observed rainfall shall appear in the corresponding time series of expansion coefficients of each mode (referred as PC in the text). It then becomes much easier to compare mode wise differences over the entire region of interest with EOF decomposition of the given data. The expansion coefficient series of the GPCP and simulated rainfall are referred in the text as PCs (observed) and PCs (simulated), respectively. Molteni *et al.* (2003) have used the singular value decomposition (SVD) modes on which the model anomalies were projected to analyze 850 hPa winds and precipitation. The time series (PC1) of

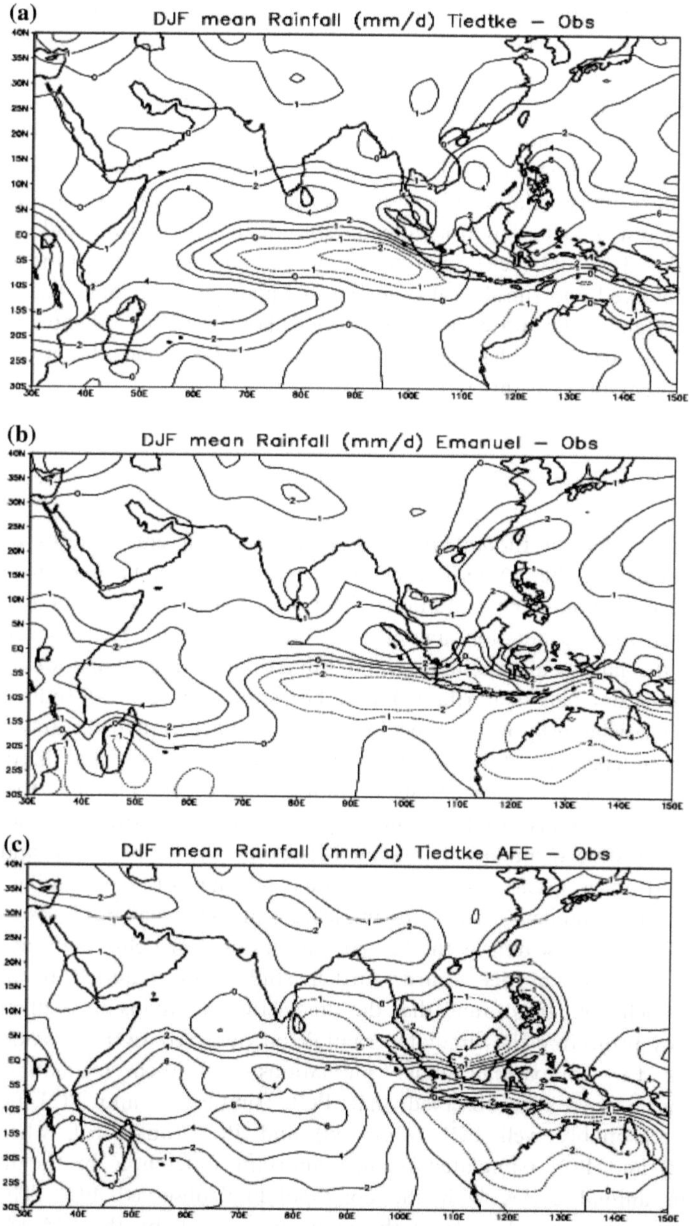

Figure 4
December-January-February (DJF) [1993–1999] mean difference plots of simulated and observed (GPCP) rainfall (mm/day): (a) Tiedtke, (b) Emanuel and (c) Tiedtke_AFE.

EOF1 of GPCP monthly rainfall from three simulations using Tiedtke, Emanuel and Tiedtke_AFE schemes are shown in Figure 5a. They represent the evolution of rainfall with strong annual oscillations in the All-India mean rainfall. The time series of PC2 (Fig. 5b) of observed and simulated rainfall, show high frequency variations between summer and winter monsoons. The simulated PC2 of each scheme does not following the PC2 of GPCP rainfall and most of the time they show an out-of-phase relationship with observations in the first 3-year simulations. This brings out a fact, that for climate, spin-up may be slower and that model roughly takes 3 annual cycles to adjust to prescribed forcing. The time series PC3 for both observed and simulation is shown in Figure 5c, and some oscillations are present throughout the entire simulation period. However, the simulated time series of PC4 (Fig. 5d) resembles closely the observed PC4 and displays the high frequency oscillations in this mode.

5. *Conclusion*

The simulations of Indian monsoon with three different cumulus parameterization schemes in the LMDZ4 general circulation model have been performed with prescribed boundary conditions. The EOF analysis of rainfall has been adopted for their evaluation against the GPCP rainfall. These schemes have simulated the large-scale circulation and rainfall over India quite well, but they produce some dry bias in northern India. The annual cycle of all-India rainfall is reproduced satisfactorily in the model for each scheme, though the July rainfall is somewhat underestimated in some years. It may be noted that Tiedtke_AFE produces greater temporal variations as compared to others, although they are not so significant as to disturb the annual cycle curve. Therefore, the entrainment/detrainment profile formulation presented here appears to be useful and may be implemented in a GCM using a mass-flux scheme for representing cumulus convection in the model. Though the implement of the given profiles is easy, there is a need for more research as the key parameters in their definition may vary one model to another. In this regard, further studies are required to produce a concrete conclusion. It is found that the time series of expansion coefficient (PCs) obtained by projecting the simulated rainfall on observed EOFs provide a better framework for comparing model simulations and their evaluation against observations. It is also observed that PC1 (simulated) and PC3 (simulated) resemble reasonably well PC1 (observed) and PC3 (observed), whereas PC2 (simulated) and PC4 (simulated) show important differences in the initial few simulation annual cycles with the corresponding observed PCs. The correct simulation of the annual cycle of the monsoon rainfall by the LMDZ4 general circulation model makes it a very promising tool for monsoon research and especially for the seasonal forecasting of Indian monsoons.

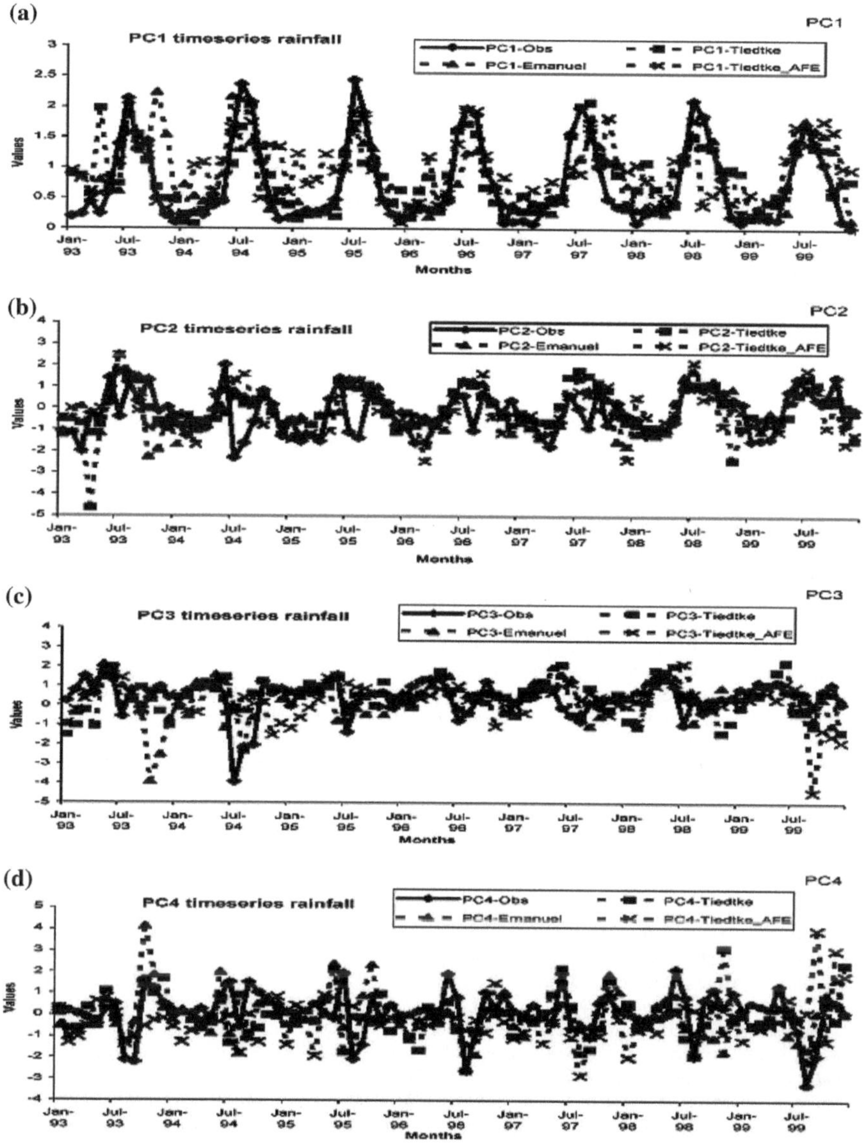

Figure 5

Time series of principal components (PCs) of GPCP monthly rainfall for [1993–1999] and of simulated rainfall: (a) PC1, (b) PC2, (c) PC3, and (d) PC4.

Acknowledgement

The authors thank one of the anonymous referees for his constructive comments and valuable suggestions which greatly improved the quality of the presentation of this paper. This research has been supported by the Council for Scientific and Industrial Research (CSIR), under the project "Adoption, Development and Evaluation of a Variable AGCM (VRAM) for Monsoon Region" under NMITLI scheme. Numerical computations were carried out at LMD during a 3-month visit of SKD under the Project 1911-2 of the Indo-French Centre for Promotion of Advanced Research /Centre Franco-Indien pour la Promotion de la Recherche Avancée (IFCPAR/CEFIPRA).

References

Arakawa, A. (2004), *The cumulus parameterization problem: Past, present and future*, J. Climate *17*, 2493–2524.

Arakawa, A. and Schubert, W. H. (1974), *Interaction of a cumulus cloud ensemble with large-scale environment: Part-1*, J. Atmos. Sci. *31*, 674–701.

Bedi, H. S. and Bindra, M. M. S. (1980), *Principal components of monsoon rainfall*, Tellus *32*, 296–298.

Eitzen, A. Z. and Randall, D. A. (1999), *Sensitivity of the simulated Asian summer monsoon to parameterized physical processes*, J. Geophys. Res. *12*, 177–12, 191.

Emanuel, K. A. and Rothman, M. Z. (1999), *Development and evaluation of a convective scheme for use in climate models*, J. Atmos. Sci. *56*, 1766–1782.

Gadgil, S. and Sajani, S. (1998), *Monsoon precipitation in the AMIP runs*, Clim. Dyn. *14*, 659–689.

Grant, A. L. M. and Brown, A. R. (1999), *A similarity hypothesis for shallow-cumulus transports*, Q. J. R. Meteor. Soc. *125*, 1913–1936.

Gregory, D. (2001), *Estimation of entrainment rate in simple models of convective clouds*, Q. J. R. Meteor. Soc. *127*, 53–72.

Hourdin, F., Musat, I., Bony, S., Braconnot, P., Codron, F., Dufresne, J-L., Fairhead, L., Filiberti, M-A., Friedlingstein, P., Grandpeix, J-Y., Krinner, G., Le Van, P., Li, Z-X., and Lott, F. (2006), *The LMDZ4 general circulation model: Climate performance and sensitivity to parameterized physics with emphasis on tropical convection*, Climate Dynamics *27*(7-8), 787–813.

Krishnamurti, T. N. and Sanjay, J. (2003), *A new approach to the cumulus parameterization issue*, Tellus A *55*(4), 275–300. doi: 10.1034/j.1600-0870.2003.00021.x

Kuo, H. L. (1974), *Further studies of the parameterization of the influence of cumulus convection of large-scale flow*, J. Atmos. Sci. *31*, 1232–1240.

Lin, C. (1999), *Some bulk properties of cumulus ensembles simulated by a cloud resolving model, Part-II: Entrainment profiles*, J. Atmos. Sci. *56*, 3736–3748.

Manabe, S., Smagorinsky, J., and Strickler, R. F. (1965), *Simulated climatology of a general circulation model with a hydrologic cycle*, Mon. Wea. Rev. *93*, 769–798.

Molteni, F., Corti, S., Ferranti, L., and Slingo, J. M. (2003), *Predictability experiments for the Asian summer monsoon: Impact of SST anomalies on interannual and intraseasonal variability*, J. Climate *16*, 4001–4021.

Neggers, R. A. J., Siebesma, A. P., and Jonker, H. J. J. (2002), *A multiparcel model for shallow cumulus convection*, J. Atmos. Sci. *59*, 1655–1668.

Ooyama, K. (1969), *Numerical simulation of the life cycle of tropical cyclone*, J. Atmos. Sci. *26*, 3–40.

Pruppacher, H. R. and Klett, J. D., *Microphysices of Clouds and Precipitation* (Kluwer Academic Publishers 1997). 954 pp.

RASMUSSON, E. M. and CARPENTER, T. H. (1983), *The relationship between eastern equatorial Pacific sea-surface temperature and rainfall over India and Sri Lanka*, Mon. Wea. Rev. *111*, 517–528.

SADOURNY, R. and LAVAL, K., *January and July Performances of LMD General Circulation Model: New Perspectives in Climate Modelling* (A. Berger, ed.), (Elsevier 1984), pp. 173–198.

SCINOCCA, J. F. and MCFARLANE, N. A. (2004), *The variability of modeled tropical precipitation*, J. Atmos. Sci. *61*(16), 1993–2015.

SIEBESMA, A. P. and CUIJPERS, J. W. M. (1995), *Evaluation of parametric assumptions for shallow cumulus convection,* J. Atmos. Sci. *52*, 650–666.

SLINGO, J. M., MOHANTY, U. C., TIEDTKE, M., and PEARCE, R. P. (1988), *Prediction of the 1979 summer monsoon onset with modified parameterization schemes,* Mon. Wea. Rev. *116*(2), 328–346.

SUD, Y. C., CHAO, W. C., and WALKER, G. K. (1992), *Role of cumulus parameterization scheme in maintaining atmospheric circulation and rainfall in the nine-layer Goddard Laboratory for Atmospheres General Circulation Model,* Mon Wea. Rev. *120*, 594–611.

TIEDTKE, M. (1989), *A comprehensive mass flux scheme for cumulus parameterization in large-scale models,* Mon. Wea. Rev. *17*, 1779–1800.

TURNER, J. S. (1962), *The starting plume in neutral surroundings*, J. Fluid Mech. *13*, 356–368.

WARNER, J. (1970), *The microstructure of cumulus cloud, part-III: The nature of the updraft*, J. Atmos. Sci. *27*, 682–688.

XIE, P., JANOWIAK, J. E., ARKIN, P. A., ADLER, R., GRUBER, A., FERRARO, R., HUFFMAN, G. J., and CURTIS, S. (2003), *GPCP pentad precipitation analysis: An experimental data set based on guage observations and satellite estimates,* J. Climate *16*, 2197–2214.

ZHANG, G. J. (1994), *Effect of cumulus convection on the simulated monsoon circulation in a general circulation model,* Mon. Wea. Rev. *122*, 2022–2038.

(Received April 4, 2006, accepted September 10, 2006)
Published Online First: June 29, 2007

Pure appl. geophys. 164 (2007) 1683–1698
0033–4553/07/091683–16
DOI 10.1007/s00024-007-0237-0

Pure and Applied Geophysics

An Objective Approach for Prediction of Daily Summer Monsoon Rainfall over Orissa (India) due to Interaction of Mesoscale and Large-scale Synoptic Systems

M. MOHAPATRA[1] and U. C. MOHANTY[2]

Abstract—Orissa State, a meteorological subdivision of India, lies on the east coast of India close to north Bay of Bengal and to the south of the normal position of the monsoon trough. The monsoon disturbances such as depressions and cyclonic storms mostly develop to the north of 15° N over the Bay of Bengal and move along the monsoon trough. As Orissa lies in the southwest sector of such disturbances, it experiences very heavy rainfall due to the interaction of these systems with mesoscale convection sometimes leading to flood. The orography due to the Eastern Ghat and other hill peaks in Orissa and environs play a significant role in this interaction. The objective of this study is to develop an objective statistical model to predict the occurrence and quantity of precipitation during the next 24 hours over specific locations of Orissa, due to monsoon disturbances over north Bay and adjoining west central Bay of Bengal based on observations to up 0300 UTC of the day. A probability of precipitation (PoP) model has been developed by applying forward stepwise regression with available surface and upper air meteorological parameters observed in and around Orissa in association with monsoon disturbances during the summer monsoon season (June-September). The PoP forecast has been converted into the deterministic occurrence/non-occurrence of precipitation forecast using the critical value of PoP. The parameters selected through stepwise regression have been considered to develop quantitative precipitation forecast (QPF) model using multiple discriminant analysis (MDA) for categorical prediction of precipitation in different ranges such as 0.1–10, 11–25, 26–50, 51–100 and > 100 mm if the occurrence of precipitation is predicted by PoP model. All the above models have been developed based on data of summer monsoon seasons of 1980–1994, and data during 1995–1998 have been used for testing the skill of the models.Considering six representative stations for six homogeneous regions in Orissa, the PoP model performs very well with percentages of correct forecast for occurrence/non-occurrence of precipitation being about 96% and 88%, respectively for developmental and independent data. The skill of the QPF model, though relatively less, is reasonable for lower ranges of precipitation. The skill of the model is limited for higher ranges of precipitation.

Key words: Monsoon disturbance, Quantitative Precipitation Forecast (QPF), Orissa.

1. Introduction

The summer monsoon rainfall over India is dominated by the semi-permanent monsoon trough which extends from west Pakistan to north Bay of Bengal across northwest India (RAO, 1976) and the westward moving synoptic scale monsoon

[1] India Meteorological Department, Mausam Bhavan, Lodi Road, New Delhi 110003, India.
[2] Centre for Atmospheric Sciences, Indian Institute of Technology, Delhi Hauz Khas, New Delhi 110016, India. E-mail: mohanty@cas.iitd.ernet.in

disturbances developing over north Bay of Bengal (MOOLEY and SHUKLA, 1989). These monsoon disturbances include depression and cyclonic storm. As per criteria of India Meteorological Department (IMD), a monsoon disturbance over the North Indian Ocean is a depression, if the wind speed is 17–27 kt, a deep depression, if the wind speed is 28–33 kt and a cyclonic storm, if the wind speed is more than or equal to 34 kt.

Considering the significance of monsoon disturbances, various attempts have been made to develop synoptic analogues for the prediction of rainfall. The simpler synoptic methods, based on classification of synoptic situations into different types, each associated with certain rainfall distribution, have been applied by many workers. A very good account on the status of quantitative precipitation forecast (QPF) models used in various countries for operational purposes has been described by BELLOCQ (1980). The recent advances in QPF research and possible future direction towards achieving improved use of QPF information in hydrological forecasting have been discussed at length by GEORGAKAKOS and HUDLOW (1984). In Indian context, PISHAROTY and ASNANI (1957) have prepared a composite picture of heavy rainfall (6.5 cm or more in 24 hours) distribution based on rainfall data of four depressions. According to them, the heavy rainfall occurs upto 1000 km ahead of the depression center and upto 400 km to the left of the depression track. MOOLEY (1973) has studied spatial distributions of mean 24-hours rainfall within 500 km of the mean position of the depression centred around 87°, 80° and 75° E longitudes. According to him, maximum rainfall is located about 150 km ahead of 24-hours mean center of depression or about 300 km from the center of depression at the beginning of the 24-hour period and between 50 and 150 km to the left of the track of depression. Considering the mean rainfall per depression day, he has shown that the mean rainfall in the left sector of the depression increases westward. DHAR and MHAISKER (1973), after compositing the rainfall distribution around 10 depressions crossing Orissa coast, have found that heavy rainfall occurs over a strip with its boundaries 40–120 km south of the track of depression. CHOWDHARY and GAIKWARD (1983) have also arrived at similar conclusions about the distribution of 24-hour rainfall around depression by using data of 540 stations in India and 27 depression cases. KRIPALANI and SINGH (1986), by analyzing data of 220 stations for July and August over 16 years period (1958–1973) have shown that the probability of occurrence of rainfall (2.5 mm or more) exceeds 0.8 over an area lying ahead of the depression and varying in extent from 3 to 5 degree latitude width and 5 to 10 degree longitude in length depending on the geographical position of the depression at the start of the 24-hour period.

All the above studies deal with various aspects of synoptic analogue techniques for the forecast of probability of precipitation (PoP) and QPF. KRUZINGA (1989) compared the forecasting of PoP over the Netherlands using an analogue technique and logistic regression. It is found that the regression method performs better than the analogue technique. GLAHN and LOWERY (1969, 1972) have used outputs from numerical weather prediction models and developed regression models to forecast PoP over different parts of the USA. KLEIN (1978) has used multiple discriminant analysis (MDA) for QPF. MOHANTY *et al.* (2001) have developed objective methods to forecast PoP and for QPF over Delhi using classical multivariate regression and discriminant analysis, respectively.

Orissa state, a meteorological subdivision of India, lies on the east coast, adjacent to north Bay of Bengal and close to the south of the normal position of the monsoon trough. The rainfall over Orissa is significantly dependent on frequency, intensity, region of occurrence and movement of the synoptic disturbances (MOHAPATRA and MOHANTY, 2004). Most of the monsoon disturbances develop over northwest Bay of Bengal and move along the monsoon trough across Gangetic West Bengal/ Orissa (MOHAPATRA and MOHANTY, 2005). In addition, there is orographic interaction due to the Eastern Ghat hill range and other hill peaks in the state and surrounding regions. Therefore the daily rainfall distribution over Orissa in association with different synoptic systems is highly complex in nature. PAEGLE (1974) has compared the forecast of PoP over different parts of the USA, derived from equations stratified with respect to the synoptic weather patterns and the equations that are not stratified. According to him, forecasts given by equations using the stratified methods are more accurate. Hence, an attempt has been made in this study to develop objective techniques to predict occurrence/non-occurrence of precipitation due to monsoon disturbances over north Bay and adjoining west central Bay of Bengal during the summer monsoon season and the quantity of precipitation over specific locations during next 24-hour period by stratifying the synoptic systems. To take into account the interaction of the mesoscale processes with the monsoon disturbances, the surface and upper air observations in and around Orissa at 0000 and 0300 UTC of the day of prediction have been considered as potential predictors out of which the actual predictors have been derived by a step-wise regression procedure. These predictors have been used to predict the occurrence/non-occurrence of precipitation over specific locations by the probabilistic approach. The same set of predictors used for the prediction of occurrence/non-occurrence of precipitation has been subjected to MDA to develop a probabilistic QPF model to predict location specific precipitation at different ranges.

Precipitation occurrence is a dichotomous variable for which numerous threat scores are available. Precipitation amount, on the other hand, is an ordinal variable (i.e., its values are defined on an ordinal scale). The measure of accuracy of forecast of such a variable should take its ordinal nature into account. Accordingly, the developed models have been validated with the independent data set. Apart from development of a prediction technique, the objective of this study is to understand the role of interaction of mesoscale convection with the monsoon disturbances over the Bay of Bengal.

2. Data and Methodology

The Figure 1(a) indicates that the maximum rainfall occurs to the south of the monsoon trough and the line of maximum rainfall passes through Orissa. It also indicates that the basic monsoon flow over Orissa is westerly.

The physiographical regions of Orissa are shown in Figure 1(b). Broadly, Orissa consists of four physiographical regions, viz. (i) coastal plain, (ii) northern upland, (iii) central river basin, and (iv) southwest hilly region. Some of the hill peaks in the northern upland and southwest hilly region are as high as 1000 m to 2000 m. Though the Eastern Ghat extends from Tamilnadu in the southwest to north Orissa in the northeast, it is most prominent in north Andhra Pradesh and south Orissa region.

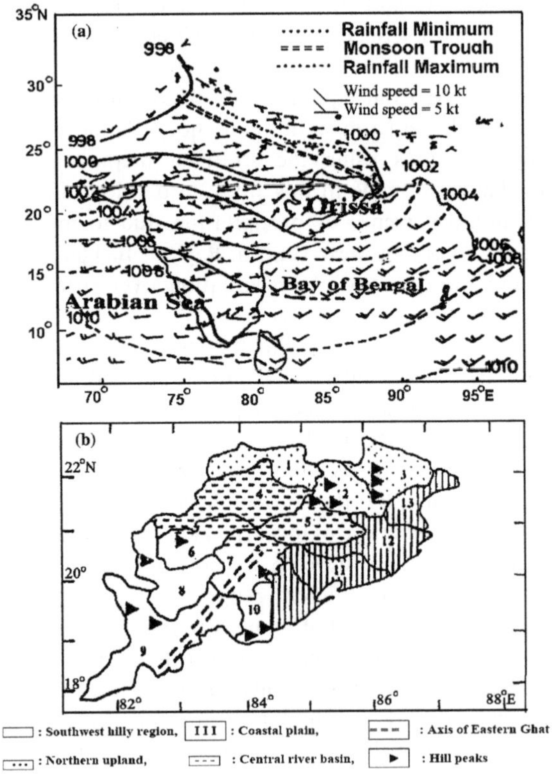

Figure 1
(a) Mean sea-level isobaric pattern (in hPa) and mean wind (in knots) in the representative month of July, plotted according to WMO code over the Indian region Source: Pathan, 1993 and (b) physiographical map of Orissa.

Due to the orographic interaction of Eastern Ghat, the western side of Eastern Ghat becomes the windward region with the basic monsoon flow as westerly in association with the synoptic disturbance over north Bay and hence gets more intense rainfall. When the disturbance lies southward, e.g., over west central Bay, the basic monsoon flow becomes easterly over Orissa and hence the western side of Eastern Ghat becomes lee side and gets less intense rainfall. SMITH (1979) has suggested three independent mechanisms of orographic rainfall, viz. (i) large-scale slope precipitation due to orographically forced vertical motion or convection triggered by smooth orographic ascent bringing the air to saturation resulting in precipitation, (ii) partial evaporation of rainfall from the pre-existing clouds before hitting low ground, and (iii) the rainfall due to orographic control of the formation of cumulonimbus clouds in a conditionally unstable air mass. KASAHARA (1980) has found that the distribution of mountain has a substantial effect on the large-scale monsoon circulation in the region.

Due to all the above reasons, the relationship of rainfall over Orissa and the monsoon disturbance over the Bay of Bengal are highly complex in nature and show large spatial variation with reference to location and intensity of the monsoon disturbance. A PoP model has been developed to predict occurrence/non-occurrence of precipitation and a probabilistic QPF model has been developed for prediction of precipitation in different ranges over six rain gauge stations (Fig. 2a) representing six homogeneous regions of Orissa as obtained by MOHAPATRA *et al.* (2003).

2.1. Development of PoP Model

The PoP model has been developed by applying a multiple regression equation with a stepwise regression technique (DRAPKER and SMITH, 1966) for the forecast of precipitation over a 24-hour period due to monsoon disturbances over the Bay of Bengal based on observed meteorological field parameters in and around Orissa. For this purpose, the daily rainfall recorded at 0300 UTC over six stations representing six homogeneous regions of Orissa (Fig. 2a) during monsoon months has been collected for the period of 1980–1998 from the India Meteorological Department (IMD). The data have been quality checked and missing data, though very few in number are filled by considering the average rainfall recorded over surrounding stations. The available surface parameters at 0300 UTC and upper air parameters at 0000 UTC observed from synoptic and radiosonde/radio wind stations in and around Orissa in association with the monsoon disturbances over north Bay of Bengal and adjoining west central Bay of Bengal have been considered as potential predictors. The regions of monsoon disturbances under consideration are shown in Figure 2(b). The data on monsoon disturbances developing over north Bay of Bengal and adjoining west central Bay of Bengal during the summer monsoon season over the period of 1980–1998 have been collected from weather reports published by IMD. The daily values of different field parameters at various levels of troposphere recorded at 0000/0300 UTC over the

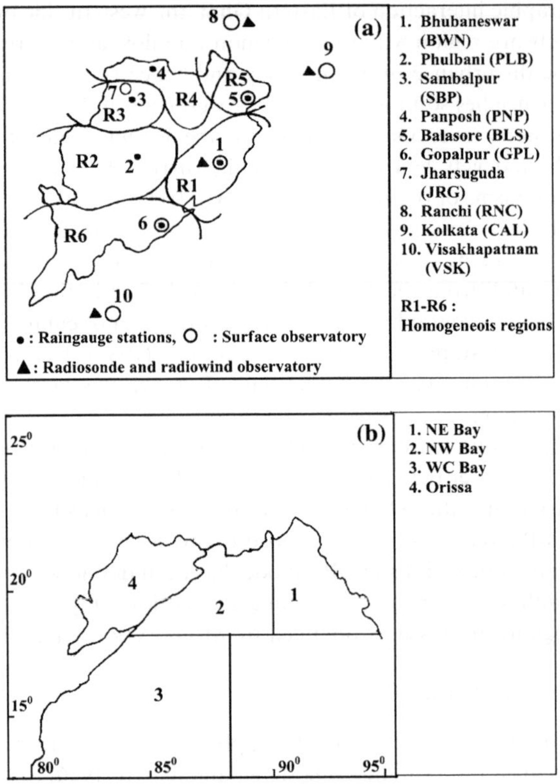

Figure 2

(a) Selected rain gauge stations and homogeneous regions of Orissa along with surface and upper air observatory stations under consideration, and (b) regions of monsoon disturbances under consideration.

stations in Orissa and its surrounding in association with the monsoon disturbances during the monsoon season for the same period of 1980–1998 have also been collected from IMD. Different surface and upper-air stations selected for the study are shown in Figure 2(a). Different field parameters considered in the study include maximum temperature, minimum temperature, dry bulb temperature, dew point, mean sea-level pressure (MSLP)/geo-potential height, zonal wind and meridional wind at surface and different standard isobaric levels in the troposphere (Table 1). The actual wind is resolved into zonal and meridional components.

As 24-hour accumulated precipitation amounts are recorded at 0300 UTC, the PoP model is initiated at 0300 UTC to generate the forecast for the next 24 hours (day 1 forecast). In the development of the PoP model, the value of predictand

Table 1

Meteorological parameters considered as potential predictors of precipitation over Orissa due to monsoon disturbances over the Bay of Bengal during the summer monsoon season

Meteorological parameters	Stations	Meteorological subdivision	Observation levels
Surface			
Maximum temperature (Tx)	Visakhapatnam	North coastal Andhra	Surface
Minimum temperature (Tn)	(VSK)	Pradesh (NCAP)	
Mean sea-level pressure	Gopalpur (GPL)	Orissa	
(MSLP)	Bhubaneswar (BWN)	Orissa	
Dry bulb temperature (T)	Balasore (BLS)	Orissa	
Dew point (Td)	Kolkata (CAL)	Gangetic West Bengal	
Zonal component of wind (u)		(GWB)	
Meridional component of	Jharsuguda (JRG)	Orissa	
wind (v)	Ranchi (RNC)	Jharkhand (JKD)	
Upper air			
Geopotential height (h)	Visakhapatnam (VSK)	NCAP	850, 700,
Dry bulb temperature (T)	Bhubaneswar (BWN)	Orissa	500, 400,
Dew point (Td)	Kolkata (CAL)	GWB	300 and 200
u and v	Ranchi (RNC)	JKD	hPa

(precipitation) is taken as 1 if the precipitation occurs and 0 if it does not. Hence, the value of predictand varies from 0 (0%) to 1 (100%). The sigma restricted parameterization is used in this stepwise regression model. The predictors are selected from the set of potential predictors as mentioned above using a forward stepwise procedure. In this criterion, the potential predictors showing a 99% level of confidence with the predictand are selected for the development of the multiple regression equation for all the specific locations under consideration except SBP (Fig. 2a). The level of confidence has been considered as 95% in the case of SBP. The level of confidence is so chosen that the variance explained by the model exceeds 50%. The correlation coefficients of the predictand with the parameters thus obtained are calculated. The parameters which do not show signification correlations have been dropped, as they may not explain any significant amount of variance. The remaining parameters have been used to develop the multiple regression model for different stations under consideration. The values of predictand are recalculated for all observations of the developmental data set. If the recalculated value of the predictand is greater than 1, it is made equal to 1 and if it is less than 0, it is made equal to 0. The recalculated values of the predictands are grouped into intervals of 0.1. For each group, the observed probability of occurrence and non-occurrence of precipitation is evaluated. The objective is to find the critical value of the predictand for decision-making so that the probability forecast can be converted into a deterministic forecast of occurrence/non-occurrence of precipitation. Thus based on this critical value, the probability

Table 2

The critical values of probability of precipitation for occurrence/non-occurrence of precipitation over specific locations of Orissa due to monsoon disturbances over the Bay of Bengal during the summer monsoon season

Raingauge station	Probability of precipitation
Phulbani (PLB)	0.72
Gopalpur (GPL)	0.80
Sambalpur (SBP)	0.60
Bhubaneswar (BWN)	0.70
Panposh (PNP)	0.53
Balasore (BLS)	0.74

forecast is converted into a deterministic forecast. The critical values for different stations under consideration are shown in Table 2.

2.2. Development of QPF Model

The parameters selected for the regression model have been considered to develop a probabilistic QPF model using MDA technique for prediction of QPF, if the occurrence of precipitation is predicted by the PoP model. For categorical prediction of precipitation in different ranges, five ranges such as 0.1–10, 11–25, 26–50, 51–100 and > 100 mm have been considered. A similar classification of precipitation is used by IMD for QPF over different river catchments on an operational basis. The MDA procedure yields (G-1) discriminant functions for the G groups, which are used to classify an event (MILLER, 1962; WILKS, 1995). The model gives a probabilistic forecast of the most likely group to which the 24-hour precipitation belongs to. Since there are five groups in the present study, the MDA procedure yields four discriminant functions of the form

$$Z_g = w_1 x_1 + w_2 x_2 + \ldots + w_m x_m,$$

where z_g are discriminant functions (scores), w_i are the discriminant weights (coefficients) and x_i are the independent variables. The interpretation of discriminant weights involves the examination of the sign and magnitude of weights. Independent variables with relatively large weights contribute more to the discriminating power of the function than the smaller ones. Thus, when the sign is ignored, each weight represents the relative contribution of its associated variable to that discriminant function. The sign merely denotes that the variable makes either a positive or negative contribution. A set of observations (predictors) is assigned to one of the groups for which the classification score is maximum. The model has been developed with *a priori* classification of probabilities. If the sample is a reflection of the true distribution in the population, *a priori* probability should be proportional to the sizes of the groups in the sample. If the unequal number of cases in different groups is only the random result of the sampling procedure, the *a priori* probability should be equal in each group. The specification of different *a priori* probabilities can greatly affect the accuracy of prediction. In the present study, the sample is the reflection of true distribution and hence *a priori* probability proportional to the sizes of the groups is more applicable. However, both equal probability and proportional probability to categories of precipitation have been used as *a priori* classification to ascertain the superior probability of classification.

The model has been developed based on data of 1980–1994 and data during 1995–1998 have been used for testing the skill of the model. The developmental data set consists of 48 days of monsoon disturbances over the Bay of Bengal, including 2 days over northeast (NE) Bay, 36 days over northwest (NW) Bay and 10 days over west central (WC) Bay off north coastal Andhra Pradesh. The test data set consists of 14 days of monsoon disturbances over the Bay of Bengal including 2 days over NE Bay, 5 days over NW Bay and 7 days over WC Bay off north coastal Andhra Pradesh. The days of monsoon disturbances over NW Bay, NE Bay and WC Bay of Bengal are only considered as these disturbances cause heavy to very heavy rainfall over Orissa. The frequency of monsoon disturbances during the period under study (1980–1998) is significantly less than the climatological average (MOHAPATRA and MOHANTY, 2004). As a result, the number of monsoon disturbance days over the regions under consideration during 1980–1998 has been significantly less. The statistical software, STATISTICA UTILITY (1994) has been applied for both the stepwise regression and the MDA. The verification has been attempted by computing various skill/threat scores based on WILKS (1995).

3. Results and Discussion

The relationship between the predictors and the location specific precipitation is presented and discussed in section 3.1. The characteristics of the developed models are discussed in section 3.2. The performance of the PoP model and QPF model evaluated with developmental and independent data sets are presented and discussed in section 3.3 and 3.4, respectively.

3.1. Relation between Selected Predictors and Precipitation

From Table 3, the precipitation over the western side of the Eastern Ghat (PLB) and the adjoining western part of the central river basin (SBP) and the western part of northern upland (PNP) is more dependent on wind at lower/middle tropospheric levels (upto 500 hPa). The precipitation in the coastal plain is more dependent on temperature and dew point at surface/lower tropospheric levels (upto 700 hPa).

The lower temperature and higher dew point over GPL at surface level is associated with rainfall over BWN. Similarly, the lower temperature and higher dew

Table 3

Correlation coefficients (CC) of the selected predictors for forecasting PoP over a particular station

Station	Predictors	CC
PLB	CAL-Tn-surface	0.294
	CAL-u-700 hPa	−0.454
	CAL-u-500 hPa	−0.407
	RNC-v-700 hPa	0.474
GPL	BWN-T-850 hPa	−0.319
	BWN-Td-850 hPa	0.363
	RNC-v-surface	0.638
SBP	RNC-Tx-surface	−0.374
	CAL-Td-700 hPa	0.468
	CAL-v-surface	−0.289
BWN	GPL-T-surface	−0.81
	GPL-Td-surface	0.71
PNP	BWN-MSLP	−0.356
	JRG-Td-surface	0.421
	VSK-u-500 hPa	0.282
BLS	GPL-Tn-surface	−0.301
	BLS-T-surface	−0.308
	GPL-Td-surface	0.781
	JRG-Td-surface	0.764
	CAL-Td-700 hPa	0.669

point over BWN at 850 hPa are associated with rainfall over GPL; the lower temperature over BLS at surface level and a higher dew point at surface level over GPL and JRG along with a higher dew point over CAL at 700 hPa with rainfall over BLS and the lower maximum temperature over RNC and a higher dew point over CAL at 700 hPa with rainfall over SBP. The above results may be due to the fact that the dew point is generally maximum along the monsoon trough and rainfall occurs around the monsoon trough, though the maximum rainfall occurs to the south of the monsoon trough. The inverse relationship between the temperature at lower levels and summer monsoon rainfall has earlier been shown by SRINIVASAN and SADASIVAN (1975). In addition to the above, the stronger southerly component of wind at surface level over RNC is favorable for rainfall over GPL due to a monsoon disturbance over the Bay. The stronger southerly component at surface is possible only if the associated monsoon trough passes to the south of RNC latitude. Hence the southward location (south of RNC and near BWN) of the monsoon trough in association with the southward location of monsoon disturbances over the Bay of Bengal, e.g., over NW and adjoining WC Bay is favorable for rainfall over GPL.

The stronger northerly component of wind at surface level over CAL is favorable for higher rainfall over SBP. It may be due to the fact that the above pattern of wind is normally associated with monsoon disturbances over the northern part of the Bay (e.g., NW and adjoining NE Bay) with the monsoon trough from the system center passing near BLS, which is favorable for rainfall over SBP. The favorable factors for rainfall over PNP include a stronger westerly component of wind over VSK at 500 hPa level along with lower MSLP over BWN and a higher dew point over JRG at surface level. The above conditions are possible only if the associated monsoon trough passes close to BWN-JRG latitude. The favorable factors for rainfall over PLB include a stronger easterly component of wind over CAL at lower and middle levels and a stronger southerly component of wind at 700 hPa over RNC. It indicates the relatively southward location of the monsoon disturbance over the northern part of the Bay of Bengal (e.g., over NW Bay) along with a stronger pressure gradient in the southwest sector and the southward location of the monsoon trough favors rainfall over PLB.

To summarize the above, the temperature, dew point and wind in the lower and mid-tropospheric levels, in association with the monsoon disturbances over the Bay of Bengal, play a significant role in rainfall over Orissa. The upper tropospheric (300 hPa and above) field parameters, in association with the monsoon disturbances over the Bay of Bengal, play an insignificant role in rainfall over Orissa, as no such field parameter figures as the predictor.

3.2. Model Characteristics

The salient features of the PoP model for prediction of occurrence/non-occurrence of precipitation are given in Table 4. The multiple correlation coefficient (MCC) of the selected predictors for the stepwise regression used in the above model with the precipitation over specific locations varies from 0.58 to 0.94. Hence, the MCCs are highly significant. The MCC is relatively higher over the coastal stations. The variance explained by the selected predictors ranges from about 33% to 89%. The explained variance is also higher over coastal Orissa. The F ratio is highly significant for all the stations. It is also higher over coastal Orissa. The MCC and variance explained are less over the western side of the Eastern Ghat, adjoining western part of the central river basin and the western side of the northern upland. It may be due to the fact that the model could not capture mesoscale convection over the region interacting with the basic flow and monsoon disturbances over the Bay of Bengal.

3.3. Performance of PoP Model

Considering the verification measures for all the stations under consideration together, the model yields very good performance for both the developmental and independent data (Table 5). However, the performance is slightly less in the case of independent data. The percentage corrects (PC) are about 96% and 88%, respectively for developmental and independent data. The Heidke skill score (HSS) is 0.75 and 0.59 for developmental and independent data, respectively. The critical success index (CSI) is very high, being 0.87 with probability of detection (POD) of 0.91 in the case of independent data. As the bias for occurrence (BIAS) is nearly 1 in

Table 4

The model characteristics for the forecast of PoP over specific locations of Orissa due to monsoon disturbances over the Bay of Bengal

Station	MCC	Variance (%)	F-ratio	
			Calculated value	Tabular value for 1% significance level
PLB	0.67	70.8	8.6	3.8
GPL	0.90	81.2	57.4	4.3
SBP	0.68	47.4	12.6	4.3
BWN	0.90	81.2	93.6	4.1
PNP	0.58	33.0	7.3	4.3
BLS	0.94	88.6	59.3	3.5

Table 5

Verification measures for the PoP model

Measure	Developmental data	Independent data
Probability of detection (POD)	0.98	0.91
False alarm rate (FAR)	0.02	0.06
Missing rate (MR)	0.02	0.08
Correct non-occurrence (C-NON)	0.76	0.70
Critical success index (CSI)	0.96	0.87
True skill score (TSS)	0.74	0.62
Heidke skill score (HSS)	0.75	0.59
Bias for occurrence (BIAS)	1.00	0.98
Percentage correct (PC)	96	88

both the cases, the model is not biased either to overforecasting or underforecasting. Hence, the developed PoP model can be very well applied for the prediction of the occurrence/non-occurrence of location specific rainfall over Orissa due to monsoon disturbances over the Bay of Bengal. However, the relatively poor performance of the model during the test period may be due to the limited period of the data based on which the model is developed. The model developed with a longer period of data may show better efficiency.

3.4. Performance of Probabilistic Model for QPF

The performance of the probabilistic model for QPF at different ranges over specific locations of Orissa due to monsoon disturbances over the Bay of Bengal is evaluated and results are presented in Table 6. Considering QPF in different ranges together, the average PC is higher with a priori proportional probability than with *a priori* equal probability of classification applied in the model for both developmental and independent data sets. The PC is relatively less with *a priori* proportional probability whereas it is relatively higher in the case of equal probability for a higher range of precipitation. The PC is slightly less in the case of independent data than in the case of developmental data. The CSI is higher with proportional probability both in the case of developmental and independent data sets for a lower range of precipitation. The CSI is significantly less for a higher range of precipitation. Considering the HSS, the model performance does not show a significant difference with respect to the type of *a priori* classification probability. The performance of the QPF model is far from satisfactory except for the lower ranges of precipitation. It may be due to the fact that the mesoscale convection leading to significant point rainfall could not be captured in the model. The meso-scale convection develops due to the interaction of monsoon disturbances over the north and west central Bay of Bengal with the basic monson flow and orography due to the Eastern Ghat and other hill peaks in Orissa and its environs. The performance of the model has further deteriorated during the independent data period. It may be due to the fact that the model is based on a limited data of 48 days of monsoon disturbances

Table 6

(a) Percentage correct (PC) measure of the QPF model

Category	Equal *a priori* probability		Proportional *a priori* probability	
	Developmental Data	Independent Data	Developmental Data	Independent Data
0.1–10 mm	40.9	37.5	63.9	50.0
11–25 mm	39.3	25.0	47.5	31.3
26–50 mm	32.7	16.7	36.7	16.7
51–100 mm	40.0	50.0	25.0	25.0
100 mm	50.0	0.0	13.6	0.0
Total	39.6	29.5	44.3	34.1

(b) Critical success index (CSI) of the QPF model

Category	Equal *a priori* probability		Proportional *a priori* probability	
	Developmental Data	Independent Data	Developmental Data	Independent Data
0.1–10 mm	0.30	0.32	0.39	0.36
11–25 mm	0.26	0.21	0.28	0.19
26–50 mm	0.21	0.08	0.22	0.08
51–100 mm	0.22	0.13	0.20	0.13
> 100 mm	0.23	0.0	0.12	0.0

(c) Heidke skill score (HSS) of the QPF model

Equal *a priori* probability		Proportional *a priori* probability	
Developmental Data	Independent Data	Developmental Data	Independent Data
0.29	0.13	0.25	0.10

over the Bay of Bengal. The use of a longer period of data for development of the model may improve the efficiency of the model.

The performance of the PoP model is considerably better than that of the QPF model, as revealed from various skill scores. This may be attributed to the fact that the precipitation amount is highly variable in space and time due to the interaction of large-scale synoptic systems such as monsoon disturbances with basic flow and orography. Another limitation may be the inadequacy of data over the Bay of Bengal and Eastern Ghat region of Orissa. However the performance of the QPF model can be further improved by further stratifying the monsoon disturbances influencing the precipitation over Orissa based on their intensity, location, movement and rate of intensification, etc. For this purpose data for a long period is very essential. Also the use of non-conventional data over the Bay of Bengal and data sparse land regions, and the use of numerical model outputs as potential predictors in addition to the conventional data for the selection of actual predictors may improve the performance of the model.

4. Conclusions

The PoP and QPF models are developed respectively for the prediction of the occurrence/non-occurrence of precipitation and amount of precipitation in different categories over specific locations in Orissa due to monsoon disturbances over the Bay of Bengal. Based on the model characteristics and the evaluation of the models with independent data sets, the following broad conclusions are drawn.

The PoP model works well for the prediction of the occurrence/non-occurrence of precipitation with PC of 88% without BIAS during the evaluation period.

The QPF model shows a relatively poor performance compared to the PoP model. Further, the QPF model is satisfactory only for the prediction of lower ranges of precipitation, The performance of the QPF model for the prediction of a higher range of precipitation (> 50 mm) is very poor, as the model could not capture mesoscale convection due to the interaction of the monsoon disturbance with basic flow and orography. The performance deteriorates during the test period probably due to the fact that the model is based on a limited data of monsoon disturbances. The use of numerical model outputs and nonconventional data sets over the Bay of Bengal and data sparse Eastern Ghat region in the mesoscale network as potential predictors for the selection of actual predictors and further stratification of large data sets of monsoon disturbances, based on their location and intensity, may improve the performance of PoP and QPF models.

The MCC of predictors with the precipitation and variance explained by them are less for the western side of the Eastern Ghat, the adjoining western part of the central river basin and the western part of northern upland as the model could not capture mesoscale convection over the region interacting with basic flow and monsoon disturbances over the Bay of Bengal.

The temperature, dew point and wind in lower and mid-tropospheric levels in association with the monsoon disturbances over the Bay of Bengal play significant roles in rainfall over Orissa. The upper tropospheric (300 hPa and above) field parameters in association with the monsoon disturbances over the Bay of Bengal play insignificant roles in rainfall over Orissa, as no such field parameter figures as the predictor.

The precipitation over the western side of the Eastern Ghat (PLB) and the adjoining western part of the central river basin (SBP) and the western part of the northern upland (PNP) is more dependent on MSLP and wind at lower/middle tropospheric levels (upto 500 hPa). The pecipitation in the coastal plain is more dependent on temperature and dew point at surface/lower tropospheric levels (upto 700 hPa).

Acknowledgement

The authors are thankful to the Director General of Meteorology, India Meteorological Department, New Delhi for his encouragement and support for this study. The authors are also thankful to the Additional Director General of Meteorology, India Meteorological Department, Pune for the supply of data to carry out the work.

References

Bellocq, A. (1980), *Operational models and quantitative precipitation forecasts for hydrological purposes and possibilities of an inter comparison*, WMO, Geneva, Switzerland.

Chowdhary, A. and Gaikward, S.D. (1983), *On some characteristic features of rainfall associated with monsoon depressions in India*, Mausam *34*, 33–42.

Dhar, O.N. and Mhaiskar, P.R. (1973), *Areal and point distribution of rainfall associated with depressions/ storms on the day of crossing the east coast of India*, Indian J. Met. Geophys. *24*, 271–278.

Draper, N.R. and Smith, H., *Applied Regression Analysis* (Wiley and Sons, New York, 1966) 407

Georgakakos, K.P. and Hudlow, M.D. (1984), *Quantitative precipitation forecast techniques for use in hydrologic forecasting*, Bull. Amer. Met. Soc. *65*, 1186–1200.

Glahn, H.R. and Lowery, D.A. (1969), *An operational method for objectively forecasting probability of precipitation*, ESSA Tech. Memo, WBTM *27*, 24 pp.

Glahn, H.R. and Lowery, D.A. (1972), *The use of model output statistics (MOS) in objective weather forecasting*, J. Appl. Met. *11*, 1203–1211.

Kasahara, A. (1980), *Influence of orography on the atmospheric general circulation*, Garp Publication series *23*, 1–49.

Klein, W. (1978), *Statistical forecast of local weather by means of model output statistics (MOS) in objective weather forecasting*, J. Appl. Met. *11*, 1203–1211.

Kripalani, R.H. and Singh, S.V. (1986), *Rainfall probabilities and amount associated with monsoon depression over India*, Mausam *37*, 111–116.

Kruzinga, S. (1989), *Statistical interpretation of ECMWF products in Dutch weather service*. In ECMWF seminar/ workshop on interpretation of NWP products, ECMWF, Reading, 360–365.

Miller, R.G. (1962), *Statistical prediction by discriminant analysis, Meteorological Monograph*, American Meteorological Society *25*, 3–14.

Mohanty, U.C., Ravi, N., and Madan, O.P. (2001), *Forecasting precipitation over Delhi during the southwest monsoon season*, Meteor. Appl. *8*, 11–21.

MOHAPATRA, M., MOHANTY, U.C. and BEHERA, S. (2003), *Spatial variability of daily rainfall over Orissa (India) during southwest summer monsoon season,* Int. J. of Climatol. *23*, 1867–1887.
MOHAPATRA, M. and MOHANTY, U.C. (2004), *Some characteristics of low pressure systems and summer monsoon rainfall over Orissa* Current Science *87*, 1245–1255.
MOHAPATRA, M. and MOHANTY, U.C. (2005), *Some characteristics of very heavy rainfall over Orissa during summer monsoon season*, J. Earth System Sci. (formerly known as Proc. Indian Academy of Sciences (Earth and Planetary Sciences) *114*, 17–36.
MOOLEY, D.A. (1973), *Some aspects of Indian monsoon depression and associated rainfall*, Mon. Wea. Rev. *101*, 271–280.
MOOLEY, D.A. and SHUKLA, J. (1989), *Main features of the westward moving low pressure systems which form over the Indian region during the summer monsoon season and their relation to the monsoon rainfall*, Mausam *40*, 137–152.
PAEGLE, J.N. (1974), *Prediction of precipitation probability based on 500 mb flow types*, J. Appl. Met. *13*, 213–220.
PATHAN, J.M. (1993), *Latitudinal variation of rainfall during the month of July in relation to the axis of the monsoon trough over India*, Mausam, *44*, 384–386.
PISHAROTY, P.R. and ASNANI, G.C. (1957), *Rainfall around monsoon depression over India,* Indian J. Met. Geophys. *8*, 15–20.
RAO, Y.P. (1976), *Southwest monsoon*, Met. Monogr. Syno. Met. *1/1976*, India Meteorological Department, 1–367.
SMITH, R.B. (1979), *The influence of mountains on the atmosphere*, Advances in Geophy. *21*, 187–230.
SRINIVASAN, V. and SADASIVAN, V. (1975), *Thermodynamic structure of the atmosphere over India during southwest monsoon season*, Mausam *26*, 169–180.
STATISTICA UTILITY (1994), *Statistica for Windows (vol. III): Statistics II, Statsoft,* Tulsa OK, 958 pp.
WILKS, D.S., *Statistical Methods in the Atmospheric Sciences* (Academic press 1995), 466 pp.

(Received April 26, 2006, accepted September 2006)
Published Online First: June 29, 2007

Pure appl. geophys. 164 (2007) 1699–1716
0033–4553/07/091699–18
DOI 10.1007/s00024-007-0240-5

Pure and Applied Geophysics

Interactive Aspects of the Indian and the African Summer Monsoon Systems

P. Sanjeeva Rao[1] and D. R. Sikka[2]

Abstract—This study addresses an understanding of the possible mutual interactions of sub-seasonal variability of the two neighboring regional monsoon systems through data analysis. The NCEP/NCAR re-analysis and OLR data for three years was used to reveal the large-scale organization of convective episodes on synoptic (~5 days) and low frequency (15–50 day) scales. It is found that synoptic scale organization over both the sectors is influenced by the eastward migration of large-scale convective episodes associated with the Madden Julian Oscillation (MJO) on the low frequency scale. The organization of convection associated with the African monsoon on the synoptic scale is influenced by the pulsatory character of lower mid-troposphere and upper troposphere wind regimes moving westward over the African sector. Over the Indian region formation of low pressure areas and depressions in the monsoon trough occur in an overlapping manner under an envelope of low frequency seasonal oscillation. We have also found some correspondence between the summer monsoon rainfall over tropical North Africa and India on a decadal basis, which would suggest a common mode of multi-decadal variability in the two monsoon systems. The study points out the need to organize simultaneous field campaigns over the Indian and the African monsoon regions so as to bring out observational features of possible interactions between the two neighboring systems, which could then be validated through modeling studies.

Key words: Monsoon, convection, clouds, inter-tropical convergence zone, Madden Julian oscillation and tropics.

1. *Introduction*

The Indian and the West African summer monsoon systems develop in proximity of each other and have similar large-scale atmospheric features controlling their behavior. The former system is more intense than the latter. The Indian monsoon also extends more northward (Lat. 25–30°N) compared to the West African monsoon (about 15°N) in their peak phase during July. This study is an attempt to compare and contrast the two systems in terms of large-scale and transient features as well as attempts to determine the influence of one with respect to the other. The

[1] Department of Science and Technology, Earth System Science Division, New Delhi 110016, India. E-mail: psanjeevarao@yahoo.com, psrao@nic.in
[2] 40, Mausam Vihar, New Delhi 110 051, India. E-mail: drsikka@yahoo.com

zonal winds based on daily and pentad average at 850, 700 and 200 hPa re-analysis (KALNAY *et al.*, 1996) of NCEP/NCAR, and visible satellite imagery from METEOSAT have been examined in this preliminary study for three years 1999, 2002 and 2003. We have relied on pentad data rather than applying any sophisticated filter to study the subseasonal fluctuations in the fields examined. In section 2 we briefly discuss the large-scale contrasting atmosphere-ocean-land features of the two monsoon systems, which may be responsible for the characteristicly large-scale monsoon circulations in the two regions. Section 3 examines the build-up of the two systems, important transients and low frequency intra-seasonal oscillations (ISO) which modulate the two systems. Section 4 broadly examines cloud clusters in respect to their spatial and temporal scales and intensity over the two regions, and the major episodes which could be discussed from the 5-day average outgoing longwave radiation (OLR) values in the three years. The data suggest the possible role of eastward moving Madden-Julian Oscillation (MJO) on low frequency (30–50 day) scale from the near-equatorial African region to the Indian longitudes in modulating monsoon activity over the Indian region. We also suspect that the modulations on the bi-weekly scale in the tropical easterly flow at 200 hPa may produce the modulations in the lower mid-tropospheric African Easterly Jet (AEJ) which control the activity of easterly waves over the African region. In section 5 we examine the possibility of a simultaneous field program in tropical Africa and India in one of the coming years for the purpose of investigating in detail the interactions among the two neighboring monsoons on a subseasonal scale. Summary and conclusions are provided in section 6.

2. Contrasts in Land – Ocean Environments and Large-scale Atmospheric Features of the Indian and the West African Summer Monsoon Systems

The geography of the two regions in terms of land and adjoining ocean features shows several contrasts in surface temperature, sea-surface temperatures (SST), vegetation cover, orographic features etc., which have an important bearing on the sensible and latent heat fluxes, surface albedo and distribution of mid-tropospheric heat sources. The contrasts in land-ocean environments to a large extent are responsible for the higher intensity and more northward reach of the Indian monsoon compared to the African monsoon. Noteworthy differences in the two monsoon systems are:

(i) Extensive cross-equatorial flow and more northward extent and depth (up to 600 hPa) of the Indian monsoon as compared to the West African monsoon, and (ii) existence of the 'Tropical Easterly Jet Stream (TEJ)' at 200 hPa along 8–15°N over the Indian longitudes with wind speeds of about 30 m/sec. The TEJ looses its identity along 40–60°E and weakens to 5–10 m/sec when it is leaving West Africa. These can be seen from Figures 1(a,b) which depict the normal July wind flow at 850 hPa and

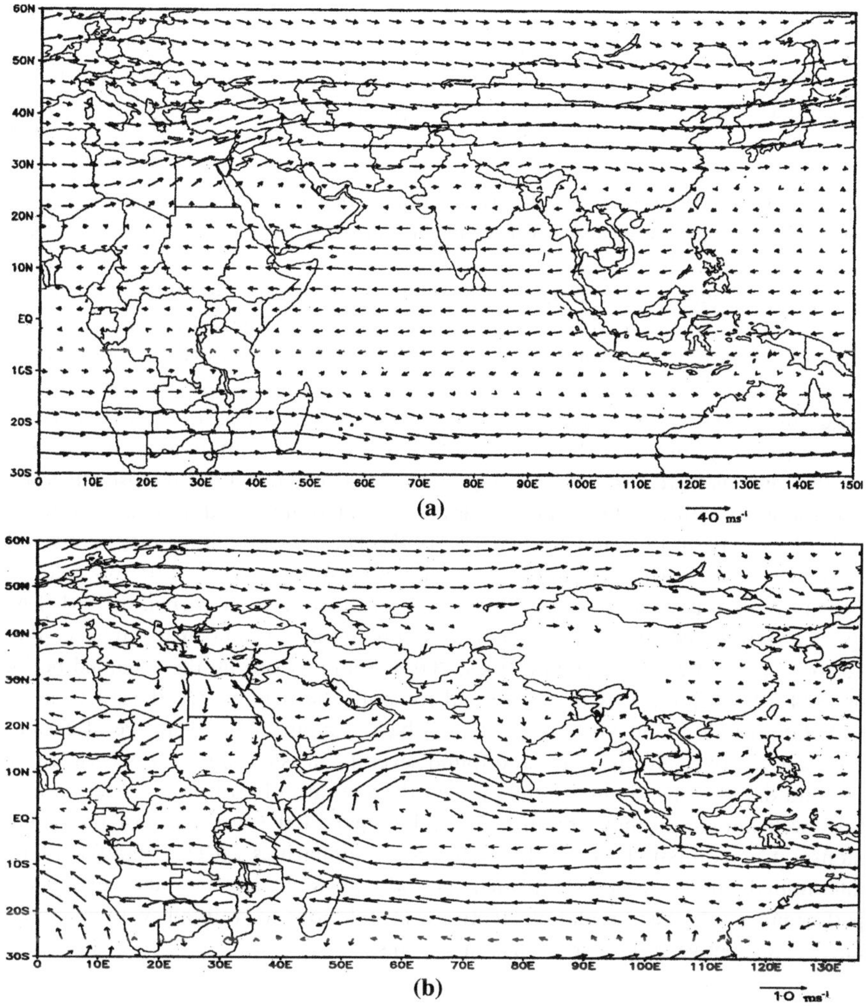

Figure 1
(a) Climatological wind pattern at 200 hPa level over the African and Indian regions during July (Source: India Meteorological Department). (b) Climatological wind pattern at 850 hPa level over the African and Indian regions during July 2002 (Source: India Meteorological Department).

200 hPa levels respectively, based on the India Meteorological Department climatological charts. Figures 1(a,b) also show the dominant presence of the monsoon trough at 850 hPa along 22–27°N over the Indian region and the ITCZ along 10–15°N over the African region. The extensive ridge at 200 hPa level stretching

from Africa to China along 27–32°N is also prominent. At 700–500 hPa (Figure not shown) levels there is a sub tropical anticyclone over the Arabian-Iranian region with its outflow directed towards the African region along 10–15°N. Perturbations in this flow (AEJ) affect the African monsoon on 5-day and biweekly scales (CARSON, 1969; OKULJA, 1970; BURPEE, 1975; REED *et al.*, 1977; PEDGLEY and KRISHNAMURTI, 1976).

NICHOLSON (1980) has broadly divided north African annual rainfall into zonally oriented belts with increasing coverage in rainfall from 50–100 cm along 10–15°N to 100–200 cm from 10°N to the equator. PARTHSARATHY *et al.* (1995) divided India into five homogeneous regions. Figures 2 (a, b) show the distribution of annual rainfall for the northern African region and India, respectively. Nearly 80 percent of the annual rainfall over India and Sahelion regions is provided by the summer monsoon period during June to September. There have been studies pertaining to the variability of African rainfall (NICHOLSON, 1997, 2000; NICHOLSON and KIM, 1997). There are many studies dealing with the Indian monsoon rainfall. Such studies over the two regions have also covered long-term inter-decadal trends and also suggested a relationship with ENSO warm episodes with reduction in rainfall on a interannual scale (SIKKA, 1980, 1999 and NICHOLSON, 1997). The two rainfall regimes fluctuate in unison to each other on a multi-decadal scale (WARD, 1998). For example the African rainfall produced −20 to −30 percent of the long-term normal in 1961–1990 epoch (NICHOLSON, 2000) and the Indian rainfall in the same epoch also registered by −5 to −7 percent of the normal (SIKKA, 2003). Similarly during the extended period of 1984–1987 rainfall was below normal for both the regions. On the inter-annual scale GRIST (2002) and GRIST *et al.* (2002) have examined the incidence of easterly wave activity over North Africa and found several consistent differences between the wet and dry seasons. Similar differences in the strength of monsoon westerlies have been well known as they are stronger and deeper in the wet years of the Indian monsoon as compared to the dry years. Such similarities on the behavior of the interannual, multi-decadal extended periods as well as wet and dry year's performance of the circulation features in relation to rainfall would suggest interactions between the two monsoon systems.

3. *Build-up of the Two Monsoon Systems and Transient Activity*

This aspect was examined in this study with respect to the onset of the westerly monsoon winds (onset of westerly winds or the cross-equatorial flow); the period in which the monsoon trough or ITCZ in the two regions reaches its maximum northward transect and the beginning of the southward march of the ITCZ in September. The monsoon advances over India gradually in a pulsatory manner and the northern-most reach of ITCZ (monsoon trough) occurs by early to mid-July. For our purpose we considered the monsoon onset in terms of the sweep and strength of

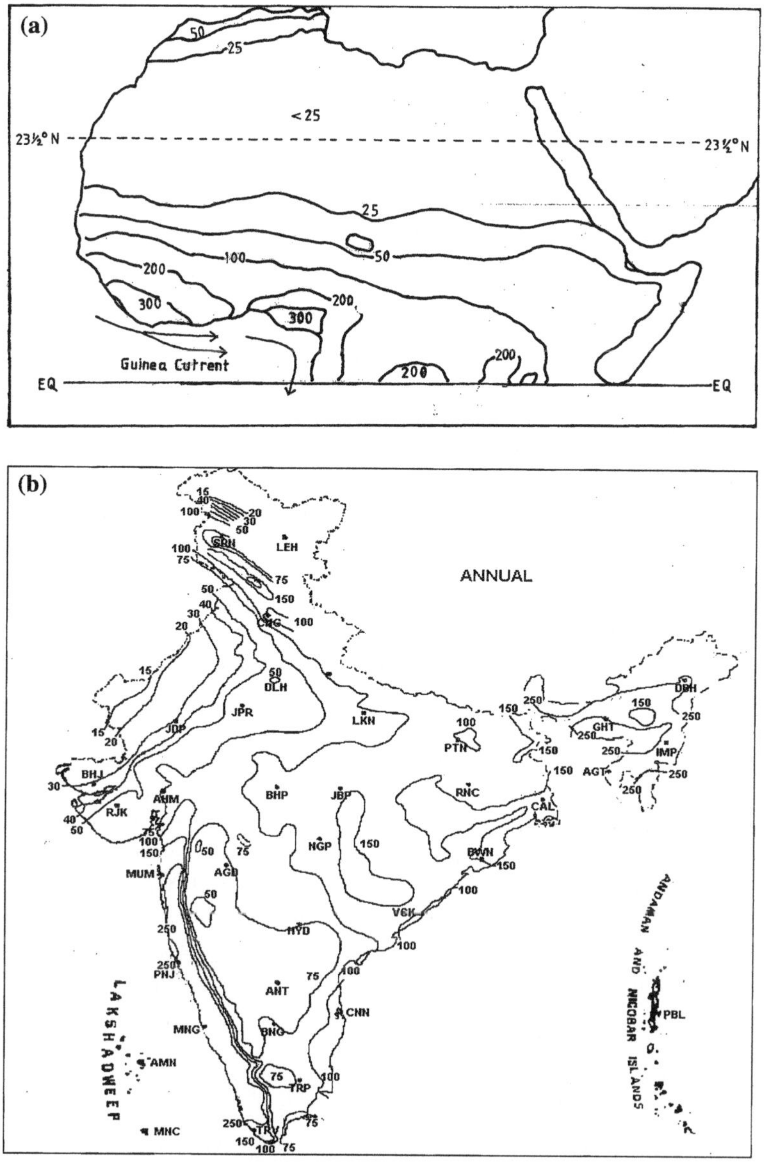

Figure 2
(a) Distribution of annual rainfall (Cm) over North Africa. (b) Distribution of annual rainfall (Cm) over India (Source: India Meteorological Department).

the westerly winds at 850 hPa. Components of the evolution of the monsoon for the two years 1999 and 2003 are given in Table 1, based on the 5-day average wind field at 850 hPa over the regions under study. In spite of the interannual variability in these events, the onset of the two systems on the whole, as shown in Table 1, is delayed by one to two weeks in the West African region compared to the Indian sector. On the contrary, the withdrawal of the monsoon occurs earlier over the African region compared to the Indian region. The intra-seasonal fluctuations in the position of the ITCZ in the two regions are dictated by three scales of atmospheric transients, viz., synoptic scale (3–5 days), quasi-biweekly scale (10–20 day) and 30–50 day low-frequency scale. Synoptic scale transients occur in the form of monsoon depressions and low pressure areas in the monsoon trough over India (SIKKA, 2006) and as mid-tropospheric easterly wave disturbances over the African region (CARSON, 1969; OKLUJA, 1970; PEDGLEY and KRISHNAMURTI, 1976; BURPEE, 1975; and REED *et al.*, 1977). There is interannual variability in the frequency of synoptic scale transients in both regions. Synoptic scale disturbances of the Indian monsoon area weaken by the time they reach 70°E and very seldom do they even move up to the Saudi Arabian coast. Easterly waves of the African region advance westward and toward the tropical North Atlantic Ocean. Thus synoptic scale transients forming in either region do not become advected from one region to other. The two monsoon systems are known to fluctuate on a biweekly scale (KRISHNAMURTI and ARDUNARY, 1982; REED *et al.*, 1977). We have examined the bi-weekly fluctuations of the two systems based on pentad average OLR and 700 hPa wind field. In this respect, no clear connection was observed between the two monsoon systems.

The Indian monsoon system oscillates prominently on a 30–50 day scale (SIKKA and GADGIL, 1980; YASUNARI, 1980; and several others since then). There is a possibility that this low frequency fluctuation of the Indian monsoon is influenced by the eastward migrating MJO, first traversing the near-equatorial African region. There is little research on the role played by MJO in modulating organized convection over the African monsoon system as most of the ITCZ over Africa is over land (from 0°E to 45°E). Also the convection remains suppressed between 45–65°E due to the prevalence of cold SST as a result of upwelling off the Somali coast. However, it is conceivable that the near-equatorial eastward propagating MJO, while

Table 1

The dates of the onset, progress and withdrawal of the monsoon over Africa and India. Onset in West Africa and India was considered at 10°N and 15°N, respectively

Sl No.	Parameter	West African Region		Indian Region	
		1999	2003	1999	2003
1	Onset of the monsoon	25 June	30 June	05 June	20 June
2	Northward reach of the monsoon	15 July	20 July	25 July	25 July
3	Beginning of the withdrawal of the monsoon	13 Sept	03 Sept	27 Sept	23 Sept

passing over tropical Africa, may modulate convection on a 30–50 day scale. This large-scale convection is suppressed while passing over 45–65°E and becomes rejuvenated over the southeast Arabian Sea (SEAS) about 70–75°E and the neighboring Bay of Bengal (80–100°E), as a result of warmer SSTs in the region. To investigate such a possibility we have examined the 5-day average position of the ITCZ (monsoon trough) over the two regions for two typical years under study. The central position of the ITCZ was determined from the turning of the wind or the latitude of the zero zonal wind. Figure 3 shows the fluctuations, on pentad basis, in the latitude of the ITCZ at 850 hPa over tropical North Africa along 10°W–10°E and 10–35°E (Central north tropical African belt) and along 75–95°E (high moisture regime of the Indian monsoon). There is a general northward seasonal migration of the central latitude of the ITCZ in all three years and for all three regions from mid-June to mid-August. Interannual variability is observed in reaching the peak northward latitude in different belts. For example, for the 75–95°E belt the maximum latitude (29°N) for the ITCZ is reached during 18–22 July, 1999 and in 2003 it was reached between 02–06 August (up to 26°N). Similarly for the westernmost sector (10°W–10°E) the most northward position (17°N) of the ITCZ was reached between 18–22 July in 1999 and in 2003 it was reached between 7–11 August (16°N). We tried to determine whether any link exists between the pentad based oscillations in the

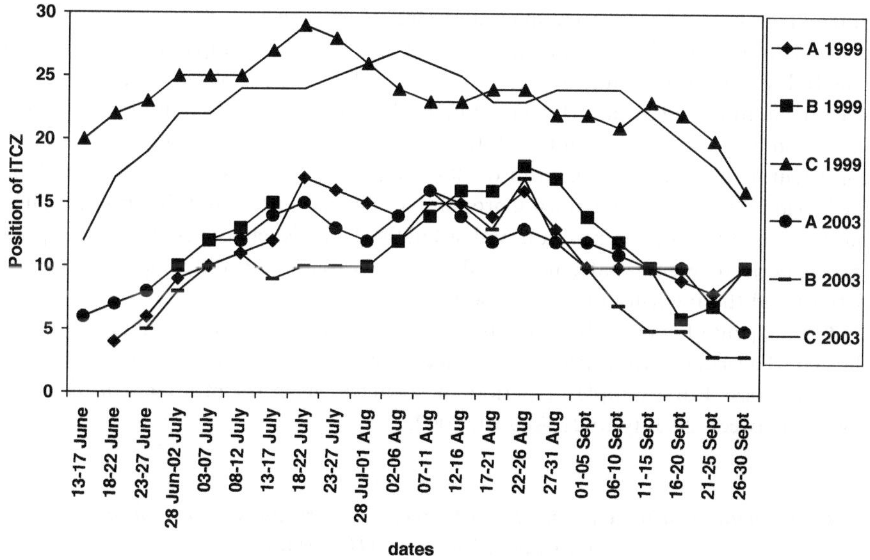

Figure 3
The position of ITCZ (Latitude °N) over two sectors (A = 10°W–10°E, B = 10°E–35°E) in tropical North Africa and one sector (C = 75°E–95°E) over India.

latitudinal position of the ITCZ over the African region with that over the Indian region. However, we did not find any correspondence between the oscillations in the two regions except that a broad seasonal northward migration in the beginning of the season and southward migration at the end of the season occur along both regions.

On the pentad scale, the fluctuations of the maximum zonal wind speed south of the ITCZ, which gives the fluctuations in the strength of the monsoon winds in the lower troposphere, for 2002 are given in Figures 4 (a,b and c) for three levels. At 850 hPa the zonal wind is westerly (monsoon westerlies) in all three sectors. At 700 hPa the zonal wind remains westerly in the Indian sector (Fig. 4c) but becomes easterly in the two African sectors (Figs. 4a,b). At 200 hPa the zonal winds are easterlies in all the sectors but are strongest over the Indian sector as compared to the two African sectors. As seen from Figure 4, apparently no clear-cut relationship is found between the pentad scale variability along different longitudinal belts except for the increase in wind strength between 20 July and 20 August, 2002 which is observed in all the three sectors. The monsoon season of 2002 was an acute drought season for India (SIKKA, 2003) and in accordance with that the weakest average wind at 850 hPa occurred in all the months along sector 75–95°E. For the other two sectors even in July and August we notice no significant differences in the wind strength, though a slight decrease is noticed in the wind strength in July 2002 for the sector 10°W–10°E which would require further study with regard to rainfall data over Nigeria and Sahel regions. Also the latitude of maximum westerly wind in different months of the two years was found to be considerably equatorward by about 4 to 6 degree latitude in the western two sectors as compared to the Indian longitudes (Figure not shown). At 200 hPa, as shown in Table 2, on the monthly scale the latitude of the easterly maximum is about 4 to 6 degrees latitude northward in the Indian region as compared to the other two longitudinal belts. This is consistent with the displacement in the 850 hPa maximum westerly winds as well as the broad rainfall pattern, ignoring orographic effects over the three longitudes. Thus on the monthly scale the position and strength of low level convergence and upper tropospheric divergent patterns would broadly explain the differences in rainfall pattern and the intensity of the monsoon in the two regions.

On the pentad average basis for the monsoon season of 2002, the fluctuations in the 200 hPa easterly wind speed match each other in the three regions, broadly on an epochal basis. For example, a period of stronger easterlies prevailed in all three regions between 20 July to 21 August, 2002.

4. Fluctuations in the Organization of Cloud Clusters and Convection as viewed from Pentad-Average OLR Data

Convection is mostly organized over both the regions in the form of cloud clusters. The cluster sizes are large with dimensions varying between 1000 to 2000 km

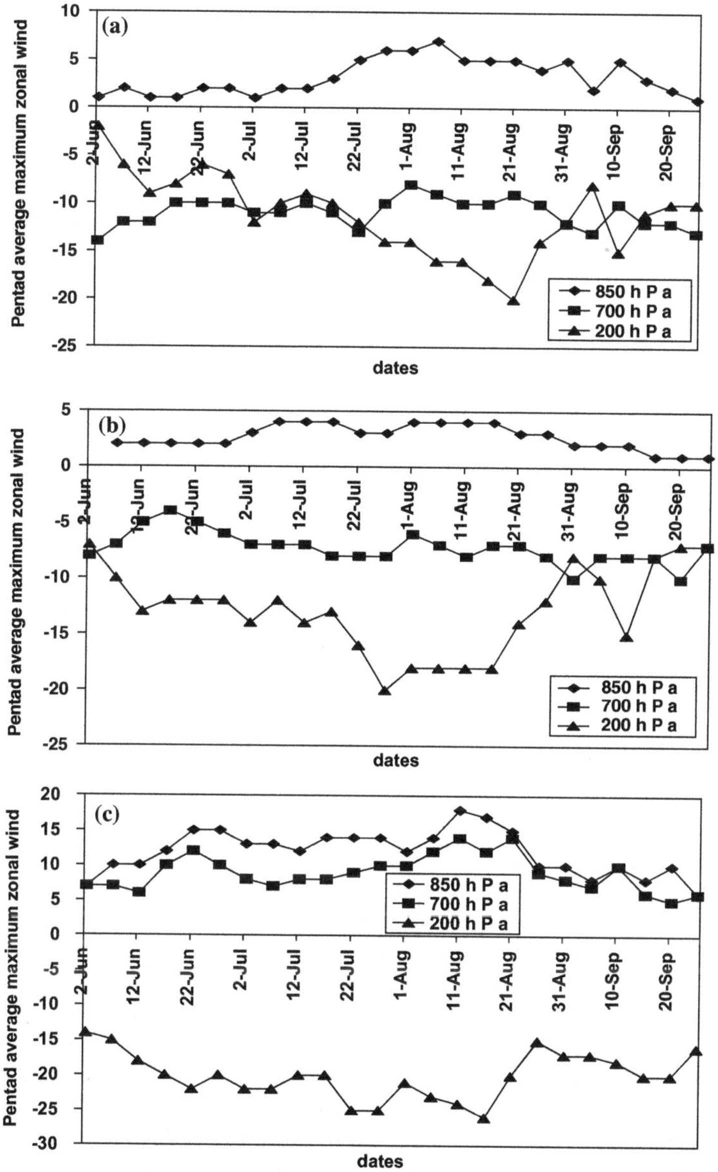

Figure 4
(a) Pentad average maximum zonal wind south of ITCZ at 850 hPa,700 hPa (westerlies positive) and 200 hPa (Easterlies negative) over 10°W–10°E (b) Pentad average maximum zonal wind south of ITCZ at 850 hPa,700 hPa (westerlies positive) and 200 hPa (Easterlies negative) over) 10°E–35°E sector during 2002. (c) Pentad average maximum zonal wind south of ITCZ at 850 hPa,700 hPa (westerlies positive) and 200 hPa (Easterlies negative) over) 75°E–95°E sector during 2002.

Table 2

Strength (m sec^{-1}) of monsoon westerlies (U-positive) at 850 hPa and upper tropospheric easterlies (U-negative) at 200 hPa along three longitudinal belts. Figures in bracket give latitude (°N) of respective maximum wind speed

Month & Year	850 hPa			200 hPa		
	10°W–10°E	10°E–35°E	75°E–95°E	10°W–10°E	10°E–35°E	75°E–95°E
June 1999	1 (6)	1 (4)	12 (16)	−5 (6)	−7 (6)	−20 (10)
July 1999	3 (8)	3 (6)	16 (13)	−16 (6)	−18 (6)	−25 (12)
August 1999	6 (10)	3 (5)	12 (13)	−18 (6)	−18 (8)	−25 (12)
September 1999	2 (6)	1 (4)	8 (11)	−14 (6)	−15 (8)	−20(10)
June 2002	2 (6)	3 (10)	12 (14)	−8(8)	−13 (8)	−18 (10)
July 2002	3 (8)	3 (12)	12 (15)	−14 (8)	−15 (8)	−21 (11)
August 2002	5 (7)	3 (8)	14 (14)	−13 (6)	−13 (6)	−22 (12)
September 2002	3 (6)	2 (5)	8 (12)	−10 (6)	−8 (6)	−17 (11)
June 2003	2 (6)	2 (4)	10 (14)	−9 (6)	−13 (16)	−20 (11)
July 2003	4 (8)	3 (5)	14 (13)	−16 (7)	−17 (8)	−22 (13)
August 2003	6 (8)	4 (6)	14 (11)	−18 (8)	−17 (10)	−20 (12)
September 2003	3 (6)	2 (5)	11 (12)	−10 (8)	−10 (7)	−20 (12)

in the east-west and 300—500 km in the south-north directions in the Indian region. The dimensions of the clusters are small in the African region. We have examined more quantitatively the average minimum OLR values in the three sectors on monthly as well as pentad scales. Table 3 presents the average OLR data on a monthly scale for three years and reveals the following.

i) On the whole OLR values are lowest along 75–95°E in all months and in all three years, and they are highest in the middle sector (10°E–35°E) which is understandable from the well known differences in the large-scale activity of the two monsoon systems.
ii) For the Indian longitudes as well as over the other two longitudinal sectors monthly average OLRs for the year 2002 are higher compared to the other two years. As mentioned the Indian monsoon season was one of acute drought in 2002, and the all-India rainfall was in deficit by nearly 50% for July 2002. This is indicated by high OLR values. However, simultaneously relatively higher OLR for the African sector is also found. This has to be examined in relation to rainfall figures over the Sahel-Nigeria region in July 2002. The higher OLR for 2002 in all the three sectors indicates that both the monsoon systems were weak on the large-scale in this year. Perhaps the build-up of the warm ENSO event in the central Pacific might have weakened the two regional monsoon systems for 2002 as a common mode with global scale ramifications.

Pentad OLR data, averaged for the three longitudinal sectors, are is shown in Table 4 for the monsoon season of 2003 as a typical case of the normal monsoon

Table 3

Monthly average 'Outgoing Long-wave Radiation' ($W\ m^{-2}$) for the three sectors under study

Month & Year	10°W–10°E	10°E–35°E	75°E–95°E
June 1999	200	220	180
July 1999	200	220	180
August 1999	200	210	180
September 1999	210	210	200
June 2002	220	240	200
July 2002	220	240	210
August 2002	220	230	190
September 2002	220	230	210
June 2003	210	240	200
July 2003	200	230	180
August 2003	210	220	200
September 2003	220	240	220

over India. The data were averaged along the latitude of the ITCZ/ monsoon trough on one hand, and in the mean equatorial zone (4°N–4°S) on the other. Over the Indian longitudes and for the zone of the monsoon trough (20–27°N), temporal fluctuations of the OLR occur as a result of the formation of low pressure areas/ depressions. There is no connection between the fluctuations along the monsoon trough/ITCZ belt over the three sectors. The Bay of Bengal disturbances on a majority of occasions do not move farther westward of 70°E along 20–25°N. Again the fluctuations in the near-equatorial belt along 10°W–10°E are only slightly modulated during most of the season and their OLR values are also higher by 20–30 Wm^{-2} compared to those over 75–95°E. This is because the West African belt is influenced by the cold SST associated with the cold ocean current flowing along the Gulf of Guinea. As a result, the signature of MJO in OLR is not clearly observed in this belt over the near-equatorial east Atlantic zone. However over the land-locked near-equatorial belt between 10°E to 35°E, from Table 4 , we notice the signature of eastward propagation of low OLR values from the central African belt (10°E–35°E) to the Indian belt (75–95°E) about 4 to 5 times in the season. Similar features were also recognized in the OLR for the other two years. On average it takes 10 days for the low OLR signal in the belt 10°W–35°E to move eastward to the belt 75–95°E. The frequency of passage of the low OLR signal corresponds to the frequency of the MJO signal on 30–50 day scale with the eastward phase speed being approximately 4–5 degree longitudes per day. The signal is clearly seen over the central part of the near-equatorial Africa (10°E–35°E). It disappears along the upwelling cold SST region off the Somali Coast and is picked up again while passing over the warm near-equatorial waters along 75–95°E. An examination of the data revealed no significant northward migration over the African land-locked region although such a migration is a characteristic feature of the Indian region (SIKKA and GADGIL, 1980; YASANURI,

Table 4

Average Pentad OLR values plus 200 (Wm^{-2}) along the central latitude of ITCZ /Monsoon Trough and near-equatorial belt during 2003

Period	Central Latitude of ITCZ/ Monsoon Trough			Near-equatorial belt		
	10°W–10°E	10°E–35°E	75°E–95°E	10°W–10°E	10°E–35°E	75°E–95°E
02 June	40	30	30	00	00	10
07 June	40	20	30	10	10	00
12 June	50	30	30	20	10	−10
17 June	60	20	50	30	30	00
22 June	60	30	00	10	20	−10
27 June	60	10	30	10	20	00
02 July	60	20	10	30	30	−20
07 July	70	30	00	20	20	00
12 July	70	40	20	10	20	10
17 July	60	40	20	00	10	00
22 July	60	30	20	20	10	10
27 July	70	40	30	30	20	−10
01 Aug	60	40	40	40	00	10
06 Aug	60	20	10	10	10	00
11 Aug	60	30	00	10	20	−10
16 Aug	70	40	30	00	00	20
21 Aug	50	20	30	10	00	10
26 Aug	60	10	40	10	10	00
31 Aug	50	20	50	10	00	−10
05 Sept	50	20	30	10	00	10
10 Sept	50	30	10	20	00	20
15 Sept	60	30	00	10	20	10
20 Sept	70	30	10	30	20	40

1980) as it is covered within the near-equatorial warm oceanic belt. We examined this aspect further with the plots published by the US Climate Diagnostic Bulletin regarding the time-longitude section (5°N–5°S) of anomalies of satellite observed OLR spanning the period May to September for the years 2003 and 2005 as depicted in Figure 5. Four eastward propagating low OLR modes are observed in the period between May to September 2003 from 30°E to 120°E. Similarly four eastward moving modes in low OLR could also be observed to pass during the same period in 2005. Similar features also are noted in several other years in the monsoon season. Figure 6 shows the plots of time longitude sections (5°N–5°S) of anomalous 200 hPa velocity potential from May to October 2003 and 2005, again after the US Climate Diagnostic Bulletin. Three to four eastward moving negative and positive pulses of velocity potential passed from 30°E to 120°E. Each pulse took about two weeks to pass over the region of our study. The eastward moving low OLR pulses and the corresponding fluctuations in the velocity potential are the signatures of the passage of the MJO signal from the near-equatorial African continent toward the near-equatorial Indian-Western North Pacific belt. A closer study is needed to firmly

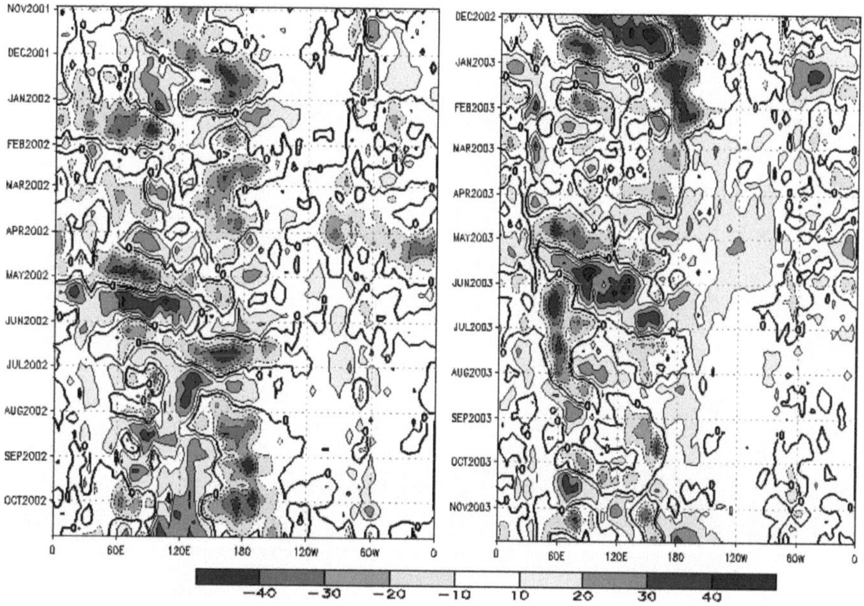

Figure 5
OLR Anomalies (Wm^{-2}) during November 2001 to November 2003. (Source: http://www.cpc.ncep.noaa.gov/products/analysis-monitoring/bulletin).

establish the connection between the eastward movement of the MJO from near-equatorial Africa toward 75–95°E. If this aspect could be firmly established, it would serve as a possible predictor for the onset of the break monsoon over India and the beginning of the revival phase after 10–15 days from near-equatorial SEAS and south Bay of Bengal. This aspect could become the focus of joint studies between the AMMA (African Monsoon Multidisciplinary Analysis) scientific community and the researchers working on the modulation of the Indian monsoon.

There also could be some relationship between the multi-decadal scale fluctuations of the summer monsoon seasonal rainfall over arid and over semi-arid parts of the Sahelian / sub-Sahelian region. Even though seasonal monsoon rainfall data over the African region on a long-term basis are not as well documented as over India, studies by NICHOLSON (1979, 1981) suggest episodes of long-term variability over the Sahel region which appear to correspond with those over India. For example, the period 1870 to 1895 was that of a wetter regime over the Sahel region and over India too it witnessed only two drought years (1873 and 1877). Again the period 1895–1920 was a drier period for the Sahel region and over India too eight monsoon droughts occurred (1896, 1899, 1901, 1904, 1905, 1911, 1918, 1920). Also the period 1965 to 1987 was a drier period for the Sahel region and over India too ten droughts occurred

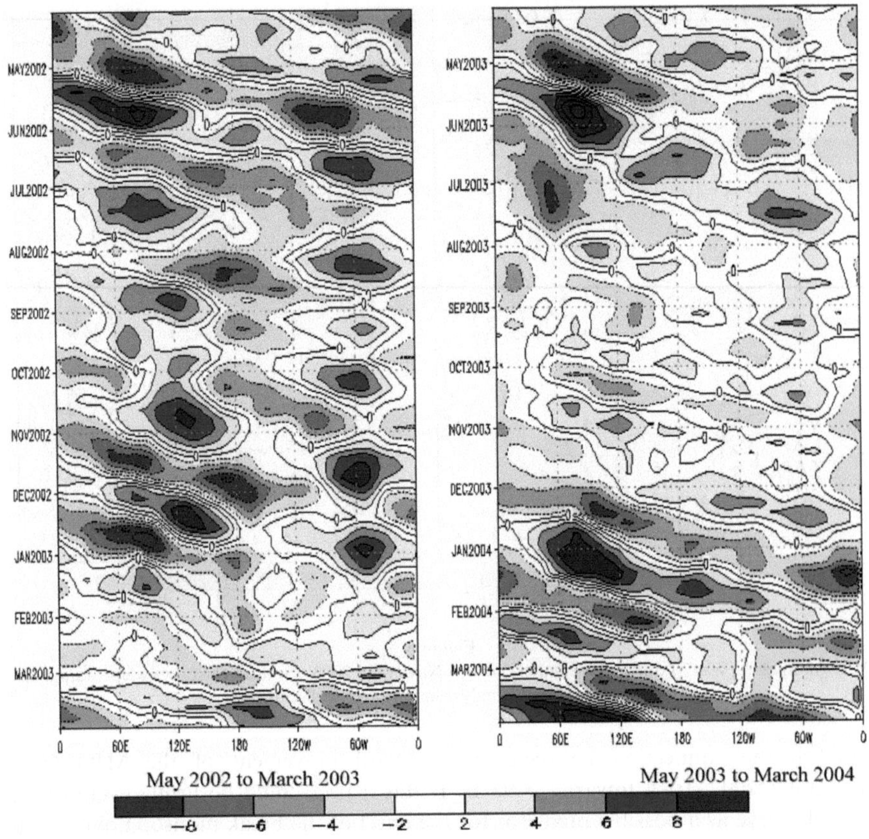

Figure 6
Time-longitude section (5°N–5°S) velocity potential anomalies at 250 hPa during May 2002 to March 2004. (Source: http://www.cpc.ncep.noaa.gov/products/analysis-monitoring/bulletin).

(1965, 1966, 1968, 1972, 1974, 1979, 1982, 1985, 1986, 1987). For the period 1930 to 1960 both monsoon systems have produced above their respective long-term normal rainfall. Thus even though there may not be year-to-year exact correspondence between the occurence of summer rains over tropical North Africa and India, yet on a long-term decadal basis both regions show some similarity. This may indicate that some global features of the coupled land-atmosphere-ocean system over tropical Africa and India fluctuate on a long-term basis in unison, which may impact the monsoon season rainfall over both regions simultaneously. If there is a common mode of long-term monsoon variability over both regions, it must be examined further by organizing joint studies under the AMMA program for the African monsoon and the Indian Climate Research Programme (ICRP) for the Indian monsoon. The common multi-decadal mode could influence the two regions for multi-decadal periods, though on an inter-annual basis the fluctuations in rainfall may not exactly coincide due to regional influences.

5. *Planned Indian Monsoon Experiments–Continental Tropical Convergence Zone*

During the period 1999 to 2003, three field programs were carried out in India under the ICRP for understanding coupled ocean-atmospheric processes of the regional monsoon system (BHAT *et al.*, 2001; SANJEEVA RAO and SIKKA, 2005). The Indian monsoon research community is currently planning another major multi-year field experiment, the 'Continental Tropical Convergence Zone (CTCZ)' for implementation during 2007–10 to understand and better predict the subseasonal fluctuations in monsoon rain episodes. The CTCZ by-and-large would cover the monsoon trough over the Indo-Gangetic plains and aims to understand the coupled land-atmosphere-ocean summer monsoon system over India in an integrated fashion. The design of the CTCZ would cover multi-scalar observational platforms covering large-scale, meso-scale and subbasin scale processes in the coupled ocean-land-biosphere-atmosphere system. Similarly the AMMA program in North Africa is being implemented to understand the coupled processes of the African monsoon system (VANDER AKKER, 2006).

In this preliminary study we have noted that a similar coupled monsoon system operates over the West African monsoon and there are indications that the two neighboring regional monsoon systems interact with each other on a low-frequency subseasonal scale. The pulsation in the TEJ over the Indian sector may influence the pulsation in the mid-tropospheric regional easterly stream over the central part of northern Africa in summer. Such pulsations may influence the generation of easterly waves in that region. Similarly, as indicated in section 3, there is a fairly consistent signal in OLR data that the low-frequency fluctuations in low OLR, associated with the eastward traveling MJO in the near-equatorial region, appears first over the central parts of the near-equatorial African continent, migrates eastward and restrengthens over the warm oceanic Indian Ocean longitudinal (75–95°E) belt. The passage of this eastward moving MJO signal could promote the beginning of a northward moving convective episode over the Indian longitudes which in turn would revive the monsoon over central India after a long break monsoon spell. As the ocean-atmosphere-land-biosphere environments of the two monsoon systems have important contrasting features, it would be of great advantage to the understanding of both the monsoon systems, if in some phase of the CTCZ experiment in India and the AMMA field program in Africa during 2007–2010 time-frame, evolution of both the monsoon systems is observed simultaneously. Such a program could be carried out independently in the two regions however a common design strategy may evolve and the field data between the two regions shared for research. Perhaps an international broad mechanism for facilitating this could be arranged through the World Climate Research Program.

6. *Summary and Conclusions*

We have attempted to discuss the common and contrasting features of the two neighboring regional summer monsoon systems viz. the West African and the Indian. The possibility of interactions among them is examined on a low-frequency subseasonal scale and it is suggested that pulsation in the TEJ over the Indian region may influence the fluctuations in the mid-tropospheric AEJ, which in turn may lead to the generation of easterly waves and the formation of associated squall lines over the region of the African monsoon system. Also the passage of organized convection under the low-frequency regime of an eastward moving MJO may influence the re-juvenation of the Indian monsoon through enhancement of organized convection in the near-equatorial warm pool of the SEAS and the Bay of Bengal. The NCEP/ NCAR reanalysis and OLR data on a pentad basis for three years, which have been used in this study, are suggestive of interactions between the West African and the Indian summer monsoon systems on low-frequency subseasonal scales. Pentad OLR data are used to describe the signature of possible eastward moving MJO on 30–50 day scale toward the near-equatorial warm pool of the SEAS and the Bay of Bengal. We have not used any time-filters for our data except the pentad averages, purposeful to examining the basic data in this preliminary attempt. We believe that the idea contained in this study may become validated by other observational studies applying sophisticated techniques as well as by examining modeling results. Our results are found to be encouraging and a more detailed study is warranted. This could be an important tool for empirically predicting the beginning of the northward moving convective episode along the 75–95°E belt over the Indian region after 10–15 days of the appearance of low OLR values along the near-equatorial (10–35°E) belt over the African region. Finally a suggestion is made that the CTCZ experiment being planned in India for 2007–2010 and the ongoing AMMA program in Africa may evolve a joint strategy of carrying out independent campaigns in one of the three years in the two regions. The data to be collected in this campaign may be focused on understanding interactions among the two neighboring regional monsoon systems; each operating under some common and some contrasting features of the coupled ocean-atmosphere-land environment.

Acknowledgement

The authors acknowledge the valuable comments of the anonymous reviewers as highly useful in improving its presentation. They are thankful to the Secretary,

Department of Science and Technology for supporting the monsoon field programs in India under the Indian Climate Research Programme. They also acknowledge the supply of some of the data used in this study by Prof. O.S.R.U. Bhanukumar, Department of Meteorology and Oceanography, Andhra University. Mr. D.R. Sikka is also grateful to Prof. J. Shukla, President, COLA/IGES, USA for stimulating him to continue pursuit of his research interest on the Indian monsoon.

References

Bhat, G.S., Gadgil, S., Harish kumar, P. V., Kalsi, S, Murty, V. S. N., Prasada Rao, C. V. K., Ramesh Babu, V., Rao, L. V. G., Rao, R. R., Ravichandran, M., Reddy, K. G., Sanjeeva rao, P., Sengupta, D., Sikka, D. R., Swain, J., and Vinayachandran, P. N. (2001), *BOBMEX—the Bay of Bengal Monsoon Experiment*, Bull. Am. Meteor. Soc. *82*, 2217–2243.

Burpee, R.W. (1975), *Some features of synoptic scale wave based on compositing analysis of GATE data*, Mon. Wea. Rev. *103*, 921–925.

Carson, T.N. (1969), *Synoptic histories of three African disturbances that developed into Atlantic hurricanes*, Mon. Wea. Rev. *97*, 256–76.

Grist, T.P. (2002), *Easterly waves over Africa. Pt. I: The seasonal cycle and contrast between wet and dry years*, Mon. Wea. Rev. *130*, 197–211.

Grist, T.P. Nicholson, S.E., Barcilon, A.I. (2002), *Easterly waves over Africa. Pt. II: Observed and modeled contrasts between wet and dry years*, Mon. Wea. Rev. *130*, 212–225.

Kalnay, E., Kanamitsu, M., Kistler, R., Collins, W., Deaven, D., Gandin, L., Iredell, M., Saha, S., White, G., Woolen, J., Zhu, Y., Chelliah, M., Ebisuzaki, W., Higgins, W., Janowiak, J., Mo, K.C., Ropelewski, C., Wang, J., Leetma, A., Reynolds, R., Jenne, R., Joseph, D. (1996), *The NCEP/NCAR 40-year reanalysis project*, Bull. Am. Meteor. Soc. *77*, 437–471.

Krishnamurti, T.N., Ardunary, (1982), *The 10–20 days westward propagating mode and break in the monsoon*, Tellus *32*, 15–26.

Nicholson, S.E. (1979), *Revised rainfall series for the West African sub-tropics*, Mon. Wea. Rev. *107*, 620–623.

Nicholson, S.E. (1980), *The nature of rainfall fluctuation in sub-tropical west Africa*, Mon. Wea. Rev., *108*, 473–487.

Nicholson, S.E. (1981), *Rainfall and atmospheric circulation during drought and wetter periods in West Africa*, Mon. Wea. Rev. *109*, 2191–2208.

Nicholson, S.E. (1997), *An analysis of the ENSO signal in tropical Atlantic and west Indian Ocean*, Int. J. Climatol, *17*, 345–375.

Nicholson, S.E. (2000), *The nature of rainfall variability over Africa in time scales of decades to millennia*, Global and Planetary Change *26*, 137–138.

Nicholson, S.E. and Kim, J. (1997), *The relationship between El Niño Southern oscillation in african Rainfall*, Int. J. Climatol. *17*, 117–135.

Okulja, P. (1970), *Synoptic flow patterns over West Africa*, Tellus, *22*, 663–680.

Pedgley, D.E. and Krishnamurti, T.N. (1976), *Structure and behavior of a monsoon cyclone over West Africa*, Mon. Wea. Rev. *104*, 149–167.

Parthasarathy, B., Munot, A., and Kothawali, D.R. (1995), *Monthly and seasonal rainfall series for All-India homogeneous regions and meteorological sub-divisions, 1871–1994*, Res. Report No. RR-065, Ind. Inst. Trop. Met., Pune, India.

Rao, Y.P. (1976), *South West Monsoon*, Met. Monograph. Synoptic Met., Ind. Met. Dept., Delhi.

Reed, R.J., Norquist, D.C., and Recker, E.C. (1977), *The structure and properties of African wave disturbance observed during phase III of GATE*, Mon. Wea. Rev. *105*, 317–333.

Sanjeeva Rao, P. and Sikka, D.R. (2005), *Intra-seasonal variability of the summer monsoon over North Indian Ocean as revealed by the BOBMEX and ARMEX Field Programs*, Pure Appl. Geophys. *162*, 1481–1510.

SIKKA, D.R. (1980), *Some aspects of the large scale fluctuations of summer monsoon in India in relation to fluctuations in the Planetary and regional scales circulation parameters*, Proc. Ind. Acad. Sci. (Earth and Planetary Sciences) *89*, 179–195.

SIKKA, D.R. (1999), *Monsoon droughts in India*, Joint COLA / CARE Tech. Report No. *2*, Centre for Ocean land and Atmosphere Studies, Centre for Application of Research on the Environment, Calverton Md. USA. 270 pp.

SIKKA, D.R. (2003), *Evaluation of monitoring and forecasting of monsoon rainfall over India and a review of monsoon drought of 2002*, Proc. Ind. Nat. Sci. Acad. *69*, 479–504.

SIKKA, D.R. (2006), *A study on the monsoon low pressure systems over the Indian region and their relationship with drought and excess seasonal monsoon rainfall*, Joint COLA / CARE Tech. Report No. 256, COLA, Md. USA.

SIKKA, D.R. and GADGIL, S. (1980), *On the maximum cloud zone and the ITCZ over the Indian longitudes during the southwest monsoon*, Mon. Wea. Rev. *108*, 1840–1853.

VANDER AKKER, ELISABETH (2006), *AMMA The international Implementation Plan* (Version 3.0), AMMA International Project Office, Paris.

WARD, M.N. (1998), *Diagnosis and short-lead time prediction of summer rainfall in tropical North Africa on inter annual and multi-decadal time scales*, J. Climate *11*, 3167–3191.

YASUNARI, T. (1980), *A quasi-stationary appearance of 30–40 day period in the cloudiness fluctuations during the summer monsoon over India*, J. Meteorol. Soc., Japan *59*, 336–354.

(Received March 16, 2006; accepted November 13, 2006)
Published Online First: June 29, 2007

Pure appl. geophys. 164 (2007) 1717–1732
0033–4553/07/091717–16
DOI 10.1007/s00024-007-0248-x

Pure and Applied Geophysics

The Preferred Structure of the Interannual Indian Monsoon Variability

DAVID M. STRAUS[1,2] and V. KRISHNAMURTHY[1,2]

Abstract—The leading empirical orthogonal function (EOF) of the June-Sept. mean, rotational horizontal wind at 850 hPa and 200 hPa (over the region 12.5°S–42.5°N, 50°E–100°E) from 56 years (1948–2003) of reanalysis (from the National Centers for Environmental Prediction) shows strong anti-cyclonic circulation at upper levels, strong Indian Ocean cross-equatorial flow and on-shore flow over western India at lower levels . The associated principal component (PC) is correlated at the 0.75 level with the seasonal mean observed Indian Monsoon rainfall (IMR). Composite differences of vertically integrated divergence (surface to 800 hPa) and vorticity (surface to 500 hPa) between "strong" years (PC-1 exceeds one standard deviation σ) and "weak" years (PC-1 less than $-\sigma$) suggest increased rising motion and storminess over the Bay of Bengal and central India. Composite difference maps of station rainfall from the India Meteorological Department (IMD) between strong years and normal years (weak years and normal years) are statistically significant over central India, with strong (weak) years associated with increased (decreased) precipitation. In both cases the maps of rainfall anomalies are of one sign throughout India. The correlation of PC-1 with global seasonal mean SST is strong and negative over the eastern equatorial Pacific, but positive in a surrounding horse-shoe like region. Significant negative correlation occurs in the northwestern Indian Ocean. The lag/lead correlation between the NINO3 SST index and PC-1 is similar to but stronger than the NINO3/IMR correlation. Modest (but significant) negative correlation is seen when NINO3 leads PC-1 (or IMR) by one-two months. Strong negative correlation is seen when PC-1 (or IMR) leads NINO3. The projections of running five-day means of horizontal rotational winds at 850 and 200 hPa onto EOF-1 (after removing the seasonal mean for each year) were pooled for strong, normal and weak years. The strong and normal year probability distribution functions (pdfs) are nearly indistinguishable, but the weak year pdf has more weight for moderate negative values and in both extreme tails and shows some hint of bi-modality.

Key words: Indian Monsoon, inter annual variability, circulation.

1. Introduction

The relationship between the seasonal mean Indian monsoon and the intra-seasonal fluctuations has been discussed extensively both from the point of view of the precipitation and the large-scale circulation (KRISHNAMURTHY and SHUKLA,

[1] Department of Climate Dynamics, George Mason University, Fairfax, Virginia, U.S.A. E-mail: straus@cola.iges.org.
[2] Center for Ocean-Land-Atmosphere Studies, Institute of Global Environment and Society, Inc., Calverton, Maryland, U.S.A.

2000; SPERBER *et al.*, 2000; GOSWAMI and AJAYA MOHAN, 2001). The basic questions that underlie much of this research are whether the seasonal mean is the residual of chaotic weather systems and their low-frequency intra-seasonal modulation, the "residual hypothesis", or whether the seasonal mean has a separate physical origin related to the slowly varying boundary conditions such as the sea-surface temperature (SST), soil moisture and snow cover, the "boundary forced hypothesis" (CHARNEY and SHUKLA, 1981; KRISHNAMURTHY and SHUKLA, 2000, hereafter KS). The well-known links of seasonal mean rainfall to Pacific SST related to the El-Niño Southern Oscillation (ENSO) as well as to winter/spring Eurasian snow cover are reviewed by WEBSTER *et al.* (1998).

The implications of this debate for potential predictability are profound, for the 'residual' hypothesis implies very little predictability for seasonal means. According to the boundary-forced hypothesis, the slowly varying land and ocean states play an important role, so there is some potential predictability in the monsoon system for the seasonal mean.

One approach taken to resolve this debate has been to seek a few large-scale seasonal mean circulation patterns that are associated with the interannual variability of the seasonal mean all-India monsoon rainfall. The existence of one or two such patterns that are highly correlated with the Indian monsoon rainfall on interannual time scales would support the boundary forced hypothesis. On the other hand, the residual hypothesis would imply that a number of patterns should be involved, for the structure of the intra-seasonal variability is rather complex, involving a number of space and time scales (as in KS and SPERBER *et al.*, 2000). GOSWAMI and AJAYA-MOHAN (2001) find the leading empirical orthogonal function (EOF) of monsoon seasonal mean 850 hPa horizontal winds from a 40-year record obtained from the reanalyses of the National Centers for Environmental Prediction/National Center for Atmospheric Research (KALNAY *et al.*, 1996; hereafter NCEP). The corresponding time series (principle component, or PC) has a correlation of 0.62 with the seasonal mean all-India Monsoon Rainfall (IMR) given by PARTHASARATHY *et al.* (1995). A similar calculation is presented by SPERBER *et al.* (2000), who show that the leading PC has a significant trend associated with it.

An alternate approach is to identify the key aspects of the seasonal mean circulation which should be related to the seasonal mean IMR on a physical basis. WEBSTER and YANG (1992) and WANG and FAN (1999) identify indices based on the dominance of the first baroclinic mode in monsoon dynamics. These indices involve the vertical shear of the zonal wind over the monsoon region. GOSWAMI *et al.* (1999) focus on the local Hadley cell associated with the monsoon to define an index based on the meridional wind shear. (Details of these indices and their correlation with the IMR will be discussed later in the paper.)

One purpose of this paper is to show that by combining the two previous approaches (EOF analysis of a single level and consideration of vertical shear), we can obtain a single, multi-level seasonal mean circulation pattern (or mode) that is closely associated with the seasonal mean IMR over a period of 56 years. Another purpose is to examine the boundary forced hypothesis by discussing the geographic distribution of observed seasonal mean Indian rainfall associated with this mode, as well as the correlation with global SST. The relationship of the intra-seasonal fluctuations to this single preferred seasonal mean mode is studied by estimating the probability distribution function (pdf) of the projections of these fluctuations on this mode for each year. In particular we study the change of the pdf from "strong" to "weak" monsoon years, these being defined from the PC of the leading mode.

Section 2 describes the data sets and EOF analysis methods, while Section 3 shows the circulation (winds, vorticity, divergence) associated with the two leading EOFs. Maps of observed Indian rainfall associated with the leading EOF are shown in Section 4, while Section 5 shows the relationship of the leading EOF to global SST. The pdfs of intra-seasonal fluctuations projected onto the leading EOF are discussed in Section 6, and Discussion and Conclusions are given in Section 7.

2. *Data and Method of Analysis*

a. Rainfall

The seasonal mean (June – September) rainfall integrated over India for the period 1948–2003 was obtained from the Indian Institute of Tropical Meteorology (PARTHASRATHY *et al.*, 1995). We refer to this as IMR, although other studies have called this the all-India rainfall. This time series was detrended in the manner described below.

Maps of rainfall were obtained from the data set created by the India Meteorological Department (IMD) using daily observations ranging up to 6329 rain gauge stations. The station data were converted into gridded data on a 1° lon. × 1° lat. grid covering land points over India for the period 1951–2003 using a suitable interpolation scheme (RAJEEVAN *et al.*, 2005). For this study, the seasonal mean and anomaly for June to September (JJAS) season were computed from the daily data.

b. Circulation Fields

The seasonal mean (JJAS) fields of horizontal wind (u,v) were obtained from the NCEP reanalysis data set for the 56–year period 1948–2003. The input to the EOF calculations (see subsection e below) consisted of the seasonal mean rotational components of (u,v) at the 850 hPa and 200 hPa levels over the region 12.5°S–42.5°N, 50°E–100°E for the 56-year period. These components were detrended prior to the EOF calculation (see subsection c). The detrending and removal of the divergent components of the wind fields were motivated by: (i) the well-known temporal inhomogeneity in the NCEP reanalysis data set, particularly in the divergent component of the circulation (KINTER *et al.*, 2004); and (ii) the artificial correlation that can be introduced by the presence of trends.

The wind components were interpolated from the original grid to a nearly equivalent Gaussian T42 grid (2.8125° lon. × ~2.8° lat.) in order to compute the rotational components.

For the composites shown in Section 3, the seasonal mean horizontal winds, vorticity and divergence were used on the 2.5° lat. × 2.5° lon. grid at the 13 levels: 1000, 925, 850, 700, 600, 500, 400, 200, 250, 200, 150, 100 and 70 hPa for the same 56-year period 1948–2003. These data were also detrended, as in subsection c.

c. Detrending

The 56-year seasonal mean time series of IMR, rotational winds at two levels, and full winds at 13 levels, were all detrended by removing the climatological mean, the linear trend and the parabolic (in time) component. These three components correspond to the first three Legendre functions in time, and so can be computed and removed in a simple manner (see STRAUS, 1983 for more details). The linear trend (not shown) of the 200 hPa winds is dominated by decreasing easterlies (increasing westerlies) near and south of the equator (near 30°N) at a rate of roughly 0.3% per year at most. At 850 hPa, the trend shows increasing easterlies and northerlies off the southwest coast of India (at the rate of roughly 0.5% per year), thus a weakening of the on-shore monsoon flow. The parabolic component is extremely weak except near the northern boundary of our domain. (A caveat here is that the trend is removed at each point whether or not it is statistically significant, so that each grid point was treated consistently.)

d. SST Data

The SST data used in this study came from the HadIsst1 (1.1 version) data set created by the Hadley Centre for Climate Prediction and Research (RAYNER *et al.*, 2003). The monthly mean SST data are provided on a 1° lon. × 1° lat. global grid and we used the data covering the period 1948–2003. The JJAS means and anomalies of SST were computed from the monthly means. The NINO3 index is defined as the areal average of SST for 5°S–5°N, 150°W–90°W.

e. EOF Analysis

For each JJAS mean, the 200 and 850 hPa fields of rotational wind (u_r, v_r) were combined into a single data vector. Each grid point was weighted by the square-root of the cosine of latitude; no other weighting was used. From the time series ($N = 56$) of these vectors, the covariance matrix over the region indicated in subsection b was constructed, and the ortho-normal set of eigenvectors (or EOFs) determined.

Projection of the time series of data vectors onto each eigenvector yielded the (dimensional) principal components (PCs).

The fields associated with the leading EOFs were computed on the basis of composites. For the full horizontal wind fields, divergence and vorticity at 13 levels (see subsection b), positive (negative) composites were obtained as the mean over all years for which the appropriate PC was greater (less) than +1.0 (−1.0) standard deviation.

3. *The Leading EOFs*

The space-time variance explained by the leading EOFs is reported in Table 1 (in terms of percent), along with the uncertainty due to sampling error. (NORTH *et al.*, 1982). The first EOF (25% explained variance) is clearly distinct from the higher order EOFs, while the significance of EOF-2 (15% explained variance) is marginal.

Differences between positive and negative composites of the full horizontal wind field associated with EOF-1 are shown in Figures 1a and 1b for the 850 and 200 hPa levels. The positive (negative) composites are defined as the mean over the 7 (12) years: 1959, 1961, 1964, 1970, 1973, 1978 and 1985 (1951, 1957, 1963, 1965, 1969, 1972, 1974, 1979, 1982, 1987, 1997, and 2002). The 10% significance level (based on a two-sided t-test) is indicated by the shading.

The lower level wind field indicates anti-cyclonic flow just north of the equator, and generally cyclonic flow in the monsoon trough region. There is an enhancement of the cross-equatorial flow in the Arabian Sea and the on-shore flow over western India. The upper level flow is dominated by a strong anti-cyclonic center at about 35°N, 65°E, leading to strong north-easterly flow over India. The three-dimensional structure of the EOF-1 composite maps is consistent with the fields of correlation of Bay of Bengal convection with upper and lower level winds presented in WANG and FAN (1999).

Table 1

EOF Variance. The space-time variance explained by the leading EOFs, in percent terms. The range of explained variance includes the uncertainty as estimated from NORTH *et al. (1981). The correlation of the corresponding PC with the IMR is given in the last column. Values significant at the 5% level are bold faced.*

	Explained Variance (%)	Range of Exp. Var (%)	Correlation with IMR
EOF-1	25.2	30–20	**0.75**
EOF-2	15.3	18–12	−0.07
EOF-3	10.8	13–9	0.15
EOF-4	6.6	8–5	0.14

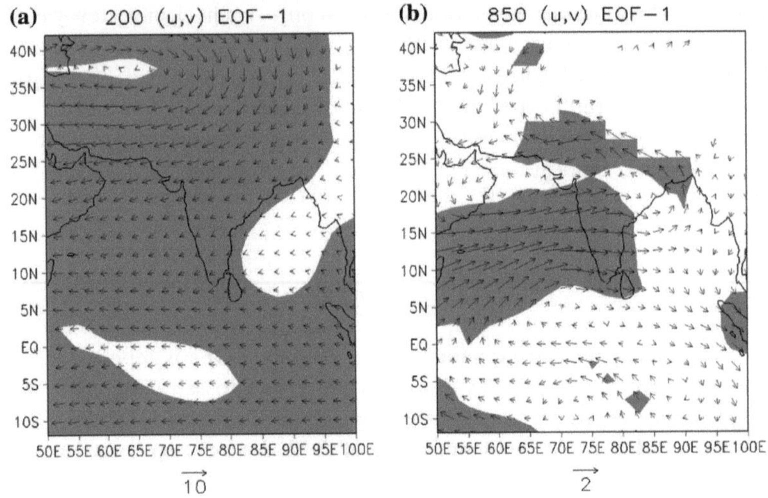

Figure 1
Composite differences of full horizontal winds for 200 hPa (a) and 850 hPa (b). Composite differences are defined as the difference between the mean of all (7) years for which PC-1 > one standard deviation (σ) and the mean of all (12) years for which PC-1 < $-\sigma$. Shading denotes 10% significance using a two-sided t test. Reference vectors show the magnitude (in m/sec) in each case.

The panels in Figure 2 show the composite differences of the mass-weighted vertical integral of: divergence from the surface to 800 hPa (Fig. 2a), vorticity from the surface to 500 hPa (Fig. 2b) and divergence from 275 to 85 hPa[1] (Fig. 2c). To the extent that the time mean vorticity integrated over the lower half of the atmosphere is a measure of the presence of cyclonic monsoon disturbances, Figure 2b suggests a storm path from the Bay of Bengal westward across India and into the Arabia Sea, consistent with Figure 1b, and consistent with the lower level divergence in Figure 2a. The upper level divergence suggests deep convection over the equator (which is confirmed by the vertical velocity composites, not shown), and the presence of upward motion over the higher elevations of Pakistan.

Composite differences of EOF-2 for the total wind field at 850 hPa and the vorticity, integrated from the surface to 500 hPa, are shown in Figures 3a and 3b. The circulation again shows enhanced cross-equatorial and on-shore Indian flow, although for a narrower region of the western Indian Ocean than was the case for EOF-1. (The area of high statistical significance is also smaller.) Over the northwestern Arabian Sea the winds are actually northerly. The lower troposphere integrated vorticity (Fig. 3b) suggests increased storminess over a region near Bangladesh and one in the northwestern Indian Ocean.

---

[1] The boundaries of the vertical integration were picked to be halfway between two analysis levels.

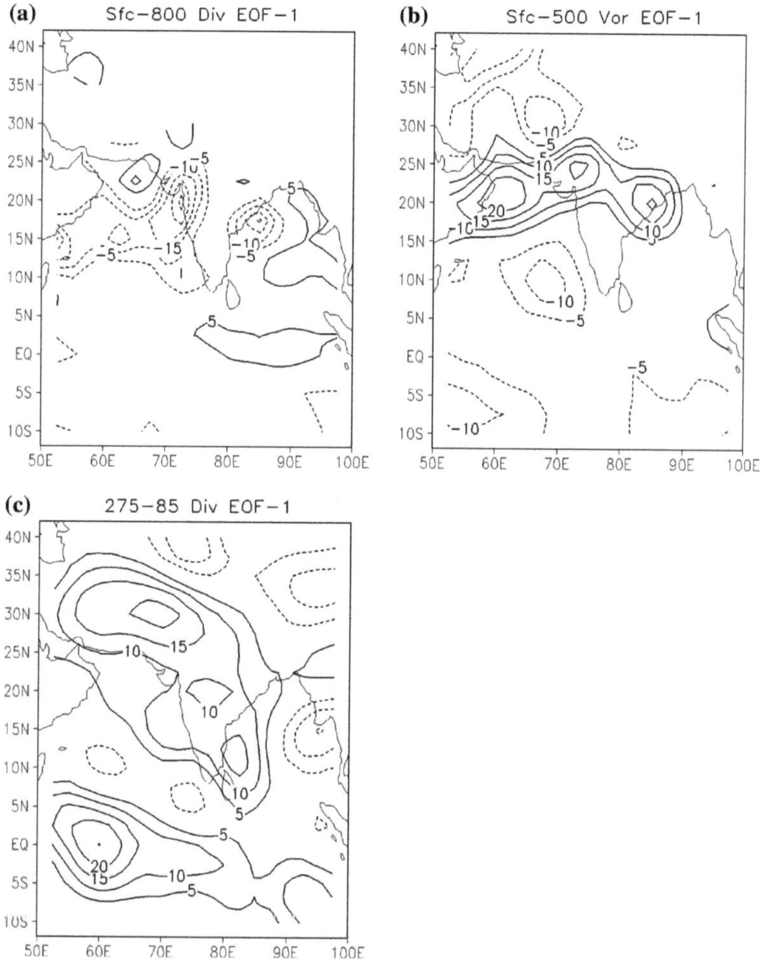

Figure 2
Composite differences of mass-weighted vertical integral of (a) Divergence (from the surface to 800 hPa), (b) vorticity (from the surface to 500 hPa), (c) divergence (from 275–85 hPa). Composite difference is the difference between the mean of all (7) years for which PC-1 > one standard deviation (σ) and the mean of all (12) years for which PC-1 < $-\sigma$. Contour interval is 1.0×10^{-4} kg/(m^2 s) in (a) and (c), 1.0×10^{-3} kg/(m^2 s) in (b). Zero contour omitted.

The leading two PCs are presented in standardized form in Figures 4a and 4b as the blue curve in each case. The standardized (and detrended) IMR is shown as the red curve. From Figure 4a it is clear that PC-1 and IMR are highly correlated (value

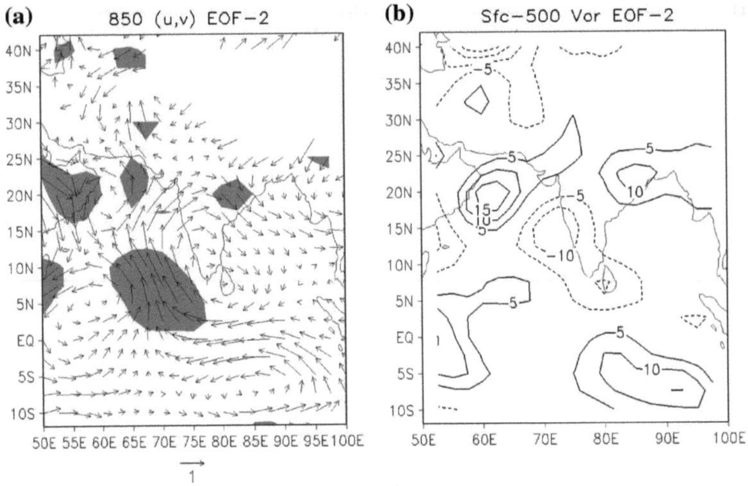

Figure 3
Composite difference of full 850 hPa horizontal wind (a) and vertical mass-weighted integral of vorticity from the surface to 500 hPa (b) for EOF-2. Composite difference is the difference between the mean of all (9) years for which PC-2 > one standard deviation (σ) and the mean of all (9) years for which PC-2 < $-\sigma$. Contour interval is 1.0×10^{-4} kg/(m^2 s) in (b). Reference vector in (a) is in m/sec. Shading denotes 10% significance using a two-sided t-test. Zero contour omitted in (b).

of 0.75); most of the extreme values of the IMR are well-matched by variations in PC-1. This is not the case for PC-2.

4. *Rainfall Composites*

Composite maps of seasonal mean rainfall anomaly over continental India were computed separately for the years in which PC-1 was greater than one standard deviation (Strong Composite), and for years in which PC-1 was less than minus one standard deviation (Weak Composite). These maps are shown in Figures 5a and 5b, along with the regions for which the composites meet the 10% significance level (based on a two-sided t-test) in Figures 5c and 5d. The strong composite shows enhanced rainfall along India (20°N–25°N), along the west coast, and to the north. The sign of the anomaly is generally the same over all of India. The weak composite shows generally reduced rainfall over the same regions, with the anomaly again being of predominantly one sign. Note that while the strong composite values are generally larger, the weak composite achieves statistical significance over a much larger region.

The structure of both composites is similar to the map of composite rainfall shown by KS, based entirely on the seasonal mean IMR for the extended period 1901–1970. The composite maps look quite different, however, from the difference

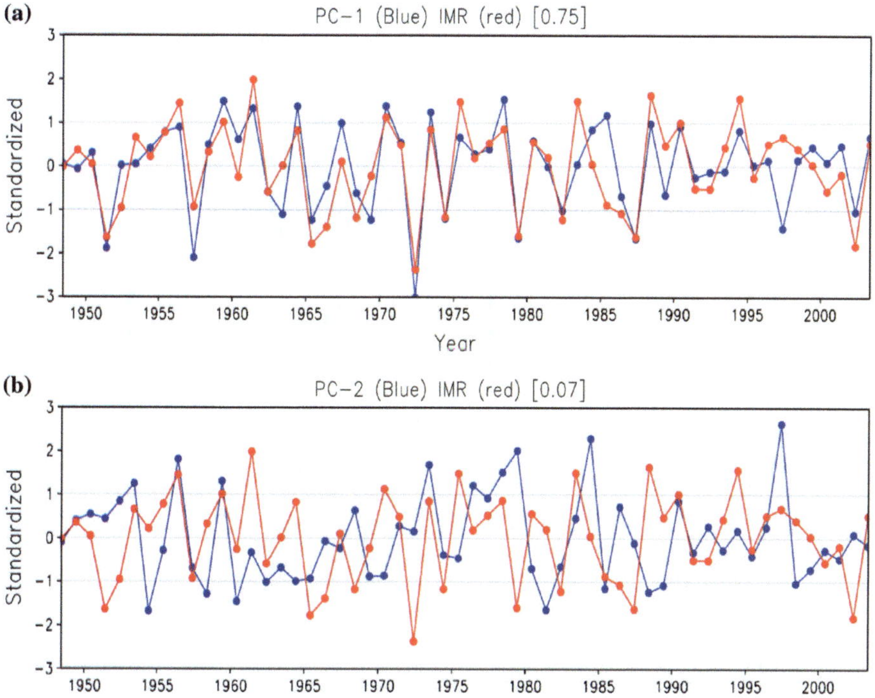

Figure 4
PC-1 (a) and PC-2 (b) shown in blue curves. IMR is shown in red curves. All series are defined for JJAS seasonal mean only, and are standardized. The correlation between the PC and IMR is shown in brackets for each panel. Thin lines show –1, 0, 1 standard deviation.

map of intra-seasonal active phase composite and break phase composite rainfall anomalies shown by KS. The latter shows much of southeastern and northern India with a rainfall anomaly of opposite sign to that of central India.

5. Global SST Relationships

The association of the leading EOF with the global distribution of seasonal mean SST can be assessed from Figure 6b, which shows the point-wise temporal correlation of SST with PC-1. (A correlation of absolute value greater than 0.30 is significant at the 2% level.) The familiar ENSO pattern is predominant in Figure 6b, along with a region of strong correlation in the Northwest Pacific. The significant negative correlation in the western Indian Ocean is indicative of two possible ways in which the atmosphere forces the ocean: (i) Strong monsoon winds lead to

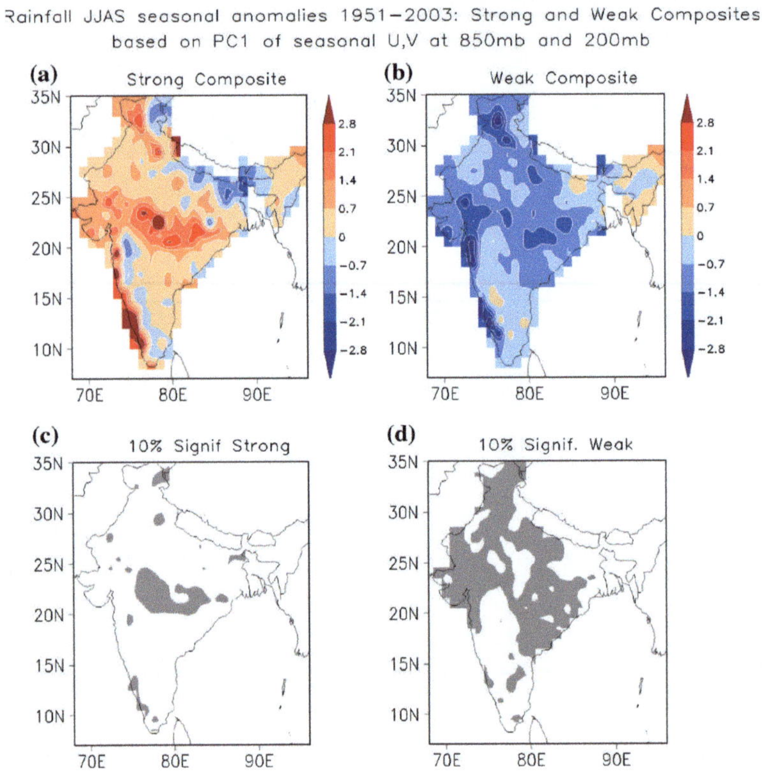

Figure 5
(a) Difference of IMD rainfall between the mean of all strong (7) years for which PC-1 > one standard deviation (σ) and the mean of all "neutral" (37) years for which $-\sigma <$ PC-1 $< \sigma$. (b) Difference of IMD rainfall between the mean of all weak (11) years for which PC-1 $< -\sigma$ and the mean of all (37) neutral years. The regions shaded in (c) and (d) show anomalies that are statistically significant at the 10% level. Units are mm/day.

evaporative cooling, and (ii) the winds force enhanced coastal upwelling. These features associated with the leading EOF are well known to be associated with the strength of the seasonal mean Indian monsoon as measured by the IMR, as discussed in WEBSTER *et al.* (1998). While the correlation pattern over the Indian Ocean may seem to involve a dipole-like structure (see e.g., SAJI *et al.*, 1999), KRISHNAMURTHY and KIRTMAN (2003) argue that SST variability in this region is linked dynamically to that in the Pacific.

The temporal lead-lag relationship between the IMR and ENSO is summarized by the green line in Figure. 6a, which shows the lag correlation of monthly anomalies of the NINO3 index with the JJAS seasonal mean anomaly of IMR.

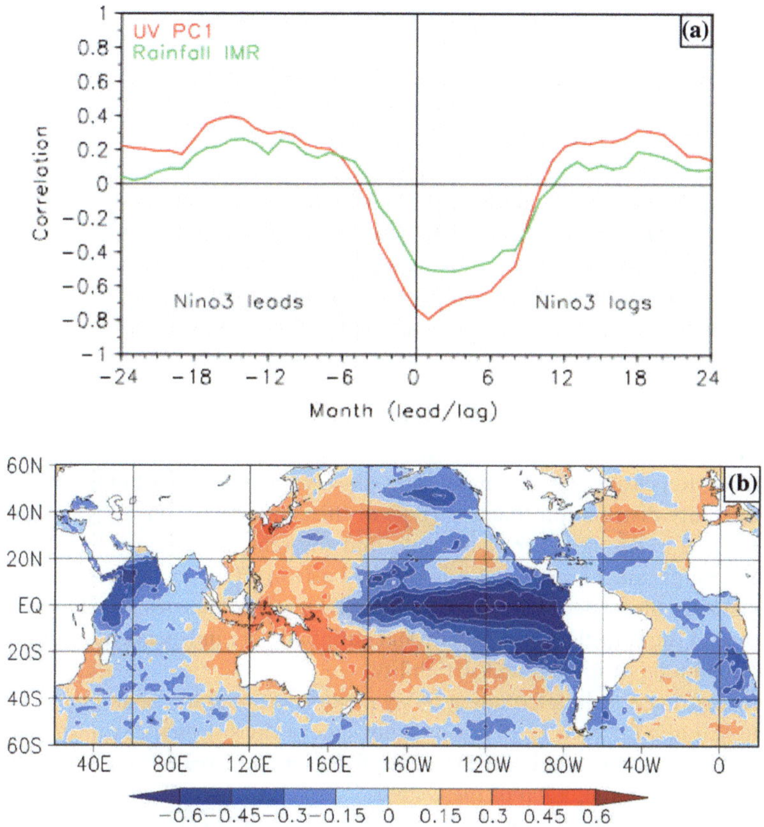

Figure 6
(a) Lag correlation of monthly anomalies of NINO3 index with PC-1 (red) and JJAS seasonal anomaly of IMR index (green) for the period 1948–2003. (b) Point correlation of PC-1 with JJAS seasonal anomaly of SST.

The overall shape of the curve, with weak positive correlation of NINO3 leading the monsoon, followed by a significant negative correlation at zero and small positive lags (monsoon leading NINO3), is quite similar to Figure 19b of WEBSTER *et al.* (1998). The red curve in Figure 6a shows the equivalent lag correlation of NINO3 with the PC of EOF-1. The strong negative correlation at zero lag indicates that the leading circulation mode is strongly SST driven, while the more modest (but significant) correlations with NINO3 leading PC-1 by one or two months indicate some predictability based on knowledge of SST alone.

6. Relationship to Intra-Seasonal Variability

The "residual" hypothesis introduced earlier states that the dominant structure of interannual variability (i.e., EOF-1) arises fundamentally from the structures of intra-seasonal variations, and that the effect of varying external forcing (e.g., SST) in a nonlinear system is to change the relative strength of different preferred intra-seasonal modes. A simple paradigm that illustrates this hypothesis was put forward by PALMER (1999), who showed that when a nonlinear system has a probability density function with several maxima (corresponding to preferred patterns), external forcing can change the relative population (or likelihood) of these maxima, leading to a link between year-to-year variations and variability within the season.

A simply way to test this paradigm in the current context is to expand each intra-seasonal state in terms of the seasonal mean EOFs, which form a set of orthogonal basis functions. The leading projection is just that component of the intra-seasonal state that has the same structure as the leading mode of interannual variability in circulation. Repeating this for many intra-seasonal states (see below), we construct the probability distribution function (pdf) of this component during years when the leading mode is strong, when it is weak, and the remaining years. The distinction among these groups of years is made on the basis of PC-1, as before. If this pdf is approximately the same for all three categories of years, that would suggest that at least some of the statistics of intra-seasonal variability are independent of the state of the leading mode of interannual variability.

We construct a series of running 5-day mean rotational horizontal wind fields at 200 and 850 hPa for each JJAS season, and remove the seasonal mean for each year separately. This yields a series of 118 states for each of 56 years (we lose two days at each of the end points). Each of these states is projected onto EOF-1 of the seasonal means, yielding 118 projection coefficients. The procedure is repeated for each year. (We find that 18.5% of the overall variance of the running pentad means is explained by the projection onto EOF-1 of the seasonal mean.) The set of $118 \times 56 = 6608$ projections is separated into three groups, depending on the value of PC-1, as described above. For each group, a separate pdf estimate is constructed using a kernel density approach (SILVERMAN, 1986).

Figure 7 shows the pdf estimates for strong years (solid line), weak years (dashed line) and the "normal" (remaining) years (dotted line). (The pdfs are normalized so that the integral under the curve is 1.0). While there is slight difference between the strong year and normal year pdfs, the weak year pdf is clearly different from the others.[2] In particular the latter shows a larger weight for moderate negative values, as well as added weight at both extreme tails. While the bi-modality suggested by

[2] The difference between the weak year pdf and either the strong or normal year pdf is statistically significant at the 1% level using the Kuiper test (PRESS *et al.*, 1989).

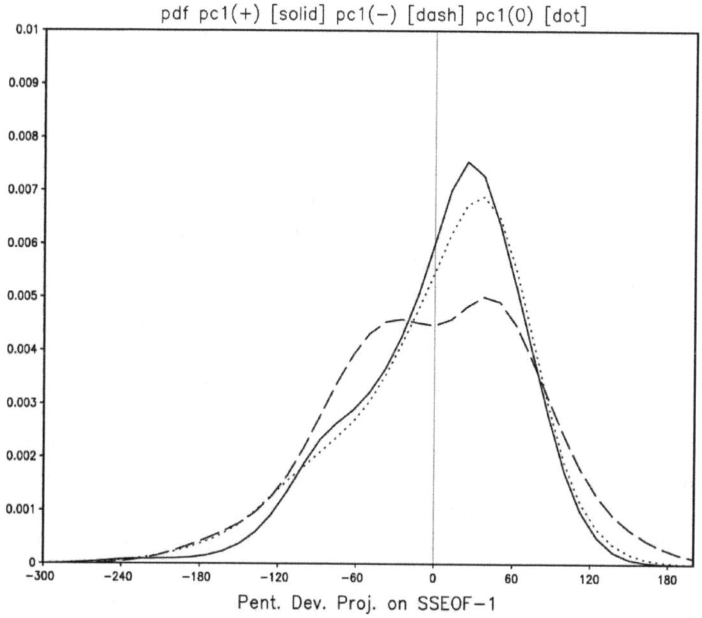

Figure 7
Probability distribution functions (pdfs) of running mean pentad projections of rotational horizontal winds at 850 and 200 hPa on EOF-1. The pdfs are computed separately for (7) strong years (PC-1 > one standard deviation σ) shown in solid line, (12) weak years (PC-1 < $-\sigma$) shown in dashed line, and (37) neutral years, shown in dotted line. The normalization is such that the area under each curve is 1.0. To obtain a dimensional number (in m/sec) from the projections on the x axis, divide by $(N)^{1/2} \sim 41$, with N = 1680 the number of components in the circulation vectors.

Figure 7 may not be statistically significant, there is a clear indication that the intra-seasonal statistics during weak years are really different than in other years.

The results suggest that during weak monsoon years the distribution of the component of intra-seasonal variability congruent with EOF-1 of the inter-annual variability is much wider than in other years. This gives partial support to Palmer's paradigm if we assume that the occurrence of a "weak" monsoon year is caused by changes in the external forcing (SST).

A more extensive analysis involving an examination of intra-seasonal variations in multiple dimensions is beyond the scope of this paper.

7. *Summary and Conclusions*

By considering the JJAS seasonal means of the rotational components of the horizontal winds at two levels (850 hPa and 200 hPa), we have shown that the leading mode of interannual Indian monsoon circulation variability (EOF-1) is highly correlated with the IMR rainfall (correlation of 0.75). This mode shows a strong anti-cyclonic circulation at upper levels and strong Indian Ocean cross-equatorial flow and on-shore flow over western India at lower levels. The monsoon trough is clearly seen in the lower level flow. Composite differences of lower level divergence and vorticity suggest increased rainfall and storminess over the Bay of Bengal and central India when this mode is strong. EOF-2 shows Indian Ocean cross-equatorial flow over a narrower region of the Indian Ocean, and the on-shore flow over western India is more southerly than for EOF-1. The correlation with IMR rainfall is indistinguishable from 0.0.

The multi-level EOF presented here may be compared with a number of other indices based on the circulation, as discussed in the Introduction and tabulated in Table 2. The PC associated with the particular EOF we present has a somewhat higher correlation (over a longer period of time) than other circulation measures. More importantly, the EOF incorporates information on the vertical shear of the zonal wind (summarizing the first baroclinic mode structure of the monsoon), the vertical shear of the meridional wind (summarizing the modified Hadley circulation) and the full horizontal structure of the low-level cross-equatorial and on-shore monsoon flow.

Composite difference maps of station IMD rainfall over India show a significant signal over central India. For both phases of the EOF-1, the maps of rainfall anomalies are of predominantly one sign over all of India. The spatial structure of the composites is different from the active/break composites of KS and more similar to their strong/weak monsoon year composites.

The point-wise correlation of PC-1 with global seasonal mean SST shows a strong negative correlation over the eastern equatorial Pacific, surrounded by a horse-shoe like pattern of positive correlation, clearly resembling the ENSO signal. Significant negative correlations are seen in the northwestern Indian Ocean. The lag/lead correlation between the NINO3 SST index and PC-1 is very similar to that between

Table 2

Monsoon Indices. Various monsoon indices and their correlation with IMR for JJAS seasonal means. A. Webster-Yang defined originally in Webster and Yang (1992); statistics above from Wang and Fan (1999). B. Monsoon-Hadley defined originally in Goswami et al. (1999); statistics above from Wang and Fan (1999). C. Westerly shear defined by Wang and Fan (1999). D. PC-1 from EOF analysis presented in Goswami and Ajaya-Mohan (2001). E. PC-4 from EOF analysis presented in Sperber et al. (2000).

Index Name	Definition	Corr (IMR)	Period
A Webster – Yang	U850 – U200 (0–20N 40–110E)	0.52	1958–1995
B Monsoon–Hadley	V850 – V200 (10–30N 70–110E)	0.64	1958–1995
C Westerly Shear	U850 – U200 (5–20N 40–80E)	0.68	1958–1995
D PC-1 (GM)	(U,V) 850 (20S–30N 40E–100E)	0.62	1958–1997
E PC-4 (S)	(U,V) 850 (20S–40N 60–120E)	0.60	1958–1997
F (this paper)	(Ur,Vr) 850, 200 (12S–42N 50–100E)	0.75	1948–2003

the NINO3 SST index and IMR, but is stronger. Both show modest (but significant) negative correlation when NINO3 leads PC-1 (or IMR) by one-two months, and stronger negative correlations when PC-1 (or IMR) leads NINO3.

The statistics of that component of intra-seasonal variability with the same spatial structure as EOF-1 has been estimated by projecting running five-day means of horizontal rotational winds at 850 and 200 hPa onto EOF-1 (after removing the seasonal mean for each year). The strong and normal year pdfs estimated from these coefficients are nearly indistinguishable, but the weak year pdf has more weight for moderate negative values and in both extreme tails.

That the correlations of PC-1 with SST (in particular with NINO3) are at least as strong (if not stronger) than the correlations of SST with IMR itself (compare Fig. 6 with Figure 19b of WEBSTER *et al.*, 1998) suggests a picture in which the slowly varying boundary forcing and large-scale seasonal mean circulation are tightly coupled (as in CHARNEY and SHUKLA, 1981). The actual amount of rainfall occurring locally or over the whole sub-continent, while strongly related to the state of the large-scale seasonal mean circulation, may have some dependence on the statistics of the intra-seasonal variability. The latter point is suggested by Figure 7, which shows the probability distribution of intra-seasonal variability is somewhat altered during low PC-1 (or dry IMR) years.

The rainfall composites (Fig. 5) and the pdfs of Fig. 7 together lend support to the conceptual picture suggested by KS: The interannual variability of the seasonal mean monsoon is a linear combination of a large-scale seasonal mean component and a statistical average of intra-seasonal variations.

Acknowledgments

This research was supported by grants from the National Science Foundation (0332910), the National Oceanic and Atmospheric Administration (NA040AR4310034), and the National Aeronautics and Space Administration (NNG04GG46G). The authors thank J. Shukla for encouraging this research.

REFERENCES

CHARNEY, J.G. and SHUKLA, J., *Predictability of monsoons*. In *Monsoon Dynamics* (eds. LIGHTHILL, J. and PEARCE, R.P.) (Cambridge University Press 1981) pp. 99–109.

GOSWAMI, B.N. and AJAYA MOHAN, R.S. (2001), *Intraseasonal oscillations and interannual variability of the Indian summer monsoon*, J. Climate *14*, 1180–1198.

GOSWAMI, B.N., KRISHNAMURTHY V., and ANNAMALI, H. (1999), *A broad-scale circulation index for the interannual variability of the Indian summer monsoon*, Quart. J. Royal Meteor. Soc. *125*, 611–633.

KALNAY, E. *et al.* (1996), *The NCEP/NCAR 40-year reanalysis project*, Bull. Am. Meteor. Soc. *77*, 437–471.

KINTER, J.L., FENNESSY, M.J., KRISHNAMURTHY, V., and MARX, L. (2004), *An evaluation of the apparent interdecadal shift in the tropical divergent circulation in the NCEP-NCAR reanalysis*, J. Climate *17*, 349–361.

KRISHNAMURTHY, V. and SHUKLA, J. (2000), *Intraseasonal and interannual variability of rainfall over India*, J. Climate *13*, 4366–4377.

KRISHNAMURTHY, V. and KIRTMAN, B.P. (2003). *Variability of the Indian Ocean: Relation to monsoon and ENSO*, Quart. J. Royal Meteor. Soc. *129*, 1623–1646.

NORTH, G.R., BELL, T.L., CAHALAN, R.F., and MOENG, F.J. (1982), *Sampling errors in the estimation of empirical orthogonal functions*, Mon. Wea. Rev. *110*, 699–706.

PALMER, T.N. (1999), *A nonlinear perspective on climate prediction.* J. Climate *12*, 575–591.

PARTHASARATHY, B., MUNOT, A., and KOTHAWALE, D.R. (1995), *All India monthly and seasonal rainfall series*: 1871–1993, Theor. Appl. Climatol. *49*, 217–224.

PRESS, W.H., TEUKOLSKY, S.A., VETTERLING, W.T., and FLANNERY, B.P. *Numerical Recipes in FORTRAN; The Art of Scientific Computing*, Second Edition (Cambrigde University Press, 1989) pp. 614–622.

RAJEEVAN, M., BHATE, J., KALE, J.D., and LAL, B. (2005), *Development of a high resolution daily gridded rainfall data for the Indian region*, Met. Monograph Climatology *22*/2005, National Climate Centre, India Meteorological Department, Pune 411005, India.

RAYNER, N.A., PARKER, D.E., HORTON, E.B., FOLLAND, C.K., ALEXANDER, L.V. ROWELL, D.P., KENT, E.C., and KAPLAN, A. (2003), *A. Global analyses of SST, sea ice and night marine air temperature since the late Nineteenth Century*, in press, J. Geophys. Res. *108*, doi:10.1029/2002JD002670.

SAJI, N.H., GOSWAMI, B.N., VINAYACHANDRAN, P.N., and YAMAGATA. T. (1999), *A dipole mode in the tropical Indian Ocean,* Nature *401*, 360–363.

SPERBER, K.R., SLINGO, J.M., and ANNAMALAI, H. (2000), *Predictability and the relationship between subseasonal and interannual variability during the Asian summer monsoon*, Quart. J. Royal Meteor. Soc. *126*, 2545–2574.

SILVERMAN, B.W., *Density Estimation for Statistics and Data Analysis* (Chapman & Hall, New York, 1986).

STRAUS, D.M. (1983), *On the Role of the Seasonal Cycle*, J. Atmos. Sci. *40*, 303–313.

WANG, B. and FAN, Z. (1999), Choice of south asian summer monsoon indices, Bull. Am. Meteor. Soc. *80*, 629–638.

WEBSTER, P.J. and YANG, S. (1992), *Monsoon and ENSO: selectively interactive systems.* Quart. J. Royal Metor. Soc. *118*, 877–926.

WEBSTER, P.J., MAGANA, V.O., PALMER, T.N., SHUKLA, J., TOMAS, R.A., YANAI, M., and YASUNARI, T. (1998), *Monsoons: Processes, predictability and the prospects for prediction*, J. Geophys. Res. *103* C7, *14*, 451–14,510.

(Received May 20, 2006, accepted November 11, 2006)
Published Online First: June 23, 2007

Pure appl. geophys. 164 (2007) 1733–1746
0033–4553/07/091733–14
DOI 10.1007/s00024-007-0239-y

Pure and Applied Geophysics

Wintertime Seasonal Scale Simulation over Western Himalaya Using RegCM3

A. P. Dimri and A. Ganju

Abstract—The Himalayan region of north India is composed of complex mountain ranges with different altitudes and orientations, causing prevailing weather conditions to be complex. Wintertime eastward moving synoptic weather systems 'Western Disturbances' (WDs) yield large amounts of precipitation over this region. Numerous micro/mesoscale circulations become generated along with prevailing weather due to surface heterogeneity and land-use variability of the Himalayan region. WDs along with these circulations may give rise to very adverse weather conditions over the region. Intraseasonal variability of surface climate over the Himalayas is studied using regional climate model (RegCM3) with 60 km resolution. A 6–month (Oct. 1999–Mar. 2000) period, as this period has received an enormous amount of precipitation in the form of snow, is considered to study surface climate variability in terms of temperature, precipitation and snow amount. Model simulations show cold bias over the Himalayan region and warm bias over northwest India. Average monthly distribution of temperature indicates that a controlled experiment could capture the areas of lowest temperature regime. Precipitation fields could be simulated only up to a certain degree of satisfaction and the influence of topographic elevation and valleys needs to be seen. RegCM3 provides a representation of resolvable atmospheric circulations that results in explaining mean variability during winter.

Key words: Snow, seasonal scale simulation, Western Himalaya.

1. Introduction

The Indian part of the Himalayas is composed of complex mountain ranges with different altitudes and orientations, which causes the prevailing weather conditions to be complex. Surface weather elements such as precipitation and temperature are intensely governed by local topography (Dimri, 2004) and local atmospheric circulations (Mohanty and Dimri, 2004) by virtue of thermodynamical and dynamical forcings. Also, numerous micro/mesoscale circulations in the narrow valleys and rugged hills are generated due to surface heterogeneity of northern Indian

Research and Development Centre, Snow and Avalanche Study Establishment, Him Parisar, Sector 37A, Chandigarh 160036, India. E-mail: apdimri@hotmail.com, apdimri@yahoo.com

Himalaya. This heterogeneity determines the precipitation and temperature pattern over the region (DIMRI and MOHANTY, 1994). Topography exerts a strong dynamical forcing on the atmospheric circulations and land surface exchanges. As a result, land atmosphere interactions substantially impact weather and climate patterns and the water and energy cycles of the climate systems (DICKINSON *et al.*, 1995; PIELKE and AVISSAR, 1990).

To study the impacts of various forcings at the regional scale, regional climate models (RCMs) are useful tools for studying mesoscale climatic processes. These previous multilayered experiments, over the united states (GIORGI *et al.*, 1993a), eastern Asia and Japan (HIRAKUCHI and GIORGI, 1995), and Europe (MARINUCCI and GIORGI, 1992; JONES *et al.*, 1995), have shown that RCMs have worked well over various domains over the globe. Also, various researchers have carried out regional climate simulation to study monsoon behavior over the Indian region. BHASKARAN *et al.* (1996) has compared seasonal simulation of the Indian Summer Monsoon with a set of three RCMs, which shows the strong orographically forced mesoscale component. Similarly, GOSWAMI and SHUKLA (1984) have studied quasi-periodic oscillation in a symmetric general circulation model and have demonstrated the impact of large-scale dynamics on the strength of the monsoon. Though numerous studies were carried out by various researchers over the Indian region they mainly pertained to the summertime monsoon phenomena and considerably less studies on multilayer integrations/simulation with RCMs during winter over the complex Himalayan region have been reported.

Therefore, in the present work an attempt is made to simulate the wintertime weather over the Indian region with a focus on mean climate conditions and intraseasonal variability. Further, most of the RCM studies have been focused on mean climate conditions and much less is known about interannual and intraseasonal variability. Therefore, the goal of this study is to provide details of intraseasonal variability. This is important for several reasons. First, the degree of similarity between modeled and observed intraseasonal precipitation variability is an important model diagnostic, as it is one way of putting sensitivity testing of a RCM to a range of meso/microscale atmospheric conditions. Second, change in precipitation has substantial socio-economic impact at the regional scale. Therefore, it is critical to evaluate when and where climate anomalies are predictable and to assess the performance of RCMs in reproducing them. Third, the intraseasonal variability of precipitation may change due to anthropogenic increases in atmospheric Greenhouse gases.

In this work (1) the mean climatic conditions and (2) the intraseasonal variability of temperature and precipitation during a particular winter season, Oct. 1999–Mar. 2000, over the western Himalayas is studied. These two issues are studied by simulating the Abdus Salam International Center for Theoretical Physics (ICTP)—RegCM3 model driven by the National Center for Environmental Prediction, US, (NCEP) observed boundary conditions. Hence, a series of regional climate simulations of six months duration over the Himalayan region is tested. The focus of the study is particularly on the Indian Himalayan region, where the heterogeneity is maximum.

In section 2, a brief description of the model and experimental design is presented. The results are discussed in section 3 and final remarks are given in section 4.

2. Model and Experimental Design

The regional climate model used in the present work is the version of RegCM developed by GIORGI *et al.* (1993a,b) with some of the updates discussed in GIORGI and SHIELDS (1999). The dynamical core of the RegCM is equivalent to the hydrostatic version of the fifth-generation Pennsylvania State University—National Center for Atmospheric Research (NCAR), US, Mesoscale Model (MM5). For the present simulation, the standard model configuration is used with 23 sigma levels, with the medium resolution PBL scheme with five levels in the lowest 1.5 km of the atmosphere, at approximately 40, 110, 310, 730 and 1400 m above surface (GIORGI and BATES, 1989). The physics parameterization employed in the simulations includes the radiative transfer package of the NCAR Community Climate Model version 3 (CCM3, KIEHL *et al.*, 1996), the nonlocal boundary scheme by HOLTSLAG *et al.* (1999) and mass flux cumulus cloud scheme of GRELL (1993).

Land-surface processes are described via Biosphere-Atmosphere Transfer Scheme or BATS (DICKINSON *et al.*, 1993). BATS is a state-of-the-art land-surface model that has been used for many years by a wide research community. It consists of a vegetation layer, three soil layers for soil water content calculations and a force restore method to calculate the temperature of a surface soil layer and a subsurface soil layer. At each model grid point a vegetation class is assigned as dependent on seasonal parameters including roughness length, maximum and minimum leaf area index, stem area index, vegetation albedo and minimum stomatal resistance. The parameter values are given by DICKINSON *et al.* (1993) for 18 land-surface classes.

In the presence of the vegetation, the temperature of canopy air and canopy foliage is calculated as demonstrated from the canopy energy balance. Sensible heat, water vapor and momentum fluxes at the surface are computed using a standard surface drag coefficient formulation based on a surface layer similarity theory. Surface evapotranspiration accounts for evaporation from the soil and the wet portion of the canopy and transpiration from the dry portion of the canopy. Ground evaporation and transpiration rates depend on the soil water content, which is a prognostic variable.

The soil hydrology calculations include predictive equations for the water content of the surface soil layer, the root zone, and a deep soil layer characterized by depths of 10 cm, 1–2 m, and 3 m, respectively. These equations account for precipitation, snowmelt, canopy foliage drip, evapotranspiration, surface runoff, infiltration below

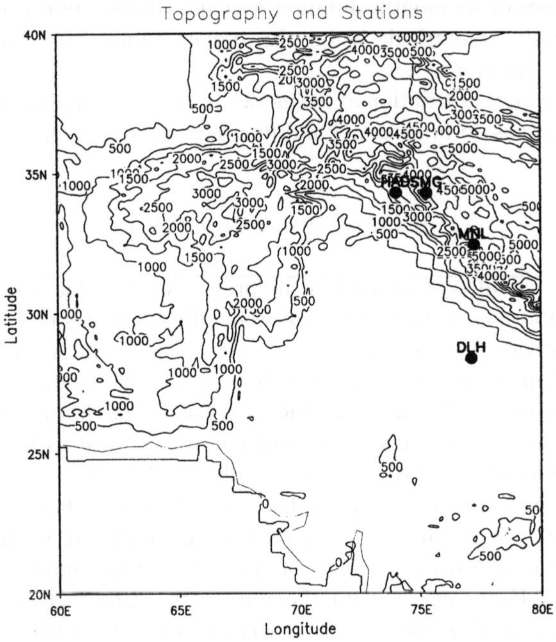

Figure 1
Domain topography (km) in the control simulation. The model grid cell size is 60 km.

the deep soil (which we refer to as base flow), and soil water movement under gravitational and capillary forces. The surface runoff rate is proportional to the precipitation + snowmelt rate and the degree of soil water saturation. Snow depth is prognostically computed from snowfall, snowmelt and sublimation.

In this paper, a control simulation is made over the domain and topography as shown in Figure 1. The computational domain is considered from the Mediterranean Sea to India using a Lambert conformal projection with grid cells of 60 km × 60 km size to understand the flow pattern of the wintertime synoptic weather system called 'Western Disturbance (WD)'. However, in the present work discussion is mainly associated with surface parameter variabilities over the Indian Himalayan and surrounding areas. The topography for control grids is obtained from a 30" (about 1 km) resolution global data set produced by the Geological U.S. Survey (USGS). The land-use distribution for the control experiments is also obtained from a 30" landuse data set produced by USGS (Loveland *et al.*, 1991). A version of this data set is already available in the form of BATS surface types. From the 30" data set we calculate the fractional cover of different surface types for each cell of the different model grids, and the grid cell is then assigned the surface type with the largest fractional cover and henceforth considering landuse and soiltype.

A 6-month simulation for the period starting from October 1, 1999, and ending on March 31, 2000, which encompasses a full winter season, is made. This particular period is chosen for the study, as an enormous amount of precipitation in the form of snow was received/recorded at SASE observatories located in the Indian part of the western Himalayas. Hence, most of the discussion pertains to surface parameter variabilities during this period. Further, lateral meteorological boundary conditions for the simulations are obtained from analyses of observations by the NCEP (KALNAY *et al.*, 1996) and therefore the model results can be directly compared with the observations for the simulated period. Soil temperatures are initialized with the temperature of the bottom model level and soil water content is initialized as a function of vegetation type (GIORGI and BATES, 1989).

3. Results

Overall evaluation of the model performance is provided within this section. A detailed analysis and comparison of control simulation with observation over the Himalayan region is presented. Observed surface air temperature and precipitation needed for the comparison with the simulated fields are obtained from 0.5° resolution global land data sets developed by the Climate Research Unit (CRU) of the University of East Anglia (NEW *et al.*, 2000) and station data from the Snow and Avalanche Study Establishment (SASE), Chandigarh, India. Results of control run simulations with observations are discussed, based on seasonal and monthly averages and variabilities of temperature, precipitation and snow depth amount. In addition to this, comparison is drawn at three stations, viz., Haddantaj (lat 34°18′43″, lon 74°02′42″. alt 3080m), Banihaltop (lat 33°31′17″, lon 75°12′00″, alt 3250m) and Manali (lat 32°16′33″, lon 77°09′03″, alt 2192m), situated in the Indian Himalayan region. These stations are chosen in such a way that they represent different climatic and geographic conditions of the region and have recorded data.

(a) Surface Air Temperature

Observed temperature (CRU) and simulated surface air temperature in control experiments averaged during the period Oct. 1999–Mar. 2000 over the region are presented in Figures 2(a,b). In addition, Figure 3(a) represents seasonal averaged observed (SASE) surface air temperature over the Indian Himalayan region. A comparison shows that the model has a warm bias over the Indian Gangetic plain and a cold bias by a few degrees over the complex mountainous region. It is likely that the model bias is artificially enhanced by a temperature overestimate/underestimate in the observed data set induced by the relatively low density of high elevation stations. Observations tend to show finer scale structure, in particular over the main mountainous ranges, which are represented well by model integrations up to a certain

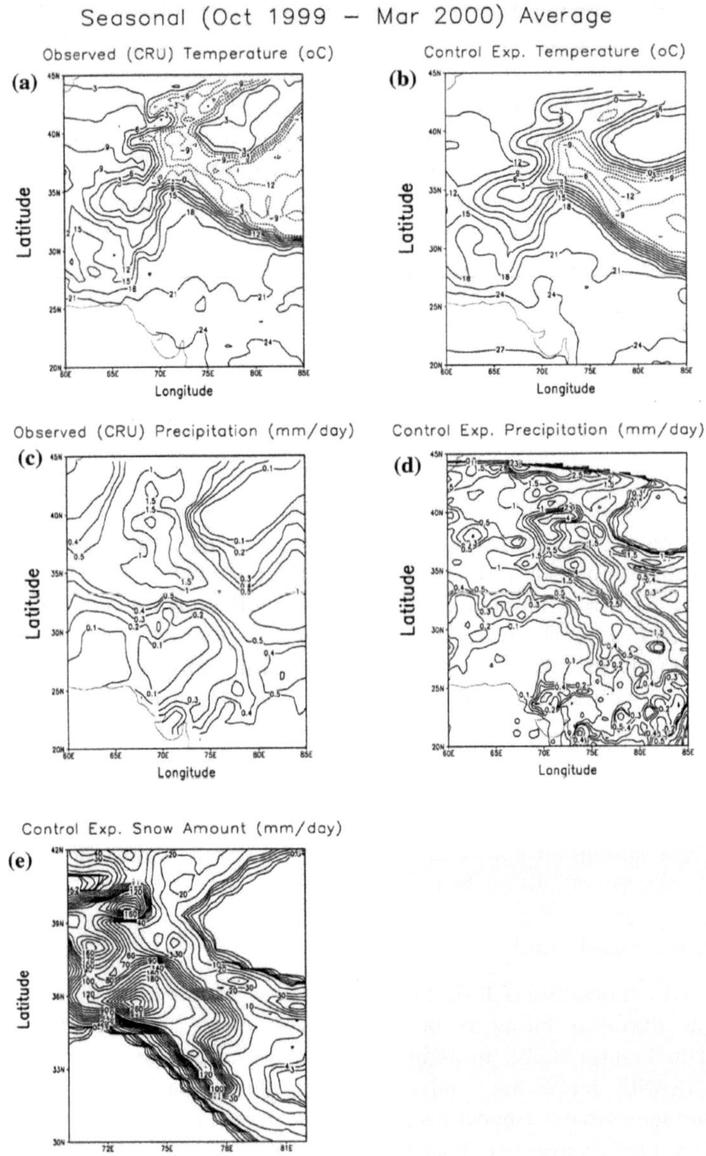

Figure 2
Seasonal (Oct 1999–Mar 2000) average:(a) Observed (CRU) and (b) simulated (control run) surface air temperature (°C); (c) observed (CRU) and (d) simulated (control run) precipitation (mm/day); (e) simulated (control run) snow amount (mm/day) over the Himalayan region.

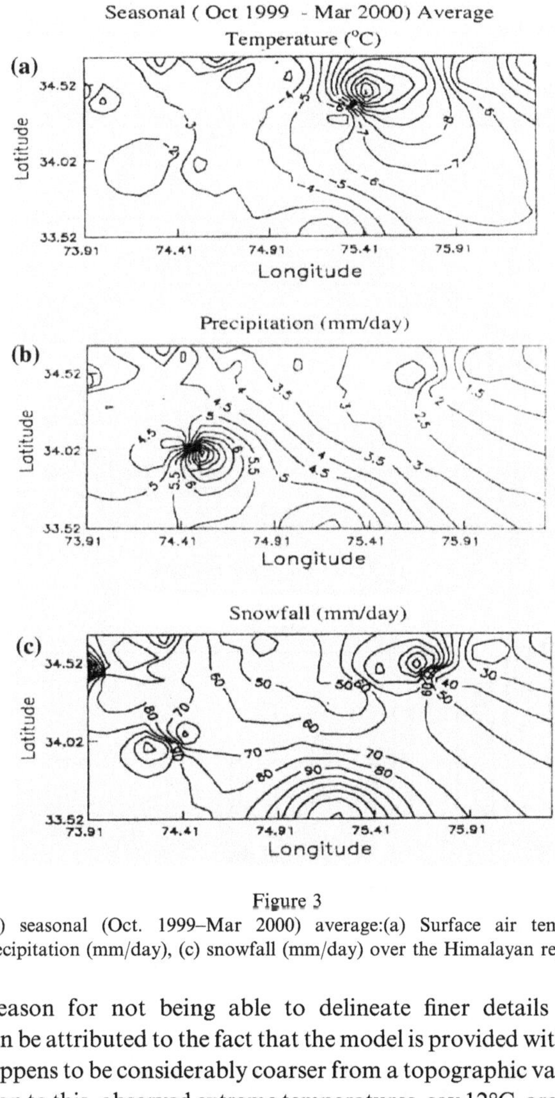

Figure 3
Observed (SASE) seasonal (Oct. 1999–Mar 2000) average:(a) Surface air temperature (°C), (b) precipitation (mm/day), (c) snowfall (mm/day) over the Himalayan region.

extent. The reason for not being able to delineate finer details of temperature distribution can be attributed to the fact that the model is provided with 1.5° resolution data, which happens to be considerably coarser from a topographic variability point of view. In addition to this, observed extreme temperatures, say 12°C, are well captured in the control experiment and the areas of lowest temperature details are well indicated over the mountainous region. Further comparison with real-time seasonal averaged observed (SASE) surface air temperature, Figure 3(a), shows that, though Figure 3(a) represents a very small area of the Himalayas, temperature distribution of the warm and cold regions is well captured by model experiments. Areas of lowest temperature are well depicted by the control experiment.

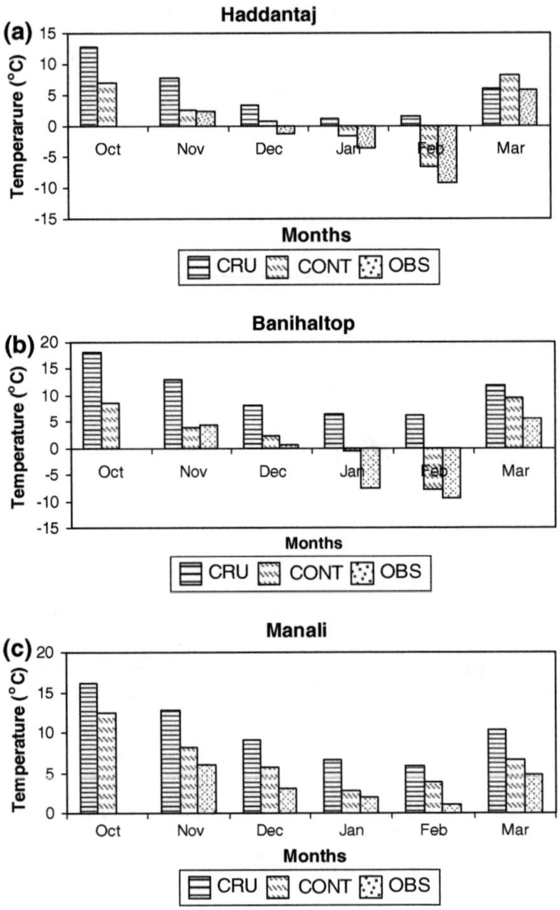

Figure 4
Comparison of monthly average surface air temperature (°C) at (a) Haddantaj, (b) Banihaltop and (c) Manali.

Further, Figures 4(a-c) represent monthly averaged observed surface air temperature of SASE and CRU and simulated surface air temperature in the control run at three of the stations located in the Indian Himalayas. Temperature value of the grid in which these stations fall is considered from CRU data and the control experiment for comparison. It should be noted that temperature data for October is not available at SASE stations. Figure 4 shows that the control experiment could capture the temporal variability and spatial variability of cold temperature over the region. In addition, control runs could produce the negative temperature distribution than the observed CRU data. It could be attributed to the fact that finer details of heterogeneity in topography and land-use are represented in the model with finer scale. Whereas in CRU observations representation of density of high elevation and low elevation stations may not be that homogeneous so that fine resolution resolvable scale circulations are not reproduced. Further, comparison shows that the model has a tendency to overestimate the cold region temperatures and underestimate the warm region temperatures by a few degrees. These biases can be attributed to the facts that they are due to the relatively low density of high elevation stations in observed data sets. Overall, the comparisons of the figures indicate that the model reproduced the observed regional temperature pattern over the Himalayan region generally well.

(b) Precipitation

Figures 2(c,d) compare seasonal averaged (October 1999 to March 2000) observed (CRU) precipitation and simulated experiments with control runs over the Himalayan region. Also, Figure 3(b) shows observed SASE precipitation. However, the control experiment could reproduce the precipitation distribution pattern, but not at all scales. However, it could generate well the precipitation amount over the northwest Indian region but over the Himalayan region the precipitation patterns are overestimated. This bias may be attributed to the fact that although in general topographically-induced cold season precipitation maxima are reproduced, the corresponding peak precipitation values are somewhat overestimated. This is evidently a problem related to the relatively coarse resolution of observed data set CRU; where over the complex topographical Himalayan region, stations of high elevations are not represented well. Nonetheless, Figures 2(c,d) indicate that model simulations are quite close to the actual precipitation amount of 0.1 to 1.0 mm/day over the western Indian region of Gujrat, Rajasthan and Punjab, whereas, over the Indian Himalayan region the amount of the precipitation is overestimated by the control run experiment. However based on this it could be stated that most of the topographically-induced precipitation is reproduced well by model simulation, therefore, overall the model captures regional topographical forcing. The reason for the model overestimate is that the dominant precipitation process is mostly of a resolvable scale nature and is induced by topographic uplift within eastward moving cyclonic systems (WDs). As a result, precipitation is mostly forced by the topographical gradients and the underrepresentation of these gradients leads to an overestimate of the precipitation maxima. These results are indeed evident from the comparison of the winter precipitation fields in Figures 2(c,d) and Figure 3(b). Small differences across the simulation are essentially due to the internal model variability (GIORGI and BI, 2000).

In addition to this, Figures 5(a-c) represent the monthly averaged observed (SASE and CRU) precipitation and simulated control experiment. Comparison

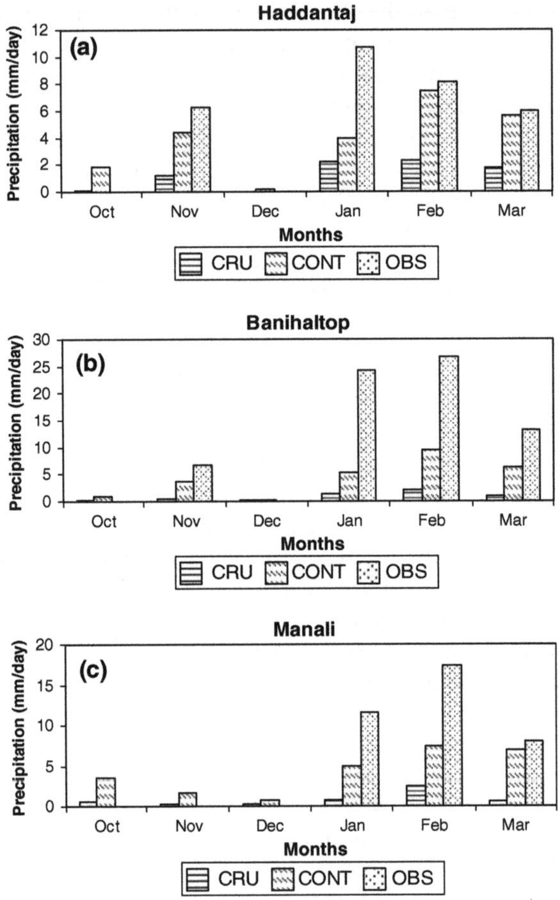

Figure 5
Comparison of monthly average precipitation (mm/day) at (a) Haddantaj, (b) Banihaltop and (c) Manali.

shows that, quantitatively, the control experiment could do better than the observed CRU data set, but still only yield a certain percent of actual precipitation recorded at Haddantaj, Banihaltop and Manali. It can be due to these possible reasons: (i) precipitation amount for control run and observed CRU data set has been considered at the grid where these station fall—it may happen then that while considering the 60 km resolution many topographical features are not well represented and (ii) the model's internal variability can play an important role.

Simulation shows high intensity of precipitation maxima lying along the orientation of the Himalayan region, which is not shown in observed (CRU) data set. This difference may lie in the fact that topography and vegetation cover is very fast changing within a km over the Himalayan region, whereas the observed data set is presented at 0.5° resolution. Due to smoothing of model topography, micro/mesoscale circulations, which are dominant in the complex topography of the Himalayas, are certainly overlooked in the 60 km simulations for obvious reasons. Apart from this, smoothing of topography and land-use in simulation experiments reduced the impact of orographic lifting due to steep gradients in the Himalayan region. During winter, most of the time orographic lifting is the predominant mechanism to modulate the precipitation amount in the complex topographical region of the Himalayas. In addition, smoothing of land-use type, particularly during winter, it is snow cover which contributes to the over/underestimate of the precipitation amount, by heat-flux exchanges due to solar radiation.

(c) Snow

Figure 2(e) shows simulated averaged (Oct. 1999–Mar. 2000) snow depth over the Himalayan region. As a reference, we also present the corresponding observed snow depths for the simulation period estimated from real time observations of SASE observatories (Figure 3(c)). Here one point to be noted, firstly, is that model calculated snow depth is in terms of the liquid water equivalent, whereas the original observed data is given in centimeters of snow. To obtain equivalent liquid water depth we scaled the snow depth by a factor of 1/3, which is roughly characteristic of the density of the aging snow (DICKINSON *et al.*, 1993). Admittedly, considerable uncertainty is implicit in this assumption because snow density mainly depends on its age, temperature, water content, etc. Second, the station density is irregular in space and it includes a relatively small number of high elevation stations. Third, observed snow depth at a station is strongly affected by processes such as snowdrift and snow sheltering by upwind obstacles, which are not included in the model. For these reasons the comparison with observations is necessarily limited in scope and mostly aims at providing qualitative indications of the model behavior.

Figure 2(e) shows that spatial variability of snow increases substantially with the resolution of the land surface, due to temperature produced such that precipitation can be in the form of snowfall over the higher peaks and rainfall over the valleys. As a result, snow tends to accumulate over the high resolution peaks and melt more effectively over the corresponding valleys over Himalayan topography. Also, limited comparison with observations indicates that the spatial scale of snow depth variability is more in line with observations. Indeed even with the limitations discussed previously, the observations clearly show that snow depth is characterized by pronounced finescale variability. In winter, not only the spatial variability of snow increases, but also do overall snow amounts over the region, i.e., snow amount varies from one region to another. This can be attributed to the inherent nonlinear nature of snow-forming processes. As the temperature threshold for snow-formation reaches, say the high elevation of a peak, snow starts accumulating. Because snow has a higher albedo than bare soil or vegetation, the overall surface albedo increases and this causes a decrease in the absorption of solar radiation at the surface. This in turn inhibits the solar warming of the surface and thus tends to cool the region and increase the lifetime of the snow pack. These feedback processes can be seen by cooling of the Himalayan region in control run and in the greater overall snow amounts.

Averaged monthly snowcover of observed (SASE) is presented in Figure 3(c). Comparison shows that spatial and temporal variability of snowfall amounts needs to be examined in a more realistic manner for assessing the model's behavior, particularly towards snow parameter. However, another feature where model observations appear at least in qualitative agreement is the seasonal evolution of the snowpack. The observations indicate that during winter snow amounts are widespread over the region, but show localized maxima in correspondence to the highest peaks.

4. Discussion and Conclusions

In this paper a regional climate model (RegCM3) is tested for its effects on the surface climate of a simulation for the Himalayan region where both topography and land-use variability are high. In this paper, six months are considered for simulations. The results might thus be sensitive to the initialization of soil variables and depend on the specific simulated year.

Control experiment shows that for temperature field the model represents some cold bias over the Himalayan region and warm bias over the northwest Indian region. However, the model could not capture the extreme temperature values, but definitely could indicate the areas of low temperatures. Also, control run could explain the temporal variability in temperature field based on a comparison with station data. In the case of precipitation field, control experiment could generate the precipitation amount over the northwest Indian region but over the Himalayan region the precipitation patterns are overestimated. In addition to this, snow-pack evolution could be generated by the simulation but has to be seen within its larger nature of complexity.

Keeping the above results in mind, an accurate simulation of the effects of surface climate does require the representation of very finescale surface processes. Also, study of subgrid scale disaggregation processes is planned in future work. In addition to this longer time simulations are planned to understand the variabilities associated with these parameters.

Acknowledgements

The first author acknowledges Professor F. Giorgi, Drs X. Bi, N. Elguindi and The Abdus Salam International Center for Theoretical Physics (ICTP), Trieste, Italy for providing necessary assistance for the above study. Also, the authors acknowledge the National Center for Environmental Prediction (NCEP), US for providing valuable data sets for accomplishing this work. The authors thank the two anonymous referees for their critical suggestions for improving the manuscript.

References

Bhaskaran, B., Jones R.G., Murphy J.M., and Noguer M. (1996), *Simulation of Indian summer monsoon using a nested regional climate model: Domain size experiments*, Climate Dyn *12*, 573–587.

Dickinson, R.E. (1995), *Land atmosphere interaction*, Rev. Geophys. *33* (Suppl.), 917–922.

Dickinson, R.E., Henderson-Sellers, A., and Kennedy P.J. (1993), *Biosphere-Atmosphere Transfer Scheme (BATS) version 1e as coupled to the NCAR Community Climate Model*, NCAR Tech. Note NCAR/TN-387 + STR, 72 pp.

Dimri, A.P. (2004), *Impact of horizontal model resolution and orography on the simulation of a western disturbance and its associated precipitation*, Meteor. Appl. *11*(2), 115–127.

Dimri, A.P. and Mohanty, U.C. (1999), *Snowfall statistics of some SASE field stations in J&K and a case study of western disturbance*, Defense Sci. J. *49*(5) 437– 445.

Giorgi, F. and Shields, C. (1999), *Test of precipitation parameterizations available in the latest version of the NCAR regional climate model (RegCM) over the continental United States*, J. Geophys. Res. *104*, 6353–6375.

Giorgi, F. and Bates, G.T (1989), *On the climatological skill of a regional model over complex terrain*, Mon. Wea. Rev. *117*, 2325–2347.

Giorgi, F., Marinucci M.R., and Bates, G.T. (1993a), *Development of a second generation regional climate model (RegCM2), Part I: Boundary layer and radiative transfer processes*, Mon. Wea. Rev. *121*, 2794–2813.

Giorgi, F. and Xi, B. (2000), *A study of internal variability of a regional climate model*, J. Geophys. Res., *105*, 29503–29521.

Giorgi, F., Marinucci M.R., De Canio, G., and Bates, G.T. (1993b), *Development of a second generation regional climate model (RegCM2), Part II: Convective processes and assimilation of lateral boundary conditions*, Mon. Wea. Rev. *121*, 2814–2832.

Goswami, B.N. and Shukla, J. (1984), *Quasi-periodic oscillations in a symmetric general circulation model*, J. Atmos. Sc. *41*, 20–37.

Grell, G.A. (1993), *Prognostic evaluation of assumptions used by cumulus parameterization*, Mon. Wea. Rev. *121*, 764–787.

Hirakuchi, H. and Giorgi, F. (1995), *Multilayer present day and $2 \times CO_2$ simulations of monsoon climate over eastern Asia and Japan with a regional climate model nested in a general circulation model*, J. Geophys. Res. *100*, 21105–21125.

Holtslag, A.A.M., de Bruijn, E.I.F., and Pan, H.L. (1999), *A high resolution air mass transformation model for short-range weather forecasting*, Mon. Wea. Rev. *118*, 1561–1575.

Jones, R.G., Murphy, J.M., and Noguer, M. (1995), *Simulation of climate change over Europe using a nested regional climate model, I. Assessment of control climate, including sensitivity to location of lateral boundaries*, Q.J.R. Meteor. Soc. *121*, 1413–1449.

Kalnay, E. *et al.* (1996), *The NMC/NCAR 40-year reanalysis project.* Bull. Am. Meterol. Soc. *77*, 437–471.

KIEHL, J.T., HACK, J.J., BONAN, G.B., BOVILLE, B.A., BRIEGLEB, B.P., WILLIAMSON, D.L. and RASCH, P.J. (1996), Description of the NCAR Community Climate Model (CCM3), NCAR Tech. Note NCAR/TN-420+STR, 152 pp.

LOVELAND, T.R., MERCHANT, J.W., OHLEN, D.O., and BROWN, J.F. (1991), *Development of a land cover characteristics database for the conterminous United States, Photogramm*, Eng. Remote Sens. *57*, 1453–1463.

MARINUCCI, M.R. and GIORGI, F. (1992), *A 2 × CO_2 climate change scenario over Europe generated using a limited area model nested in a general circulation model, 1. Present-day seasonal climate simulation*, J. Geophys. Res. *97*, 9989–10009.

MOHANTY, U.C. and DIMRI, A.P. (2004), *Location specific prediction of probability of occurrence and quantity of precipitation over western Himalayas*, Weather Forecasting *19* (3), pp. 520–533.

NEW, M.G., HULME, M., and JONES, P.D. (2000), *Representing twentieth century space time climate variability, Part II: Development of a 1901–96 monthly grid of terrestrial surface climate*, J. Climate *13*, 2217–2238.

PIELKE, R. and AVISSAR, R. (1990), *Influence of landscape structure on local and regional climate*, Landscape Ecol. *4*, 133–155.

(Received April 25, 2006, accepted September 16, 2006)
Published Online First: June 23, 2007

Pure appl. geophys. 164 (2007) 1747–1764
0033–4553/07/091747–18
DOI 10.1007/s00024-007-0229-0

Pure and Applied Geophysics

A Seasonal Statistical Evaluation of COAMPS® over the Arabian Gulf Region

Rebecca E. Eager,[1] Sethu Raman,[1] Teddy R. Holt,[2] Douglas Westphal,[2] Jeffrey Reid,[2] Jason Nachamkin,[2] Ming Liu,[2] and Abdulla Al Mandoos[3]

Abstract—A statistical evaluation of the Coupled Ocean/Atmosphere Mesoscale Prediction System (COAMPS®) was performed over the Arabian Gulf region for the period, 1 August to 5 October, 2004. Verification skill scores of bias and root-mean-square error were estimated for surface variables and for vertical profiles to investigate any diurnal variations. The model predictions of boundary-layer heights are compared with the observations at Abu Dhabi, United Arab Emirates. The Middle East presents challenges to numerical weather prediction due to complex land-ocean-land mesoscale processes. An independent data set of surface measurements from 50 stations in the UAE was available from the Department of Water Resources Studies, Abu Dhabi for model verification. The results indicate a diurnal variation in the model errors. The errors are small considering the magnitudes of the observed variables. Errors in the coastal region can be attributed to the differences in the timing of the onset of sea and land breeze circulations in the simulations as compared to the observations. Errors are relatively smaller in the offshore locations.

Key words: Statistical evaluation, UAE, COAMPS, verification.

1. Introduction

The Coupled Ocean/Atmosphere Mesoscale Prediction System (COAMPS®) (Hodur, 1997) developed by the U.S. Naval Research Laboratory, provided weather and dust forecasts for the Arabian Gulf region in support of the United Arab Emirates Unified Aerosol Experiment (2004) (UAE^2). Verification of COAMPS® has been done for winter months in the Mediterranean (Nachamkin and Hodur, 2000) and in the Middle East (Shi *et al.*, 2004). Shi *et al.* (2004) performed a model verification of COAMPS® for January-March 1991 on a 15-km grid centered over Iraq. They found bias errors of near zero at the analysis time for levels below

COAMPS® is a registered trademark of the Naval Research Laboratory.

[1] Department of Marine, Earth, and Atmospheric Sciences, North Carolina State University, Raleigh, North Carolina, U.S.A.
[2] Naval Research Laboratory, Monterey, California, U.S.A.
[3] Department of Atmospheric Studies, Ministry of Presidential Affairs, United Arab Emirates.

200 hPa. At the 12-hour forecast below 200 hPa there was a negative temperature bias error of 0.2–0.7°C, a negative geopotential height bias of 3 m to 9 m, and a positive wind bias of 0.5 m s^{-1}, which are similar to the bias errors of the models listed in WHITE *et al.* (1999). NACHAMKIN and HODUR (2000) found the wind and temperature fields to be underpredicted and relative humidity to be overpredicted, with magnitudes similar to the bias errors in SHI *et al.* (2004) and WHITE *et al.* (1999).

This paper describes a statistical evaluation of 48-hour COAMPS® forecasts performed in the Arabian Gulf region during the summer and early fall, from 1 August to 5 October, 2004. Section 2 discusses the model, available data, verification methods and gives a brief synoptic overview of the period; Section 3 discusses the verification of surface variables across the region; Section 4 includes the model verification for upper-level variables; and Section 5 gives a summary of the results. Throughout the paper, both UTC and local time (LT; UTC + 4 hours) are used when referring to forecast valid time, because mesoscale circulations driven by land-sea temperature contrasts are the dominant weather features during most of the verification period.

2. Model, Verification Methods, and Data

2.1. Model Description

The atmospheric component of COAMPS® with non-hydrostatic dynamics was used in this study. The atmospheric model contains physical parameterizations for subgrid scale mixing, sub-grid scale convection, short- and longwave radiation, and explicit moist physics. A full description of the model physics and the equations can be found in CHEN *et al.* (2003) and on the COAMPS® website (http://www.nrlmry.navy.mil/coamps-web/web/home). Three one-way nested grids are used (Fig. 1). The outer grid is 92 × 68 grid points with 81-km resolution, grid 2 is 127 × 109 grid points with 27-km resolution, and grid 3 is 181 × 181 grid points with 9-km resolution. The domain has 30 vertical sigma levels, with increased vertical resolution (50 to 100 m) in the lower levels. For this study the 9-km grid domain is used for verification. Boundary conditions are provided by the Navy Operational Global Atmospheric Prediction System (NOGAPS) model, as are the first-guess fields for the first model run. Subsequent model runs use the previous COAMPS® 12-hr forecast for the first-guess field. Data assimilation was performed using multivariate optimum interpolation analysis every 12 hours at 0000 UTC and 1200 UTC.

2.2. Observed Data

Three data sets are used for verification of surface variables. The first data set is the ADP data set, consisting of all the observations contained within the 9-km model

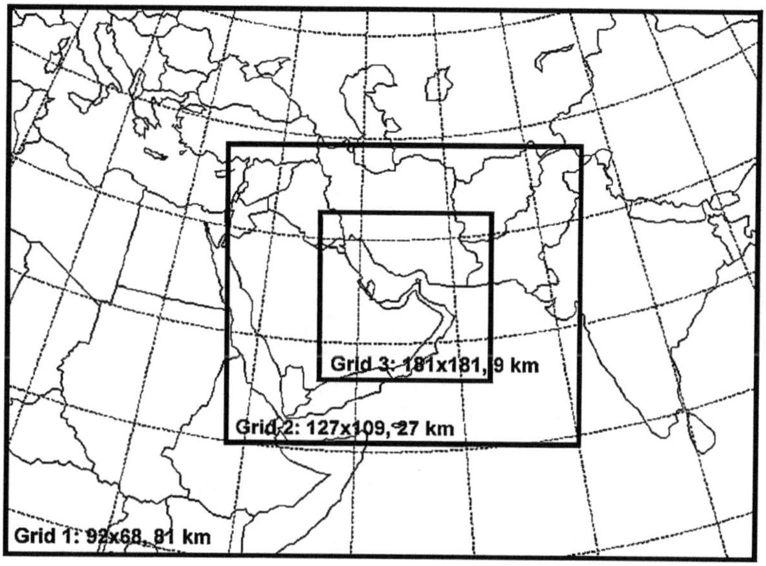

Figure 1
COAMPS® model triple nested domain with grid points and horizontal resolution given for each nest.

grid that were available for model analysis, including surface station, radiosonde, buoy, and satellite data. This data set undergoes a complete data quality process (BAKER, 1992) prior to assimilation into the model. The second data set, the UAE data set, is the independent data set not used in the analysis, consisting of the Department of Water Resources Studies (DWRS) surface weather station data. The status of the quality control on the DWRS data is unknown. Thirty-six DWRS stations are located in the interior of the UAE, nine are coastal stations, and seven are on islands. Sixteen stations are excluded for the entire period for one of two reasons: (1) model terrain height for the station was more than 100 m different than the actual terrain height, or (2) the model designated the location of a coastal or island station as water instead of land. Single hours of data at a station are excluded from the calculations of skill scores when any one of the variables was not present; for example, if the wind speed was missing at a particular hour, the other variables at that hour were not included. Locations of the stations used in the model verification are given in Figure 2. The third data set, designated as BOTH, consisted of the combined ADP and UAE data sets. Visibility is not measured at the DWRS stations and therefore model surface visibility is only verified against the ADP data set. The ADP data set is also used for verification of the upper-level variables. Radiosonde soundings from Abu Dhabi, available from the Department of Atmospheric Science at the University of Wyoming from their website (http://weather.uwyo.edu/upperair/

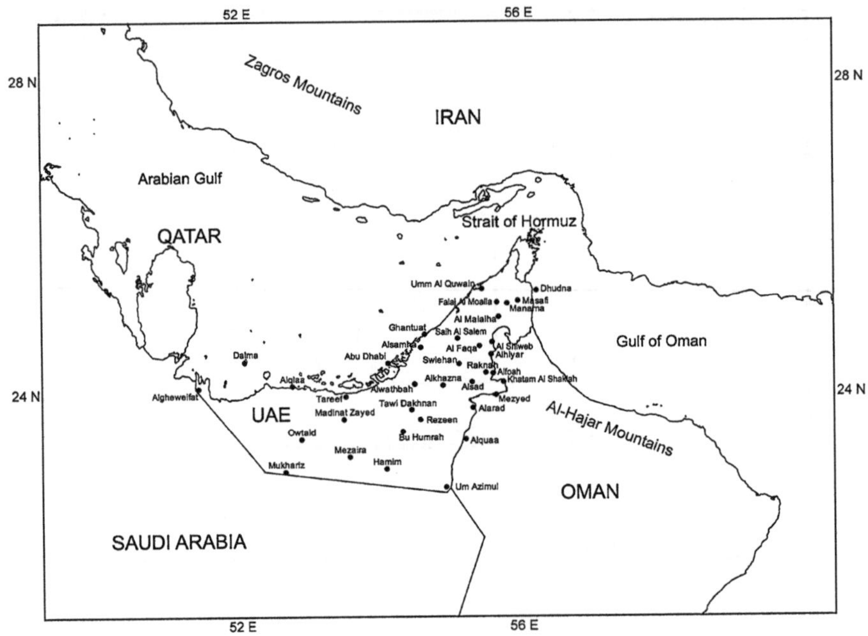

Figure 2
Surface meteorological stations included in the verification of surface parameters. These stations are operated by the UAE Department of Water Resources.

sounding.html) are used in the verification of model boundary layer heights. Radiosonde data are available every 12 hours at 0000 UTC (0400 LT) and 1200 UTC (1600 LT) for 1 August to 5 October, 2004 except for seven soundings.

2.3. Verification Techniques

Verification skill scores of bias error and root-mean-square (RMS) error are estimated for air temperature, dew point temperature, wind speed, wind direction, surface visibility, and geopotential height. COAMPS® model output of dew point depression is converted to dew point temperature. Visibility, in kilometers, is estimated from the COAMPS® output of surface dust concentration (*sfdscc*), with units of mg m^{-3}, using the following relation:

$$\text{visibility (km)} = \frac{0.16}{(sfdscc)^2}. \tag{6.1}$$

This equation is derived from the Koschmeider Equation (SEINFELD, 1986). The bias error measures the model's tendency to underpredict or overpredict a value. The

RMS error measures the magnitude of the model error. The bias error is estimated for any given variable x, a s

$$\text{bias}\,(x) = \frac{1}{N}\sum_{n=1}^{N}\left(x_n^f - x_n^o\right), \tag{6.2}$$

and the RMS error using

$$\text{rmse}\,(x) = \left[\frac{1}{N}\sum_{n=1}^{N}\left(x_n^f - x_n^o\right)^2\right]^{1/2}, \tag{6.3}$$

where N is the number of locations, $x_n{}^f$ is the value of the forecast variable and $x_n{}^o$ is the value of the observed variable. Bias and RMS errors are estimated for each model-observation pair and then averaged over all available pairs.

2.4. Verification Procedure

A cold start model simulation was begun on 25 July, 2004 with 48-hour forecasts produced every 12 hours through 1200 UTC 5 October, 2004. Model output was stored every three hours. The statistical evaluation period for the 9-km domain is from 0000 UTC 1 August to 1200 UTC 5 October, 2004. The model performance skill scores for the region are calculated every 6 hours for all surface variables for both the 0000 UTC (0400 LT) and 1200 UTC (1600 LT) model initialization times. Bias and RMS errors are also determined for upper-level variables, and the model predicted boundary-layer height is verified against observations at Abu Dhabi.

2.5. Synoptic Overview

During most of the verification period, the area of interest is dominated by the southwest monsoon, which causes light northwesterly winds of 2 to 4 m s^{-1} over the Arabian Gulf. Mesoscale features, such as the sea and land breeze circulations, are the main influences on the weather during the monsoon. The surface winds near the coast can increase to approximately 10 m s^{-1} due to the sea and land breeze circulations. Observed hodographs from several stations throughout the UAE region are available in EAGER (2005). Occasionally, stronger winds due to synoptic features over the Gulf overwhelm the coastal circulations; however, these periods only last for about one to two days. The summer monsoon ended between 23 and 27 September, 2004, followed by a tropical disturbance that moved westward across the northern Arabian Sea.

The typical diurnal range of air temperature is from about 32° to 40°C at island and coastal locations, and 26° to 40°C at inland sites. The dew point temperature ranges from 20° to 28°C offshore, 18° to 26°C at coastal stations, and 8° to 16°C at inland stations, with more moist conditions at night. The wind speed varies from about 8 to 10 m s^{-1} for all stations (EAGER, 2005).

3. Surface Verification

The model temperature bias and RMS errors for the 0000 UTC (0400 LT) and 1200 UTC (1600 LT) initialization runs are shown in Figures 3a and 3b. For both times, an average of 32 model-observation pairs are used to estimate the bias and RMS errors. COAMPS® generally underpredicts the temperature at each time period with a 0 to 1°C cold bias in the late morning (1000 LT) and afternoon (1600 LT) and a 2.5°C cold bias during the night (2200 LT) and early morning (0400 LT). When verified against the BOTH data set, the nighttime cold bias decreases to values

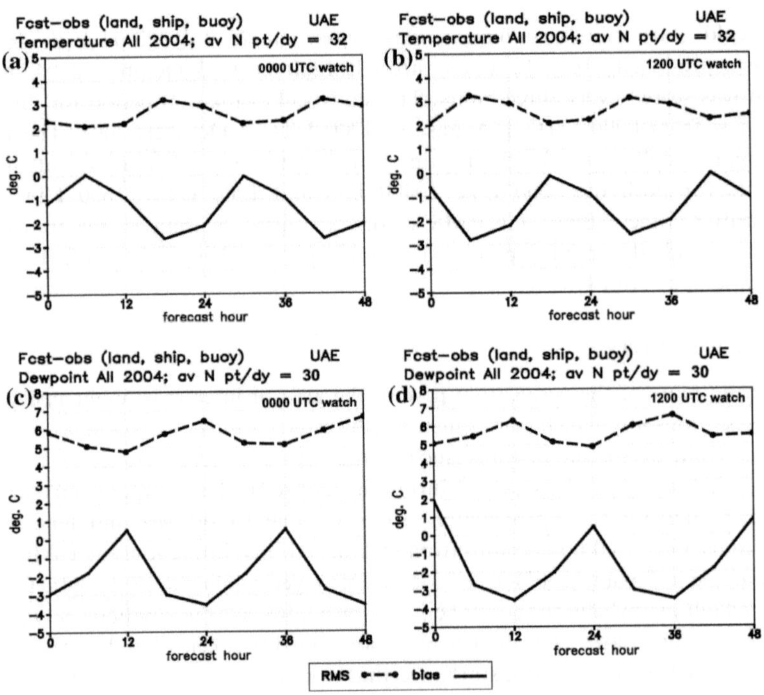

Figure 3
The 2 m air temperature RMS and bias errors at (a) 0000 UTC (0400 LT) and (b) 1200 UTC (1600 LT) initialization times. The 2 m dew point temperature RMS and bias errors at (c) 0000 UTC (0400 LT) and (d) 1200 UTC (1600 LT) initialization times. UAE data set was used for verification. Watch refers to model initialization time.

between 0°C and –2°C. As mentioned, typical daytime maximum temperature is approximately 40°C and minimum around 26°C inland. This translates to about 2.5% in the daytime and 8% during night time for bias errors. The RMS errors show little diurnal variation and vary from about 5% during nights to about 8% during the day. It is interesting to note the reversal in magnitudes of errors between the bias and the RMS errors. These errors are possibly caused by errors in the model prediction of sea breeze and land breeze circulations.

Model dew point temperature bias and RMS errors are shown in Figures 3c and 3d for the 0000 UTC and 1200 UTC (0400 LT and 1600 LT) initialization times. An average of 30 pairs per day are used. COAMPS® has a 2 to 4°C dry bias during the night (2200 LT) and morning hours (0400–1000 LT). The model bias in the afternoon (1600 LT) is very slightly moist. When COAMPS® is verified against BOTH, the model dew point bias improves at all forecast time periods, but still ranges between a 0°C to +1°C in the afternoon to a bias of about –2°C at night and in the early morning. When compared to typical observed values of dew point temperatures of 18 to 26°C at the coast, the bias errors are of the order of 5 to 10%. The RMS errors are relatively larger (about 20%), possibly due to low moisture conditions of the atmosphere in this location.

The wind speed bias and RMS errors are shown in Figures 4a and 4b for the 0000 UTC and 1200 UTC (0400 LT and 1600 LT) model initialization time. An average of 32 pairs per day is available for verification. For both the 0000 UTC (0400 LT) and 1200 UTC (1600 LT) initialization times, the model winds are 0.5–1 m s^{-1} too fast in the morning hours (0400–1000 LT) and 0.5 m s^{-1} too slow during the afternoon (1600 LT) and night (2200 LT). The model wind speed bias error slightly improves to –0.5 m s^{-1} to 0.5 m s^{-1} when verified against the BOTH data set. The most improvement in the forecasts occurs during the late morning (1000 LT; not shown). RMS errors are about 2 m s^{-1}. Typical observed near-surface wind speeds are of the order of 10 m s^{-1} due to sea and land breeze circulations (EAGER, 2005), indicating an RMS error of 20%. Some of this error can be attributed to observational errors related to averaging time and instrument response and model errors related to horizontal inhomogeneity and prediction of the onset of sea and land breezes.

The model wind direction bias and RMS errors are plotted in Figures 4c and 4d for the 0000 UTC (0400 LT) and 1200 UTC (1600 LT) initialization times. An average of 32 grid points per day was available to estimate the bias and RMS errors. The wind direction errors are estimated accounting for the zero crossing. For both model initialization times there is a 10–20° bias error in the forecasts valid in the night (2200 LT) and morning (0400 LT and 1000 LT). For the afternoon (1600 LT) forecast integration times there are no wind direction bias errors. The wind direction bias improves to 0° to 10° at both initialization times for the forecasts verified against BOTH (not shown). The RMS errors vary from 70° to 80°. and are rather large. It is interesting to note that the bias is small and the RMS error is large. The possible reason for this difference is that the model is not

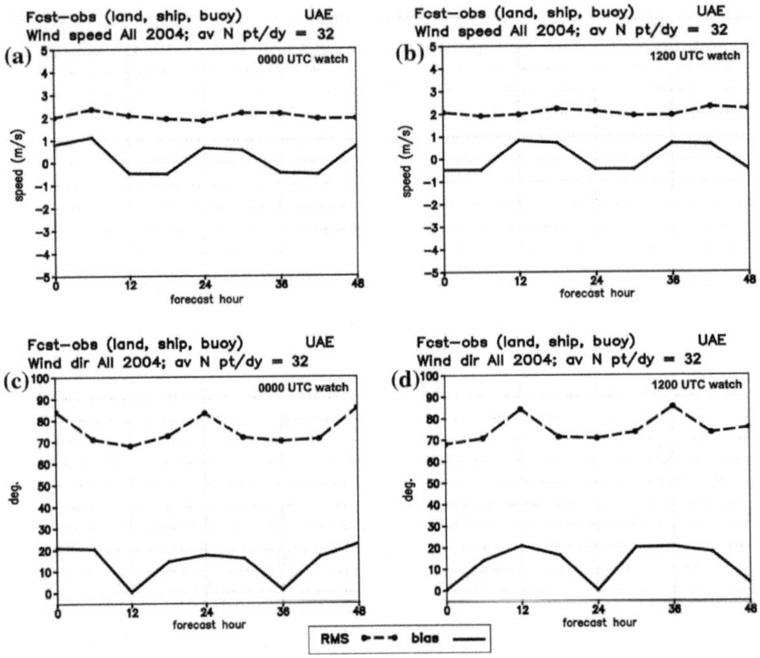

Figure 4
Wind speed RMS and bias errors at (a) 0000 UTC (0400 LT) and (b) 1200 UTC (1600 LT) initialization times. Wind direction RMS and bias error at (c) 0000 UTC (0400 LT) and (d) 1200 UTC (1600 LT) initialization times. UAE data set was used for verification. Watch refers to model initialization time.

able to accurately predict the location of the sea and land breeze fronts with the result that at a given location the model forecast would be southerly (offshore) winds and the observed winds northerly (onshore). This produces an error of 180°. However, while estimating the bias error, the errors cancel out giving a small net value. Predicting the circulation and the sea breeze and land breeze fronts has always been a challenge for the mesoscale models.

Visibility is not measured at the DWRS stations, but was available in the ADP data set. An average of 38 model-observation pairs is available per day. The model visibility bias and RMS errors are shown in Figure 5. The bias at the initialization time improves from 4 km in the 0000 UTC (0400 LT) runs to 3 km in the 1200 UTC (1600 LT) runs. The visibility bias error decreases in the afternoon, and increases during the morning and night. The RMS error is between 6–7 km at all forecast integration times. The large bias error may be due to the inherent problem of model visibility versus observed visibility that arises from the visibility observations typically being less than the model-determined values, particularly for clear sky

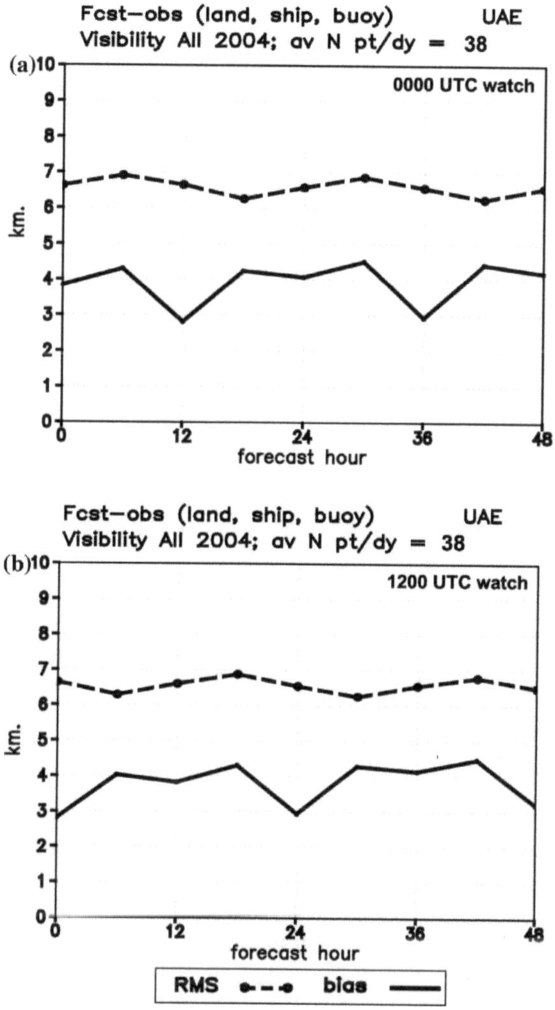

Figure 5
Visibility RMS and bias errors at (a) 0000 UTC (0400 LT) and (b) 1200 UTC (1600 LT) initialization times. ADP data set was used for verification. Watch refers to model initialization time.

conditions. In addition, observations of visibility are typically given qualitatively instead of exact values as determined by the model.

Summaries of the bias and RMS error values calculated using the UAE data set at each forecast valid time are given in Table 1 (0000 UTC simulations) and Table 2 (1200 UTC simulations) for temperature, dew point temperature, wind direction, and

Table 1

Summary of RMS and bias error values for COAMPS® verified against the UAE data set for the 0000 UTC (0400 LT) initialized simulations

	Temperature (°C)		Dewpoint (°C)		Wind Speed (m s^{-1})		Wind Direction (°)	
Tau	RMS	Bias	RMS	Bias	RMS	Bias	RMS	Bias
0	2.2	−1.2	5.7	−3.0	1.9	0.8	82.4	20.8
6	2.0	0	4.7	−1.8	2.3	1.1	67.0	20.0
12	2.1	−0.9	4.6	0.6	2.0	−0.5	66.3	0.4
18	3.1	−2.7	5.6	−3.0	1.9	−0.5	70.6	14.5
24	2.8	−2.2	6.3	−3.7	1.8	0.6	80.8	17.1
30	2.1	0	4.9	−2.0	2.1	0.5	69.5	15.5
36	2.2	−0.9	4.9	0.6	2.1	−0.5	68.1	0.7
42	3.1	−2.7	5.7	−2.8	1.9	−0.5	69.4	16.3
48	2.8	−2.1	6.5	−3.7	1.9	0.7	83.6	22.2

Table 2

Summary of RMS and bias error values for COAMPS® verified against the UAE data set for the 1200 UTC (1600 LT) initialized simulations

	Temperature (°C)		Dewpoint (°C)		Wind Speed (m s^{-1})		Wind Direction (°)	
Tau	RMS	Bias	RMS	Bias	RMS	Bias	RMS	Bias
0	2.0	−0.6	4.9	1.9	2.0	−0.5	66.0	0.4
6	3.2	−2.8	5.3	−2.7	1.9	−0.5	68.5	13.9
12	2.8	−2.2	6.2	−3.6	1.9	0.8	82.5	20.8
18	2.0	−0.1	4.8	−2.1	2.2	0.7	68.0	16.1
24	2.2	−0.9	4.7	0.5	2.1	−0.4	68.5	−0.2
30	3.1	−2.6	5.9	−3.0	1.9	−0.5	71.2	19.8
36	2.8	−2.0	6.5	−3.6	1.9	0.7	83.5	20.0
42	2.2	0	5.2	−1.9	2.2	0.6	70.3	17.5
48	2.3	−1.0	5.3	1.0	2.1	−0.5	72.9	3.4

wind speed. The model bias and RMS errors are smallest at forecast integration times in the afternoon, and largest in the early morning (0400 LT). Air temperature RMS errors are between 2°C and 3.25°C and the bias errors are between −3°C and 0°C. The daytime temperature bias is typically underpredicted by 2.5% and by 7.7% during the night. The RMS error during the day is about 5% of the temperature and about 11.5% during the night. The dew point temperature RMS errors are 4.6°C to 6.5°C and bias errors are −3.7°C to 1.9°C. The dew point temperature is overpredicted by about 1.9% during the day and underpredicted by 16.7% during the night at coastal stations. The RMS error is about 19.2% during the daytime and 33.3% during the nighttime. The wind speed errors are small, with RMS errors of up to 2.5 m s^{-1} and bias errors between -0.5 and 1 m s^{-1}. The wind direction RMS errors are between 65° and 85° with bias errors of 0° to 25°.

RMS and bias error values at individual stations are different than the errors summed over all the stations. On the island station used for verification, the temperature bias was smaller during the night and the dew point temperature bias was positive during all forecast integration times, both contrasting to the errors calculated for all the stations. COAMPS® predicts the dew point temperature better at this offshore location than at stations on the coast or inland. Stations along the coast show similar bias and RMS error tendencies as the errors for all the stations. Inland stations show similar temperature bias errors as for all the stations. The dew point temperature and wind speed bias errors vary between the inland and coastal stations (EAGER, 2005).

4. Comparison of Vertical Profiles

Bias and RMS errors are calculated every 12 hours for the 0000 and 1200 UTC (0400 and 1600 LT) soundings using 11 available observed soundings from the ADP data set. Geopotential height, temperature, dew point temperature, wind direction, and wind speed are verified at pressure levels of 1000, 925, 850, 700, 500, 400, 300, 250, 200, 150, and 100 hPa. Model vertical soundings at Abu Dhabi are also verified against observations to determine the ability of the model to predict the boundary-layer height.

4.1. Vertical Verification—All Stations

The model temperature bias and RMS errors in the vertical soundings are shown in Figure 6. The smallest RMS error of 1–2°C occurs at the 700 hPa level while the largest RMS error is at the 100 hPa level (12°C). COAMPS® performs well in the lower layers with only a slight warm bias. Above 850 hPa there is a cold bias that increases with height and forecast integration time.

The model dew point temperature bias in the upper levels is shown in Figure 7. The RMS error is between 5°C and 8°C from 1000 to 700 hPa, and error values increase with forecast integration time. The model dew point temperature has a moist bias at all levels, which increases with height and forecast valid time. Below 700 hPa the bias is 0°C to 4°C too moist, but above this level the dew point bias increases to a maximum of 14°C too moist at the 48-hr forecast. The moist dew point temperature bias and cold air temperature bias means that the model dew point depression is too small and that too much moisture is predicted at all levels. Larger errors in higher altitudes could be due to the problems associated with the model initialization and boundary conditions.

The geopotential height bias error and RMS error are shown in Figure 8. The smallest RMS error (15–30 m) is at 1000 hPa. The largest RMS error is at the 200 hPa level, where it increases from 40 m in the 0-hr forecast to 80 m in the 48-hr

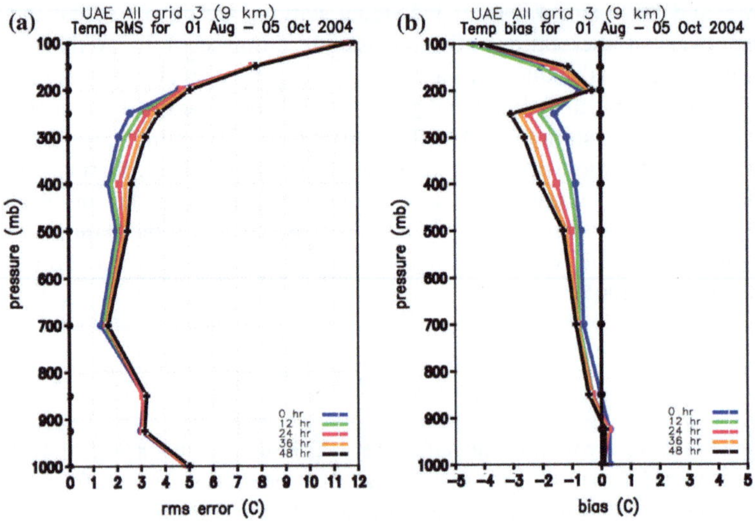

Figure 6
Temperature RMS and bias errors for all vertical soundings available on grid 3.

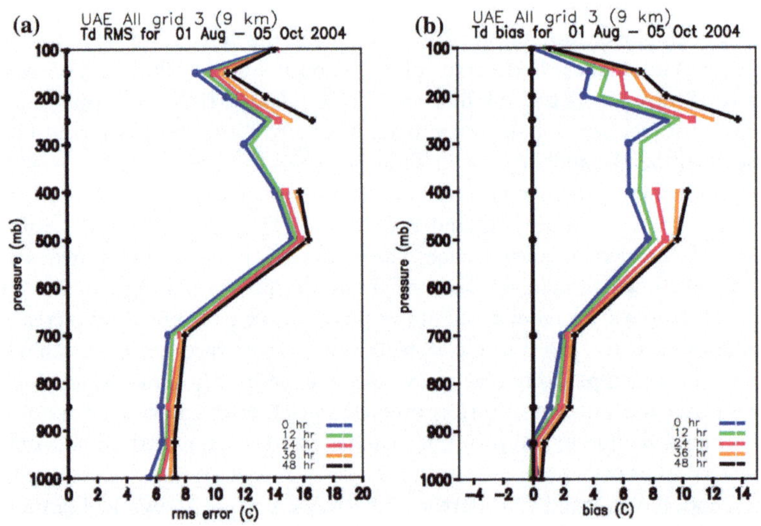

Figure 7
Dew point temperature RMS and bias errors for all vertical soundings available on grid 3.

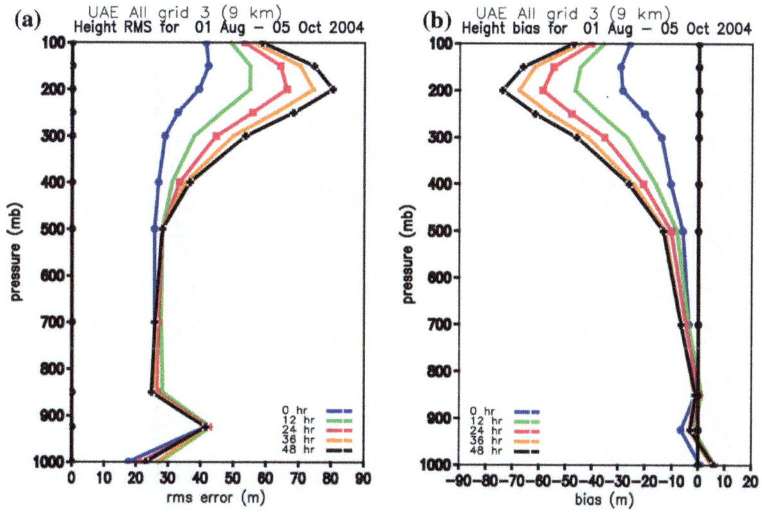

Figure 8
Geopotential height RMS and bias error for all vertical soundings available on grid 3.

forecast. The geopotential height is initialized well with a bias of ± 10 m for all levels up to the 400 hPa level and -10 to -30 m above 400 hPa. The model geopotential height bias increases with each forecast valid time, most notably at 500 hPa and above, where the bias error increases to a maximum of about −75 m.

COAMPS® upper-level wind speed bias error and RMS error are shown in Figure 9. An RMS error of 2 m s^{-1} occurs at 1000 hPa and increases with height to 6 m s^{-1} at 200 hPa. RMS errors increase in each subsequent forecast time. The model wind speed bias indicates an underprediction of 0–1.5 m s^{-1}. From 250–150 hPa, the model overpredicts the wind speed by up to 1 m s^{-1}. Wind speeds are 0–1.5 m s^{-1} too slow from the surface to 250 hPa.

COAMPS® upper-level wind direction bias and RMS error are shown in Figure 10. The RMS errors decrease with height, from an error of 60° at 1000 hPa to an error of 40° at 100 hPa. RMS errors increase in each subsequent forecast integration time. The wind direction bias error is about 0°–10° from 1000 hPa to 200 hPa. Above 200 hPa, the wind direction bias error is −5° to −15°.

4.2. Vertical Verification – Boundary-Layer Height

The model derived boundary-layer height is compared with observed soundings at Abu Dhabi, a coastal site. The observed boundary-layer height is determined by the location of the base of the elevated inversion using the virtual potential temperature. The boundary-layer depth is derived in COAMPS® based on the

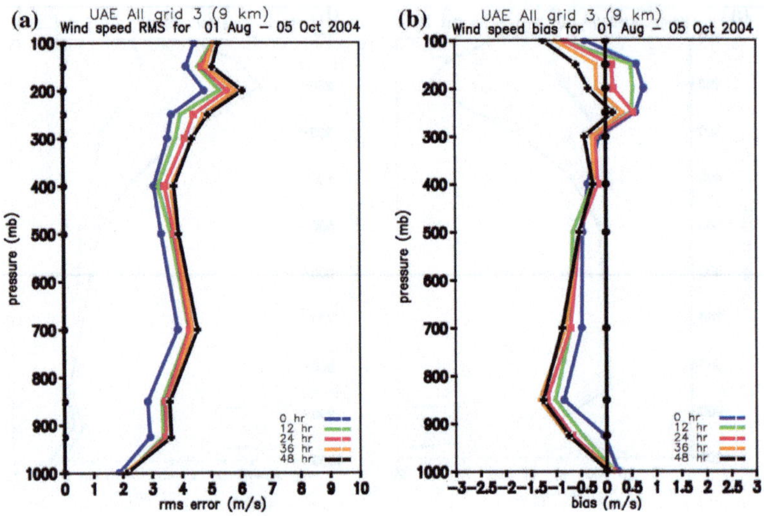

Figure 9
Wind speed RMS and bias error for all vertical soundings available on grid 3.

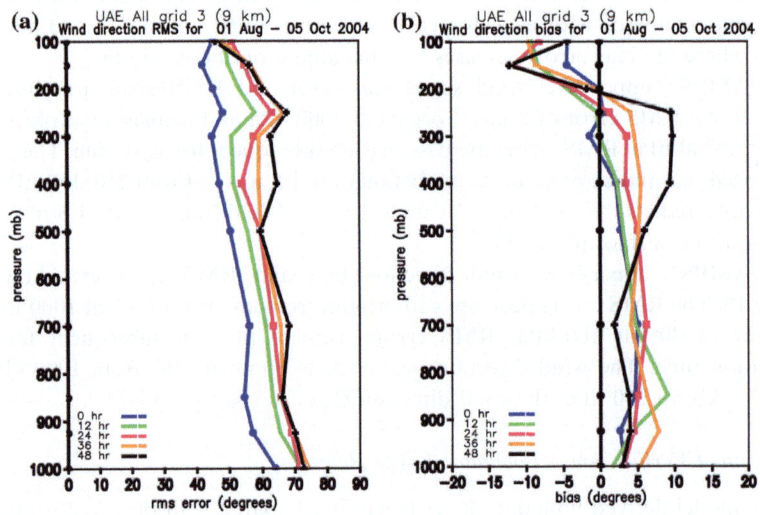

Figure 10
Wind direction RMS and bias error for all vertical soundings available on grid 3.

Richardson number (*Ri*), and is defined as the lowest level at which *Ri* exceeds a value of 0.5 (CHEN *et al.*, 2003). The model-determined boundary-layer height is compared with the boundary-layer height as determined from the model potential temperature profile. This also shows how the potential temperature inversion that signifies the top of the boundary layer was resolved in the model. The dates of 3 and 24 August and 15 September were chosen to illustrate the differences in the two methods of estimating the boundary-layer height.

On both 3 and 24 August, 2004, the boundary layer height was better predicted by the model potential temperature profile as compared to the *Ri* method. The observed boundary layer height on 3 August, 2004 at 1200 UTC (1600 LT) was 676 m. The boundary-layer height as determined by the *Ri* method was 2240 m, while the height estimated from the model potential temperature was 676 m. The observed boundary-layer height on 24 August, 2004 at 1200 UTC (1600 LT) is 1510 m. The model boundary layer height determined by the *Ri* method was 2257 m, but when estimated from the model potential temperature profile, the boundary layer height is 1489 m, a value similar to the observed boundary layer height as shown in Figure 11.

Better agreement between the observed boundary-layer height and the model-determined height using the *Ri* method is found on a few days. For example, on 15 September, 2004, the observed boundary-layer height at Abu Dhabi was 540 m at 1200 UTC (1600 LT). The model boundary-layer height determined by the *Ri* method was 520 m, while the value estimated from the model potential temperature profile was 731 m.

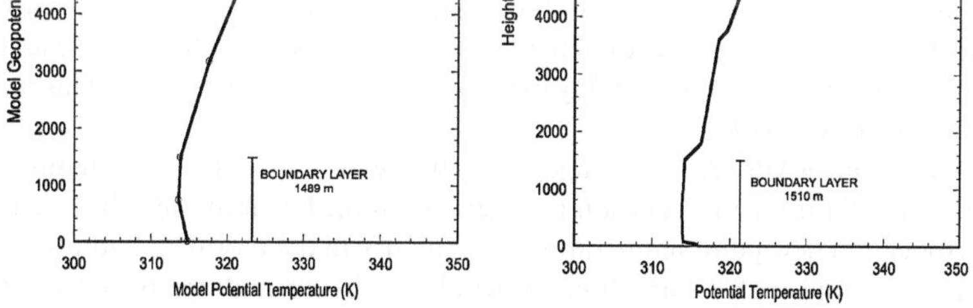

Figure 11
(a) Modeled and (b) observed potential temperature soundings for 1200 UTC (1600 LT) 24 August, 2004. Model sounding is from the initialization time.

The coarse vertical resolution of the model is a problem at higher altitudes (above 1 km) which makes comparisons of the *Ri* method and the potential temperature method difficult; however, the *Ri* method also poses problems at altitudes below 1 km, where there is higher vertical resolution. Some of the overprediction by COAMPS® of the boundary-layer height may be due to the location of the site of the radiosonde station. Abu Dhabi is located on the coast, and therefore slight differences between the modeled and observed wind direction would cause the air mass over Abu Dhabi to be very different: dry continental air due to southerly winds or maritime air due to northerly winds. This in turn may lead to errors in the boundary layer height due to the different stabilities in the modeled and observed boundary-layer air mass.

5. *Summary and Conclusions*

A statistical evaluation of COAMPS® using the verification skill scores of bias and RMS error was performed for the 1 August to 5 October, 2004 time period. This time period included the days of the UAE^2 experiment. Model output fields of 2 m air temperature, 2 m dew point temperature, 10 m wind speed, and 10 m wind direction were verified against three data sets. Surface visibility was verified against the ADP dataset only because no measurements were available in the UAE data set. RMS and bias errors were calculated for modeled upper-level variables of temperature, dew point temperature, geopotential height, wind speed, and wind direction. The predicted height of the boundary layer was verified against soundings at Abu Dhabi.

The initial analysis and model forecast errors are well within the bounds of mesoscale models, and are similar to the RMS and bias errors shown in SHI *et al.* (2004) and WHITE *et al.* (1999) for the winter season. In the verification of the vertical profiles, bias and RMS errors do not grow significantly with time as compared to the values at the initial time (0-hour forecast). The study location is in a region of weak synoptic forcing during the summer. Forecast errors grow slowly with time because of the mesoscale features rather than the synoptic features. The mesoscale features in the Arabian Gulf region are driven by diurnal variations. This may lead to the forecasts valid at similar times of day exhibiting similar forecast errors. This makes it ideal for model validation.

For bias and RMS errors averaged over the region, model 2 m air temperatures are cooler at all forecast integration times except in the late morning when there is no bias. Model 2 m dew point temperatures are too dry in the morning, and too moist in the afternoon and night. The observed wind speeds are underpredicted during the afternoon and night and overpredicted during the morning. Differences in bias errors over land and over water will cause differences in the model temperature gradient, which in turn could produce differences in wind speed. At night the air temperatures are too cool, and the modeled temperature gradient between land and water is too small. This results in a smaller modeled pressure gradient and weaker winds. Later during the night, the model temperature over land has continued to cool and is now much cooler than the sea-surface temperature, resulting in a large temperature difference between land and water. This in turn causes the larger wind speed bias. In the late morning there is no temperature bias, so there should also be minimal bias in the modeled horizontal temperature gradient between land and water. In the afternoon there is a cold temperature bias, resulting in a small temperature difference between land and water. This causes a smaller modeled pressure gradient and lighter modeled wind speeds, as shown in the negative wind speed bias.

The wind direction bias is 10–20° at most forecast times, with little bias in the afternoon. Individual stations show a wind direction bias of 35–70°, which could be due to incorrect predictions of the timing and development of the sea and land breeze circulations. The 3–4 km surface visibility bias indicates that COAMPS® is not predicting as much aerosol loading in the near-surface layer as that which is actually there. This may be due to the lighter simulated wind speeds in the afternoon not causing enough aerosols to be lifted or remain suspended. This error may have important practical implications for real-time forecasting in desert regions. The model boundary-layer height error at Abu Dhabi may be due to the method used to calculate the boundary-layer height, and errors may decrease by estimating the boundary-layer height from the modeled potential temperature. In summary, in spite of the biases and errors discussed here, the model does well in predicting the diurnal variations over water and over land, such as the development of the sea and land breezes and the thermal internal boundary layers. Large RMS errors in wind direction indicate that the model is not able to predict the time and location of the sea and land breeze fronts. This is a problem with the present mesoscale models and needs to be addressed through improved surface and land use parameterizations.

Acknowledgements

This research was funded by the Office of Naval Research, grant number N00173-03–1-G902. We would especially like to thank Quentin Saulter of ONR 35 for funding this research. The authors thank the Department of Atmospheric Studies, Ministry of Presidential Affairs (formerly the Department of Water Resources Studies, Office of His Highness the President in Abu Dhabi) for allowing us to use their meteorological data for this research. We also are grateful to the staff of the State Climate Office of North Carolina for providing computer lab and other support. In addition, the authors express appreciation to the anonymous reviewers whose comments improved the content of this paper.

References

Baker, Nancy L. (1992), *Quality control for the Navy operational atmospheric database*, Wea. Forecasting *7*, 250–261.

Chen, S., Cummings, J., Doyle, J., Hodur, R., Holt, T., Liou, C.S., Liu, M., Ridout, J., Schmidt, J., Thompson, W., Mirin, A., and Sugiyama, G. (2003), *COAMPS^{TM} version 3 model description – General theory and equations*, Naval Research Laboratory Technical Report, NRL/PU7500-04-448, 141 pp.

Eager, Rebecca E. (2005), *An investigation of the coastal circulations and aerosol transport in the Arabian Gulf region*, M.S. Thesis, 537 pp., North Carolina State Univ., Raleigh, 10 June. http://www.lib.ncsu.edu/theses/available/etd-07152005–143028/.

Hodur, R.M. (1997), *The Naval Research Laboratory's Coupled Ocean/Atmosphere Mesoscale Prediction System (COAMPS)*, Mon. Wea. Rev. *135*, 1414–1430.

Nachamkin, Jason E., and Hodur, Richard M. (2000), *Verification of short-term forecasts from the Navy COAMPS over the Mediterranean.* Preprints, 15 Conf. *Probability and Statistics in the Atmospheric Sciences*, Asheville, NC, Amer. Meteor. Soc., *5*.4.

Seinfeld, John, *Atmospheric Chemistry and Physics of Air Pollution* (Wiley-Interscience, 1986).

Shi, Jainn J., Chang, Simon W., Holt, Teddy R., Hogan, Timothy F., and Westphal, Douglas L. (2004), *A meteorological reanalysis of the 1991 Gulf War*, Mon. Wea. Rev. *132*, 623–640.

White, B.G., Paegle, J., Steenburgh, W.J., Horel, J.D., Swanson, R.T., Cook, L.K., Onton, D.J., and Miles, J.G. (1999), *Short-term forecast validation of six models*, Wea. Forecasting *14*, 84–108.

(Received March 29, 2006, accepted September 13, 2006)
Published Online First: June 23, 2007

Pure appl. geophys. 164 (2007) 1765–1785
0033–4553/07/091765–21
DOI 10.1007/s00024-007-0230-7

Pure and Applied Geophysics

Influence of Pacific on Southern Indian Ocean Rossby Waves

B. H. Vaid,[1] C. Gnanaseelan,[1,2] P. S. Polito,[3] and P. S. Salvekar[1]

Abstract—The sea-surface height anomalies derived from Simple Ocean Data Assimilation (SODA) during 1958–2001, Topex/Poseidon satellite during 1993–2001 and the SODA heat content anomalies (125 m depth) during 1958–2001 are filtered into annual and biennial Rossby wave components using a two-dimensional Finite Impulse Response filter. The filtered Rossby wave components (both annual and biennial) in the southern Pacific and Indian Oceans have considerable strength and variability. The propagation of annual and biennial Rossby waves in the Indonesian through-flow region [12.5°S–7.5°S] of the Indian Ocean is in phase with the southern Pacific Ocean waves. So it is speculated that the Pacific Ocean influences the Indian Ocean, especially through the region 17.5°S to 7.5°S and thus the southern Pacific Rossby waves could be an unexplored contributor to the Indian Ocean Rossby waves. We also carried out Fast Fourier Transform (FFT) and wavelet analysis on the tide gauge sea-level data along the Australian coast to support our claim. Filtered annual and biennial components of SODA heat content anomalies (125 m depth) also support these findings.

Key words: Variability, Rossby waves, Indonesian through-flow.

1. Introduction

Wyrtki (1961) provided the first reported description of the Indonesian through-flow by combining observations from several resources. A number of studies on Indonesian through-flow have since been attempted to advance the understanding of the ocean and atmosphere. Several model studies have assessed the role of the Indonesian through-flow on the mean global ocean circulation (Hirst and Godfrey, 1993; Schiller *et al.*, 1998). Observational studies have suggested that the Indonesian through-flow transport oscillates in phase with El Niño Southern Oscillation (ENSO): namely, larger Indonesian through-flow transport during La Niña conditions and smaller Indonesian through-flow transport during El Niño phases (Gordon and Fine, 1996; Meyers, 1996). Godfrey (1996) provided a comprehensive review of the current knowledge of the effect of the Indonesian through-flow on ocean circulation and heat exchange with the atmosphere. They

[1] Indian Institute of Tropical Meteorology, Pune, 411008, India.
[2] Department of Meteorology, Florida State University, Tallahassee, FL 32306.
[3] Instituto Oceanográfico da Universidade de São Paulo (IOUSP), Brazil.

noted that Indonesian through-flow has considerable impact on global climate. The vast movement of waters from the Pacific to the Indian Ocean through the Indonesian seas plays an important role in a range of climate phenomena. An estimated 17 million cubic meters of tropical water per second is transferred from the Pacific to the Indian Ocean through the Indonesian seas. This through-flow has numerous interconnected climate ramifications (L-DEO SCIENTISTS, 1993).

The knowledge of the Indonesian through-flow is very important to the regional climate and environmental variability in the maritime continent, as it is an oceanic teleconnection of ENSO. BAQUERO-BERNAL *et al.* (2002) studied the Indian Ocean Dipole (IOD)-ENSO relation by conducting sensitivity experiments including and excluding the ENSO influences in their Coupled General Circulation Model (CGCM) simulations. Their CGCM includes both the Indian and Pacific Oceans. The influence of ENSO is excluded in their no-ENSO CGCM simulation by replacing sea-surface temperature simulated through changes in surface heat flux or Indian Ocean circulation (e.g., VENZKE *et al.*, 2000; SCHILLER *et al.*, 2000). Rossby waves in the southern tropical Indian Ocean are first investigated by WOODBERRY *et al.* (1989) using a one-and-a-half-layer reduced gravity ocean model. In a subsequent study PERIGAUD and DELECLUSE (1992) showed that the wind stress is weak in the interior of the basin where the waves reach maximum amplitude (at 90°E) and suggested that the waves are produced from coastal winds along the western shore of Australia. Together with observational analyses, BAQUERO-BERNAL *et al.* (2002) found no evidence for an ENSO-independent oscillatory dipole mode in the Indian Ocean sea-surface temperature. HUANG and SHUKLA (2005) aim to identify the physical mechanisms responsible for the teleconnections over the tropical Indo-Pacific region, focusing the major role of the oceans. The ENSO phase strongly influences the thermocline depth in the Indonesian through-flow region and in the southeastern Indian Ocean (BRACCO *et al.*, 2004), which further strengthens the role of Indonesian through-flow in the Indian Ocean. The idea of energy 'leakage' from the Pacific to the Indian Ocean is first described by CLARKE (1991). He determined that 10% of the energy flux that reaches the western equatorial Pacific due to propagating long Rossby waves makes it into the Indian Ocean. In exploring the unrealistic IOD in the Commonwealth Scientific and Industrial Research Organisation (CSIRO) Mark 3 model, CAI *et al.* (2005a) found that the unrealistic representation of the model Java-Timor geometry resulted in an artificial pathway for Pacific upwelling Rossby waves to reach the model Java coast, triggering an IOD event about 6 to 9 months after the ENSO peak. In short, they showed some coarse-resolution coupled GCMs with a large opening in the Indonesian seaways simulate a positive IOD event after the mature phase of El Niño in the Pacific (CAI *et al.*, 2005a). Their findings highlighted the importance of proper topographic and geographic representation of the Indonesian region in modeling the variability of the Indian Ocean. The dynamics of El Niño Southern Oscillation (ENSO) cycles has previously been interpreted in terms of discharge and recharge of mass and heat in the equatorial Pacific. Some of the ENSO discharge/recharge signals transmit into the Indian Ocean, although the pathways are yet to be fully established. However CAI *et al.* (2005b), highlighted the importance of an unnoticed subtropical north-Pacific pathway and its involvement in the transmission of ENSO signals to the Indian Ocean. They proposed that signals in the Indian Ocean are coming from the northern Pacific, then via coastal Kelvin waves, into the Indian Ocean. Their analysis confirms that ENSO discharge/recharge arrives mainly at the central western Australia coast and the energy would then propagate southward as a coastal Kelvin wave. WIJFFELS and MEYERS (2004) find that the response to wind forcing over the Pacific is transmitted to the Indian Ocean primarily through the coastal waveguide of the Australian Coast.

In view of the above, there is always a question whether the southern Pacific waves have any influence on the Indian Ocean Rossby waves. Therefore it is worthwhile to examine the Rossby wave propagations (annual and biennial periods) in the southern Pacific and Indian Oceans and understand their relationships. In the present study, we establish a possible relationship between the Indian Ocean and the Pacific Ocean Rossby waves. In other words the Pacific to Indian Ocean connection is investigated with respect to its effect on southern tropical Indian Ocean Rossby waves. We also studied the associated oceanic processes in these two basins.

2. Data and Methodology

The sea surface height (SSH) data (from January/1958 – December/2001) derived from Simple Ocean Data Assimilation (SODA) product (CARTON *et al.*, 2000) is used in the analysis. The assimilation product used for SODA is based on Geophysical Fluid Dynamics Laboratory Modular Ocean Model 2.2. Area of study is 35°E–120°E, 20°S–10°N. The Topex/Poseidon altimeter now provides more than a decade of sea-surface height anomalies measurement with unprecedented accuracy of 2 cm and the RMS (Root-Mean-Square) difference between Topex/Poseidon and SODA SSHA for 9 years (1993–2001) annual climatology is observed to be less than 1.2 cm (for area averaged grid points). The spatial map (for all the grid points) showed that RMS difference is high (more than 5 cm) over the Somali coast while in most of the Indian Ocean the difference is less than 3 cm. For studying the influence of Indonesian through-flow on biennial Rossby waves and annual Rossby waves propagation, the SSH anomalies data over the region 120°E–72°W, 12.5°S–7.5°S was taken into account. Surface winds derived from National Center for Environmental Prediction (NCEP) and National Center for Atmospheric Research (NCAR) reanalysis data product (1958–2001) (KALNAY *et al.*, 1996) were taken to calculate wind stress and wind stress curl anomalies. The tide gauge sea-level data for the stations BOOBY IS (lat. = 10 36.0S, long. = 141 55.0E), DARWIN (lat. = 12

28.0S, long. = 130 51.0E) and BROOME (lat. = 18 00.0S, long. = 122 13.0E) along the Australian coast was also used in the study.

The two-dimensional Finite Impulse Response (FIR) filter, which is a band-pass filtering technique, is used to filter the SODA SSH anomalies into biennial and annual Rossby wave components. The Finite Impulse Response (FIR) filters are based on convolution of two sequences, the original data η_0, and the filter f, resulting in the filtered data η_f. The statistical significance of these signals is measured by the amplitude, A, fractional variance V, and the signal-to-noise ratio S/N estimated for each component. If no statistically significant wave signal is present for a given spectral band, the filtered signal will have very low amplitude, explain little or nothing of the total variance, and have a signal-to-noise ratio nearly zero. In the present case $\eta_0(x,t)$ is a function of longitude and time; therefore the filter $f(i,j)$ is a function of longitudinal lag i and temporal lag j. The filtered matrix $\eta_f(x,t)$ is obtained from

$$\eta_f(x,t) = \sum_{i=-m}^{m} \sum_{j=-n}^{n} \eta_0(x+i, t+j) f(i,j). \tag{1}$$

The size of the filter is controlled by the parameters m and n. The value of n is set *a priori* to be approximately one period and m is adjusted automatically within the filtering routine to be approximately one wavelength. For the band-pass filtering of the annual and biennial Rossby wave components (say η_{12}, η_{24}, respectively) the filter $f(i,j)$ is a Gaussian tapered cosinusoidal surface

$$f(i,j) = \frac{e^{-1/2((\pi i/m)^2 + (\pi j/n)^2)}}{N} \cos\left(\frac{2\pi i}{L} - \frac{2\pi j}{T}\right) + M, \tag{2}$$

with M and N such that

$$\begin{cases} \sum_{i=-m}^{m} \sum_{j=-n}^{n} |f(i,j)| = 1, \\ \sum_{i=-m}^{m} \sum_{j=-n}^{n} f(i,j) = 0, \end{cases},$$

where L and T are the wavelength and period of the approximate center of each pass band. T is set to 365 and 730 days, respectively and L is such that $L = CpT$. The Gaussian part of Equation (2) works as a smoother while the cosine part limits the power of the response to a specific area in the 2-D spectrum. This can be understood as a 2-D version of the tapered cosine window functions (Blackman, Hamming, Hanning, etc.) used in classical 1-D spectral analysis, whose main advantages are the same: to minimize phase distortion and to reduce the amplitude of side-lobes (leaking). A thorough description of FIR filters and its performance used in this study is given in POLITO and CORNILLON (1997), POLITO et al. (2000), POLITO and LIU (2003), and CHOWDARY et al. (2006).

The FIR filter method offers an alternative, where each filter component covers a finite range of periods and wavelengths, a spectral band, instead of a spectral line. Amplitudes, periods and wavelengths can vary within certain thresholds, with the period chosen *a priori*, centered within the bandwidth. No functional shape is assumed and the data are processed simultaneously in the time and longitude domains. This makes the filter design process more intuitive thus easier to adapt to oceanographic applications in which two out of the three parameters *cp*, *T* and *L* are known to be within a certain range. The filter parameters are adjusted to bracket the spectral bands related to previously observed physical processes. This is an advantage to classical Fourier analysis and complex empirical orthogonal functions, since both methods do not allow for such adjustment.

3. Results and Discussion

The region between 12°S–7°S is very important in respect to Indonesian through-flow point of view (MURTUGUDDE *et al.*, 1998). It was shown that the climatological effect of the Indonesian through-flow is to warm the Indian Ocean and to cool the Pacific Ocean (MURTUGUDDE *et al.*, 1998). In the present study the effect of the Pacific on annual and biennial Rossby waves is studied by applying the FIR filter on SODA SSH anomalies [during 1958–2001] for both the Indian Ocean and the Pacific Ocean separately. Figure 1 shows the propagation of annual Rossby waves in the Indian Ocean from 24.5°S to 10.5°N. Strong signals are observed along 10.5°S, starting from the Indonesian through-flow region in January and progressing westward with the expected latitude-dependent phase speed. The phase speed along 10.5°S for annual Rossby waves is found to be about −18 cm/sec. The phase speeds of annual Rossby waves are estimated using the Radon Transform method (POLITO *et al.*, 2000). The phase speed difference occasionally forces the fronts to break up in smaller, eddy-like features. Both fronts and eddies propagate with the local Rossby wave phase speed and therefore are considered as Rossby waves. Figure 2 shows the propagation of annual Rossby waves (mm) in the Indian Ocean (averaged over 12.5°S–7.5°S) and Pacific Ocean (17.5°S–7.5°S) from 1958–2001. The longitudinal gap between the Indian Ocean and the Pacific Ocean seen in Figure 2 is due to the fact that data over shallow areas (depth $H \leq 1000\,\mathrm{m}$), small islands ($x \leq 3°$) and zonally enclosed water bodies are excluded to facilitate filtering. Strong signals are observed between 12.5°S–7.5°S in the Indian Ocean, starting from the Indonesian through-flow region along 120°E in January/February and progressing westward with the expected latitude-dependent phase speed [Fig. 2(a)]. In most years during the study period, these annual Rossby waves in the Indian Ocean are seen to be in phase with the Pacific Ocean Rossby waves. It is clear from the past studies that energy leaked from the Pacific influences the southern Indian Ocean Rossby waves. But we in this study hypothesize a new possibility of the energy leakage through the

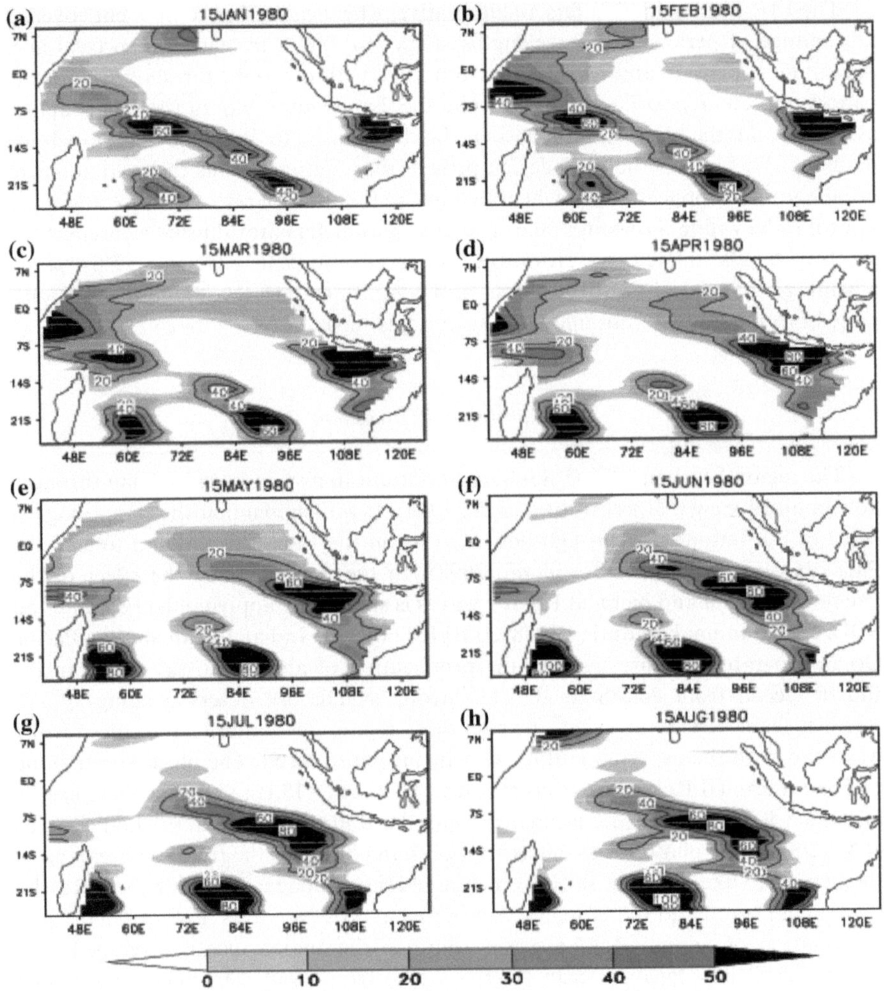

Figure 1
Propagation of annual Rossby waves (mm) in 1980.

Indonesian through- flow region. Figures 3 and 4 show wind stress anomalies and wind stress curl anomalies averaged over 12.5°S–7.5°S, respectively. To support our findings, we analyzed local wind forcing at the same time (the time when annual Rossby waves in the Indian Ocean began propagating westward before the annual Rossby waves in the Pacific reached the boundary), and in most cases it is observed that the role of wind forcing to generate the Indian Ocean annual Rossby waves is

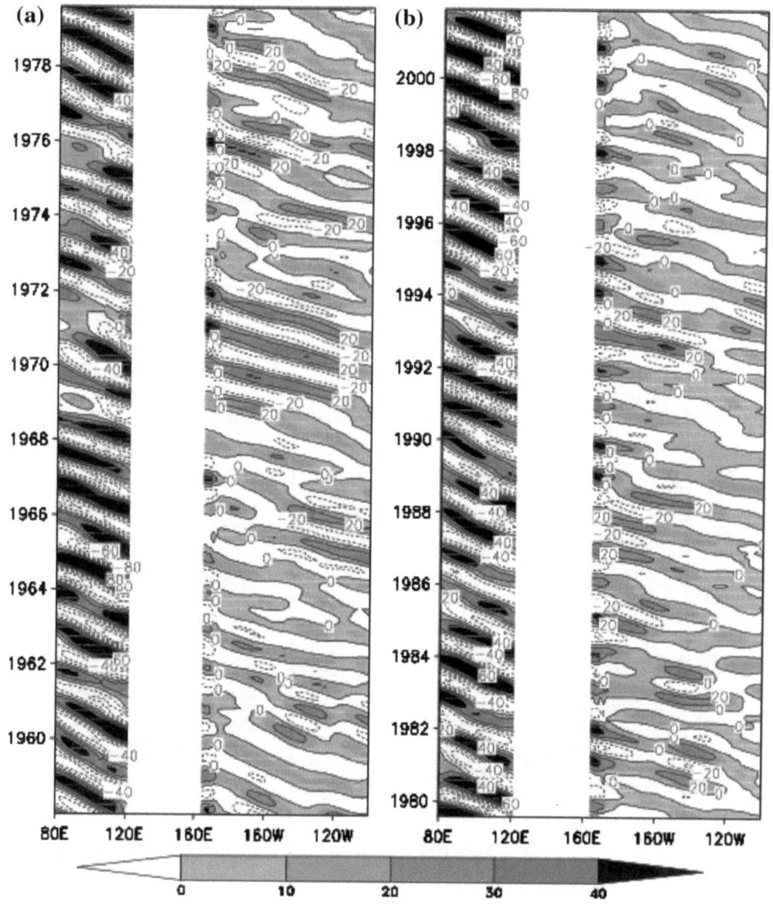

Figure 2
Propagation of annual Rossby waves (mm) in the Indian Ocean [averaged over 12.5°S–7.5°S] and Pacific Ocean [averaged over 17.5°S–7.5°S].

seen to be considerably less (Figs. 3 and 4). However, in year 2000, a considerable value of wind stress curl in the southern Indian Ocean was observed which in fact shows that there could be some local generation of Rossby waves in the Indian Ocean itself or the local forcing might strengthen the Indian Ocean Rossby waves. Also Rossby wave signals in the Pacific are observed to be very small, almost within 20 mm, but at the same time the waves in the Indian Ocean are very large (about 50 mm); this questions the sources of extra energy. Apart from the southern Pacific waves, the waves from the northern Pacific entering the Indian Ocean near the

ENSO impinge on the western boundary and move equatorward along the raypath of Kelvin-Munk waves, and reflect as equatorial Kelvin waves. En-route to the equator, these waves are reinforced by wind stress anomalies associated with ENSO. The reflected equatorial Kelvin waves impinge on the Australasian continent and move poleward along the northwestern Australia coast as coastally-trapped waves, radiating Rossby waves into the south Indian Ocean. Study by POTEMRA *et al.* (2001) and CAI *et al.* (2005b) placed emphasis on equatorial Pacific Rossby waves and their connection to southern tropical Indian Ocean Rossby waves. However, the present study focuses on southern Pacific Rossby waves as they make their way into the Indian Ocean. The process, specifically how these waves are getting around Australia, is still not explained. A southern Pacific Rossby wave hits the east coast of Australia, short Rossby waves are reflected and two coastal waves are generated, one propagates north (counter-clockwise) around Australia, the other propagates south (clockwise) around Australia. It is not clear that either one will make it around Australia into the Indian Ocean, particularly the northern one, given that the New Guinea land mass extends to the equator. But we cannot test which one makes it around Australia into the Indian Ocean using our technique due to the restrictions in the FIR filter. We speculate that it makes its way along the northwest Australian coast at 10°S. Figure 5 shows the Topex/Poseidon SSHA (bicubically interpolated) averaged over 12.5°S to 10.5°S. The Rossby wave propagation seen in Figure 5 supports our speculation that Rossby wave signals form the Pacific Ocean enter into the Indian Ocean through the Indonesian through-flow region. To further explore this, we analyzed tide gauge sea-level data along the Australian coast during 1993–2002 for station BOOBY IS (lat. = 10 36.0S, long. = 141 55.0E), DARWIN (lat. = 12 28.0S, long. = 130 51.0E) and BROOME (lat. = 18 00.0S, long. = 122 13.0E). The FFT power spectra for the selected locations are shown in Fig. 6 (X-axis represents periods in month, Y-axis represents power of the signal). FFT spectrum of tide gauge sea-level data for all three stations (BOOBY IS, DARWIN, BROOME) showed dominances of 1 to 2 year signals. Significant power is found active for 1 to 2 year signals for the selected location, which in fact supports our speculations. In other words, the continuity of signals (of 1 to 2 year) from BOOBY IS, DARWIN to BROOME is evident. Consequently the southern Pacific Rossby waves could be an unexplored contributor to the Indian Ocean Rossby waves. Since an advantage of wavelet analysis is to obtain information about time when variability will become intense during the period chosen. In the present study we also carried out a wavelet analysis to explore the dominance of signals for all three selected locations along the Australian coast. The wavelet analysis has been carried out with ion research system team wavelet software. The detailed information about the wavelet analysis is given by TORRENCE and COMPO (1998). Fig. 7(a) shows BOOBY sea level (Tide Gauge) and Fig. 7(b) shows the wavelet power spectrum. The power has been scaled by the global wavelet spectrum (at right). The cross-hatched region is the cone of influence, where zero padding has reduced the variance. Figure 7(b) gives a three-dimensional

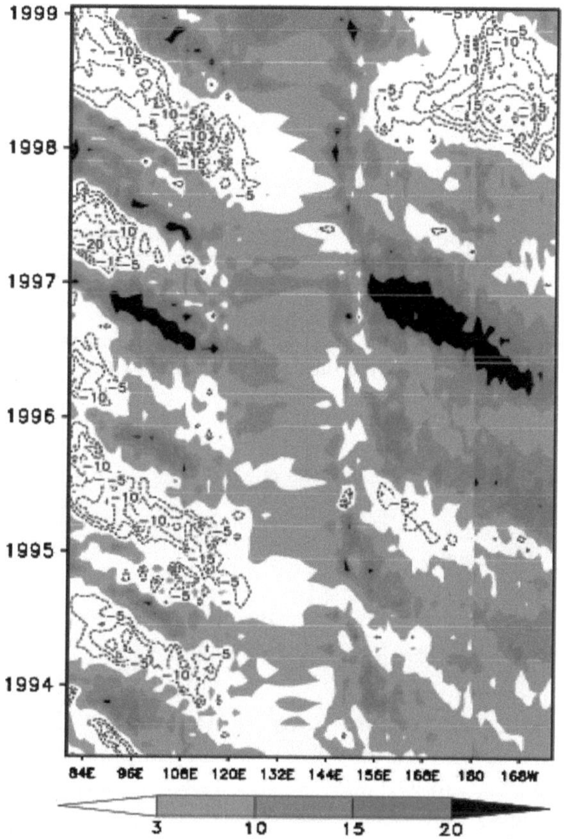

Figure 5
Time longitude plot of T/P SSHA (bicubically interpolated) averaged over 12.5°S–10.5°S.

plot, x-axis is the information of time (1 to 120 months correspond to January 1993 to December 2002), y-axis is information about the period of the signal and shading gives the information of the power. Contour with dark line symbolizes that the particular signal is within the 10% significance level (90% confidence level). The power, (absolute value)^2 gives information on the relative power at a certain scale and a certain time. The wavelet has the advantage of incorporating a wave of a certain period, as well as being finite in extent. In fact, the wavelet shown in Fig. 7 (called the Morlet wavelet) is nothing more than a sine wave multiplied by a Gaussian envelope. A detailed analysis of BOOBY sea level showed that 1 to 2 year variability was intense. A similar analysis has been carried out for the rest of the selected locations DARWIN (Fig. 8) and BROOME (Fig. 9). The power spectrums

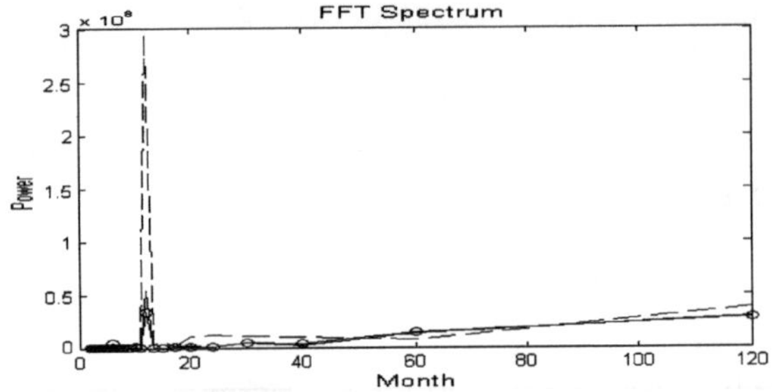

Figure 6
FFT spectrum, dash line represents BOOBY IS (lat. = 10 36.0S, long. = 141 55.0E), solid line represents DARWIN (lat. = 12 28.0S, long. = 130 51.0E) and circle represents BROOME (lat. = 18 00.0S, long. = 122 13.0E).

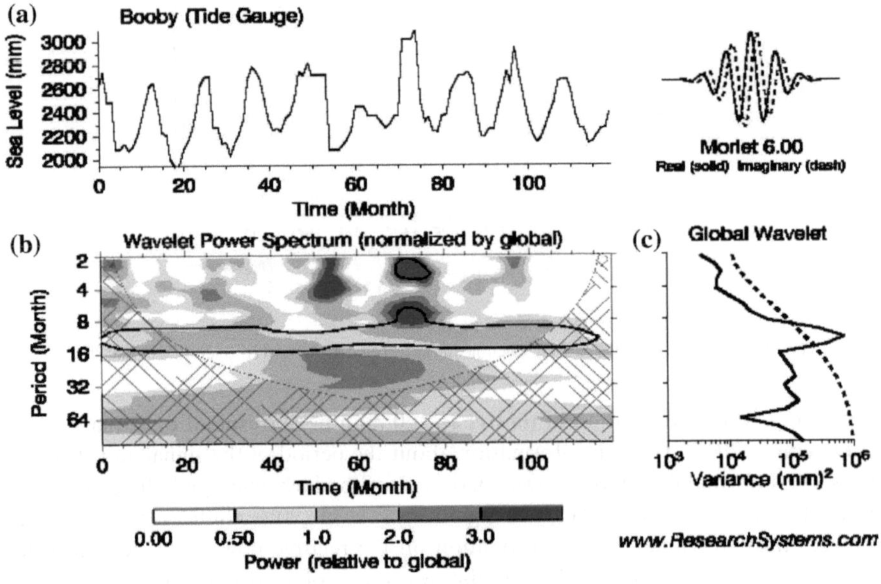

Figure 7
(a) Booby sea level (Tide Gauge). (b) The wavelet power spectrum. The power has been scaled by the global wavelet spectrum (at right). The cross-hatched region is the cone of influence, where zero padding has reduced the variance. Black contour is the 10% significance level (90% confidence level), using a red-noise (autoregressive lag1) background spectrum. (c) The global wavelet power spectrum (black line). The dashed line is the significance for the global wavelet spectrum, assuming the same significance level and background spectrum as in (b).

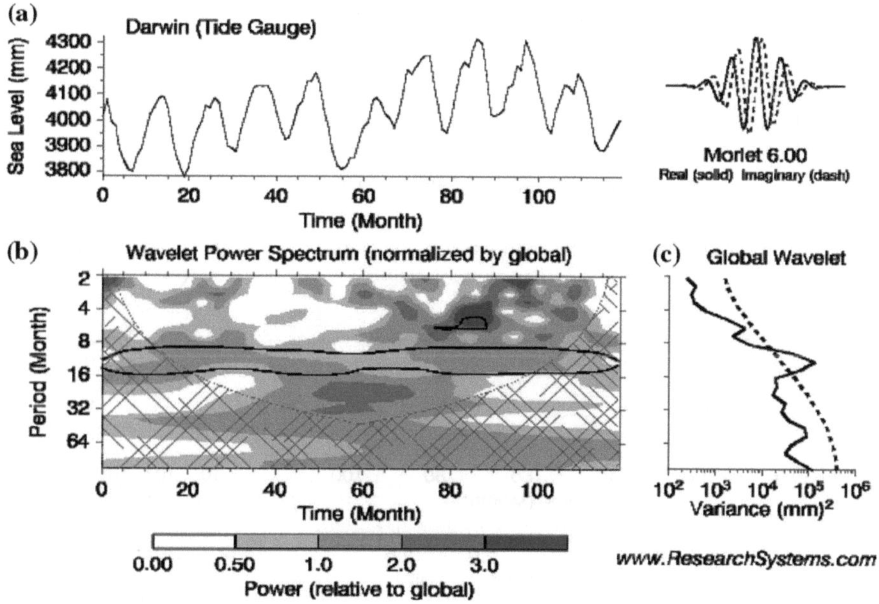

Figure 8
Same as Figure 7, but for DARWIN Sea level (Tide Gauge).

of the wavelet for the selected locations (BOOBY, DARWIN and BROOME) are dominant for the period of 1 to 2 years. Thus the propagation of annual and biennial Rossby waves in the Indonesian through-flow region [12.5°S–7.5°S] of the Indian Ocean is in phase with the southern Pacific Ocean waves. However, more detailed model studies are required to establish a stronger relation, which will be carried out in a future study.

To further strengthen the study, annual Rossby waves are filtered out for both the Indian Ocean and the Pacific Ocean from Topex/Poseidon SSH anomalies data [1993 to 2001]. Similar results are observed in the annual Rossby waves propagation in Topex/Poseidon SSH anomalies (Figure not shown), as observed in the SODA SSH anomalies. Moreover this phenomenon is not only found in the surface but also in the subsurface. This was reflected in the SODA heat content anomalies of 125 m depth (Fig. 10) in the Indian Ocean and Pacific Ocean.

GNANASEELAN *et al.* (2006) did a rather extensive analysis of the near equatorial biennial Rossby waves observed in the Topex/Poseidon altimetry data. They proposed that downwelling signals in the west, and upwelling signals in the east formed in 1994 and 1997 give a seesaw gradient to the thermocline in the equatorial Indian Ocean. Downwelling biennial Rossby waves are dominant along two

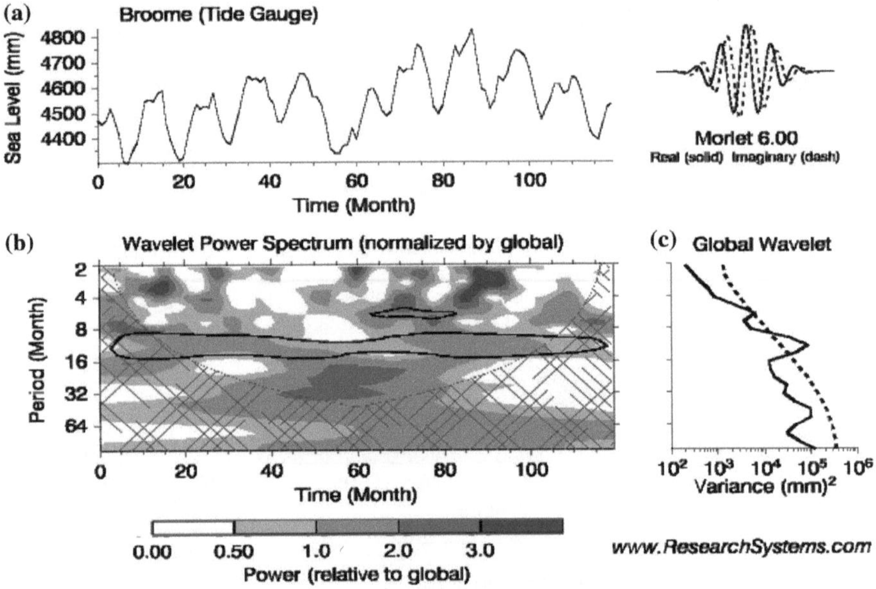

Figure 9
Same as Figure 7, but for BROOME Sea level (Tide Gauge).

latitudinal belts, 3.5°S–1.5°N (peak around 1.5°S) and 12.5°S–7.5°S (peak around 10.5°S). Downwelling biennial Rossby waves along 1.5°S started propagating westward from the eastern boundary in July/August, 1993 and 1996, more than one year prior to the formation of the Indian Ocean dipole events. These strong downwelling signals reach the western equatorial Indian Ocean during the peak dipole time. According to them, the biennial Rossby wave signals are triggered by the wind stress curl in the eastern boundary. The biennial signals along 10.5°S apparently affect the SST propagation and are observed to be the main dynamical mechanisms that drive the SST anomalies during dipole years. However, no mechanism is proposed by GNANASEELAN *et al.* (2006) for the generation of downwelling biennial Rossby waves in the region 12.5°S–7.5°S. In this study, we tried to examine the biennial Rossby wave propagation in the Indian Ocean and Pacific Ocean by applying the FIR filter in both regions (ocean basins) separately. Figure 11 shows the geographical distribution of biennial Rossby waves in the Indian Ocean from 24.5°S to 10.5°N from January to August 1980. The propagation of biennial Rossby waves is evident in the panels of Fig. 11. Strong signals can be seen along 10.5°S, starting from the Indonesian through-flow region in January and progressing westward. Phase speed of biennial Rossby waves along 10.5°S is found to be −11 cm/sec. The phase speeds of biennial Rossby waves are estimated using the Radon Transform

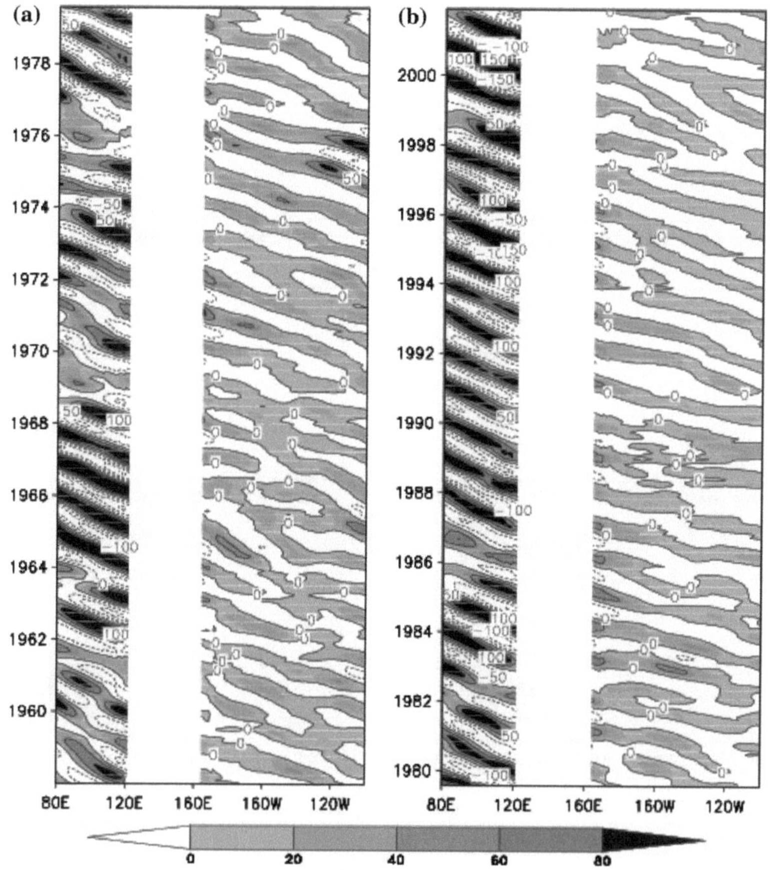

Figure 10
Propagation of filtered annual Rossby waves from SODA heat content anomalies of 125 m depth (°C m) in the Indian Ocean [averaged over 12.5°S–7.5°S] and Pacific Ocean averaged over 17.5°S–7.5°S].

method (POLITO *et al.*, 2000). A similar pattern is observed throughout the study period. Figure 12 shows the propagation of biennial Rossby waves (mm) in the Indian Ocean (averaged over 12.5°S–7.5°S) and the Pacific Ocean (averaged over 17.5°S–7.5°S) from 1958–2001. The longitudinal gap between Indian Ocean and Pacific Ocean seen in the figure is due to the fact that data over shallow areas (depth $H \leq 1000$ m), small islands ($x \leq 3°$) and zonally enclosed water bodies are excluded to facilitate filtering. From Fig. 12 it can be observed that biennial Rossby wave propagation in this region of the Indian Ocean is influenced by the Pacific Ocean. Strong biennial Rossby waves over 12.5°S to 7.5°S are seen to be responsible for westward propagation in the sea-surface temperature anomalies over 12.5°S to 7.5°S

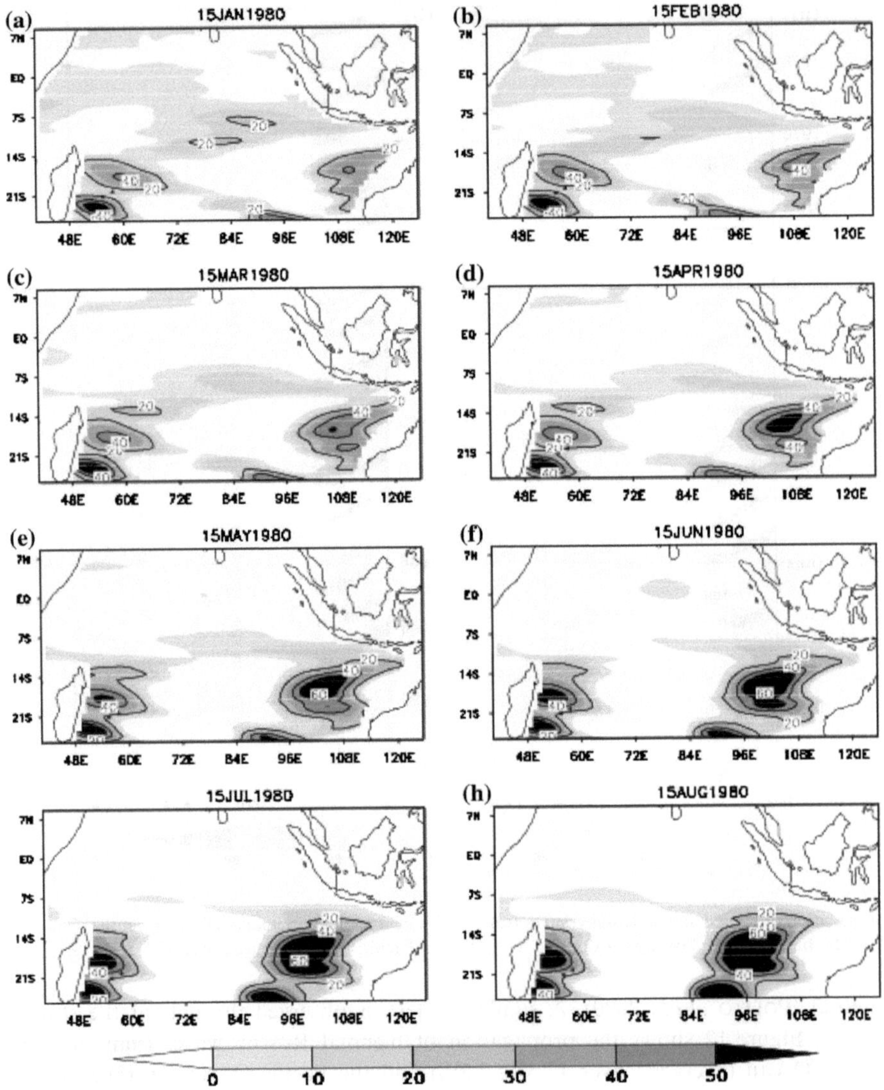

Figure 11
Propagation of biennial Rossby waves (mm) in 1980.

during the Indian Ocean Dipole (IOD) years (GNANASEELAN *et al.*, 2006). Moreover, an extensive study of biennial Rossby waves using longer term series SODA SSHA data by VAID *et al.* (2006), also showed the importance of biennial Rossby waves

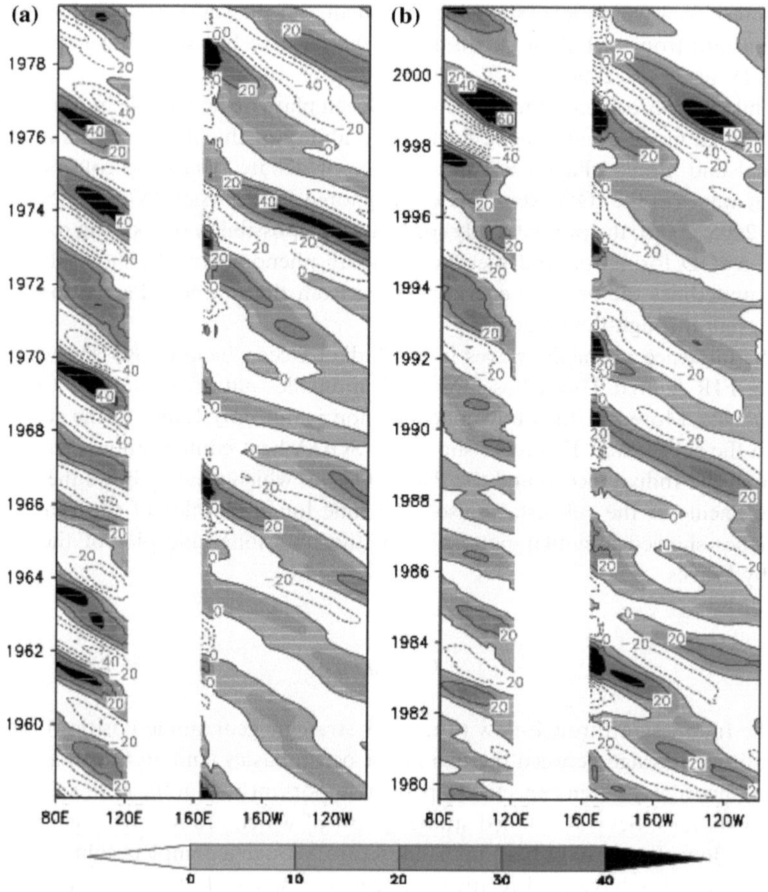

Figure 12
Propagation of biennial Rossby waves (mm) in the Indian Ocean [averaged over 12.5°S–7.5°S] and Pacific Ocean [averaged over 17.5°S–7.5°S].

during IOD. They performed a composite analysis of biennial Rossby waves for different cases (for example: composite of 1961, 1994 and 1997 [the recent strongest IOD years], composite of 1965, 1976 and 1986 [El Niño years], composite of 1961, 1977 and 1994 [positive IOD years], composite of 1972, 1982 and 1997, [positive IOD years coinciding with El Niño years], composite of 1959, 1970 and 1996 [negative IOD] years and composite of 1969 and 1992 [negative IOD years coinciding with El Niño years] over the Indian Ocean and showed the importance of biennial Rossby waves during IOD. From the composite map analysis of the biennial Rossby wave signals, a dipole pattern was clearly observed in the equatorial Indian Ocean with the wave crest occupying the western Indian Ocean and the wave trough occupying the southeastern tropical Indian Ocean during the positive Indian Ocean Dipole (IOD) events (VAID *et al.*, 2006).

While there is evidence that the Indian Ocean plays an active role in some of these dipole/zonal mode events, the self–sustained ability of this mode and its impact on regional and global climate is still a matter of debate. Since the role of biennial Rossby waves in the IOD event cannot be ignored (GNANASEELAN *et al.*, 2006; VAID *et al.*, 2005, 2006), the present study speculates the possibility of a strong relationship between IOD formation and the Pacific Ocean phenomenon. In short, the present study underlines the possible energy leakage from the southern Pacific through the Indonesian through-flow region.

The influence of Pacific waves in the Indian Ocean biennial Rossby waves was seen in FIR filtered Topex/Poseidon SSH anomalies data from 1993–2001 (Figure not shown). Moreover the above phenomenon is not only found in the surface but also in the subsurface. Figure 13 shows the SODA heat content anomalies of upper 125 m in the Indian Ocean and the Pacific Ocean, which clearly shows the influence of the Pacific in the subsurface also. The time longitude plot of the heat content anomalies showed a similar pattern as of the time longitude plot of the biennial Rossby Waves.

4. *Summary*

The Indonesian through-flow (due to its strategic geographic position), the only low-latitude passage between the two major ocean basins (Indian Ocean and Pacific Ocean), has always been expected to play an important role in the ocean circulations and regulations. Several detailed studies (most of which are referred) regarding this through-flow already have been in the literature, though a complete solution is yet to be addressed. We have used relatively longer time series of SODA assimilative SSH anomalies data to examine the possible link between the southern Indian Ocean and Pacific Ocean Rossby Waves. This extends the time record available in conventional observational data sets (e.g., Topex/Poseidon satellite data, etc.). The filtered Rossby wave components (both annual and biennial) in the southern Pacific and Indian Ocean have considerable strength and variability. The study speculates that the Rossby waves found in the southern Indian Ocean are partially influenced by the Pacific Ocean Rossby waves. To support our speculations we analyzed the tide gauge sea-level data for the station BOOBY IS (lat. = 10 36.0S, long. = 141 55.0E), DARWIN (lat. = 12 28.0S, long. = 130 51.0E) and BROOME (lat. = 18 00.0S, long. = 122 13.0E) along the Australian coast. The FFT and wavelet analysis performed for the three selected locations along the coast of Australia showed that the annual and biennial Rossby wave in the Indonesian through-flow region [12.5°S–7.5°S] of the Indian Ocean is in phase with the southern Pacific Ocean waves. Thus

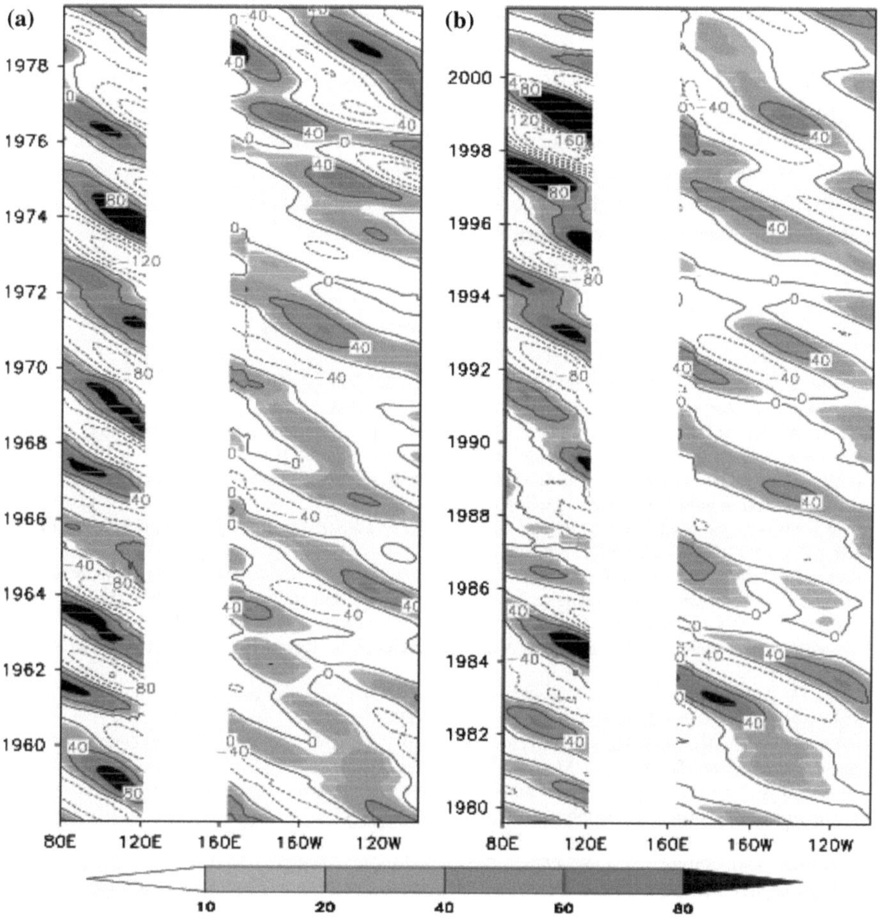

Figure 13
Propagation of filtered biennial Rossby waves from SODA heat content anomalies of 125 m depth (°C m) in the Indian Ocean [averaged over 12.5°S–7.5°S] and Pacific Ocean [averaged over 17.5°S–7.5°S].

the southern Pacific Rossby waves could be considered as an unexplored contributor to the Indian Ocean Rossby waves. Filtered annual and biennial components of the SODA heat content anomalies (125 m depth) also showed similar results.

Acknowledgement

We thank Director, IITM for providing the infrastructure required for the study. The financial support was provided by Department of Ocean Development,

Govt. of India through DOD-INDOMOD project. We acknowledge Drs. Gary Meyer, D. Sengupta, Patrick Caldwell, Daryl Metters, and A. Parekh, J. S. Chowdary and S. Mahapatra for the scientific discussions and help. The authors are thankful to the T/P team and AVISO altimetrie, SODA, NCEP/NCAR, University of Hawaii sea-level center, NTC Baseline array of tide gauges from the Bureau of Meteorology and wavelet by ion research system. We thank the editor and reviewers for their valuable comments and suggestions, which helped us to enchance the manuscript considerably.

References

Baquero-Bernal, A., Latif, M., and Legutke, S. (2002), *On dipole like variability of sea- surface temperature in the tropical Indian Ocean*, J. Clim. *15*, 1358–1368.

Bracco, A., Kucharski, F., Kallummal, R., and Molteni, F. (2004), *Internal variability, external forcing and climate trends in multi-decadal AGCM ensembles*, Clim. Dyn. *23*, 659–678.

Cai, W., Hendon, H. H., and Meyers, G. (2005a), *Indian Ocean dipole-like variability in the CSIRO Mark 3 coupled climate model*, J. Climate *18*, 1449–1468.

Cai, W., Meyers, G., and Shi, G. (2005b), *Transmission of ENSO signal to the Indian Ocean*, Geophys. Res. Lett. *32*, L05616, doi: 10.1029/2004GL021736.

Carton, J. A., Chepurin, G., Cao, X., and Giese, B. (2000), *A Simple Ocean Data Assimilation analysis in the global upper ocean 1950 95, Part I: Methodology*, J. Phys. Oceanogr. *30*, 294 309.

Chowdary, J. S., Gnanaseelan, C., Vaid, B. H., and Salvekar, P. S. (2006), *Changing trends in the Tropical Indian Ocean SST during La Niña years*, Geophys. Res. Lett. *33*, L18610, doi:10.1029/2006GL026707.

Clarke, A. J. (1991), *On the reflection and transmission of low-frequency energy at the irregular western Pacific Ocean boundary*, J. Geophys. Res. *96*, 3289–3305.

Gnanaseelan, C., Vaid, B. H., and Polito, P. S. (2006), *Rossby waves in the Indian Ocean and their impact on Indian Ocean Dipole*, J. Mar. Res. (submitted).

Godfrey, J. S. (1996), *The effect of the Indonesian through-flow on ocean circulation and heat exchange with the atmosphere: A review*, J. Geophys. Res. *101*, 12 217–12 237.

Gordon, A. L. and Fine, R. A. (1996), *Pathways of water between the Pacific and Indian Oceans in the Indonesian Seas,* Nature *379*, 146–149.

Hirst, A. C. and Godfrey, J. S. (1993), *The role of Indonesian through-flow in a global ocean GCM,* J. Phys. Oceanogr. *23*, 1057–1086.

Huang, B. and Shukla, J. (2005), *On the mechanisms of the interannual variability in the tropical Indian Ocean, Part I: The role of remote forcing from the tropical Pacific*, COLA Technical Report.

Kalnay *et al.* (1996), *The NMC/NCAR 40-Year Reanalysis Project*, Bull. Am. Meteorol. Soc., *77*, 437–471.

L-Deo Scientists Lead Exploration in Indonesia Seas (1993), Columbia University Record, November 12, vol. 19, no. 10.

Meyers, G. (1996), *Variation of Indonesian through-flow and El Niño-Southern oscillation,* J. Geophys. Res. *101*, 12, 255–12, 263.

Murtugudde, R., Busalacchi, A. J., and Beauchamp, J. (1998), *Seasonal to interannual effects of the Indonesian through-flow on the tropical Indo-Pacific Basin,* J. Geophys. Res. *103*, 21425–21441.

Perigaud, C. and Delecluse, P. (1992). *Annual sea-level variations in the southern tropical Indian Ocean from GEOSAT and shallow water simulations*, J. Geophys. Res. *97*, 20169–20178.

Potemra, J. T. (2001), *Contribution of equatorial Pacific winds to southern tropical Indian Ocean Rossby waves,* J. Geophys. Res. *106*, 2407–2422.

Polito, P. S. and Cornillon, P. (1997), *Long baroclinic Rossby waves detected by TOPEX/POSEIDON*, J. Geophys. Res. *102*, 3,215–3,235.

POLITO, P. S., SATO, O. T., and LIU, W. T. (2000), *Characterization and validation of the heat storage variability from TOPEX/POSEIDON at four oceanographic sites,* J. Geophys. Res. *105*, C7, 16911–16921.

POLITO, P. S. and LIU, W. T. (2003), *Global characterization of Rossby waves at several spectral bands,* J. Geophys. Res. *108*, C1, 3018, doi: 0.1029/2000JC000607.

SAJI, N. and YAMAGATA, T. (2003), *Structure of SST and surface wind variability during Indian Ocean Dipole mode events: Coads observations*, J. Clim. *16*, 2735–2751.

SCHILLER, A., GODFREY, J. S., MCINTOSH, P. C., MEYERS, G., and WIJFFELS, S. E. (1998), *Seasonal near-surface dynamics and thermodynamics of the Indian Ocean and Indonesian through-flow in a global ocean general circulation model,* J. Phys. Oceanogr. *28*, 2288–2312.

SCHILLER, A., GODFREY, J. S., MCINTOSH, P. C., MEYERS, G., and FIEDLER, R. (2000), *The interannual dynamics and thermodynamics of the Indo-Pacific Oceans,* J. Phys. Oceanogr. *30*, 987–1012.

TORRENCE, C. and COMPO, G. P. (1998), *A Practical Guide to Wavelet Analysis,* Bull. Amer. Meteor. Soc. *79*, 61–78.

VAID, B. H., GNANASEELAN, C., POLITO, P. S., and SALVEKAR, P. S. (2005), *Role of Biennial Rossby waves in the Indian Ocean Dipole Formation*, WGNE Report, research activities in Atmospheric and Ocean Modelling, No. 2–47.

VAID, B. H., GNANASEELAN, C., POLITO, P. S., and SALVEKAR, P. S. (2006), *Influence of El Niño on the Biennial and Annual Rossby waves propagation in the Indian Ocean with special emphasis on Indian Ocean Dipole*, IITM Research Report RR-109.

VENZKE, S., LATIF, M., and VILWOCK, A. (2000), *The coupled GCM ECHO-2, II, Indian Ocean response,* J. Clim. *13*, 1371–1383.

VERSCHELL, M., KINDLE, J., and HOWDEN, S. (1995), *Effects of Indo-Pacific through-flow on the upper tropical Pacific and Indian Ocean*, J. Geophys. Res. *100*, 18409–18420.

WIJFFELS, S. E. and MEYERS, G. (2004), *An intersection of oceanic waveguides: Variability in the Indonesian through-flow region*, J. Phys. Oceanogr. *34*, 1232–1253.

WOODBERRY, K. E., LUTHER, M. E., and OBRIEN, J. J. (1989), *The wind-driven circulation in the southern tropical Indian Ocean*, J. Geophys. Res. *94*, 17985–18002.

WYRTKI, K. (1961), *Physical oceanography of the Southeast Asian waters, scientific results of maritime investigations of the South China Sea and Gulf of Thailand 1959–1961,* NAGA Report *2,* Scripps Institute of Oceanography, 195 pp.

(Received March 17, 2006, accepted September 26, 2006)
Published Online First: June 19, 2007

D. Land Surface Processes

Pure appl. geophys. 164 (2007) 1789–1809
0033–4553/07/091789–21
DOI 10.1007/s00024-007-0241-4

Pure and Applied Geophysics

A Simple Reclassification Method for Correcting Uncertainty in Land Use/Land Cover Data Sets Used with Land Surface Models

JOSEPH G. ALFIERI,[1] DEV NIYOGI,[1,2] MARGARET A. LEMONE,[3] FEI CHEN,[3] and SOULEYMANE FALL[1]

Abstract—With increasing computational resources, environmental models are run at finer grid spacing to resolve the land surface characteristics. The land use/land cover (LULC) data sets input into land surface models are used to assign various default parameters from a look-up tables. The objective of this study is to assess the potential uncertainty in the LULC data and to present a reclassification method for improving the accuracy of LULC data sets. The study focuses on the Southern Great Plains and specifically the Walnut River Watershed in southeastern Kansas, USA. The uncertainty analysis is conducted using two data sets: The National Land Cover Dataset 1992 (NLCD 92) and the Gap Analysis Program (GAP) data set, and a reclassification logic tree. A comparison of these data sets showed that they do not agree for approximately 27% of the watershed. Moreover, an accuracy assessment of these two data sets indicated that neither had an overall accuracy as high as 80%. Using the relationships between land-surface characteristics and LULC, a reclassification of the watershed was conducted using a logical model. This model iteratively reclassified the uncertain pixels according to their surface characteristics. The model utilized normalized difference vegetation index (NDVI) measurements during April and July 2003, elevation, and slope. The reclassification yielded a revised LULC dataset that was substantially improved. The overall accuracy of the revised data set was nearly 93%. The study results suggest: (i) as models adopt finer grid spacings, the uncertainty in the LULC data will become significant; (ii) assimilating NDVI into the land-surface models can reduce the uncertainty due to LULC assignment; (iii) the standard LULC data sets must be used with caution when the focus is on local scale; and (iv) reclassification is a valuable means of improving the accuracy of LULC data sets prior to applying them to local issues or phenomena.

Key words: Land use, land cover, land surface modeling, NDVI, land-surface characteristics, surface heterogeneity.

1. *Introduction*

An accurate description of land-surface characteristics is important to a wide array of applications. Some examples include managing natural resources (LIU *et al.*, 2005) and local development (YANG and LO, 2002), understanding the relationship

[1] Department of Agronomy, Purdue University, 915 W. State Street, West Lafayette, Indiana 47907. E-mail: dniyogi@purdue.edu
[2] Department of Earth and Atmospheric Sciences, Purdue University, West Lafayette, Indiana 47907.
[3] National Center for Atmospheric Research, Boulder, Colorado 80307.

between development and economic growth (LAMBIN *et al.*, 2003), and assessing risk for anthropogenic or natural hazards such as landslides (DHAKAL *et al.*, 1999). An accurate land use/land cover (LULC) dataset is also important to both understanding and modeling the exchange of mass and energy between the land surface and the atmosphere (BALLESTER *et al.*, 2003; ISHIZUKA *et al.*, 2005; KOTHAVALA *et al.*, 2005; REITHMAIER *et al.*, 2006), describing the deposition and distribution of atmospheric pollutants (CHOI *et al.*, 2005; NIYOGI *et al.*, 2004, 2006), and characterizing the role of biogeophysical and biogeochemical processes in regional atmospheric processes (WALKO *et al.*, 2000).

Since land-surface characteristics impact environmental processes on micro- (NIYOGI *et al.*, 2006), meso- (HOLT *et al.*, 2006), regional (PIELKE *et al.*, 2002), and global (FEDDEMA *et al.*, 2005) scales, land-surface characteristics exert an important control on the surface-atmosphere exchange and atmospheric processes leading to a broad range of environmental phenomena. For example, LEMONE *et al.* (2002; 2006) have linked land-surface characteristics with the partition of the surface energy balance and the evolution of the atmospheric boundary layer. Similarly, land cover and terrain conditions have been linked to the development of mesoscale circulations (ANTHES, 1984; SEGAL *et al.*, 1988), the initiation of convective storms (PIELKE, 2001; HOLT *et al.*, 2006), and the distribution of atmospheric precipitation (PIELKE *et al.*, 2007).

In order to fully describe or correctly model the role of the land surface in atmospheric or environmental processes, an accurate representation of the characteristics of the LULC is needed. ALAPATY *et al.* (1997) and NIYOGI *et al.* (1999) analyzed the impact of uncertainty in the surface characteristics and concluded that errors in the surface characterization propagate into the modeling of both the surface energy budget and the evolution of the boundary layer. While LULC data sets, which have been developed to represent a generalized regional scale, may provide an adequate representation on that scale, small, yet potentially important, local surface features may not be represented (WARDLOW and EGBERT, 2003; WICKHAM *et al.*, 2005).

In this study, we suggest that such limitations of the LULC data sets should be considered when conducting any study of land-atmosphere processes. Additionally, we propose a simple reclassification method to correct ambiguous LULC classifications. We focus on the Southern Great Plains (SGP) of the United States; specifically, the reclassification test case is the Walnut River Watershed (WRW) located in southeastern Kansas. This watershed was selected because it contains many of the LULC and surface characteristics typical of the SGP as a whole. For example, the WRW contains a mixture of tall- and short-grass prairie, agricultural fields, and developed (urban) areas. In addition, this watershed contains both expanses of gently rolling hills and steeper, more rugged terrain (LEMONE *et al.*, 2000).

This paper presents both a comparison of two LULC data sets, the National Land Cover Dataset 1992 (NLCD 92) and the Gap Analysis Program (GAP) data sets, and a reclassification method to enhance the accuracy of these data sets. The reclassification method utilizes a decision tree logical model based on the relationships between surface characteristics and LULC classification to reclassify those data points with an inaccurate or uncertain classification within the NLCD 92 or GAP data set. For example, the normalized difference vegetation index (NDVI) of an uncertain pixel is used to clarify whether that pixel is correctly categorized as water, developed, or vegetated land cover.

The following section describes the WRW and the LULC data sets used in this analysis. Section 3 discusses the methods used in the preliminary analyses and the reclassification process. Section 4 presents the results of the analyses and section 5 presents the conclusions. A follow-up study will evaluate the influence of surface features on boundary layer evolution and mesoscale processes.

2. Site Description and Data Sets

2.1. Walnut River Watershed

The Walnut River Watershed is located in southeastern Kansas between the city of Wichita to the west and the Flint Hills to the east. The watershed encompasses an area of approximately 5000 km^2 with a maximum extent of approximately 100 km from north to south and approximately 60 km from east to west (SONG and WESELY, 2003). The watershed is characterized by east-west gradients in precipitation, geology, and vegetation cover (LEMONE *et al.*, 2000). The western side of the WRW, which consists of gently rolling hills rising out of the watershed, is dominated by agricultural use including winter wheat, sorghum, and soybeans with some urban encroachment. The eastern side of the WRW is characterized by steeper slopes along the edge of the Flint Hills. In addition, this portion of the WRW is characterized primarily by grassland environments with a mixture of both tall– and short–grass prairie species (LEMONE *et al.*, 2000), some of which is used for pasture (Fig. 1).

2.2. National Land Cover Data Set 1992

The National Land Cover Data set 1992 dataset is a consistent, generalized, 30 m resolution LULC dataset for the contiguous United States (VOGELMANN *et al.*, 2001). Using Landsat thematic mapper imagery captured primarily during the period from 1990 to 1993, NLCD 92 was developed using a two-step classification process. In the first step, an unsupervised classification was applied to the Landsat imagery and each cluster was assigned one or more land-use categories according to the Anderson Level II scheme (ANDERSON *et al.*, 1976) shown in Table 1. In the second step, the confounded clusters, i.e., clusters that may be assigned to multiple land use classes, were refined using logical models based on ancillary data such as terrain, population density, and soil characteristics (VOGELMANN *et al.*, 2001). A recent analysis of the NLCD 92 dataset

Table 1

(Contd.)

NLCD 92	GAP	Simplified	NLCD 92	GAP	Simplified
Deciduous For	Maple-Basswood For.	Wooded	Row Crops	Cropland	Cropland
Evergreen For	Oak-Hickory For.		Small Grains		
Mixed For	Post Oak-Blackjack Oak For.		Fallow		
	Pecan Floodplain For.		Orchards/Vineyards		
	Ash-Elm-Hackberry Floodplain				
	For. Cottonwood Floodplain For.				
	Mixed Oak Floodplain For.				
	Bur Oak Floodplain Wood.				
	Mixed Oak Ravine Wood.				
	Post Oak-Blackjack Oak Wood.				
	Cottonwood Floodplain Wood.				
	Evergreen For.-Disturbed Land				
	Deciduous For.-Mined Land				
	Maple Floodplain For.				
	Deciduous Wood.				
			Woody Wetlands	Grass Playa Lake	Wetland
			Herbaceous Wetland	Salt Marsh/Prairie	
				Spikerush Playa Lake	
				Playa Lake	
				Low or Wet Prairie	
				Freshwater Marsh	
				Bulrush Marsh	
				Cattail Marsh	
				Wetland	
				Weedy Marsh	

accuracy of slightly more than 87% and agree with the NLCD 92 dataset for approximately 68% of the land area of Kansas. As with the NLCD 92 data set, grasslands are the most accurately classified LULC type and wetlands are the least accurately classified LULC type (WARDLOW and EGBERT, 2003).

2.4. Additional Data Sets

Aerial photographs taken as a part of National Agricultural Imagery Program (NAIP) were used as ground truth during the validation stage of the analysis. These photographs, which have a 2 m resolution, were collected during 2003. Two Landsat images collected on clear sky days during April and July 2002 were also used to calculate both NDVI and its distribution. Finally, the 1″ (approximately 30 m) National Elevation Data set (NED) digital elevation model (DEM) was used to determine the terrain characteristics including elevation, slope, and aspect employed in the analysis.

3. Reclassification Method

3.1. Simplification of the Data Sets and Preliminary Analyses

The NLCD 92 and GAP data sets are not directly comparable since they do not utilize the same classification scheme. Thus, the first step of the analysis was to map each of the categories in the NLCD 92 and GAP data sets to a single, simplified classification scheme (Table 1). This simplified scheme consists of seven categories representing each of the broad categories of the Anderson land cover scheme used with the NLDC 92 data set. The categories used to develop the simplified data sets for the study area were: Water, Urban, Barren, Wooded, Grassland, Cropland, and Wetland (Table 2).

The resulting LULC data sets, which are shown in Figure 2a and 2b, were then assessed for accuracy through a pixel-by-pixel comparison. The accuracy assessment used 500 randomly selected points distributed throughout the catchment; the locations of each of these random points were the same for all accuracy assessments. The ground truth for the accuracy assessment was interpreted from the 2003 NAIP imagery. The accuracy assessments for both the NLCD 92 and GAP data sets are summarized in Table 3. The Producer's Accuracy, which provides an estimate of how well the data were classified during the development of the data set, is defined as the ratio of the number of pixels within a given category that is correctly classified to the total number of pixels in that category based on ground truth. The User's Accuracy is the ratio of the number of pixels within a given category that is correctly classified to the total number of pixels classed into that category. Overall Accuracy is defined as the total number of pixels that is correctly classified to the total number of pixels in the dataset (LILLESAND *et al.*, 2004).

Table 2

The description of the seven classification categories utilized to simplify both the NLCD 92 and GAP data sets

Classification	Description
Water	Open water such as lakes and rivers; perennial snow and ice are also included
Urban	Developed areas including residential, such as single-family homes, commercial, and industrial areas; recreational grasslands, such as parks and golf courses, are also included
Barren	Areas perennially devoid of vegetation, such as strip mines, gravel pits, and quarries, and areas sparsely vegetated due to disturbance.
Wooded	Wooded areas including deciduous, evergreen, and mixed forest
Grassland	Grassland, shrubland, and herbaceous vegetation; pasture and hayfields are also included
Cropland	Row crops, small grains, such as wheat, and fallow fields
Wetland	Herbaceous and woody vegetation in areas that have periodically saturated soils or are covered in water

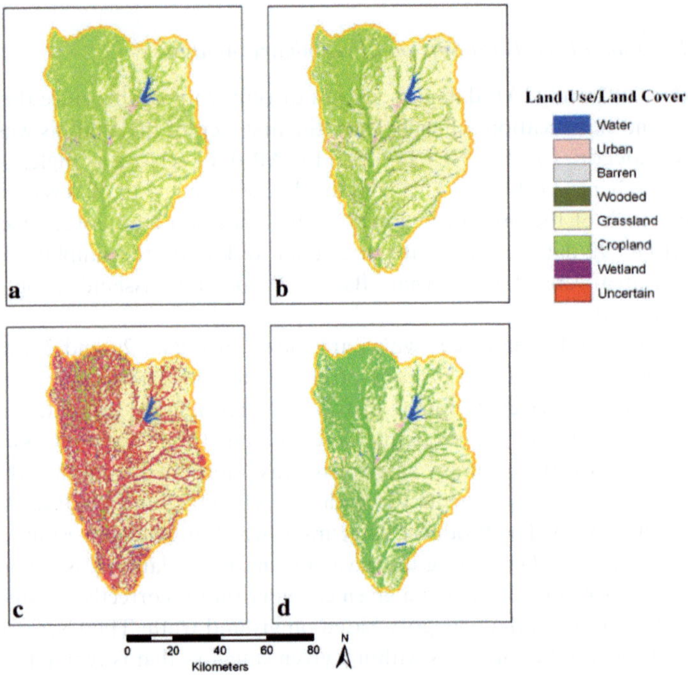

Figure 2

The simplified NLCD 92 (**a**) and GAP (**b**) data sets are shown for the Walnut River Watershed. The areas of uncertainty between these data sets (**c**) are shown. The revised LULC dataset (**d**) is also shown.

Table 3

The summarized accuracy assessment for each of the land use/land cover data sets

Classification	Water	Urban	Barren	Wooded	Grassland	Cropland	Wetland
				NLDC 92			
Producer's	69.8	73.5	100.0	15.4	89.9	68.7	33.3
User's	88.6	73.5	50.0	71.4	64.0	72.0	12.5
Overall	**68.7**						
				GAP			
Producer's	77.8	82.4	0.0	64.6	92.4	73.7	66.7
User's	94.6	93.3	0.0	72.4	73.0	90.1	15.4
Overall	**79.2**						
				Reclassified			
Producer's	84.4	82.4	0.0	90.8	97.5	94.4	66.7
User's	100.0	96.6	0.0	89.0	89.0	93.0	100.0
Overall	**92.9**						

3.2. Relationships between Surface Characteristics and Land Use/Land Cover

Histograms of the NDVI distribution for the simplified GAP dataset were generated for each of the LULC classifications. The NDVI values for each of the LULC classifications, as well as the complete data set, were sorted into 20 bins. The number of pixels within each bin was then normalized by the total number of pixels and plotted. Normalization enables a comparative analysis of the NDVI distributions for the various LULC classifications, while using the complete simplified GAP dataset facilitated the analysis for misclassification errors while clarifying the change in NDVI with time.

Using only those pixels for which both NLCD 92 and GAP data sets agree, the relationships of the land surface characteristics, including elevation, slope, and aspect, with the percent coverage of each LULC classification were developed. To accomplish this task, the WRW was first divided into slices based on a given land surface characteristic (Table 4). For example, the study region was sliced into seven elevation levels ranging from the lowest elevation with the WRW (330 m) to the highest (510 m) in nominally 20 m increments. Next, the percent coverage of each LULC classification was calculated for each slice. Finally, by plotting the percent coverage of each LULC classification as a function of the surface characteristics, relationships were developed that could be used during the reclassification process.

3.3. Reclassification Model

A five-step reclassification method (Fig. 3; Table 5) was developed to reclassify uncertain pixels. This method used a logical model that focused on four surface

Table 4

The criteria of slicing or subdividing the Walnut River Watershed when calculating the relationships between these surface characteristics and land use/land cover classification

Data Type	Nominal Data Range	Number of Slices	Slice Interval	Comments
Elevation	330 to 510 m	9	20 m	—
Slope	0° to 30°	7	0.0° to 0.5° 0.5° to 1.0° 1.0° to 1.5° 1.5° to 2.5° 2.5° to 3.5° 3.5° to 6.0° 6.0 to 30°	the slice intervals were selected based on a standard deviation partition of the data
Aspect	Flat and 0 to 360°	9	45°	each slice is centered on a defined direction; for example, the North slice include aspects from 337.5° to 22.5° with due north being 0°; flat has an aspect of −1.

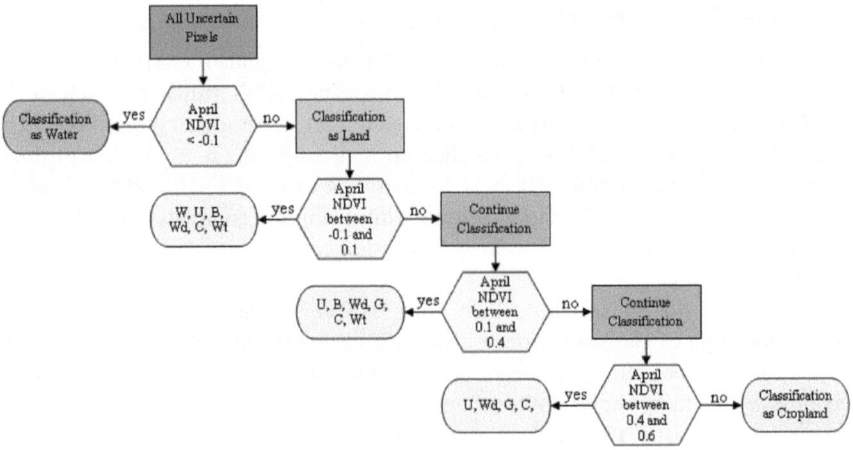

Figure 3
A flowchart showing the classification decisions made during the first step of the reclassification process is shown. Beginning with all of the uncertain pixels in the Walnut River Watershed, the NDVI during April is used as the basis for this initial step of the reclassification process. The symbol key is as follows: W – Water, U – Urban, B – Barren, Wd – Wooded, G –Grassland, C – Cropland, and Wt – Wetland.

characteristics: NDVI during April, NDVI during July, elevation, and slope. These four factors were selected because each showed a strong relationship of the LULC categories; for example, a low or negative NDVI is associated with the Water and Urban classes while a high positive NDVI is often associated with the Wooded. (The

Table 5

The conditional statements used in this research for reclassifying the uncertain data based on each of the four surface characteristics are shown. The symbol key is as follows: W – Water, U – Urban, B – Barren, Wd – Wooded, G – Grassland, C – Cropland, and Wt – Wetland

	April NDVI	July NDVI	Elevation	Slope
All Uncertain Pixels	NDVI < −0.1 Water			
	−0.1 < NDVI < 0.1 W, U, B, Wd, C, Wt	NDVI < −0.1 Water		
		−0.1 < NDVI < 0.0 U, B	Elevation < 420 m Urban 420 m < Elevation Barren	
		0.0 < NDVI < 0.2 U, B, C	Elevation < 420 m U, C	Can Not Be Differentiated
			420 m < Elevation B, C	Slope < 18° Cropland 18° > Slope Barren
		0.2 < NDVI < 0.4 U, C, Wt	Elevation < 380 m U, C, Wt	Slope < 3° U, C 3° < Slope Wetland
			380 m < Elevation < 420m U, C	Can Not Be Differentiated
			420 m < Elevation Cropland	
		0.4 < NDVI < 0.7 Wd, C, Wt	Elevation < 380 m Wd, C, Wt	Slope < 5° Cropland 5° < Slope Wd, Wt
			380 m < Elevation < 440 m Wd, C	Slope < 5° Cropland 5° < Slope Wooded
			440 m < Elevation Cropland	
		0.7 < NDVI Wooded		
	0.1 < NDVI < 0.4 U, B, Wd, G, C, Wt	NDVI < 0.0 Urban		
		0.0 < NDVI < 0.3 U, B, C	Elevation < 420 m U, C	Can Not Be Differentiated

Table 5

(*Contd.*)

April NDVI	July NDVI	Elevation	Slope
		420 m < Elevation B, C	Slope < 18° Cropland 18 > Slope Barren
	0.3 < NDVI < 0.5 U, Wd, G, C, Wt	Elevation < 380 m U, Wd, C, Wt	Slope < 3° U, Wd, C 3° < Slope Wetland
		380 m < Elevation < 420 m G, C	Slope < 1° Cropland 1° > Slope Grassland
		420 m < Elevation Grassland	
	0.5 < NDVI < 0.7 Wd, G, Wt	Elevation < 380 m Wd, Wt	Slope < 3° Wooded 3° < Slope Wetland
		380 m < Elevation Grassland	
	0.7 < NDVI Wd, G	Elevation < 380 m Wetland	
		380 m < Elevation Grassland	
0.4 < NDVI < 0.6 U, Wd, G, C	NDVI < 0.0 Urban		
	0.0 < NDVI < 0.3 U, C		
	0.3 < NDVI < 0.5 U, G, C	Elevation < 420 m U, C	Can Not Be Differentiated
		420 m < Elevation Grassland	
	0.5 < NDVI < 0.7 Wd, G, C	Elevation < 420 m Wd, C	Slope < 5° Cropland 5° < Slope Wooded
		420 m < Elevation Grassland	
	0.7 < NDVI Wd, G	Elevation < 420 m Wooded	
		420 m < Elevation Grassland	
0.6 < NDVI Cropland			

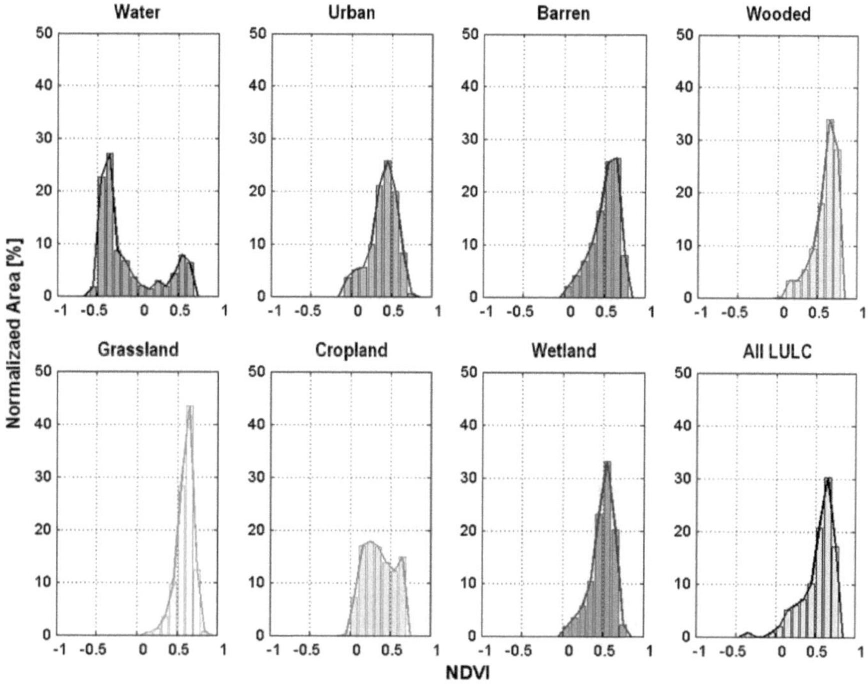

Figure 4
Histograms showing the distribution of NDVI for each land use/land cover classification provide evidence for both misclassification error and a potential method for improving the overall accuracy of a land use/land cover dataset for the Walnut River Watershed.

relationship between of the LULC categories and the various surface characteristics used in the logical model are discussed further below.) The logic model is a decision tree functioning analogously to a dichotomous key and focusing on each surface characteristic in turn to reclassify the uncertain pixels (FRIEDL and BRODLEY, 1997; DEFRIES and CHAN, 2000).

While the specific conditional statements are provided in Table 5, the five-step reclassification process can be summarized as follows. In the first step of the reclassification process, the data are first partitioned as either a water or land-based LULC type based on the NDVI during April 2002 (Fig. 3). Those pixels with an uncertain classification that had an NDVI during April less than −0.1 were reclassified as Water. Since nearly 90% of the pixels that have a known classification and an NDVI during April greater than 0.6 are classified as Cropland, those pixels with an NDVI value greater than 0.6 were reclassified as Cropland. The pixels that were not classified as Water or Cropland were then divided into three subsets

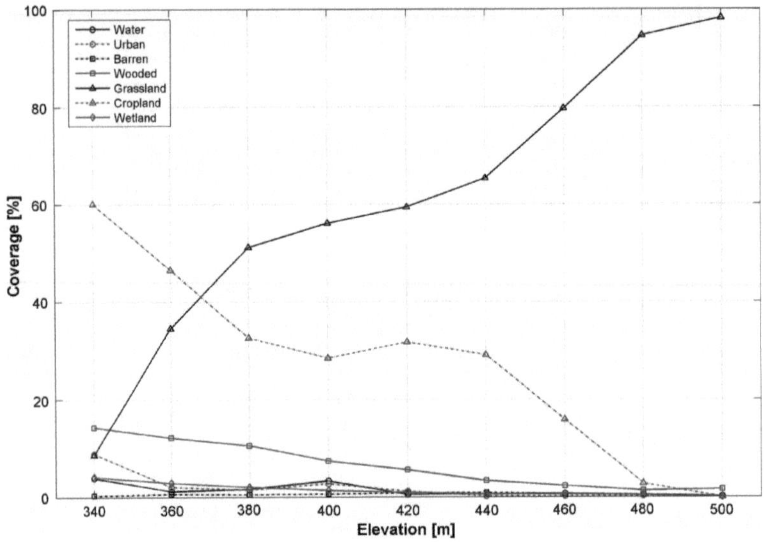

Figure 5
The change in the percent coverage with elevation for each of the land use/land cover types demonstrates the dominance of Grassland and Cropland throughout the Walnut River Watershed.

depending on their NDVI during April. Each of these subsets represents a unique grouping of potential LULC classifications. Those subsets are further refined in subsequent steps.

Using the NDVI during July as a basis, the classification of the three subsets of pixels that could not be classified uniquely in the previous step was further refined in the second step of the reclassification process. This resulted in each of the subsets from step one being further subdivided into four smaller, more refined subsets. For example, the subset of uncertain pixels with an NDVI during April between −0.1 and 0.1 was refined into four subsets, based on the NDVI for July 2002, including a subset containing NDVI values between −0.1 and 0.0, a subset containing NDVI values between 0.0 and 0.2, a subset containing NDVI values between 0.2 and 0.4, and a subset containing NDVI values between 0.4 and 0.7. Approximately 10% of the uncertain pixels were reclassified during this step.

In the third step of the reclassification process, each of the twelve subsets generated during the previous step was further refined, based on elevation. Using elevation as the criterion, the subsets developed previously were further refined and subdivided into a total of 14 subsets. Additionally, it was possible to reduce the size of eight of the twelve subsets developed in the previous two steps by uniquely reclassifying at least a portion of the pixels contained within those subsets.

The fourth step of the reclassification process utilized slope as the partition criterion. This step allowed for more than 99.9% of the uncertain pixels to be classified into a unique LULC classification. The 307 pixels that could not be classified based on the decision tree were reclassified based on the classification of the majority of their nearest neighbors defined here as a 5×5 grid centered on the pixel to be classified.

To illustrate the reclassification process, consider an uncertain pixel within the riparian zone with NDVI = 0.2 during April, NDVI = 0.6 during July, elevation = 375 m, and slope = 2°. In the first step of the reclassification process, this pixel would be partitioned into the subset for NDVI during April with a range from 0.1 to 0.4. This subset included all of the land-based LULC classifications. In the second step, the pixel would be partitioned into the subset for NDVI during July ranging from 0.5 to 0.7, thereby reducing the potential LULC classifications by half. In the third step of the reclassification process, which used elevation as the decision criterion, the pixel would be partitioned into the subset for elevations less than 380 m. This subset included both the Wooded and Wetland LULC types. In the final step, which used slope as the decision criterion, this pixel would be reclassified as Wooded.

Once the reclassification process was completed, an accuracy assessment was conducted using the same method as used with the simplified NLCD 92 and GAP data sets. This analysis showed an overall accuracy of nearly 93% (Table 3). This is an improvement in the overall accuracy of 24.2% and 13.7% compared to the NLCD 92 and GAP data sets, respectively.

4. Results

4.1. Simplification of the Data Sets and Preliminary Analyses

While the classification represented by the two simplified data sets (Figs. 2a and 2b) appears reasonable, the accuracy assessments of these data sets (Table 3) indicate that they contained errors. The accuracy assessments indicated that the NLCD 92 overall accuracy was approximately 68.7%. For NLCD 92, the classification with the highest Producer's Accuracy was associated with the Barren and Grassland categories while the highest User's Accuracy was associated with the Water classification. The classification of Wetland was particularly weak for the NLCD 92 data set.

Similarly, the accuracy assessment for the GAP dataset indicated an overall accuracy of 79.2% for the WRW. The classification categories with the highest accuracy were Grassland and Water for Producer's and User's Accuracy, respectively. Like the NLCD 92 data set, the GAP dataset was least accurate for the Wetland category. Moreover, a more accurate representation of all of the classification categories is reflected in an improved overall accuracy of more than

10% as compared to the NLCD 92 data set. However, while the GAP dataset is more accurate than the NLCD 92 data set, it still contains significant errors.

The comparison of the simplified NLCD 92 and GAP data sets conducted through an overlay process demonstrated where the two data sets agreed and where the LULC classification was uncertain (Fig. 2c). This analysis indicated that the two LULC data sets agree in the classification of approximately 73% of the area of the WRW. Further analysis of the 27% of the WRW with an uncertain classification, i.e., the two data sets did not agree, suggests that the differences are located primarily in one of the following three geographic regions:

1. Along the transitions from large areas of contiguous LULC, specifically the Water and Urban classifications, with another LULC classification. This is, at least in part, reflective of the changing boundaries of these LULC types with time.
2. Within riparian zones where clusters of trees, as well as grassland and cropland environments, were often misclassified. This may be the result of mixed pixels containing multiple LULC types, the application of smoothing algorithms during the development of these two data sets, or signature confusion such that the classification algorithms used were unable to differentiate the correct LULC type.
3. In areas containing a mixture of LULC. In areas of mixed agricultural and grassland environments there are often differences between the NLCD 92 and GAP classification. This is most likely due to either mixed pixels or signature confusion.

4.2. Relationships between Surface Characteristics and Land Use/Land Cover

Analysis of the NDVI histograms – Figure 4 shows the NDVI during April - indicates both a degree of misclassification error and a cause for that error. These plots also suggest that NDVI may be used as a part of the reclassification process aimed at eliminating that error to produce a more accurate representation of the WRW. Specifically, the analysis demonstrated that both the absolute value of the NDVI during April and July and its change over that period were linked with their LULC classification. As such, NDVI could be used to reclassify uncertain pixels, and assimilating NDVI within land-surface models can assist in minimizing errors.

Examples of potential misclassification can be seen in the Water histogram. The histogram for April and July both show a bimodal distribution with approximately 30% of the pixels classified as Water in the GAP dataset contained within the second peak and having an NDVI greater than 0. The position and shape of the second peak is consistent with the NDVI distribution of Wetland. Finally, the majority of the pixels contained within the second peak are located near rivers, lakes, or small ponds within the WRW. Together, these results suggest those pixels may have been misclassified. These results also suggest that reclassifying pixels categorized as Water but having a positive NDVI would result in a more accurate data set.

When the percent coverage of a given LULC classification was plotted as a function of a land-surface characteristic, several strong relationships were found between elevation and LULC coverage. This is particularly interesting given the ongoing discussion regarding whether mesoscale circulations in the SGP are a function of vegetation cover (WEAVER and AVISSAR, 2001) or topography (DORAN and ZHONG, 2001). Using 20 m elevation slices spanning the WRW, the percent coverage of each LULC classification was determined. Plotting the percent coverage for each LULC classification (Fig. 5) revealed several important relationships. In addition to demonstrating the dominance of the Grassland and Cropland classifications throughout the WRW, it showed clearly that Cropland was the most prevalent LULC classification at lower elevations while Grassland was most common at the higher elevations. Indeed, at the highest elevations above 480 m, the percent coverage of Grassland approaches 100%. Additionally, there is a nearly linear, inverse relationship between the percent coverage of the Wooded classification and elevation. Finally, it was found that Water, Wetland, and especially Urban were confined mainly to the lowest elevations.

A similar analysis relating LULC classification with slope also yielded strong relationships. For example, Cropland is confined primarily to the flattest portions of the WRW. For slopes less than 0.5°, Cropland is the dominant LULC classification with a percent coverage of nearly 47%; the percent coverage drops to less than 10% when the slope exceeds 4.5° and less than 3% when the slope exceeds 6°. Additionally, the steepest slopes were dominated by either Grassland at the higher elevation or Wooded areas in riparian zones. The steepest slopes in the WRW are found along the shorelines and channel cuts of lakes, streams, and rivers. Finally, while Urban LULC was most common when the slope was less than 2°, Water, Barren, and Wetland classifications were somewhat more prevalent at slopes greater than 4.5°; however, it must be noted that in all cases these LULC classifications constitute a small percentage of the total land cover.

The relationship of LULC classification with aspect was more ambiguous. No specific LULC classification or subset of classifications could be uniquely associated with any given aspect. The LULC classifications were evenly distributed among all aspects. As such, aspect would not provide a useful relationship for the reclassification of uncertain LULC classifications within the Walnut River Watershed.

4.3. Reclassification Model

The results of the reclassification (Fig. 2d) indicate a substantially improved classification of the LULC within the WRW. The accuracy assessment, which was conducted using the same methods as used with the simplified NLCD 92 and GAP data sets, yielded an overall accuracy of 92.9% for the revised data set. This is more than 13% greater than the GAP dataset and nearly 25% greater than the overall accuracy of the NLCD 92 dataset (Table 3).

In particular, the revised data sets showed strong improvements in the classification of all LULC categories as compared to the NLCD 92 data set. The revised dataset also showed strong improvements in the classification of the Wooded, Grassland, and Cropland environments as compared to the GAP data set. For example, the Producer's Accuracy for the Wooded LULC classification increased from 64.6% to 90.8% while the User's Accuracy increased from 72.4% to 89.0%. Similarly, the User's Accuracy increased from 73.0% and 90.1% to 89.0% and 93.0% for the Grassland and Cropland classifications, respectively.

5. *Conclusions*

The comparison of the NLCD 92 and GAP data sets demonstrates significant differences; these two dataset agree for only 73% of the Walnut River Watershed. Moreover, accuracy assessments of these data sets indicate an overall accuracy of less than 80% in both cases. The results strongly suggest that these data sets should be used with caution when they are applied to local areas. We offer a reclassification approach using a logical model that could greatly improve the accuracy of these LULC data sets when the focus is on a local scale.

The reclassification process is based on several strong relationships between land-surface characteristics and LULC classification that are evident from this research. In particular, NDVI, elevation, and slope demonstrated useful relationships. For example, based on this research, Cropland is confined to flat or gently sloping areas while Grassland is more prevalent in areas of steeper slopes and higher elevation. Similarly, developed land use, i.e., the Urban classification, is greatest at the lowest elevations within the WRW where the slope is minimal. In contrast to these land surface characteristics, aspect appeared to have no influence on LULC classification.

The relationships of LULC classification and surface characteristics evident from this research not only indicate a strong linkage between LULC and the land surface, but they also suggest the reclassification method tested in this research. That method, a logical model that acts like a dichotomous key, uses surface characteristics to reclassify uncertain LULC classifications to produce a more accurate representation of the WRW. Indeed, the reclassification method resulted in improvements in the overall accuracy of the LULC dataset by as much as 24.2% when compared with the NLCD 92 data set.

While the results of this research are unique for the WRW, the strong relationship between LULC type and surface features is broadly applicable. As a result, by developing relationships of the type used here for other local areas, the reclassification method applied here also could be applied to those areas.

As modelers adopt finer grid spacing for land surface models, the impacts of the surface specifications become increasingly more important. As a result, an accurate LULC dataset also becomes increasingly more important. This is because uncertainty in LULC could introduce errors into model simulations that could propagate into the results. The magnitude of the influence of the uncertainty in LULC data sets on land surface model output is the focus of ongoing research and will be presented in a follow up paper. Study results suggest that assimilating NDVI into the land surface models can reduce the uncertainty due to LULC assignment.

Acknowledgments

The authors would like to acknowledge Dr. Steve Nelson and NSF (ATM 0233780; ATM 0236885; ATM 0296159), Dr. Jared Entin and NASA–THP (NNG05GQ47G), Drs. Jared Entin and Garik Gutman and NASA-IDS (NNG04GL61G), the Joint Center for Satellite Data Assimilation, and NASA Land Use Land Cover Change Program for support of the research. The authors would also like to thank Profs. Chris Johannsen and Melba Crawford at Purdue University for their insights and helpful review of this manuscript. The authors would also like to thank the three anonymous reviewers for their comments and suggestions that greatly improved the quality and clarity of the manuscript. Finally, the authors would like to thank Megan O'Connell for her assistance during this research.

References

Alapaty, K., Raman, S., and Niyogi, D. (1997), *Uncertainty in the specification of surface characteristics: A study of prediction errors in the Boundary Layer*, Boundary-Layer Meteor. *82*, 475–502.

Anderson, J.F., Hardy, E.E., Roach, J.T., and Witmer, R.E. (1976), *A land use and land cover classification system for use with remote sensor data*, Geolog. Survey Prof. Paper 964. United States Geological Survey, Washington D.C., 41 pp.

Anthes, R. (1984), *Enhancement of convection by mesoscale variability in vegetation coverage in semi-arid regions*, J. Climate Appl. Meteor. *23*, 541–554.

Ballester, M.V.R., Victoria, D.D., Krusche, A.V., Coburn, R., Victoria, R.L., Richey, J.E., Logsdon, M.G., Mayorga, E., and Matricardi, E. (2003), *A remote sensing/GIS-based physical template to understand the biogeochemistry of the Ji-Parana River basin (Western Amazonia)*, Remote Sensing of Environ. *87*, 429–445.

Choi, B.Y., Yun, S.T., Yu, S.Y., Lee, P.K., Park, S.S., Chae, G.T., and Mayer, B. (2005), *Hydrochemistry of urban groundwater in Seoul, South Korea: Effects of land-use and pollutant recharge*, Environ. Geology *48*, 979–990.

DeFries, R.S. and Chan, J.C.-W. (2000), *Multiple criteria for evaluating machine learning algorithms for land cover classification from satellite data*, Remote Sensing of Environ. *74*, 503–515.

Dhakal, A.S., Amanda, T., and Aniya, M. (1999), *Landslide hazard mapping and the application of GIS in the Kulekhani watershed, Nepal*, Mountain Res. Develop. *19*, 3–16.

DORAN, J.C. and ZHONG, S. (2001). *Comments on "Atmospheric disturbance caused by human modification of the landscape"*, Bull. Am. Meteor. Soc. *82*, 277–279.

EGBERT, S.L., PETERSON, D.L., STEWART, A.M., LAUVER, C.L., BLODGETT, C.F., PRICE, K.P., and MARTINKO, E.A. (2001), *The Kansas Gap Land Cover Map: Final Report*, Kansas Biological Survey Report 98. [http://maps.kansasgis.org/kgcc/docs/datadocs/gap_finalrep.pdf.]

FEDDEMA, J.J., OLESON, K.W., BONAN, G.B., MEARNS, L.O., BUJA, L.F., MEEHL, F.A., and WASHINGTON, W.M. (2005), *The importance of land-cover change in simulating future climates*, Science *310*, 1674–1676.

FRIEDL, M.A. and BRODLEY, C.E. (1997), *Decision tree classification of land cover from remotely sensed data*, Remote Sensing of Environ. *61*, 399–409.

HOLT, T., NIYOGI, D., CHEN, F., LEMONE, M.A., MANNING, K., and QURESHI A.L. (2006), *Effect of land-atmosphere interactions on the IHOP 24-25 May 2002 convection case*, Mon. Wea. Rev. *134*, 113–133.

ISHIZUKA, S., ISWANDI, A., NAKAJIMA, Y., YONEMURA, S., SUDO, S, TSURUTA, H., and MURDIYARSO, D. (2005), *The variation of Greenhouse emissions from soils of various land-use/cover types in Jambi province, Indonesia*, Nutrient Cycling in Agro-ecosystems *71*, 17–32.

KOTHAVALA, Z., ARAIN, M.A., BLACK, T.A., and VERSEGHY, D. (2005), *The simulation of energy, water vapor and carbon dioxide fluxes over common crops by the Canadian land surface scheme (CLASS)*, Agricultural and Forest Meteor. *133*, 89–108.

LAMBIN, E.F., GEIST, H.J., and LEPERS, E. (2003), *Dynamics of land-use and land-cover change in tropical regions*, Ann. Rev. Environ. Resources *28*, 205–241.

LILLESAND, T.M., KIEFER, R.W., and CHIPMAN, J.W. (2004), *Remote Sensing and Image Interpretation*. (John Wiley and Sons, New York, 2004).

LIU, Y.S., WANG, D.W., GAO, J., and DENG, W. (2005), *Land use/cover changes, the environment and water resources in northeast China*, Environ. Management *36*, 691–701.

LEMONE, M.A., GROSSMAN, R.L., COULTER, R.L., WESLEY, M.L., KLAZURA, G.E., POULOS, G.S., BLUMEN, W., LUNDQUIST, J.K., CUENCA, R.H., KELLY, S.F., BRANDES, E.A., ONCLEY, S.P., MCMILLEN, R.T., and HICKS, B.B. (2000), *Land-atmosphere interaction research, early results, and opportunities in the Walnut River Watershed in southeast Kansas: CASES and ABLE*, Bull. Am. Meteor. Soc. *81*, 757–779.

LEMONE, M.A., GROSSMAN, R. L., MCMILLEN, R., LIOU, K.-N., OU, S.C., MCKEEN, S., ANGEVINE, W., IKEDA, K., and CHEN, F. (2002), *CASES-97: Late-morning warming and moistening of the convective boundary layer over the Walnut River watershed*, Boundary-Layer Meteor. *104*, 1–52.

LEMONE, M.A., CHEN, F., ALFIERI, J.G., TAWANI, M., GEERTS, B., MIAO, Q., GROSSMAN, R.L., and COULTER, R.L. (2006), *Influence of land cover, soil moisture, and terrain on the horizontal distribution of sensible and latent heat fluxes and boundary-layer structure*, J. Hydrometeor., in press.

NIYOGI, D., RAMAN, S., and ALAPATY, K. (1999), *Uncertainty in specification of surface characteristics, Part 2: Hierarchy of interaction explicit statistical analysis*, Boundary-Layer Meteor. *91*, 341–366.

NIYOGI D., CHANG, H., SAXENA, V. K., HOLT, T., ALAPATY, K., BOOKER, F., CHEN, F., DAVIS, K.J., HOLBEN, B., MATSUI, T., MEYERS, T., OECHEL, W.C., PIELKE, R.A., SR., WELLS, R., WILSON, K., and XUE, Y.K. (2004), *Direct observations of the Effects of Aerosol loading on Net Ecosystem CO_2 Exchanges over Different Landscapes*, Geophys. Res. Lett. *31*, L20506, doi:10.1029/2004GL020915.

NIYOGI, D., ALAPATY, K., PHILLIPS, S., and ANEJA, V. (2006), *Considering ecological formulations for estimating deposition velocity in air quality models*, Internat. J. Global Environ. Issues *6*, 270–284.

PIELKE, R.A., SR. (2001), *Influence of the spatial distribution of vegetation and soils on the prediction of cumulus convective rainfall*, Rev. Geophys. *39*, 151–177.

PIELKE, R.A., SR., MARLAND, G., BETTS, R.A., CHASE, T.N., EASTMAN, J.L., NILES, J.O., NIYOGI, D., and RUNNING, S. (2002), *The influence of land-use change and landscape dynamics on the climate system: Relevance to climate change policy beyond the radiative effect of Greenhouse gases*, Philosoph. Transact. Roy. Soc. London *360*, 1705–1719.

PIELKE, R.A. SR., ADEGOKE, J., BELTRAN-PRZEKURAT, A., HIEMSTRA C.A., LIN, J., NAIR, U.S., NIYOGI, D., and NOBIS, T.E. (2007), *An overview of regional land use and land cover impacts on rainfall*, Tellus B, *59*, 587–601.

REITHMAIER, L.M., GOCKEDE, M., MARKKANEN, T., KNOHL, A., CHURKINA, G., REBMANN, C., BUCHMANN, N., and FOKEN, T. (2006), *Use of remotely sensed land use classification for a better evaluation of micrometeorological flux measurement sites,* Theoretical Appl. Climatology *84,* 219–233.

SCOTT, J.M., DAVIS, F., CSUTI, B., BUTTERFIELD, B., GROVES, C., ANDERSON, H., CAICCO, S., D'ERCHIA, F., EDWARDS, T.C., ULLIMAN, J., and WRIGHT, G. (1993), *Gap Analysis: A Geographic Approach to Protection of Biological Diversity* (The Wildlife Society, Washington, D.C., 1993).

SEGAL, M., AVISSAR, R., MCCUMBER, M., and PIELKE, R. (1988), *Evaluation of vegetation effects on the generation and modification of mesoscale circulations,* J. Atmos. Sci. *45,* 2268–2292.

SONG, J., and WESELY, M.L. (2003), *On comparison of modeled surface flux variations to aircraft observations,* Agricultural and Forest Meteor. *117,* 159–171.

VOGELMANN, J.E., HOWARD, S.M., YANG, L., LARSON, C.R., WYLIE, B.K., and VAN DRIEL, N. (2001), *Completion of the 1990s national land cover dataset for the conterminous United States from Landsat thematic mapper data and ancillary data sources,* Photogrammetric Engin. Remote Sensing *67,* 650–662.

WARDLOW, B.D. and EGBERT, S.L. (2003), *A state-level comparative analysis of the GAP and NLCD land-cover data sets,* Photogrammetric Engin. Remote Sensing *69,* 1387–1397.

WALKO, R.L, BAND, L.E, BARON, J., KITTEL, T.G.F., LAMMERS, R. LEE, T.J., OJIMA, D., PIELKE, R.A., TAYLOR, C., TAGUE, C., TREMBACK, C.J., and VIDALE, P.L. (2000), *Coupled atmosphere-biophysics-hydrology models for environmental modeling,* J. Appl. Meteor. *39,* 931–944.

WEAVER, C. P. and AVISSAR, R. (2001), *Atmospheric disturbances caused by human modification of the landscape,* Bull. Am. Meteor. Soc. *82,* 269–281.

WICKHAM, J.D., STEHMAN, S.V., SMITH, J.H., and YANG, L. (2005), *Thematic accuracy of the 1992 national land-cover data for the western United States,* Remote Sensing of Environ. *91,* 452–468.

YANG, X. and LO, C.P. (2002), *Using time series of satellite imagery to detect land use and land cover changes in the Atlanta, Georgia metropolitan area,* Internat. J. Remote Sensing *23,* 1775–1798.

(Received May 31, 2006, accepted September 11, 2006)
Published Online First: July 4, 2007

Pure appl. geophys. 164 (2007) 1811–1837
0033–4553/07/091811–27
DOI 10.1007/s00024-007-0236-1

Pure and Applied Geophysics

Analysis of Weak Wind Stable Conditions from the Observations of the Land Surface Processes Experiment at Anand in India

ADITI and MAITHILI SHARAN

Abstract—The observations in weak wind stable conditions are scarce. The present study examines the observations from the Land Surface Processes Experiment (LASPEX) conducted at Anand, (Gujarat, India) during the year 1997–1998 to study the characteristics of surface layer under weak wind stable conditions. The observed surface fluxes are compared with those computed using Monin-Obukhov (M-O) similarity theory. The upper air observations and regional climatology are used to justify the persistence of weak wind conditions at Anand. The frequency of occurrence of weak wind stable conditions is observed to be around 67%. In 86% of the cases under weak wind conditions, bulk Richardson number (R_{iB}) is found to be larger than 0.2. The magnitude of surface fluxes computed from M-O similarity theory is shown to be smaller in comparison to those based on the observations in weak wind stable conditions. Surface fluxes computed using the empirical relations for the eddy diffusivities and drag and heat exchange coefficients are found to be comparable with those based on M-O similarity theory however these fluxes are under-predicted in comparison to the observations. The traditional M-O similarity theory is not able to simulate the observed fluxes well in weak wind stable conditions at Anand.

Key words: Bulk Richardson number, eddy diffusivities, stable conditions, surface fluxes, surface layer, weak wind.

1. Introduction

The surface layer is an active link between the earth's surface and free atmosphere and the flow in this layer has a turbulent character which transports heat, momentum, moisture and pollutant both horizontally and vertically (SORBJAN, 1989; GARRATT, 1992; KAIMAL and FINNIGAN, 1994). Spatial distribution of surface fluxes significantly influences the global atmospheric fields, thus increasing the need to simulate land-surface processes in atmospheric models.

Many experimental studies on atmospheric surface layer are reported in literature (WYNGAARD *et al.*, 1971; WYNGAARD and COTE, 1971, 1972; BUSINGER *et al.*, 1971; KAIMAL and WYNGAARD, 1990). In addition, a number of boundary layer experiments like Wangara (CLARKE *et al.*, 1971), Kansas (IZUMI, 1971), Plume validation experiment, Electric Power Research Institute (EPRI, 1981) Kincaid,

Centre for Atmospheric Sciences, Indian Institute of Technology, Delhi, Hauz Khas, New Delhi 110 016, India.

Cabauw (1987), Hydrological Atmosphere Pilot Experiment (HAPEX) (ANDRE *et al.*, 1990; GOUTEROBE *et al.*, 1997), First ISLSCP (International Satellite Land Surface Climatology Project) Field Experiment (FIFE) (SELLERS *et al.*, 1992), Cooperative Atmosphere Surface Exchange Study (CASES-99), LITFASS-2003 experiment in Lidenberg (Germany) in May/June 2003 and the Urban Boundary Layer Experiment ESCOMPTE 2004 have been conducted with various objectives to provide a better insight into land-surface processes. All of these have a common objective of studying the surface fluxes and their parameterizations required for development of atmospheric models. A few experiments conducted over the Indian sub-continent include the coastal boundary layer studies over Trivandrum (PRAKASH *et al.*, 1992, 1993; KUNHIKRISHNAN *et al.*, 1993; RAMACHANDRAN *et al.*, 1994), the monsoon boundary layer experiment (MONTBLEX-1990) conducted during the south-west summer monsoon season over the monsoon trough region of the Gangetic plains (GOEL and SRIVASTAVA, 1990; KUSUMA *et al.*, 1991, 1995; KUSUMA and NARASIMHA, 1996; KAILAS and GOEL, 1996) and Land Surface Processes Experiment (LASPEX) (VERNEKAR *et al.*, 2003) in the semi-arid region of western India to generate surface data set to parameterize the land surface processes in numerical models. Most of these field experiments conducted to date are performed in strong and moderate wind conditions. As a part of the diffusion experiment in low wind conditions, conducted at the Indian Institute of Technology Delhi in February 1991, an attempt has been made to analyze the turbulence characteristics over land in the tropics in weak wind conditions (AGGARWAL *et al.*, 1995; YADAV *et al.*, 1996, 2003). However, the observations in weak wind stable conditions are scarce.

Low wind stable conditions occur in the surface layer for a considerable period of time in most parts of the world and are associated with high air pollution potential. They assume greater significance in a tropical environment such as India because they occur frequently during the winter season and are crucial for air pollution episodes (SHARAN *et al.*, 1995, 1996a; SHARAN and GOPALAKRISHNAN, 2003). The knowledge of atmospheric boundary layer processes is limited under weak wind stable conditions and thus it is necessary to examine the boundary/surface layer characteristics under these conditions.

The surface layer characteristics have been studied (SHARAN *et al.*, 2003b) using EPRI and CASES-99 observations. However, turbulence observations were not available in EPRI data set whereas very few observations were found in CASES-99 in weak wind stable conditions. Thus, extensive data are required to analyze the surface layer characteristics in weak wind stable conditions. The aim of the present study is to analyze the observations available from the LASPEX to examine characteristics of the surface layer in weak wind stable conditions.

The objective of the paper is two fold: (i) To examine the persistence of weak wind conditions at Anand, and (ii) to analyze the surface layer characteristics in weak and strong wind stable conditions.

2. Experiment and Data Analysis

2.1. Observational Site

LASPEX was conducted in the Sabarmati River basin (Fig. 1) of the Gujarat state in western India with an objective to collect surface and subsurface atmospheric hydrological database against which parameterized models for land-surface processes can be tested for improvement and further development. The Sabarmati river basin lies in the semi-arid part of India and provides contrasting meteorological conditions from one season to another as well as a variety of weather conditions on the scale of a week within a season. The main experiment was conducted from January 1997 to February 1998. Five experimental platforms were chosen, namely Anand (central station), Sanad, Arnej, Khanda and Derol. These stations are at a distance of about 60–100 km apart. The meteorological data of Anand during the year 1997 are used in the present study.

Anand lies in the middle of the Gujarat state with latitudes 22° 35′ N and longitudes 72° 55′ E and with an elevation of 45.1 m above mean sea level. The climate is of a semi-arid type and the type of soil is loamy (SHEIKH, 1997). The mean wind speed is found to be quite low in the range of 0.63 to 1.8 m s^{-1} (PANDEY *et al.*, 2001).

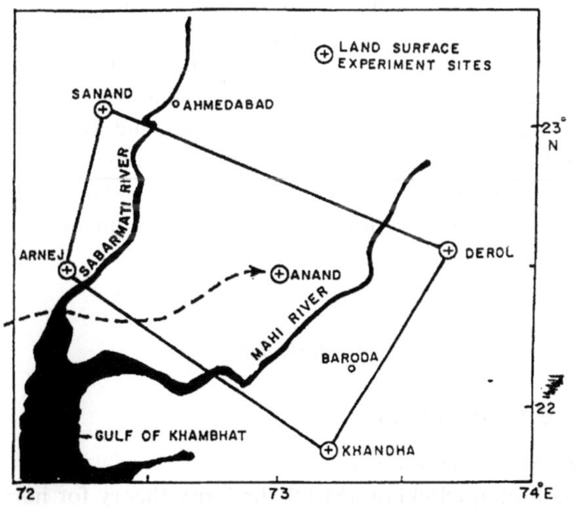

Figure 1
Location of Anand—One of the experimental sites of LASPEX (VERNEKAR *et al.*, 2003).

2.2. Instrumentation and Data Collection

The experimental setup used for the study consisted of slow-response meteorological sensors for measuring wind direction, wind speed, temperature and humidity at 1, 2, 4 and 8 m height on a tower. The slow response meteorological sensors are used to measure the mean values whereas the fast response sensors can take measurements with very fine resolution, which makes them well suited for turbulence measurements. The turbulence data were collected by a fast-response Metek sonic anemometer, with a sampling interval of 10 Hz at 9.5 m. The Metek sonic anemometer has a built-in processor that analyzes the wind and temperature data and evaluates the surface layer parameters such as momentum flux, sensible heat flux and friction velocity using eddy correlation technique Radiosonde ascents were taken by the India Meteorological Department using low ascent rate balloons at synoptic hours 0530, 0830, 1130, 1430, 1730 IST (Indian Standard Time) every day for 5 days (13–17) in each month of 1997.

The experiment was conducted from January 1997 to February 1998. During the year 1997, the slow data at 1, 2, 4 and 8 m levels from the tower were available for the months of February, March, April, May, June, July, August, October, November and December and the turbulence data were available during the months of April, May, June, July, August, October and December. However, in the year 1998 the slow data were available for the months of January and February whereas the turbulence data were available for the months of May, August and November. In the present study we have utilized only those months (April, May, June, July, August, October and December during 1997) for which both fast and slow data are available. The slow data are primarily analyzed for stable conditions during night-time by looking for (i) positive potential temperature gradient and (ii) a smooth variation of wind. Based on the above criteria, the data during the months of May, June, October and December in the year 1997 are selected for further analysis.

The screened data are further classified into strong and weak winds based on the surface wind speed (U_s) measured at 8-m level on the tower and geostrophic wind speed (U_g) observed from the Radiosonde observations. The wind at the 850–hPa level from Radiosonde observations is considered as an approximation to the geostrophic wind.

3. Methodology

3.1. Calculation of Surface Fluxes

(a) Similarity theory in terms of z/L

Based on the Monin-Obukhov (M-O) similarity theory for horizontally homogeneous and stationary surface layer, the nondimensional wind and temperature profiles are given by

$$\frac{kz}{u_*}\frac{\partial U}{\partial z} = \phi_m\left(\frac{z}{L}\right), \tag{1}$$

$$\frac{kz}{\theta_*}\frac{\partial \theta}{\partial z} = \phi_h\left(\frac{z}{L}\right), \tag{2}$$

where k is the von Karman constant, U is the mean wind speed, θ is the potential temperature, u_* is the frictional velocity, θ_* is the frictional temperature scale, ϕ_m and ϕ_h are the similarity functions for momentum and heat, L is the Obukhov length scale and z/L is the stability parameter defined as

$$\frac{z}{L} = \frac{kg}{\overline{\theta}}\frac{z\theta_*}{u_*^2}, \tag{3}$$

in which g is the acceleration due to gravity and $\overline{\theta}$ is mean potential temperature. Integrating Eqs. (1) and (2) from z_0 to z in the surface layer, we obtain

$$U(z) = \frac{u_*}{k}\left[\ln\left(\frac{z}{z_0}\right) - \psi_m\left(\frac{z}{L}, \frac{z_0}{L}\right)\right], \tag{4}$$

$$\theta(z) - \theta_0 = \frac{\theta_*}{k}\left[\ln\left(\frac{z}{z_0}\right) - \psi_h\left(\frac{z}{L}, \frac{z_0}{L}\right)\right], \tag{5}$$

where z_0 is the surface roughness length and θ_0 is the potential temperature at z_0. Ψ_m and Ψ_h are integrated similarity functions defined as

$$\psi_m\left(\frac{z}{L}, \frac{z_0}{L}\right) = \int_{\frac{z_0}{L}}^{\frac{z}{L}} [1 - \phi_m(\varsigma')]\frac{d\varsigma'}{\varsigma'}, \tag{6}$$

$$\psi_h\left(\frac{z}{L}, \frac{z_0}{L}\right) = \int_{\frac{z_0}{L}}^{\frac{z}{L}} [1 - \phi_h(\varsigma')]\frac{d\varsigma'}{\varsigma'}, \tag{7}$$

where $\varsigma' = z'/L$. The surface in this study is assumed to be smooth and accordingly the roughness lengths for momentum and heat are taken to be equal to z_0.

The application of linear functional forms for ϕ_m and ϕ_h is normally limited (SHARAN *et al.*, 2003a) to moderate to strong wind stable conditions. In the present study we propose to use the functional forms for ϕ_m and ϕ_h proposed by BELJAARS and HOLTSLAG (1991). Based on these, Ψ_m and Ψ_h are given by:

$$\psi_m = -a\left(\frac{z}{L} - \frac{z_0}{L}\right) - y + y_o, \tag{8}$$

$$\psi_h = -\left(1 + \frac{2}{3}a\frac{z}{L}\right)^{3/2} + \left(1 + \frac{2}{3}a\frac{z_0}{L}\right)^{3/2} - y + y_o, \tag{9}$$

where

$$y = b\left(\frac{z}{L} - \frac{c}{d}\right)\exp\left(-d\frac{z}{L}\right), \tag{10}$$

$$y_o = b\left(\frac{z_0}{L} - \frac{c}{d}\right)\exp\left(-d\frac{z_0}{L}\right). \tag{11}$$

Equations (3), (4) and (5) along with Eqs. (8) and (9) are solved numerically to compute u_*, θ_*, L and hence the fluxes.

(b) Fluxes in terms of bulk Richardson number (R_{iB})

Another stability parameter R_{iB} is readily available from the routine meteorological observations. It is defined as

$$R_{iB} = \frac{g}{\bar{\theta}}\frac{(\theta - \theta_0)(z - z_0)}{U^2}. \tag{12}$$

where the wind speed at z_0 is taken to be zero. The potential temperature at z_0 may be obtained by extrapolating θ from tower observations. Equation (12) expresses R_{iB} in terms of known variables U, θ, θ_0 and z.

R_{iB} can be obtained as a function of z/L from Eqs. (3), (4) and (5) as

$$R_{iB} = \frac{z - z_0}{L}\left[\ln\left(\frac{z}{z_0}\right) - \psi_h\left(\frac{z}{L}, \frac{z_0}{L}\right)\right]\left[\ln\left(\frac{z}{z_0}\right) - \psi_m\left(\frac{z}{L}, \frac{z_0}{L}\right)\right]^{-2}, \tag{13}$$

in which Ψ_m and Ψ_h are defined in Eqs. (8) and (9). The estimated value of R_{iB} from the observations in Eq. (12) can be used in Eq. (13) to calculate z/L numerically using the bisection method (PRESS *et al.*, 1986). Once z/L is known, u_* and θ_* are computed from Eqs. (4) and (5).

3.2. Calculation of Eddy Diffusivities of Momentum and Heat

Eddy diffusivity coefficients (K_M and K_H) can be obtained from the computed values of u_* and θ_* (ARYA, 1988) through

$$K_M = \frac{u_*^2}{\left(\frac{\partial U}{\partial z}\right)}, \tag{14}$$

$$K_H = \frac{u_*\theta_*}{\left(\frac{\partial \theta}{\partial z}\right)}. \tag{15}$$

3.3. Drag and Heat Exchange Coefficients

The surface fluxes are usually expressed in terms of dimensionless drag (C_D) and heat exchange (C_H) coefficients. According to M-O similarity theory, these coefficients are some universal functions of z/z_0 and z/L (ARYA, 1988). The coefficients C_D and C_H as functions of z/L can be expressed as

$$C_D = \frac{k^2}{\left(\ln\left(\frac{z}{z_0}\right) - \psi_m\left(\frac{z}{L}, \frac{z_0}{L}\right)\right)^2}, \tag{16}$$

$$C_H = \frac{k^2}{\left(\ln\left(\frac{z}{z_0}\right) - \psi_m\left(\frac{z}{L}, \frac{z_0}{L}\right)\right)\left(\ln\left(\frac{z}{z_0}\right) - \psi_h\left(\frac{z}{L}, \frac{z_0}{L}\right)\right)}. \tag{17}$$

In general, z/L is not available from routine meteorological measurements whereas R_{iB} can readily be obtained from them. Thus, R_{iB} can be used to obtain z/L, which in turn can be utilized for computing the surface fluxes.

Table 1a

Surface wind speed at Ahemdabad

Month	Time (IST)	Number of days with wind speed ($U \geq 17$ ms^{-1})	Number of days with wind speed ($5.5 \leq U < 17$ ms^{-1})	Number of days with wind speed ($0.27 \leq U < 5.5$ ms^{-1})	Number of days with wind speed ($0 \leq U < 0.27$ ms^{-1})
January	0830	0	1	22	7
	1730	0	2	23	0
February	0830	0	1	20	7
	1730	0	3	25	0
March	0830	0	1	24	6
	1730	0	4	26	1
April	0830	0	3	25	2
	1730	0	5	24	1
May	0830	0	4	26	1
	1730	0	9	22	0
June	0830	0	3	25	2
	1730	0	11	28	1
July	0830	0	2	27	2
	1730	0	7	23	1
August	0830	0	1	27	3
	1730	0	2	27	2
September	0830	0	2	25	3
	1730	0	1	26	3
October	0830	0	0	21	10
	1730	0	1	25	5
November	0830	0	1	23	6
	1730	0	0	24	6
December	0830	0	2	23	6
	1730	0	1	26	4

4. Results and Discussions

4.1. Observational Analysis for Justification of Weak Wind Conditions at Anand

The objective of the present study is to analyze the observations from LASPEX at Anand for weak wind stable conditions. To justify the persistence of weak wind conditions, we first briefly discuss the climatological features at Anand.

4.1.1. Synoptic conditions at Anand

Anand lies between Ahemdabad and Baroda (Fig. 1). The surface wind speed values of these two stations at 0830 IST and 1730 IST are given in Tables 1a & 1b. These values are based on the observations from 1951 to 1980 (Climatological Tables in India 1951–1980, India Meteorological Department). The wind speed data can be divided into four groups (i) 17 m s^{-1} or more; (ii) 5.5–17 m s^{-1}; (iii) 0.27–5 m s^{-1}; and (iv) 0 m s^{-1} or calm winds. For both Ahemdabad and Baroda, maximum number of days (more than 70% at both the times 0830 IST and 1730 IST) reports the wind

Table 1b

Surface wind speed at Baroda

Month	Time (IST)	Number of days with wind speed ($U \geq 17$ ms^{-1})	Number of days with wind speed ($5.5 \leq U<17$ ms^{-1})	Number of days with wind speed ($0.27 \leq U<5.5$ ms^{-1})	Number of days with wind speed ($0 \leq U<0.27$ ms^{-1})
January	0830	0	0	19	12
	1730	0	0	25	6
February	0830	0	0	17	11
	1730	0	0	24	4
March	0830	0	0	19	12
	1730	0	0	27	4
April	0830	0	0	22	8
	1730	0	0	27	3
May	0830	0	0	28	3
	1730	0	1	29	1
June	0830	0	0	28	2
	1730	0	2	27	1
July	0830	0	0	27	4
	1730	0	1	27	3
August	0830	0	0	26	5
	1730	0	1	27	3
September	0830	0	0	23	7
	1730	0	0	25	5
October	0830	0	0	17	14
	1730	0	0	20	11
November	0830	0	0	17	13
	1730	0	0	18	12
December	0830	0	0	20	11
	1730	0	0	21	10

speed values between 0.27–5 m s^{-1}, thus indicating the prevalence of weak wind conditions.

This is further supported by the fact that the southwest monsoon withdraws from the northwestern parts of the country by mid-September. Thereafter a high pressure area establishes itself that becomes most intense during December-January months (Fig. 2). The pressure gradient over the entire region (comprising states of Punjab, Haryana, Rajasthan, Gujarat and parts of Madhya Pradesh, Uttar Pradesh) becomes

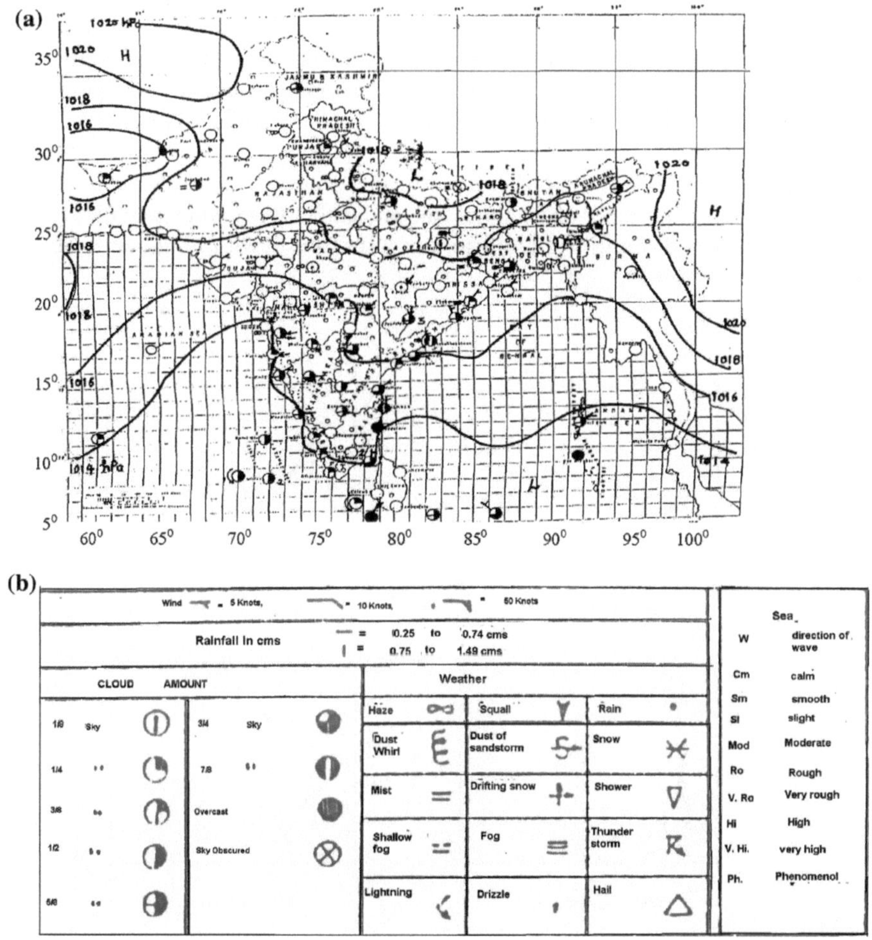

Figure 2

Indian Daily Weather Report, (a) Weather Map at 0830 IST on Saturday November 1, 1997 and (b) Symbols in Figure 2a.

very weak. The surface wind or the upper wind within a kilometer of the surface is usually weak. Thus, the months from October to January can be categorized as 'weak wind regime' months.

On the basis of the above facts it may be assumed that weak wind conditions persist at Anand. To support further, we analyze the observations available at Anand.

4.1.2. Analysis of upper air observations

We have analyzed the wind speed values available from the Radiosonde observations for the months of May, June, October and December, selected for analysis in section 2.2. These observations are available at five synoptic hours only for five days (13 to 17) in each month of 1997. The winds are considered weak when $U_g < 4\,\mathrm{ms}^{-1}$ (Estournel and Guedalia, 1987; Tjemkes and Duyenkerke, 1989; Sharan *et al.*, 1995).

The examination of wind speed profiles for these selected months reveals that U_g is less than 4 m s^{-1} for most of the days, thus indicating weak wind conditions. The wind profiles for a day selected in different months are shown in Figures 3(a–d).

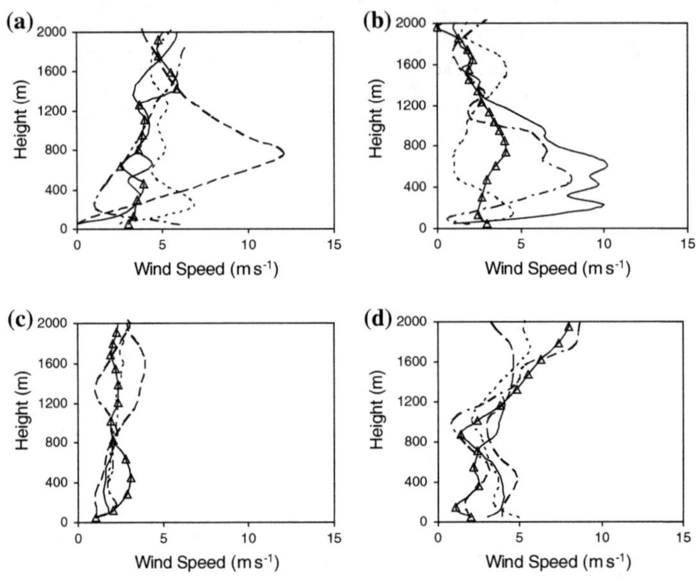

Figure 3
Variation of wind speed with height above ground for a day in the month of (a) May 16 and (b) June 13, (c) October 13, (d) December 15 in the year 1997 at five different hours. (—— 0530 IST, — — — 0830 IST, – – – 1130 IST, – · – · – 1430 IST, —Δ— 1730 IST).

4.1.3. Analysis of surface winds from tower observations

To analyze further, we examine the hourly surface wind speeds (U_S) at 8 m level in the year 1997 at 0530 IST, 0830 IST, 1730 IST and 2330 IST. It is found that for all the hours U_S is observed to be less than 2 m s^{-1} for the maximum number of days (Figs. 4a–d) in most of the months, thus indicating the occurrence of weak wind conditions. During the months of November and December the number of days with U_S < 2 ms^{-1} at 1730 IST is slightly more than those at 2330 IST. For these days, U_S at 2330 IST was in the range 2.0–2.5 m s^{-1}.

The wind roses are plotted using the hourly wind speed at 8 m level available from the tower for each month. The wind speed values are divided into different intervals such as 0–0.5 m s^{-1}, 0.5–2 m s^{-1}, 2–4 m s^{-1}, and 4–6 m s^{-1}. The wind speed values lying in the interval 0–0.5 m s^{-1} are assumed as calm winds and the frequency of occurrence of wind speed less than 2 m s^{-1} is obtained by combining the percentage of wind speed values lying in the intervals 0–0.5 m s^{-1} and 0.5–2 m s^{-1}. It is found that except June and July, the frequency of occurrence of wind speed less than 2 m s^{-1} is more than 50%. This supports our hypothesis of the persistence of

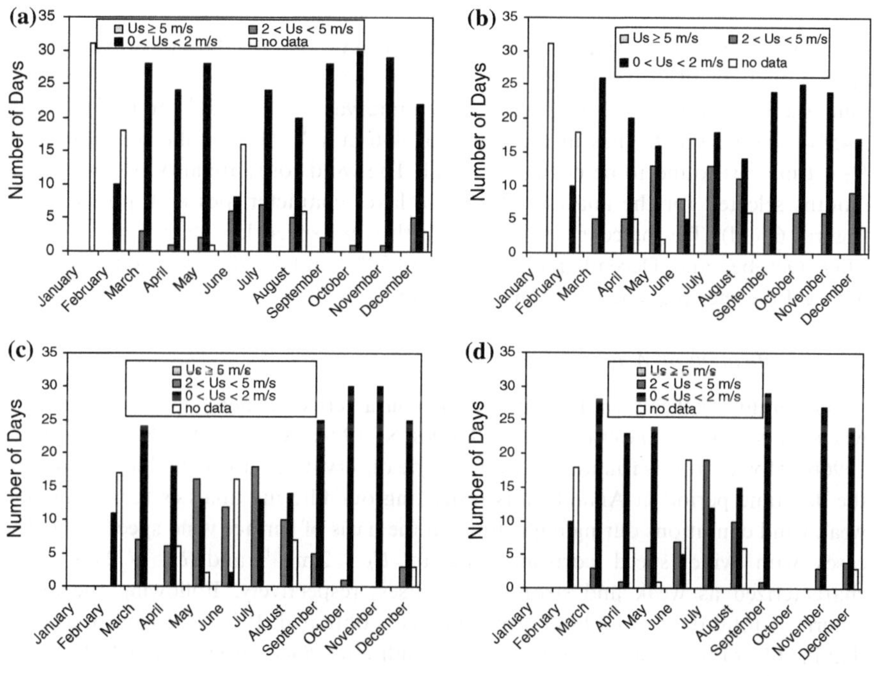

Figure 4
Histograms of surface wind speed at 8 m at Anand for February-December 1997 at (a) 0530 IST, (b) 0830 IST, (c) 1730 IST, and (d) 2330 IST.

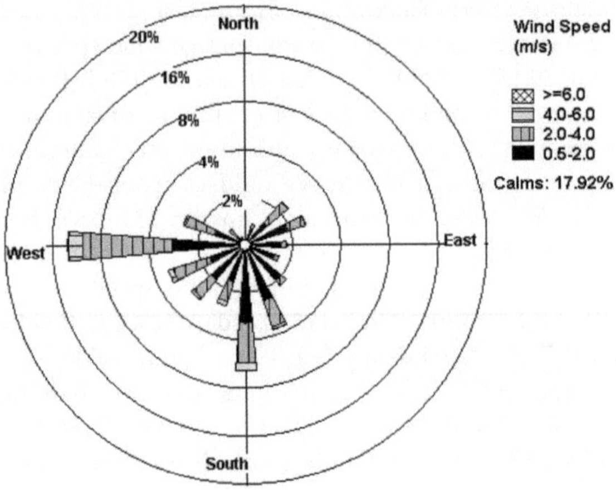

Figure 5
Wind rose for the period of one year (February-December 1997) at Anand (range of calms 0.0–0.5 m s^{-1}).

low wind conditions at Anand. A wind rose is also plotted for the entire year (Fig. 5) and on the basis of its frequency distribution table it is concluded that for 67.50% of time, the U_S is less than 2 m s^{-1}; it is in the interval of 2–4 m s^{-1} for 16.76% time, 3.45% of time the U_S lies in the range of 4–6 m s^{-1}. The number of cases with $U_S < 6\,\mathrm{ms}^{-1}$ is found to be negligibly small. The wind roses are also plotted for the months selected for the analysis of surface layer characteristics and are shown in Figures 6(a–d). The percentage of winds with wind speed less than 2 m s^{-1} during May, October and December is 55.80%, 88.20% and 71.90% respectively, while during June the frequency of occurrence is only 33.80%.

4.2. Classification of Winds

Generally the weak wind conditions are characterized either when $U_g < 4\,\mathrm{ms}^{-1}$ or when $U_S < 2\,\mathrm{ms}^{-1}$ (SEGENDORF and DICKSON, 1974; ARYA, 1995; SHARAN *et al.*, 1996b). However, it is found that Radiosonde observations are available only during the day-time period at Anand, thus restricting our filtering analysis for strong and weak wind conditions during night-time on the basis of surface wind speed only. The cases with wind speed corresponding to $U_S \leq 2\,\mathrm{ms}^{-1}$ and $U_S > 2\,\mathrm{ms}^{-1}$ are characterized as weak and strong wind cases, respectively. Following the above criteria, a total data of 84 hours for 9 nights is obtained. It is clear from Figure 7a that in 86% of the cases under weak wind conditions, R_{iB} is larger than 0.2. R_{iB} is also used to characterize the strong and weak wind conditions (SHARAN *et al.*, 2003b). The points corresponding to $R_{iB} \geq 0.2$ with $U_S > 2\,\mathrm{ms}^{-1}$ and $R_{iB} < 0.2$ with

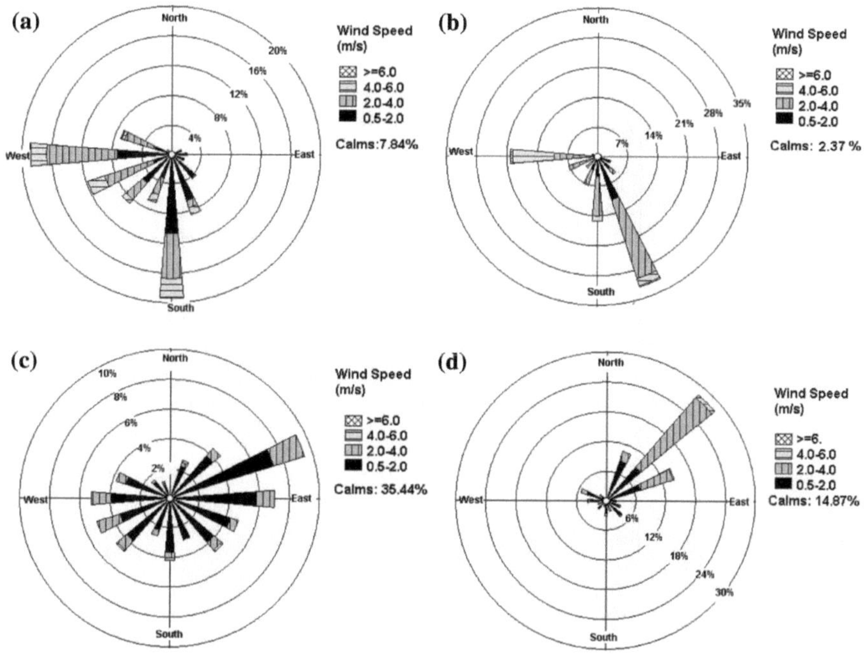

Figure 6
Wind roses for the month of (a) May, (b) June, (c) October, and (d) December (range of calms is 0.0–0.5 m s^{-1}).

$U_S < 2\,\text{ms}^{-1}$ are neglected. Accordingly, we have selected 77 hours (Table 2) out of which 8 hours correspond to strong wind and the remaining 69 hours to weak wind conditions.

An empirical relation between U_s and R_{iB} (Fig. 7a) is obtained by fitting a curve through the values in the range $0.01 < R_{iB} < 3.6$ using Microsoft Excel version (5.0.2195) and is given as

$$U_s = a1(R_{iB})^p, \tag{18}$$

in which $a1 = 0.7412$ and $p = -0.4937$ with the coefficient of determination $R^2 = 0.6778$.

Corresponding to the data given in Figure 7a, a scattered diagram between the observed u_* and R_{iB} is also plotted (Fig. 7b). An empirical relation is fitted between u_* and R_{iB} in the range $0.01 < R_{iB} < 3.6$ using Microsoft Excel version (5.0.2195) and is given as

$$u_* = a2(R_{iB})^q, \tag{19}$$

in which $a2 = 0.0732$ and $q = -0.5271$ with $R^2 = 0.357$.

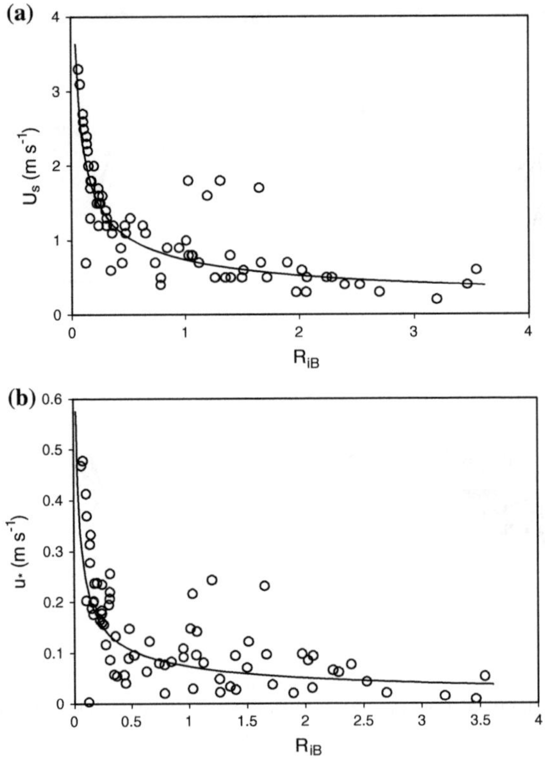

Figure 7
Scattered diagram of (a) surface wind speed (U_s) and (b) friction velocity (u_*) with bulk Richardson number (R_{iB}). Solid line is obtained from the relations $U_s = 0.7412\, R_{iB}^{-0.4937}$ in (a) and $u_* = 0.732\, R_{iB}^{-0.5271}$ in (b), respectively. Symbols (o o o) indicate the observational points.

To justify further the persistence of strong and weak wind conditions for the selected strong and weak wind days given in Table 2, we examine the geostrophic wind speed from the wind profiles available from the Radiosonde observations for these selected days. The wind profiles for the days corresponding to strong wind conditions, i.e., June 16, 17 and 18 are not available nonetheless the examination of wind profile of June 15th reveals $U_g < 4\,ms^{-1}$. However, the upper air observations for these days were examined on the neighboring stations (Ahemdabad and Baroda) and for the majority of hours U_g was found to be less than 4 m s^{-1}. Thus, these cases may not be explicitly falling within the criteria of strong wind conditions.

For weak wind conditions, three days viz. 15–16 May, 13–14 October and 14–15 December are chosen on the basis of the examination of wind profiles available from the Radiosonde observations. The days May 16th, October 13th and December 15th

Table 2

Cases filtered for the analysis in stable conditions

Case	Date	Hours (IST)
Weak	13 May 1997	2000, 2100, 2200, 2300,2400
Weak	14 May 1997	0100, 0200, 0400, 0500
Weak	14 May 1997	2000, 2100, 2200, 2300, 2400
Weak	15 May 1997	0100, 0200, 0300, 0400, 0500, 0600
Weak	15 may 1997	1900, 2000, 2100, 2300, 2400
Weak	16 May 1997	0100, 0200, 0400, 0500, 0600
Weak	16 May 1997	2100, 2200, 2300, 2400
Weak	17 May 1997	0100, 0200, 0300, 0400, 0500
Weak	17 May 1997	1900, 2000, 2100, 2200, 2300, 2400
Weak	18 May 1997	0100, 0200, 0300, 0400, 0500, 0600
Strong	16 June 1997	2000, 2100
Strong	17 June 1997	0100, 0200, 0300
Strong	18 June 1997	2000, 2100, 2200
Weak	13 October 1997	1900, 2100, 2200, 2300, 2400
Weak	14 October 1997	0100, 0200, 0300, 0400, 0500
Weak	14 December 1997	2000, 2100, 2200, 2300, 2400
Weak	15 December 1997	0200, 0400, 0500

are already considered as weak wind days on the basis of an analysis of upper air observations in section 4.1.2. Figure 8 represents the variation of wind speed with height for May 15th, October 14th and December 14th. It is clear from Figure 8 that U_g is less than 4 m s^{-1} for May 15th and October 14th for most of the hours. However, there is some discrepancy on December 14th which may be due to the following facts observed from the IDWR of Gujarat state.

1. The pressure gradient over Gujarat is reduced, as is evident between the pressure difference between Jodhpur and Bombay at 1730 IST.
2. A western disturbance affected the northwestern parts of India (i.e., Punjab, Haryana, Rajasthan, Western U.P.) probably on 13 December and moved eastward on 14 December.
3. Pressure in general shows a rising trend from 14 to 15 December. As the winds at surface or lower levels are weak under high pressure, we again get weak wind conditions on 15 December.

4.3. Surface Layer Characteristics

The surface layer parameters are calculated using hourly averaged wind and temperature at 9.5 m level and temperature at the surface. The hourly averaged surface temperatures are obtained by extrapolating the temperature profile from the tower and the surface roughness length is taken to be 0.1 m (Murthy *et al.*, 2004) for Anand. The value of the von Karman constant is taken as 0.4. Eqs. (3), (4) and (5)

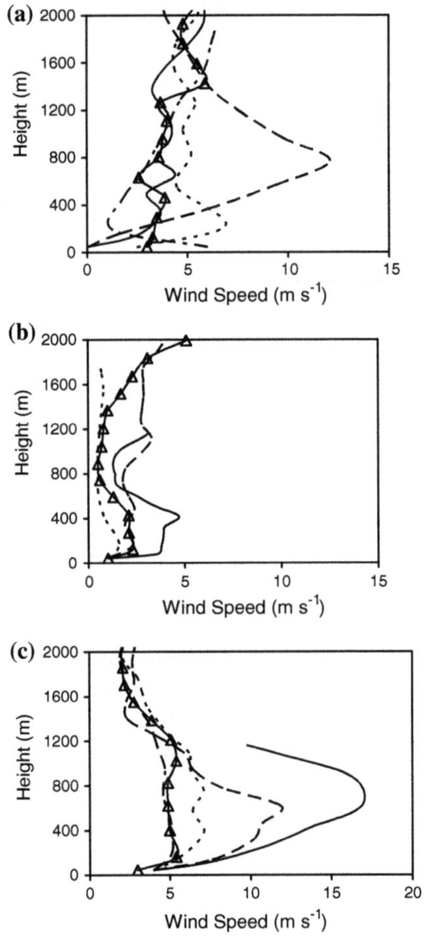

Figure 8
Variation of wind speed with height above ground for (a) May 15th 1997, (b) October 14th and (c) December 14th 1997 at five different hours (—— 0530 IST, — — — 0830 IST, – – – 1130 IST, – · – · – 1430 IST, —Δ— 1730 IST).

are used for computing u_*, θ_* and L, and Eq. (12) is utilized for computing R_{iB}. The eddy diffusivity coefficients of momentum (K_M) and heat (K_H) are obtained using calculated u_* and θ_*. The drag (C_D) and heat (C_H) exchange coefficients are computed using Eqs. (16)–(17) as functions of z/L. The results are discussed in the subsequent sections.

4.3.1. Variation of wind, temperature and R_{iB}

Evolution of hourly wind speed at the 9.5 m level, temperature at the 1 m level and R_{iB} are shown in Figs. 9 (a–c), respectively for a typical case of strong wind (16–17 June) and a weak wind case (13–14 October). R_{iB} is computed using Eq. (12) and it is found less than 0.2 during the period in the strong wind case whereas it is considerably greater than 0.2 in the weak wind case, indicating a strongly stratified surface layer (Fig. 9c).

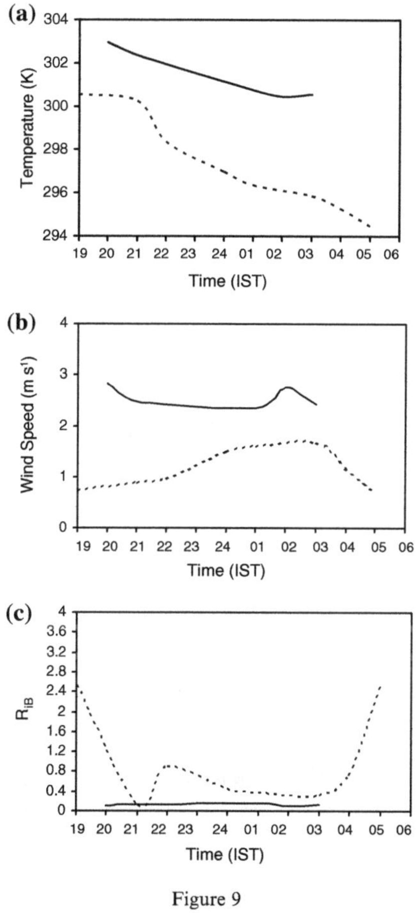

Figure 9
Hourly variation of (a) temperature at 1 m level, (b) wind speed at 9.5 m level, and (c) R_{iB} in strong wind (June 16–17——) and weak wind (October 13–14 -----).

4.3.2. Eddy diffusivity of momentum and heat

The eddy diffusivity coefficients of momentum (K_M) and heat (K_H) are obtained from Eqs. (14) and (15) using the computed values of u_* and θ_*. Figures 10 (a–b) represent the variation of eddy diffusivities (K_M and K_H) with R_{iB}. The magnitudes of K_M and K_H are approximately in the range 0.034 to 0.131 m^2 s^{-1} and 0.033 to 0.130 m^2 s^{-1}, respectively in strong wind conditions, whereas they approximately vary between 1.33×10^{-4} to 0.06 m^2 s^{-1} and 3.2×10^{-5} to 0.05 m^2 s^{-1}, respectively in weak wind conditions. Thus, the magnitudes of K_M and K_H are smaller in weak wind conditions compared to strong wind conditions. Further, K_M and K_H can be computed from the empirical relations proposed by SHARAN *et al.* (2003b) in stable conditions as:

$$K_M = b1(R_{iB})^r, \tag{20}$$

$$K_H = b2(R_{iB})^s. \tag{21}$$

where $b1 = 2.4 \times 10^{-3}$, $b2 = 1.4 \times 10^{-3}$, $r = -1.11$ and $s = -1.22$.

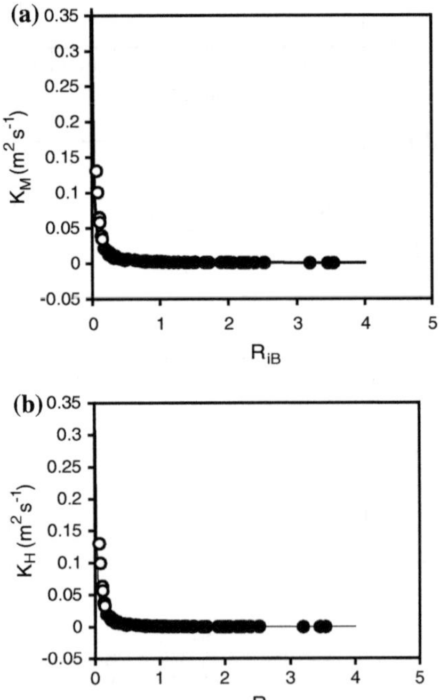

Figure 10

Variation of (a) K_M and (b) K_H with R_{iB} in strong and weak wind conditions. Strong wind ○ ○ ○; weak wind ●●●. Solid line is obtained from the relations $K_M = 2.4 \times 10^{-3} R_{iB}^{-1.11}$ and $K_H = 1.4 \times 10^{-3} R_{iB}^{-1.22}$, respectively.

The values of K_M and K_H obtained from the empirical relations are found to be in the range .02 to .06 m^2 s^{-1} and .016 to .047 m^2 s^{-1}, respectively in strong wind conditions whereas they vary between 5.0×10^{-4} to .03 $m^2 s^{-1}$ and 2.0×10^{-4} to .02 $m^2 s^{-1}$, respectively in weak wind conditions. It is found that the values of K_M and K_H computed using these empirical relationships are close (Figs. 10a and 10b) to those computed from Eqs. (12) and (13).

4.3.3. *Drag and heat exchange coefficients*

The drag (C_D) and heat (C_H) exchange coefficients are computed using Eqs. (16) and (17) as functions of z/L. The variation of C_D and C_H with R_{iB} is shown in Figures 11a and 11b, respectively. The magnitudes of C_D and C_H are approximately in the range 1.5×10^{-3} to 3.0×10^{-3} and 1.4×10^{-3} to 3.6×10^{-3}, respectively in strong wind conditions whereas they approximately vary between 2.5×10^{-5} to 2.4×10^{-3}

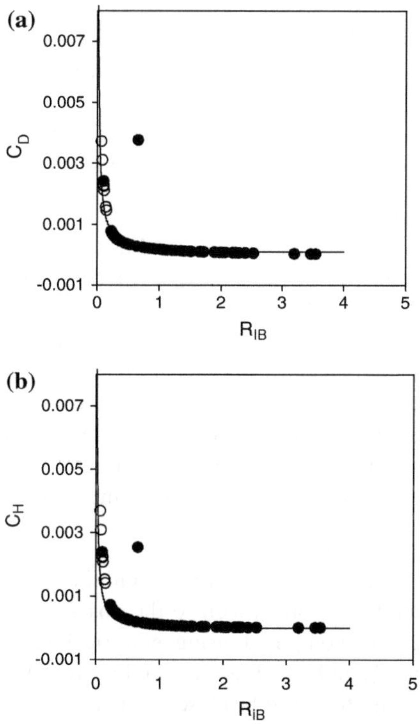

Figure 11

Variation of (a) C_D and (b) C_H with R_{iB} in strong and weak wind conditions. Strong wind ○ ○ ○; weak wind • • •. Solid line is obtained from the relations $C_D = 2 \times 10^{-4} R_{iB}^{-0.87}$ and $C_H = 2 \times 10^{-4} R_{iB}^{-0.97}$, respectively.

and 6.1×10^{-6} to 2.3×10^{-3}, respectively in weak wind conditions. Thus, the magnitudes of C_D and C_H are relatively smaller in weak wind conditions compared to those under strong wind conditions. The empirical relations proposed by SHARAN *et al.*, (2003b) for stable conditions can also be used to compute C_D and C_H as

$$C_D = c1(R_{iB})^t, \tag{22}$$

$$C_H = c2(R_{iB})^u. \tag{23}$$

where $c1 = 2.0 \times 10^{-4}$, $c2 = 2.0 \times 10^{-4}$, $t = -0.87$ and $u = -0.97$. The values of the coefficients C_D and C_H obtained from these relations are found to be in the range 1.2×10^{-3} to 2.7×10^{-3} and 8×10^{-4} to 2.0×10^{-3}, respectively in strong wind conditions whereas they vary approximately in the range 1.0×10^{-4} to 1.5×10^{-3} and 2.6×10^{-5} to 1.0×10^{-3}, respectively in weak wind conditions. It is clear from Figures 11a and 11b that the values of the coefficients C_D and C_H obtained from these relations are close to those computed from Eqs. (16) and (17).

4.3.4. Validation of surface fluxes

The surface layer parameters such as u_* and heat flux are computed using Eqs. (3), (4) and (5) in both strong and weak wind conditions. The entire data set considered here is divided in two groups corresponding to $R_{iB} < 0.2$ and $R_{iB} \geq 0.2$. The points with $R_{iB} \geq 0.2$ correspond to weak wind stable conditions.

(i) Surface fluxes in strong wind stable conditions

For $R_{iB} < 0.2$, the calculated values of u_* obtained from similarity theory are under predicted in comparison to observations (Fig. 12a). However, the computed heat flux is relatively closer to the observations (Fig. 12b). This is known that the computed values of u_* and heat flux for $R_{iB} < 0.2$ corresponding to strong wind conditions are close to observations (SHARAN *et al.*, 2003b). The discrepancy here is attributed to the smaller magnitude of U_g as discussed in section 4.2. These cases may not be explicitly falling within the criteria of strong wind conditions and thus are not included within the further analysis. In the remaining part we restrict our analysis with data points corresponding to $R_{iB} \geq 0.2$, implying weak wind stable conditions.

(ii) Surface fluxes in weak wind stable conditions

For $R_{iB} \geq 0.2$ the magnitudes of both u_* and heat flux obtained using the similarity theory are smaller in comparison to the observations (Figs. 13a and 13b). In addition, the fluxes are computed using empirical relations for K_M and K_H (Eqs. (20) and (21)) and C_D and C_H (Eqs. (22) and (23)). The fluxes computed using these empirical relations are found to be close to each other, as well as close to those computed using the similarity theory (Figs. 13a and 13b). The Normalized Mean Square Error (NMSE) and Fractional Bias (FB) for the similarity theory (Eqs. (3), (4) and (5)), K_M and K_H method (Eqs. (20) and (21)) and C_D and C_H method (Eqs. (22) and (23)) under weak wind conditions are given in Table 3.

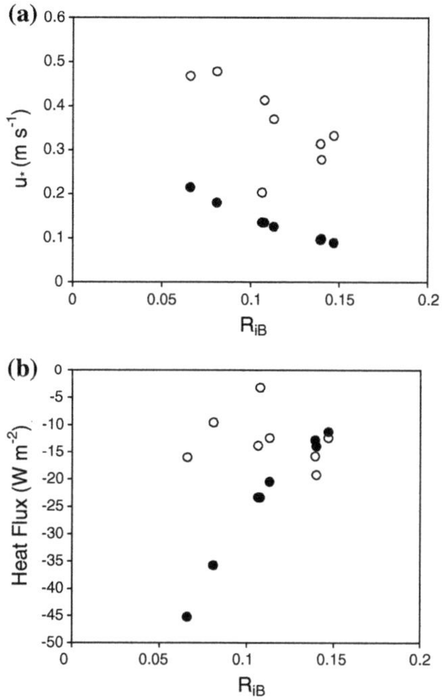

Figure 12
Variation of surface (a) friction velocity and (b) heat flux with R_{iB} in strong wind conditions (computed • • •; observations ○ ○ ○).

To analyze further, the fluxes are computed and compared with observations in three different seasons: summer, autumn and winter in weak wind stable conditions. During each season, computed frictional velocity is smaller than that observed (Figs. 14a, c and e). The heat flux is overestimated during summer (Fig. 14b) and winter (Fig. 14f) seasons. In autumn (Fig. 14d) the computed heat flux is somewhat closer to the observations. The statistical errors such as NMSE and FB are computed with the calculated and observed u_* and heat flux in different seasons and are given in Table 4. The statistical measures reveal that the computed values of u_* and heat flux are close to the observations in autumn and winter seasons. The relatively large values of NMSE and FB for u_* and heat flux indicate their overprediction in the summer season.

Thus we conclude that the computed fluxes show deviations from the observations in weak wind stable conditions. One of the possible reasons for deviations of fluxes from the observations is that the turbulent fluxes are measured at 9.5 m level which may be outside the surface layer in weak winds. NIEUWSTADT (1984) pointed out that in stable conditions the vertical motion is restricted. Consequently the height

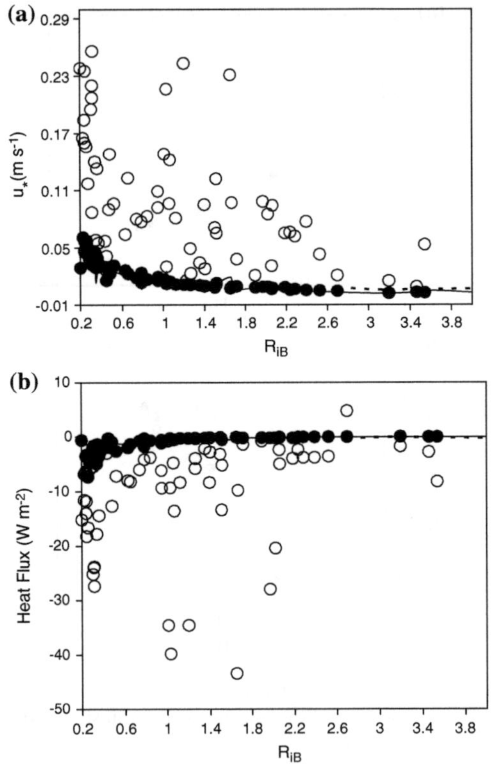

Figure 13

Variation of surface (a) friction velocity and (b) heat flux with R_{iB} in weak wind conditions (similarity theory • • •; empirical relations for K_M and K_H - - - - ; empirical relations for C_D and C_H ——; observations ○ ○ ○).

of the boundary layer and thus of the surface layer is lowered. The use of the local similarity theory above the surface layer was suggested (NIEUWSTADT, 1984). Accordingly, the similarity theory based on the local approach may be used to calculate the surface fluxes. MAHRT (1998) discussed the surface layer in very stable conditions and pointed out that the local similarity theory works better than the traditional M-O similarity theory. Further, it is desirable to validate the fluxes with the turbulence observations at a lower level within the surface layer as and when the data become available.

Another possible reason for deviations of observed fluxes from those based on the observations may be the surface heterogeneities. However, Anand falls in the middle Gujarat agroclimatic zone and the type of soil is loamy sand (PANDEY *et al.*, 2001). In

Table 3

Normalized Mean Square Error (NMSE) and Fractional Bias (FB) for u_* and heat flux for different methods under weak wind conditions

Method of calculation of u_* and heat flux	Statistical Measures	u_*	Heat Flux
Similarity theory with Beljaars and Holtslag, (1991) functions	NMSE FB	3.59 1.21	8.66 1.33
K_M, K_H method	NMSE FB	4.30 1.24	5.29 1.55
C_D, C_H method	NMSE FB	5.83 1.44	5.28 1.82

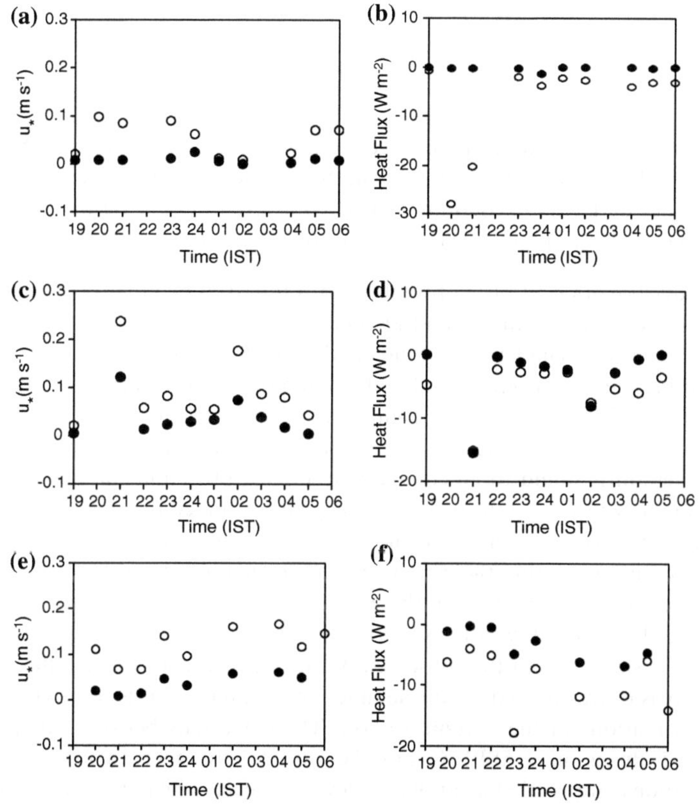

Figure 14

Comparison of surface friction velocity and heat flux for three days in different seasons (a & b) 15–16 May, (c & d) 13–14 October, (e & f) 14–15 December (computed • • •, observed ○ ○ ○).

Table 4

Normalized Mean Square Error (NMSE) and Fractional Bias (FB) for u_* and heat flux in different seasons under weak wind conditions

Parameter	Summer		Autumn		Winter	
	NMSE	FB	NMSE	FB	NMSE	FB
u_*	4.64	1.30	1.17	0.84	1.64	1.05
Heat flux	8.44	1.89	0.47	0.46	1.26	0.87

view of the relatively homogeneous terrain features at Anand, the role of surface heterogeneity is not expected to explain fully the deviations of fluxes from the observations. However, one needs to examine further this aspect.

5. *Conclusions*

The study was undertaken to analyze the observational data from LASPEX over Anand for the weak wind stable conditions and to study the surface layer characteristics under these conditions. The winds are found to be weak for most of the time in the selected period. The main features observed from the present study are:

1. The frequency of occurrence of weak wind is found to be more than 77% in most months except May, August and December.
2. In 86% of the cases under weak wind conditions, R_{iB} is found to be larger than 0.2.
3. The surface fluxes in weak wind stable conditions are computed utilizing three methods: (i) similarity theory, (ii) empirical relations for eddy diffusivities and (iii) empirical relations for drag and heat exchange coefficients. The performance of all these method is found to be comparable to each other but the observations are underpredicted by all of them. Thus, it may be concluded that the traditional M-O similarity theory is not able to simulate the observed fluxes well in weak wind stable conditions during the experiment conducted in a tropical environment.
4. The fluxes are also computed in three different seasons viz., summer, autumn and winter and compared with the observations. It is found that computed frictional velocity is smaller than that observed during each season whereas the computed heat flux is overestimated during summer and winter and it is somewhat closer to the observations in the autumn season. The deviations between computed and observed fluxes in LASPEX may be due to the fact that the turbulence observations are available at 9.5 m level, a level which may be outside the surface layer. Under such conditions the use of local similarity theory (NIEUWSTADT, 1984; MAHRT, 1998) may give better results.
5. Another possible reason for the deviations between observed and computed values of fluxes may be the poor response of wind measuring instruments during light

winds (SHARAN *et al.*, 1996b). The sonic anemometer used for wind measurements is a fast responding and sensitive instrument. It is desirable to improve upon the reliability and accuracy of the meteorological information given by it under weak wind conditions.

The observations in the weak wind stable conditions are very scarce. Thus it is necessary to design a surface and boundary layer experiment in weak wind stable conditions taken care of some of the points mentioned in this study.

Acknowledgements

The authors wish to thank Gr. Capt. O. P. Madan for his assistance in synoptic data analysis. The authors would like to thank Prof. Sethu Raman and Mr. D. R. Sikka for their critical review of the manuscript. The authors also acknowledge appreciatively Dr. Sanjeeva Rao of Department of Science and Technology, India Meteorological Department and Indian Institute of Tropical Meteorology (IITM) Pune for providing the data for the present study. The authors also extend gratitude to the reviewers for their valuable comments.

REFERENCES

AGGARWAL, P., YADAV, A. K., GULATI, A., RAMAN, S., RAO, S., SINGH, M. P., NIGAM, S., and REDDY, N. (1995), *Surface layer turbulence processes in low wind speeds over land*, Atmos. Environ. *29*, 2089–2098.

ANDRE, C., BOUGEAULT, P., and GOURTORBE, J. P. (1990), *Regional estimates of heat and evaporation fluxes over non-homogeneous terrain.* Examples from the HAPBEX-MOBILIHY Programme, Boundary-Layer Meteor. *50*, 77–108.

ARYA, S. P. S., *Introduction to Micrometeorology* (Academic Press, 1988) 307 pp.

ARYA, S. P. S. (1995), *Modelling and parameterization of near-source diffusion in weak winds*, J. Appl. Meteor. *34*, 1112–1122.

BELJAARS, A. C. M. and HOLTSLAG, A. A. M. (1991), *Flux parameterization over land surfaces for atmospheric models*, J. Appl. Meteor. *30*, 327–341.

BUSINGER, J. A., WYNGAARD, J. C., IZUMI, Y., and BRADLEY, E. F. (1971), *Flux profile relationships in the atmospheric surface layer*, J. Atmos. Sci. *28*, 181–189.

CLARKE, R.H., DYER, A.J., BROOK, R.R., REID, D. G., and TROUP, A.J. (1971), *The Wangara Experiment: Boundary Layer Data*, Div. of Meteor. Physics Tech. Paper No. *19*, Commonwealth Scientific and Industrial Research Organization (CSIRO), Australia. 341 pp.

ESTOURNEL, C., and GUEDALIA, D. (1987), *A new parameterization of eddy diffusivities for nocturnal boundary-layer modelling*, Boundary-Layer Meteor. *39*, 191–203.

GARRATT, J.R., *The Atmospheric Boundary Layer* (Cambridge University Press, U.K. 1992) 316 pp.

GOEL, N. and SRIVASTVA, H. N. (1990), *Monsoon trough boundary layer experiment (MONTBLEX)*, Bull. Amer. Meteor. Soc. *71*, 1594–1600.

GOUTORBE, J-P., LEBEL, T., DOLMAN, A.J., GASH, KABAT, P., KERR, Y. H., MONTENY, B. A., PRINCE, S.D., STRICKER, J. N. M., TINGA, A., and WALLACE, J. S. (1997), *An Overview of HAPEX-Sahel: A study in climate and desertification*, J. Hydrol. 188–189, 4–7.

IZUMI, Y. (1971), *Kansas 1968 Field Program Data Report*, Air Force Cambridge Res. Lab. AFCRL-72-0041. Environ. Res. Paper No. 379, Hanscom AFB, MA.79 pp.

KAILAS, S. V. and GOEL, M. (1996), *Planning MONTBLEX- An Overview*, Proc. Indian Acad. Sci. (Earth and Planet . Sci.) *5*, 209–225.

KAIMAL, J. C. and FINNIGAN, J.J., *Atmospheric Boundary Layer Flows- Their Structure and Measurements* (Oxford University Press, Oxford 1994) 289 pp.

KAIMAL, J. C. and WYNGAARD, J. C. (1990), *The Kansas and Minnesota experiments*, Boundary-Layer Meteor. *50*, 31–47.

KUSUMA, G. R. and NARASIMHA, R. (1996), *Estimation of drag coefficient at low wind speeds over the monsoon trough land region during MONTBLEX-90*, Geophys. Res. Lett. *23*, 2617–2620.

KUSUMA, G. R., RAMAN, S., and PRABHU, A. (1991), *Boundary layer heights over the monsoon trough region during active and break phases*, Boundary-Layer Meteor. *57*, 129–138.

KUSUMA, G. R., RAMAN, S., and PRABHU, A., and NARASIMHA, R. (1995), *Turbulent heat flux variation over the monsoon trough region during MONTBLEX-90*, Atmos. Environ. *29*, 2113–2129.

KUNHIKRISHNAN, P. K., SEN GUPTA, K., RAMACHANDRAN, R., PRAKASH, J. W. J., and NAIR, K. N. (1993), *Structure of the thermal internal boundary layer over thumba, India*, Ann. Geophys. 11, 52–60.

MAHRT, L., SUN, J., WILLIAM, B., DELANY, T., and ONCLEY, S. (1998), *Nocturnal boundary layer regimes*, Boundary-Layer Meteor. *88*, 255–278.

MURTHY, B.S., PARASNIS, S.S., and EK, M. (2004), *Interactions of the land surface with atmospheric boundary layer: Case studies from LASPEX*, Current Sci. *86*, 25–30.

NIEUWSTADT, F.T.M. (1984), *The turbulent structure of the stable nocturnal boundary layer*, J. Atmos. Sci. *41*, 2202–2216.

PANDEY, V., KUMAR, M., and SHEKH, A.M. (2001), *Agroclimatic features of LASPEX sites*, J. Agrometeorology *3*, 39–55.

PRAKASH, J. W. J., RAMACHANDRAN, R., NAIR, K. N., SEN GUPTA, K., and KUNHIKRISHNAN, P. K. (1992), *On the structure of sea-breeze fronts observed near the coastlibe of Thumba, India*, Boundary-Layer Meteor. *59*, 111–124.

PRAKASH, J. W. J., RAMACHANDRAN, R., NAIR, K. N., SEN GUPTA, K., and KUNHIKRISHNAN, P. K. (1993), *On the spectral behavior of atmospheric boundary layer parameters at Thumba, India*, Quart. J. Roy. Meteor. Soc. *119*, 187–192.

PRESS, W. H., FLANNERY, B. P., TEUKOLSKY, S. A., and VETTERLING, W. T., *Numerical Recipes: The Art of Scientific Computing* (Cambridge University Press, 1986) 818 pp.

RAMACHANDRAN, R., PRAKASH, J.W. J., SEN GUPTA, K., NAIR, K. N., and KUNHIKRISHNAN, P. K. (1994), *Variability of surface roughness and turbulence intensities at a coastal site in India*, Boundary-Layer Meteor. *70*, 385–400.

SEGENDORF, J. F. and DICKSON, C. R. (1974), *Diffusion under low wind speed inversion conditions*, NOAA Technical Memo-ERL-ARL-52, Air Resources Laboratories, Silver Spring.

SELLERS, P. J., HALL, F. G., ASRAR, G., STREBEL, D. E., and MURPHY, R. E. (1992), *An Overview of the First International Sattelite Land Surface Climatology Project (ISLSCP) Field Experiment (FIFE)*, J. Geophy. Res. *97*, 18345–18373.

SHARAN MAITHILI, GOPALAKRISHNAN, S. G., MCNIDER, R. T., and SINGH, M. P. (1996a), Bhopal gas leak: A numerical investigation of prevailing meteorological conditions, J. App. Meteor. *3*, 1637–1656.

SHARAN MAITHILI and GOPALAKRISHNAN, S. G. (2003), *Mathematical Modelling of Diffusion and Transport of Pollutants in the Atmospheric Boundary Layer*, Pure Appl. Geophys. *160*, 357–394.

SHARAN MAITHILI, MCNIDER, R. T., GOPALAKRISHNAN, S. G., and SINGH, M. P (1995), *Bhopal gas leak: A numerical simulation of episodic dispersion*, Atmos. Environ. *29*, 2061–2070.

SHARAN MAITHILI, RAMA KRISHNA, T.V.B.P.S., and ADITI. (2003a), *On the bulk Richardson Number and flux-profile relations in an atmospheric surface layer under weak wind stable conditions*, Atmos. Environ. *37*, 3681–3691.

SHARAN MAITHILI, RAMA KRISHNA, T.V.B.P.S., and ADITI (2003b), *Surface Layer Characteristics in the Stable Boundary Layer with Strong and Weak Winds*, Boundary-Layer Meteor. *108*, 257–288.

SHARAN MAITHILI, SINGH, M. P., YADAV, A. K., AGGARWAL, P., and NIGAM, S. (1996b), *A Mathematical Model for Dispersion of Air Pollutants in Low Wind Conditions*, Atmos. Environ. *30*, 1209–1220.

SHEIKH, A.M. (1997), *Agroclimatic Atlas of Gujarat, Gujarat Agricultural University, Anand, India.* (AQ: please update the publisher?)

SORBJAN, Z., *Structure of the Atmospheric Boundary Layer* (Prentice Hall, Englewood Cliffs, NJ 1989) 317 pp.

TJEMKES, S. A. and DUYNKERKE, P. G. (1989), *The nocturnal boundary layer: Model Calculations Compared with Observations*, J. Appl. Meteor. *28*, 161–175.

VERNEKAR, K. G., SINHA, S., SADANI, L.K., SIVARAMKRISHNAN, S., PARASNIS, S.S., BRIJ MOHAN, SAXENA, S., DHARAMRAJ, T., PATIL, M. N., PILLAI, J. S., MURTHY, B. S., DEBAJE, S. B., and BAGAVATSINGH, A. (2003), *An overview of the Land Surface Processes Experiment (LASPEX) over a semi-arid region of India*, Boundary-Layer Meteor. *106*, 561–572.

WYNGAARD, J. C. and COTE, O. R. (1971), *The budget of turbulent kinetic energy and temperature variance in the atmospheric surface layer*, J. Atmos. Sci. *28*, 190–201.

WYNGAARD, J. C., COTE, O. R., and IZUMI, Y. (1971), *Local free convection, similarity and budgets of shear stress and heat flux*, J. Atmos. Sci. *28*, 1171–1182.

WYNGAARD, J. C. and COTE, O. R. (1972), *Cospectral similarity in the atmospheric surface layer*, Quart. J. Roy. Meteor. Soc. *98*, 590–603.

YADAV, A.K., RAMAN, S., and SHARAN MAITHILI (1996), *Surface layer turbulence and spectra and eddy dissipation during low winds in tropics*. Boundary-Layer Meteor. *79*, 205–223.

YADAV, A.K., RAMAN, S., and NIYOGI, D. (2003), *A note on estimation of eddy diffusivity and dissipation length in low winds over a tropical urban terrain*, Pure Appl. Geophys. *160*, 395–404.

(Received June 26, 2006, accepted October 26, 2006)
Published Online First: June 19, 2007

Pure appl. geophys. 164 (2007) 1839–1859
0033–4553/07/091839–21
DOI 10.1007/s00024-007-0232-5

Pure and Applied Geophysics

Simulation of Atmospheric Boundary Layer Characteristics during Indian Summer Monsoon using Observations from Monsoon Trough Boundary Layer Experiment at Jodhpur, India

T. V. B. P. S. RAMA KRISHNA[1] and MAITHILI SHARAN[2]

Abstract—The diurnal structure of the boundary layer during Indian summer monsoon period is studied using a one-dimensional meteorological boundary layer model and the observations collected from the Monsoon Trough Boundary Layer Experiment conducted in 1990 at Jodhpur, India. The model was initialized with the observed temperature profiles at 0530 LST on 17 July, 1990 at Jodhpur and was run for 26 hours. The study is carried out with a geostrophic wind speed of 9.5 m s^{-1} corresponding to the strong wind simulation. The mean thermodynamic and wind structure simulated by the model are in good agreement with those observed from 30 m tower. The computed surface layer characteristics such as the surface fluxes, TKE and standard deviations of velocity components are found to be reasonably in good agreement with those based on turbulence measurements. The shear and buoyancy budget computed from the model are also compared with the turbulence measurements. The integrated cooling budget in the nocturnal boundary layer is examined.

Key words: Surface fluxes, turbulence measurements, shear, buoyancy, turbulent kinetic energy, integrated cooling budget.

1. Introduction

The atmospheric boundary layer (ABL) is the lowest region of the atmosphere which experiences surface effects through vertical exchanges of heat, momentum and moisture (PANOFSKY and DUTTON, 1984) and is thus of crucial dynamical and thermodynamical importance. The primary mechanism for exchanging these properties is turbulence, which arises from instability and once initiated tends to maintain itself by horizontal and vertical shears of wind and by thermal convection in which surface properties play an important role (SIKKA and NARASIMHA, 1997). Many of the air pollution activities take place within this layer and thus it is important to examine the structure of the ABL.

Since the work of LETTAU (1932) atmospheric turbulence in the ABL has been extensively studied in the last seven decades from theoretical and modelling

[1] EIRA Division, National Environmental Engineering Research Institute, Nehru Marg, Nagpur – 440 020, India.
[2] Centre for Atmospheric Sciences, Indian Institute of Technology Delhi, Hauz Khas, New Delhi, 110 016, India. E-mail: mathilis@cas.iitd.ernet.in

considerations and through field observations and laboratory measurements (GARRATT, 1982; HOLT and RAMAN, 1988; BELJAARS and HOLTSLAG, 1991; BETTS and BELJAARS, 1993). These studies have often been driven by practical problems such as the dispersion of pollutants, flow past obstacles and micro-climatic variations within growing crops and forest canopies (GARRATT, 1978; PANOFSKY and DUTTON, 1984). The turbulence processes in the ABL have been investigated from several field experiments conducted in mid-latitudes (e.g., Kansas—1968, Minnesota—1973, Kincaid—1980-81, Oklahoma—1983, CASES—1999 at Leon), Wangara experiment—1967 in Australia, Cabauw—1978–1979 experiment in The Netherlands, Land Surface Processes Experiment (LASPEX)—1999 at Anand in India, Urban boundary layer field experiment over Marseille in France in 2001 and LITFASS-2003 experiment at Falkenberg in Germany.

The MONsoon Trough Boundary Layer EXperiment (MONTBLEX) was conducted in 1990 in India to study the structure of the ABL during the Indian summer monsoon (southwest monsoon) season, its variation in time and space, role of eddy fluxes in the maintenance of monsoon trough and energetics of the monsoon trough (SIKKA and NARASIMHA, 1997). The experiment was conducted at Delhi, Varanasi, Jodhpur, Kharagpur and Calcutta over land and over the Bay of Bengal. SIKKA and NARASIMHA (1997) have thoroughly discussed the importance of the monsoon trough, its location and characteristics during the monsoon season.

Several modelling and observational studies of ABL were carried out in the Indian atmospheric conditions. The boundary layer processes under different atmospheric stability conditions were studied through modelling or observations made from several field experiments like MONsoon EXperiment (MONEX)-1979 and routine meteorological data (HOLT and RAMAN, 1985, 1988; MANJU KUMARI, 1985; VERNEKAR *et al.*, 1991; GAMO *et al.*, 1994) and IIT Diffusion experiment (AGGARWAL *et al.*, 1995; YADAV *et al.*, 1996). The boundary layer turbulence was studied using the measurements obtained from the meteorological tower (RAMAN *et al.*, 1990; AGGARWAL *et al.*, 1995; YADAV *et al.*, 1996) and through monostatic sodars (SINGAL *et al.*, 1982; 1993). A high resolution primitive equation one-dimensional model has been used to study the ABL over the monsoon trough region by POTTY *et al.* (1997) and it found that Turbulent Kinetic Energy (TKE) increases from the eastern side of the monsoon trough to the western sector. RAO *et al.* (1997) have compared the simulations of one-dimensional mean structure of the monsoon trough boundary layer with e-ℓ and e-ε closure models with the observations obtained for Delhi and Calcutta during the MONTBLEX-90. Two models with different schemes for parameterizing turbulent fluxes have been used with Kytoon data obtained for Kharagpur during the MONTBLEX-90 for understanding the turbulent structure of the ABL by CHATTERJEE *et al.* (1997). The radiative and turbulence processes in the Stable Boundary Layer (SBL) under weak wind conditions have been examined using a one-dimensional mesoscale meteorological model by GOPALAKRISHNAN *et al.* (1998). SATYANARAYANA *et al.* (2000) have studied the atmospheric boundary layer characteristics at Anand, India using land-surface processes experimental data sets. RAMA KRISHNA *et al.* (2003) have analyzed the mean structure of the Nocturnal Boundary Layer (NBL) under strong and weak wind conditions using the extensive measurements collected during the plume validation experiment conducted by the Electric Power Research Institute at Kincaid (USA). The effect of gentle slopes on a weak and strong-wind NBL was investigated (GOPALAKRISHNAN *et al.*, 2005) using a one-dimensional model, with a simple correction term to account for the terrain effects, very similar to the one used by BROST and WYNGAARD (1978).

Many of the above studies show the role of turbulence and radiation within the ABL. The objective of the present study is to examine the evolution of ABL in terms of mean thermodynamic and turbulence structure; shear and buoyancy budgets and turbulent and radiative cooling budgets within the ABL during the summer monsoon season in India. The diurnal structure of the ABL is analyzed using a one-dimensional mesoscale meteorological boundary layer model and the meteorological data collected during the MONTBLEX-90 at Jodhpur, India. The details of the data used are given in the following section and the description of the numerical model is given in section 3. The results are discussed in section 4 and finally conclusions are presented in section 5.

2. *Observational Meteorological Data*

The meteorological data collected from the MONTBLEX conducted in 1990 at Jodhpur (26.3°N; 73°E), India is used in the present study. The location of Jodhpur with respect to the monsoon trough is shown in Figure 1. The monitored data comprises (i) the turbulence data from Sonic and Gill anemometers, (ii) the slow response data from the 30-m micro-meteorological tower and (iii) the upper air data from Radio-Sonde (RS)/Rawin-Sonde (RW). The details of the data and instruments are given in Table 1.

The fast response/turbulence data comprising the components of wind speed and temperature were collected using the sonic anemometer at 4-m level and the wind speed was also measured using the Gill anemometer at 15 m level. Fast temperature was measured at 2 and 4 m levels using Resistance Thermometer Devices (RTDs). The data were recorded at the rate of 8 samples per second continuously for a sampling period of 15 minutes at pre-determined intervals (RAO, 1997). The slow response data of wind speed, wind direction and temperature were measured at four levels 1, 8, 15 and 30 m on 30 m tower (RUDRA KUMAR *et al.*, 1997). The upper air data comprising the temperature, pressure, wind speed and wind direction was obtained using RS/ RW. The extensive measurements were collected during the

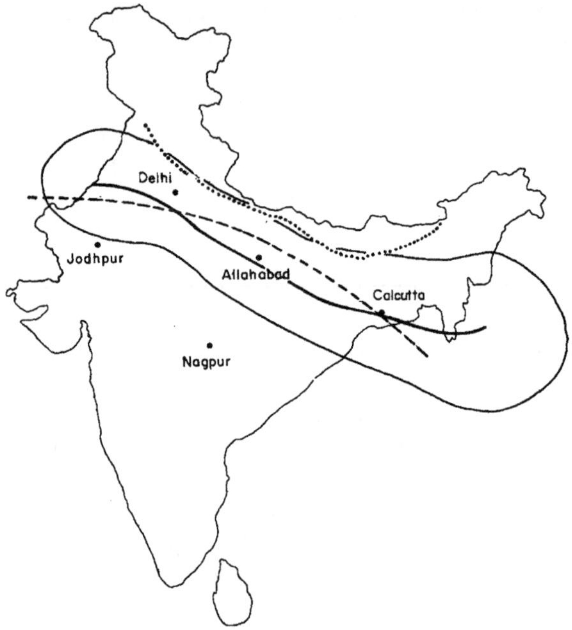

Figure 1
Map showing the location of Jodhpur with respect to monsoon trough. Monsoon trough positions during active (- - - - -) and weak (........) phases with region of one standard deviation (enveloping continuous line) around the mean. [Source: PAUL and SIKKA (1976)].

Intensive Observation Period (IOP) from 17–21 July, 1990 at Jodhpur (RUDRA KUMAR *et al.*, 1997).

The meteorological data for the period from 29 June, 1990 to 30 September, 1990 has been analyzed for choosing the appropriate data for simulating the structure of the diurnal boundary layer during the monsoon period. In general, the geostropic wind is calculated from the network of pressure measurements. In the absence of such measurements, the wind speed at 850 hpa ($\approx$ 1.2 km) level is assumed as the geostrophic wind speed in the present study. Several past studies (GOPALAKRISHNAN *et al.*, 1998; SHARAN *et al.*, 2003) have assumed the geostrophic wind speed to be the wind speed at 850 hpa level. Thus the mean flow at 850 hPa ($\approx$ 1.2 km) was chosen in isolating the weak wind from the strong wind data. A wind speed of less than 4 m s^{-1} at 850 hPa is classified as a weak forcing (RAMA KRISHNA *et al.*, 2003; SHARAN *et al.*, 2003). It was found that no weak wind conditions prevailed during the stable conditions in nighttime during the above period. One case during 17–18 July, 1990 with a geostrophic wind speed of 9.5 m s^{-1} was selected for the present study on the basis of the following criteria: (i) persistence of clear sky conditions; (ii) continuous cooling of near-surface temperatures during the

Table 1

Meteorological data collected during MONTBLEX-90 at Jodhpur (RUDRA KUMAR et al. 1997)

(a) Slow response data

Parameter	Instrument	Levels	Height (m)	Accuracy
Wind speed (m s^{-1})	3 cup anemometer	4	1, 8, 15, 30	1.5–2%
Wind direction (deg.)	Wind vane	4	1, 8, 15, 30	± 3 degrees
Temperature (°C)	Platinum Resistance Thermometer Devices	4	1, 8, 15, 30	± 0.2%
Relative humidity	Humicap, Vaisälä	3	1, 4, 30	± 2%

(b) Fast response data

Parameter	Instrument	Levels	Height (m)	Accuracy
Components of wind speed (m s^{-1})	Sonic	1	4	± 0.05 m s^{-1}
Temperature (°C)	Sonic	1	4	± 1%
Wind speed (m s^{-1})	Gill	1	15	± 1%
Fast temperature (°C)	RTDs	2	2, 4	± 0.2%

(c) Upper air data

Parameter	Instrument	Accuracy	Levels in vertical
Temperature (°C)	Radio sonde/ Rawin sonde	± 0.5°C	At intervals of 50 hPa up to 200 hPa; 20–30 hPa during IOP
Pressure	-do-	± 1 hPa	
Humidity	-do-	± 5% up to troposphere	
Wind speed	-do-	± 1 m s^{-1} below 10 m s^{-1} ± 10% above 10 m s^{-1}	
Wind direction	-do-	± 5% for wind speed above 25 m s^{-1} ± 10% for lower wind speeds	

night; (iii) the temperature difference between the observations at 0000 LST and 0600 LST at 850 hPa is less than 1 K during these individual nights, so that the advection can be neglected (TJEMKES and DUYENKERKE, 1989); (iv) fair weather conditions without any disturbing weather phenomenon. The selected day incidentally falls during the IOP at Jodhpur during the MONTBLEX-90. The synoptic weather situation during this period was western disturbances moving across Himalayas (north of Jodhpur). Though these disturbances exist, no precipitation was reported at Jodhpur during this period. Jodhpur falls in the western side of the monsoon trough, which is of semi-arid climate where the monsoon trough is generally of undisturbed type.

3. Numerical Model

An improved version of the hydrostatic mesoscale meteorological model originally developed by PIELKE (1974) is used in the present study. The current version includes a TKE mixing length closure for boundary layer parameterization (SHARAN and GOPALAKRISHNAN, 1997), a layer-by-layer emissivity based longwave parameterization scheme (GOPALAKRISHNAN *et al.*, 1998; RAMA KRISHNA *et al.*, 2003) and a surface layer formulation based on BELJAARS and HOLTSLAG (1991). Some of the major aspects of the model are described here.

3.1 Model Equations

The following is the set of equations governing a horizontally homogeneous flow in the boundary layer:

$$\frac{\partial u}{\partial t} = fv - fv_g + \frac{\partial}{\partial z}\left[K_M \frac{\partial u}{\partial z}\right], \tag{1}$$

$$\frac{\partial v}{\partial t} = -fu + fu_g + \frac{\partial}{\partial z}\left[K_M \frac{\partial v}{\partial z}\right], \tag{2}$$

$$\frac{\partial \theta}{\partial t} = \frac{\partial}{\partial z}\left[K_H \frac{\partial \theta}{\partial z}\right] + F_\theta, \tag{3}$$

$$\frac{\partial q}{\partial t} = \frac{\partial}{\partial z}\left[K_H \frac{\partial q}{\partial z}\right], \tag{4}$$

$$\frac{\partial E}{\partial t} = K_M\left(\left[\frac{\partial u}{\partial z}\right]^2 + \left[\frac{\partial v}{\partial z}\right]^2\right) - \frac{g}{\theta}\left(K_H \frac{\partial \theta}{\partial z}\right) + C\frac{\partial}{\partial z}\left[K_M \frac{\partial E}{\partial z}\right] - \frac{[0.2E]^{3/2}}{\ell}, \tag{5}$$

where u and v are the velocity components of the wind along east-west and north-south directions, respectively; u_g and v_g are the components of the geostrophic wind; θ is the potential temperature; q is the specific humidity; f is the Coriolis parameter; K_M and K_H are, respectively, the vertical exchange coefficients of momentum and heat, related to the Turbulence Kinetic Energy (E) by MONIN and YAGLOM (1971) dimensional argument (i.e., $K_M = K_H = 0.4\,\ell E^{1/2}$); g is the acceleration due to gravity; C is a constant taken to be 1. The mixing length, ℓ is determined in terms of a simple diagnostic relationship (BLACKADAR, 1962; SHARAN and GOPALAKRISHNAN, 1997); F_θ is the source term that includes contributions from short- and longwave heating/cooling processes and is expressed as

$$F_\theta = \frac{1}{\rho C_p}\frac{\partial R_N}{\partial z} \tag{6}$$

in which R_N is net radiative flux, ρ is the density of air and C_p is the specific heat capacity at constant pressure. The shortwave radiation parameterization is discussed in MAHRER and PIELKE, (1977). Many authors (e.g., MCNIDER and PIELKE, 1981) adopted SASAMORI's (1972) radiation scheme in an isothermal atmosphere for its simplicity and computational efficiency for the parameterization of longwave fluxes in the NBL. However in the present study, a scheme based on emissivity approach (MAHRER and PIELKE, 1977) in which water vapor and carbon dioxide are considered as principle emitters of longwave radiation (GOPALAKRISHNAN *et al.*, 1998) is adopted.

Equations (1) to (5) were approximated by a version of Crank Nicholson scheme (PIELKE, 2002). Instead of the usual equal weighing at the present time steps, the new time step is overweighed. In the present study, the weighing factor has been taken as 0.75 (PAEGLE *et al.*, 1976).

3.2. Initial and Boundary Conditions

In order to solve the equations (1) to (5), initial conditions for the field variables and their values at the upper and lower boundaries are required. The observed temperature and dew point temperature profiles obtained from RS/RW measurements at Jodhpur site were used to initialize the model up to a height of 6623 m and are shown in Figure 2.

The one-dimensional and three-dimensional versions of the mesoscale meteorological model were used with local closure for eddy diffusivities (MCNIDER *et al.*, 1988). The details of the three-dimensional model can be found in SHARAN *et al.* (1996). The surface temperatures were computed using the surface energy balance. Both versions of the model were run with the same initial conditions. The comparison of potential temperature profiles at 00 LST (Fig. 3a) and at 06 LST (Fig. 3b) show that temperature advection is reasonably negligible. As a result of this, the one-dimensional version of the model has been used in the numerical simulations described hereafter. The TKE-mixing length closure is used to parameterize the eddy diffusivities in the one-dimensional model and the initial condition for TKE was set to 0.001 $m^2\ s^{-2}$. The one-dimensional model is based

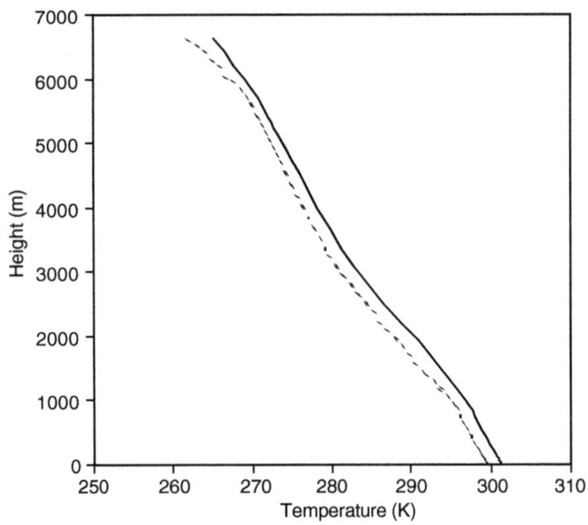

Figure 2
Initial temperature (——) and dew point temperature (.........) profiles used for simulation.

on the TKE-mixing length closure to parameterize the eddy diffusivities, an updated radiation scheme and BELJAARS and HOLTSLAG (1991) profiles for parameterizing the nondimensional temperature and wind profiles in the surface layer. Since the boundary layer schemes are more advanced in the one-dimensional model, we have adopted the one-dimensional model in the present study.

The lower and upper boundary conditions required to solve the equations for the mean variables (equations (1)–(4)) were obtained from the observations. The temperature at the surface at each time step was obtained by interpolating the hourly surface temperatures. The temperature at the surface is extrapolated from the temperatures obtained from tower data (temperature at 1, 8, 15, 30 m levels). A no slip condition was assumed for the components of velocity at the ground. The surface specific humidity at the lower boundary was prescribed initially from the observations and held constant during the simulation. At the upper boundary, the geostrophic wind, the temperature and the specific humidity were prescribed initially from the upper air data and were held constant throughout the simulation and TKE was set to zero. The surface albedo of 0.2 and the surface roughness of 15 cm are considered. The surface roughness length computed from the observations varies from 0.5 to 26 cm on different hours of different days (RAO, 1997). An average value of 15 cm has been taken in the analysis. The surface-cooling rate is prescribed from observations and the surface specific humidity is taken as

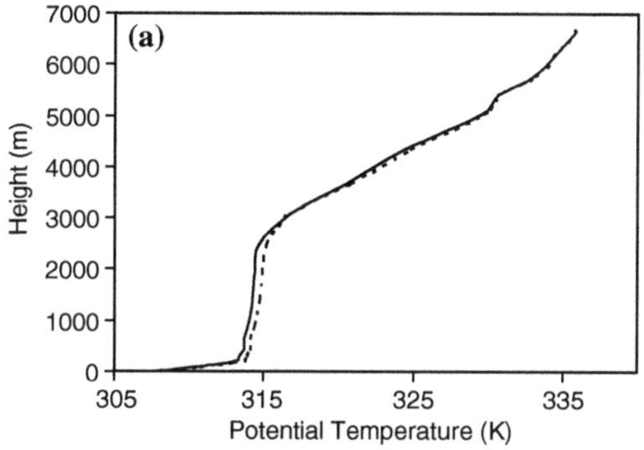

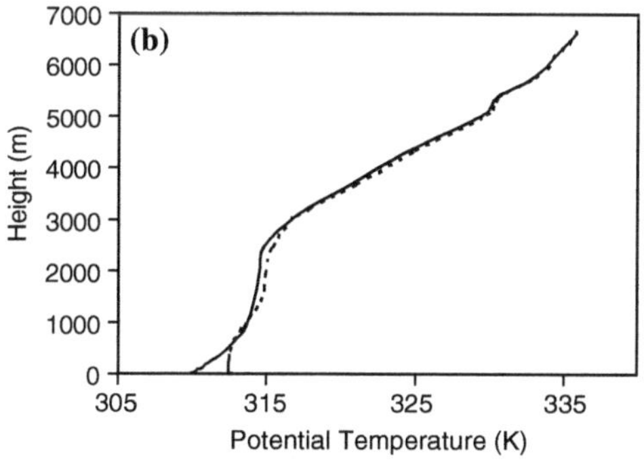

Figure 3
Comparison of potential temperature profiles at (a) 0000 LST and (b) 0600 LST. 1-D model (—); 3-D model (........).

0.0073 kg·kg^{-1}. The time step used for integration is 30 s and the model is run with 22 levels in the vertical up to a height of 6623 m. The values of input parameters used in the numerical simulations are given in Table 2.

Table 2

List of input parameters used in the numerical simulations

List of input parameters	Date: 17–18 July, 1990
Day of (Julian) Year	119
Surface pressure (hPa)	973.6
Surface temperature (K)	302.15
Mean wind speed (m s^{-1})	9.5
Mean wind direction (°)	345
Simulation start time (LST)	0530
Total simulation time (hours)	26
Model levels in meters	2, 10, 25, 50, 80, 120, 170, 220, 315, 438, 650, 916, 1416, 1940, 2488, 3067, 3681, 4335, 5039 5400, 5800, 6623

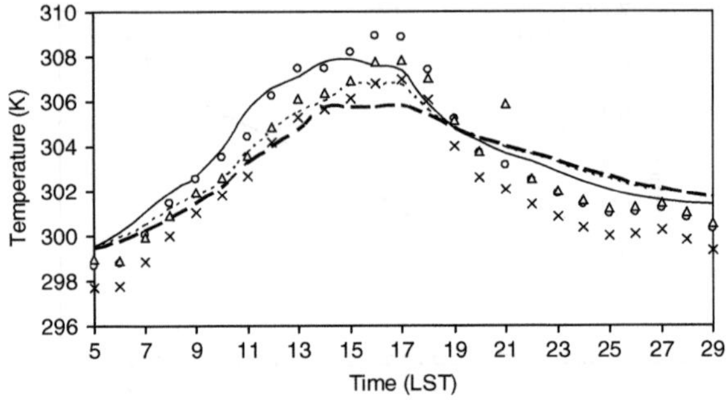

Figure 4
Diurnal variation of temperature at three levels at Jodhpur during MONTBLEX-90. Simulated: 2 m ——; 15 m ....; 30 m – – – – –; Observed: 2 m ○ ○ ○○; 15 m × × ××; 30 m ΔΔΔΔ.

The lower boundary condition required to solve Eq. (5) for TKE was obtained from the relationship (HOLT and RAMAN, 1988)

$$E = \begin{cases} 3.75u_*^2; & z/L > 0 \\ 3.75u_*^2 + 0.2w_*^2 + (-z/L)^{2/3}u_*^2; & z/L < 0 \end{cases}, \qquad (7)$$

where u_* is the surface friction velocity, w_* is the convective velocity scale and L is the Obukhov length.

4. *Results and Discussion*

For the simulations, the model was initialized at 0530 LST on 17 July 1990 and was run for 26 hours. The simulated results are compared with those based on the measurements obtained from the MONTBLEX-90 at Jodhpur.

4.1. *Evolution of Temperature and Wind Speed*

Figure 4 illustrates the diurnal variation of temperatures simulated by the model and observed at 2, 15 and 30 m levels on a 30 m micrometeorological tower. The model simulated temperatures at 15 m and 30 m were obtained from the interpolation of temperatures at model levels 10, 25 and 50 m. It is seen from the observations that the temperature at 15 m level is less than the temperature at 30 m during both daytime and nighttime. This would imply that there is considerable stability between 15 m and 30 m at all hours which is not expected. The temperature at surface is found to be more than the temperature at 15 m level at all hours, implying slightly unstable conditions even during night time immediately above ground.

We have thoroughly examined in detail the actual measured values of the temperatures by the sensors at 1 m, 2 m, 8 m, 15 m and 30 m levels. The observations show a superadiabatic lapse rate between 1 to 2 m, a small inversion between 2 and 8 m, substantial fall in temperature between 8 and 15 m and again inversion between 15 to 30 m at all hours. This is not physically realistic behavior and it appears that the sensors are not working properly at 8 and 15 m levels. We have examined the data for the period July 13–30, 1990 and found the similar behavior suggesting the malfunctioning of the sensors at 8 and 15 m levels. Thus, for the further analysis, we have considered the data at 2 and 30 m levels.

The observed diurnal amplitudes between 0500 to 1700 hours at 2 m, 15 m and 30 m are 10.2, 9.3 and 8.9 K, respectively whereas the corresponding computed amplitudes are 8.4, 7.2 and 6.3 K. The computed values are somewhat lower than those observed and the differences between computed and observed values increase from 0800 to 1600 hours during daytime. The computed values are considerably higher than those observed during nighttime at 2 m. This shows that the model is producing relatively less warming during the daytime and reduced cooling at nighttime compared to observations. The environment of the station is semi-desert (sandy soil) and as such it is expected that the daytime warming and nighttime cooling both would be faster. The model is not able to account for this aspect. Apparently, it would require a sensitivity analysis for tuning the parameters to account this near-surface difference between observed and computed temperatures.

As we have suspected the sensor at 15 m level, the computed temperatures are compared with those observed at 30 m level. Figure 4 reveals that the model is able to capture the observed behavior very well at all hours at 30 m level. Simulated results also show that the temperature decreases during the daytime with the height whereas it increases during nighttime implying that the predicted results are consistent with those expected in unstable and stable atmosphere close to the surface.

The diurnal variation of wind speed simulated by the model and the corresponding observations at 2 and 30 m levels is shown in Figure 5. The predicted winds are found to be close to those measured at 2 m level, whereas the model

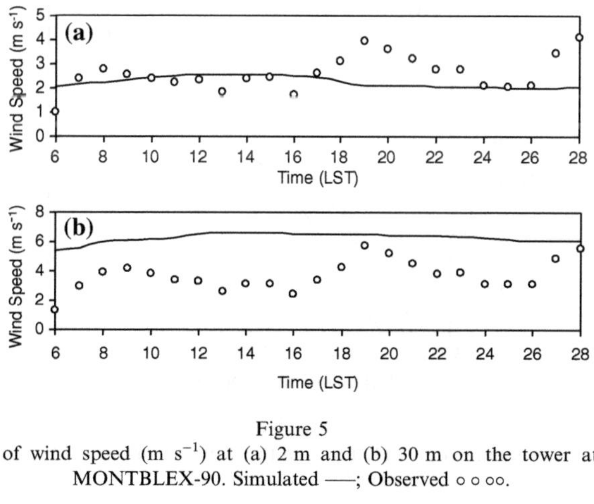

Figure 5
Diurnal variation of wind speed (m s^{-1}) at (a) 2 m and (b) 30 m on the tower at Jodhpur during MONTBLEX-90. Simulated —; Observed o o oo.

overpredicts the wind speed at 30 m level. The overprediction of wind speed at 30 m level may be due to the uncertainties associated with the specification of geostrophic wind and surface roughness length. An oscillatory behavior is seen in the observations but the simulated winds do not show these oscillations. The oscillations in the observations are found with a period of 6 h which is much smaller than the period $2\pi/f$ of an inertial oscillation (RAMA KRISHNA *et al.*, 2003).

Figure 6 shows the profiles of simulated and observed potential temperatures at 2240 LST (Fig. 6a) and at 0510 LST (Fig. 6b). The predicted potential temperatures are reasonably close to the corresponding observed profile at 2240 LST, whereas at 0510 LST the model overpredicts the temperatures. This may be due to the excessive cooling as the model uses a prescribed cooling rate through observed surface temperatures. The mixed layer shown by the model simulated temperatures may be

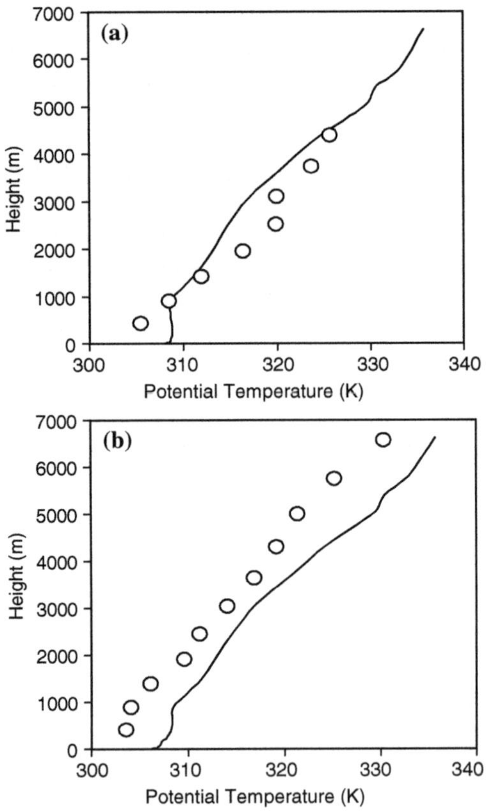

Figure 6
Profiles of potential temperature at (a) 2240 LST and (b) 0510 LST during 17–18 July, 1990 at Jodhpur. Simulated —; Observed o o oo.

due to the slightly unstable conditions at the surface and slightly stable conditions at the elevated layers as shown in Figure 4. The substantial cooling indicated by the observations may be due to cold advection. The potential temperature profiles are compared only at two hours due to availability of observed data only at these hours. Both predicted and observed values show cooling during the night.

4.2. *Surface Layer Characteristics*

The surface layer characteristics such as surface friction velocity, kinematic heat flux and TKE simulated by the model are compared with those observed from turbulence measurements.

(a) Surface fluxes

The diurnal variation of the simulated surface friction velocity (u_*) and the kinematic heat flux and the corresponding measured fluxes are shown in Figures 7a and 7b, respectively. The turbulence measurements for u_* are obtained from the sonic anemometer at 4-m level and also from Gill anemometer at 15-m level. The heat flux measurements are obtained from the sonic anemometer at 4-m level. The model simulated u_* values are closer to those obtained from Gill anemometer than from sonic anemometer (Fig. 7a).

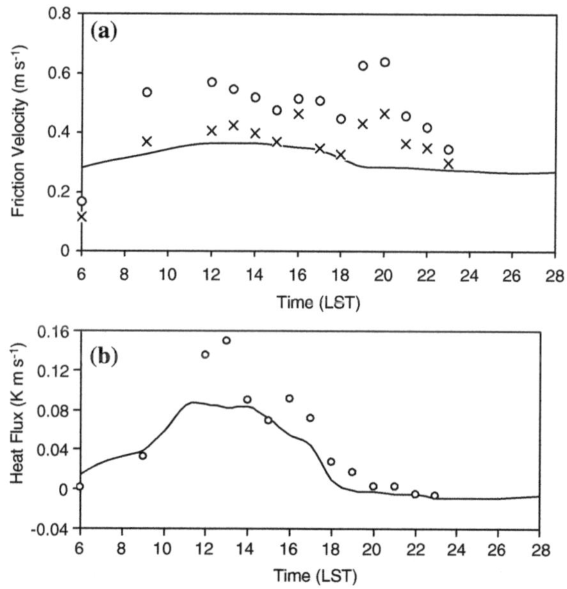

Figure 7
Diurnal variation of (a) surface friction velocity and (b) kinematic heat flux during 17–18 July, 1990 at Jodhpur. Simulated —; Observed ∘ ∘ ∘∘ (Sonic), × × ×× (Gill anemometer).

There is a reasonably good agreement between the simulated and the observed heat fluxes (Fig. 7b) except in a couple of hours around noon. It may be concluded that the model is able to produce the surface fluxes reasonably well.

(b) TKE

The TKE is one of the most important variables in micrometeorology because it is a measure of intensity of turbulence. It is directly related to the momentum, heat and moisture transport throughout the boundary layer. TKE is also sometimes used as a starting point for approximation of turbulent diffusivities (STULL, 1988).

The TKE is estimated from the turbulence measurements as:

$$\mathrm{TKE} = 0.5(\overline{u'^2} + \overline{v'^2} + \overline{w'^2}). \tag{8}$$

The model simulated TKE at the surface is compared (Fig. 8) with that obtained from the turbulence measurements at 4-m level. The model overpredicts TKE during the daytime convective conditions whereas the simulated TKE is close to that observed during nighttime stable conditions. The model failed to simulate the peak values of TKE compared to observations as the model predicted winds near the surface also show no peak values (Fig. 5a).

4.3. Shear and Buoyancy Budget

The shear and buoyancy budgets in the NBL up to a height (h_t) where TKE is 5% of that at the surface are computed. Integrating the TKE equation (RAMA KRISHNA *et al.*, 2003) with respect to z from 0 to h_t, we obtain

$$S = \frac{1}{h_t}\int_0^{h_t} K_M\left[\left(\frac{\partial u}{\partial z}\right)^2 + \left(\frac{\partial v}{\partial z}\right)^2\right]dz, \tag{9}$$

$$B = -\frac{g}{h_t}\int_0^{h_t}\left[\frac{K_H}{\theta}\frac{\partial \theta}{\partial z}\right]dz. \tag{10}$$

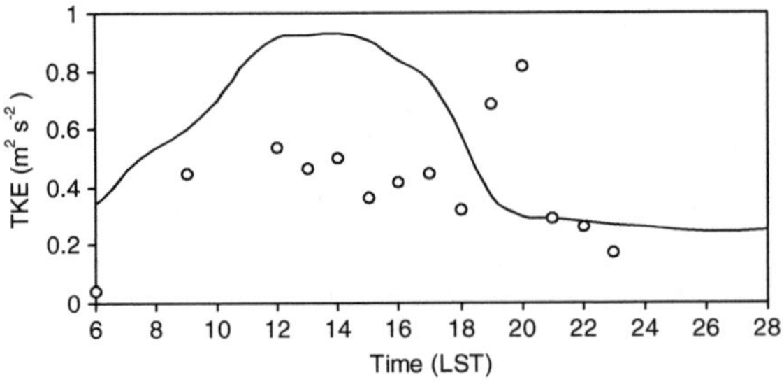

Figure 8

Diurnal variation of TKE during 17–18 July, 1990 at Jodhpur. Simulated —; Observed o o oo.

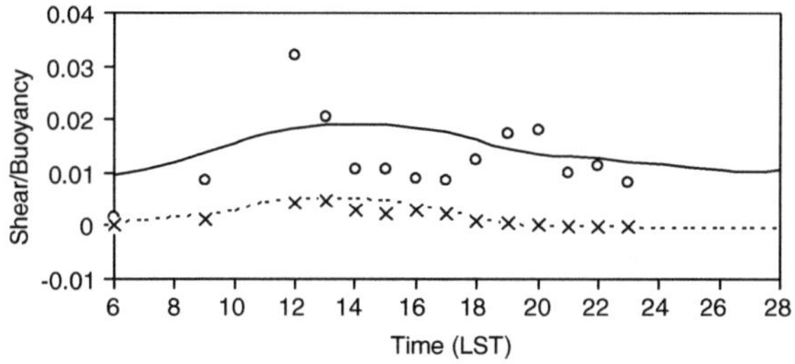

Figure 9
The shear and buoyancy budget in the SBL. Simulated shear ———; buoyancy ....; Observed shear ○ ○ ○ ○; buoyancy × × × ×.

Relations (9) and (10) refer to the shear and buoyancy production terms, respectively. The negative value of B represents the buoyancy consumption. The relative contribution of the shear and buoyancy to the total TKE is examined. Integrals in (9) and (10) are evaluated numerically (PRESS *et al.*, 1986). The simulated values of shear production and buoyancy consumption in the SBL are found to be closed (Fig. 9) to those obtained from turbulence measurements. The shear production is found to dominate over the buoyancy consumption in the SBL. The shear and buoyancy budgets are in accordance with those obtained in an earlier study (RAMA KRISHNA *et al.*, 2003).

4.4. Standard Deviations of Velocity Components

The turbulence in terms of standard deviations of velocity components, namely, σ_u, σ_v and σ_w are computed using several formulae available in literature for unstable/convective as well as stable conditions.

4.4.1. Unstable/convective conditions

The expressions for σ's in unstable conditions are given by ARYA (1999)

$$\sigma_{u,v} = u_*\left(12 - 0.5\frac{z_i}{L}\right)^{1/3}, \tag{11a}$$

$$\sigma_w = 1.3u_*\left(1 - 3\frac{z}{L}\right)^{1/3}, \tag{11b}$$

in which z_i is the PBL thickness.

4.4.2. Stable conditions

The expressions for σ_u and σ_v (HANNA, 1982) and σ_w (NIEUWSTADT, 1984) in stable conditions are given by

$$\sigma_u = 2u_*\left(1 - \frac{z}{h_t}\right), \tag{12a}$$

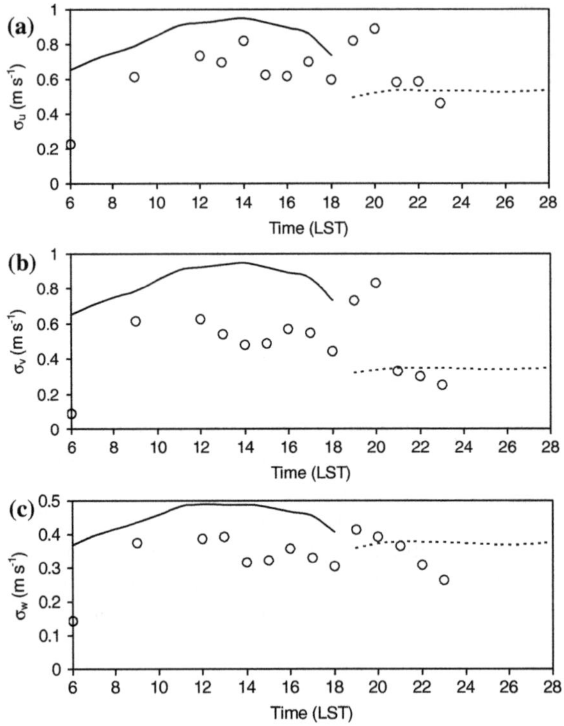

Figure 10
Diurnal variation of (a) σ_u, (b) σ_v and (c) σ_w during 17–18 July, 1990 at Jodhpur. Computed values in unstable (—) and in stable (.........) conditions; Observed oooooo.

$$\sigma_v = 1.3u_*\left(1 - \frac{z}{h_t}\right), \tag{12b}$$

$$\sigma_w = 1.4u_*\left(1 - \frac{z}{h_t}\right)^{3/4} \tag{12c}$$

in which h_t is the height of the turbulent boundary layer.

Computed values of σ's are calculated from the relations (11–12) using the model simulated values of u_*, z_i and h_t. The computed values of σ_u (Fig. 10a), σ_v (Fig. 10b) and σ_w (Fig. 10c) are found to be reasonably in good agreement with those based on turbulence measurements. It is observed that the computed values of σ_u, σ_v and σ_w are slightly higher in unstable conditions, whereas they are reasonably close to the turbulence measurements in stable conditions. Figure 10 shows that the turbulence increases during daytime, reaching a maximum value around noon and decreasing thereby towards sunset. After sunset, the turbulence is maintained at almost a constant value throughout the night in NBL.

4.6. Integrated Cooling Budget

The integrated cooling budget is examined within the surface inversion layer. The individual contribution of turbulent and radiative cooling within the inversion layer of depth h is obtained by integrating the thermodynamic energy equation (3) taking into consideration that the turbulent heat flux vanishes at the top of the inversion layer, we obtain

$$\overline{\left(\frac{\partial\theta}{\partial t}\right)} = \frac{1}{h}\overline{(w'\theta')}|_0 + \frac{R_N|_h - R_N|_0}{\rho C_p h}, \tag{13}$$

where $\overline{(w'\theta')}|_0$ is the turbulent heat flux at the surface, $R_N|_h$ is the net radiative flux at $z = h$ and $R_N|_0$ is the corresponding flux at the surface.

The inversion depth in an ideal fair weather NBL is determined as the height above the surface at which the local potential temperature gradient (Γ) vanishes. For computational purposes, it is taken to be the height where Γ is approximately equal to a pre-assigned small value δ (GOPALAKRISHNAN *et al.*, 1998). The value of δ is taken to be 3.5×10^{-3} K m^{-1} as used by ANDRE and MAHRT (1982).

The first term on the right-hand side of Eq. (13) represents the integrated turbulent cooling and the second term gives the radiative cooling. It may be noted that the heat flux at the surface contributes to the turbulent cooling.

The role of turbulent and radiative cooling in the NBL is examined under strong wind with a geostrophic wind speed of 9.5 m s^{-1} (actual simulation made for the present study as given in Table 2) and under weak wind with a geostrophic wind speed of 4 m s^{-1} (assumed for weak wind). The weak wind simulation is carried out with the same initial and boundary conditions along with the input data as given in Table 2 only to find out the role of turbulent and radiative cooling within the NBL when wind becomes weak.

Figure 11 shows that the turbulent cooling dominates over the radiative cooling under strong wind conditions, whereas the radiative cooling is found to be dominating over the turbulent cooling when wind becomes weak. However, the difference (in magnitude) between the radiative cooling and turbulent cooling in the weak wind NBL is not that significant as that in the case of strong wind NBL. This result is in conformity with the earlier findings (GOPALAKRISHNAN *et al.*, 1998; RAMA KRISHNA *et al.*, 2003) in which it was shown that turbulent cooling dominates over radiative cooling in a strong wind NBL, whereas the radiative cooling dominates over turbulent cooling in weak wind NBL.

Based on the above results it may be concluded that the one-dimensional mesoscale meteorological model used in the present study is able to reasonably predict the diurnal variation of the atmospheric boundary layer. The one-dimensional model used in the present study assumes that the flow is horizontally homogeneous and thus, the effect of advection is neglected, which is ascertained by comparing the potential temperature profiles predicted using the 1-D and 3-D

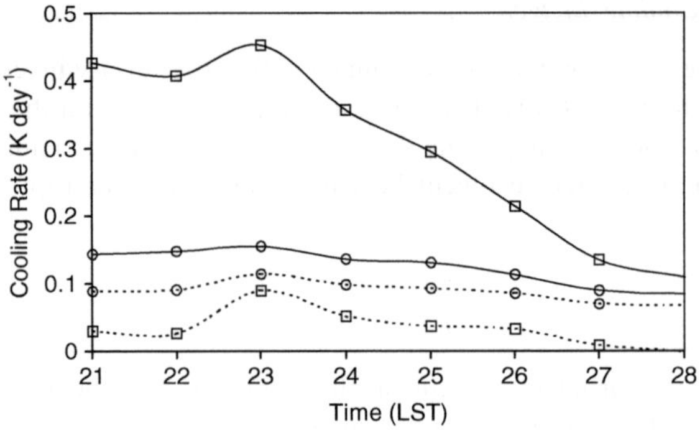

Figure 11
The integrated cooling budget in the NBL. Strong wind case: Turbulent cooling –□–□–, Radiative cooling –o–o–; Weak wind case: Turbulent cooling ---□---□---, Radiative cooling ---o---o---.

models. The role of advection may be examined using a three-dimensional model further. The hourly surface temperatures are prescribed externally from observations and the surface temperature at each time step is obtained by linearly interpolating the temperature at two consecutive hours. The surface energy balance may be used for computing the surface temperatures in the future.

The one-dimensional model used in the present study takes into account all the physical aspects including the (i) improved parameterization of the surface layer using BELJAARS and HOLTSLAG (1991) profiles for parameterizing the nondimensional temperature and wind profiles in the surface layer, (ii) TKE for turbulent parameterization and (iii) an updated radiation parameterization scheme for parameterizing the longwave radiation. Such a detailed model has been used to simulate the boundary layer features observed during the experiment. In addition, the terrain features at Jodhpur are relatively flat. Thus the application of the one-dimensional model is justified for understanding the diurnal variation of boundary layer parameters in the present study. At present the three-dimensional model is available which is based on Blackadar's formulation for turbulence parameterization (BLACKADAR, 1962) rather than TKE and it uses the Bussinger's linear profile for parameterizing the surface fluxes (BUSINGER *et al.*, 1971) and a radiation scheme based on isothermal atmosphere. However, with updating of the three-dimensional mesoscale model, it will be used to analyze the observations.

5. Conclusions

The diurnal structure of the boundary layer during the Indian summer monsoon period is studied using a one-dimensional meteorological boundary layer model and the observations collected from the MONTBLEX conducted in 1990 at Jodhpur,

India. The model was initialized with the observed temperature profiles at 0530 LST on 17 July, 1990 at Jodhpur and was run for 26 hours. The study is carried out with a geostrophic wind speed of 9.5 m s^{-1} corresponding to the strong wind simulation. The mean thermodynamic and wind structure simulated by the model is compared with the observations obtained from 30 m tower and upper air data. The model simulations are in good agreement with the observations. The computed surface layer characteristics such as the surface fluxes, TKE and standard deviations of velocity components are found to be reasonably in good agreement with those based on turbulence measurements. The shear and buoyancy budgets computed from the model are also compared with the turbulence measurements. The integrated cooling budget in the NBL is given.

Acknowledgments

The authors wish to thank the financial support provided by the Department of Science and Technology, Govt. of India for carrying out this work. The authors had a fruitful discussion related to the observations with Dr. D.R. Sikka. One of the authors TVBPSRK would like to thank Dr. S. Devotta, Director, National Environmental Engineering Research Institute, Nagpur for his encouragement. The authors extend appreciation to the anonymous reviewers for the valuable comments and suggestions in improving the manuscript.

REFERENCES

AGGARWAL, P., YADAV, A.K., GULATI, A., RAMAN, S., RAO, S., SINGH, M.P., NIGAM, S., and REDDY, N. (1995), *Surface layer turbulence processes in low wind speeds over land*, Atmos. Environ. *29*, 2089–2098.

ANDRE, J.C. and MAHRT, L. (1982), *The nocturnal surface inversion and influence of clear-air radiational cooling*, J. Atmos. Sci. *39*, 864–878.

ARYA, S.P., *Air Pollution Meteorology and Dispersion* (Oxford University Press, Oxford 1999).

BELJAARS, A. C. M. and HOLTSLAG, A. A. M. (1991), *Flux parameterization over land surfaces for atmospheric models*, J. Appl. Meteor. *30*, 327–341.

BETTS, A.K. and BELJAARS, A.C.M. (1993), *Estimating effective roughness length for heat and momentum from FIFE data*, Atmos. Res. *30*, 251–261.

BLACKADAR, A. K. (1962), *The vertical distribution of wind and turbulent exchange in neutral atmosphere*, J. Geophys. Res. *67*, 3095–3102.

BROST, R.A. and WYNGAARD, J.C. (1978), *A model study of the stably stratified planetary boundary layer*, J. Atmos. Sci. *35*, 1427–1440.

BUSINGER, J.A., WYNGAARD, J.C., IZUMI, Y., and BRADLEY, E. F. (1971), *Flux-profile relationships in the atmospheric surface layer*, J. Atmos. Sci. *28*, 181–189.

CHATTERJEE, M., SEN, G.K., and SINHA, D.K., *Understanding the turbulent structure of the atmospheric boundary layer: A diagnostic approach*. In *The Monsoon Trough Boundary Layer* (R. Narasimha *et al.*, eds.) (Indian Academy of Sciences, Bangalore 1997) pp. 393 – 410.

GAMO, M., GOYAL, P., MANJU, KUMARI MOHANTY, U.C., and SINGH, M.P. (1994), *Mixed layer characteristics as related to the monsoon climate of New Delhi*, Boundary-Layer Meteor. *67*, 213–227.

GARRATT, J.R. (1978), *Flux profile relations above tall vegetation*, Q. J. Roy. Meteor. Soc. *104*, 199–211.

GARRATT, J.R. (1982), *Observations in the nocturnal boundary layer*, Boundary-Layer Meteorol. *22*, 21–48.

GOPALAKRISHNAN, S.G., SHARAN, M., MCNIDER, R.T., and SINGH, M.P. (1998), *Study of radiative and turbulent processes in the stable boundary layer under weak wind conditions*, J. Atmos. Sci. *55*, 954–960.

GOPALAKRISHNAN, S.G., FREEDMAN, F., SHARAN, M., and RAMA KRISHNA, T.V.B.P.S. (2005), *A model study of the strong and weak wind, stably stratified nocturnal boundary layer: Influence of gentle slopes*, Pure Appl. Geophys. *162*, 1795–1809.

HANNA S.R., *Applications in air pollution modelling*. In *Atmospheric Turbulence and Air Pollution Modeling* (F.T.M. Nieuwstadt and H. Van Dop eds.), (Riedel 1982), pp. 275–310.

HOLT, T. and RAMAN, S. (1985), *Air craft and ship observations of the mean structure of the marine boundary layer over Arabian Sea during MONEX-79*, Boundary-Layer Meteor. *33*, 259–282.

HOLT, T. and RAMAN, S. (1988), *A review and comparative evaluation of multilevel boundary layer parameterizations for first-order and turbulent kinetic energy closure schemes*, Rev. Geophys. *26*, 4, 761–780.

LETTAU, H., *Atmosphaerische Turbulenz* (Leipzig, Akademische Verlagsgesellschaft 1932).

MAHRER, Y. and PIELKE, R.A. (1977), *A numerical study of the airflow over irregular terrain*, Beiträge zur Physik der Atmosphäre *50*, 98–113.

MANJU KUMARI (1985), *Diurnal variation of mean mixing depths in different months at Delhi*, Mausam *36*, 71–74.

MCNIDER, R.T. and PIELKE, R.A. (1981), *Diurnal boundary- layer development over sloping terrain*, J. Atmos. Sci. *38*, 2198–2212.

MCNIDER, R.T., MORAN, M.D., and PIELKE, R.A. (1988), *Influence of diurnal and inertial boundary layer oscillations on long-range dispersion*, Atmos. Environ. *22*, 2445–2462.

MONIN, A.S. and YAGLOM, A.M., *Statistical Fluid Mechanics*, vol. I (MIT Press, Cambridge, Mass 1971).

NIEUWSTADT, F.T.M. (1984), *The turbulent structure of the stable nocturnal boundary layer*, J. Atmos. Sci. *41*, 2202–2216.

PAEGLE, J., ZDUNKOWSKI, W.G., and WEICH, B.M. (1976), *Implicit differencing of predictive equations of the planetary boundary layer*, Mon. Wea. Rev. *104*, 1321–1324.

PANOFSKY, H.A. and DUTTON, J.A., *Atmospheric Turbulence*, Wiley (Inter Science, New York 1984).

PAUL, D.K. and SIKKA, D.R. (1976), *Extended range forecasting-categorization of weather charts, Part I: Monsoon sea-level pressure field*, Project report no. ERF/1, 34 pp., Indian Institute of Tropical Meteorology, Pune.

PIELKE, R.A. (1974), *A three-dimensional numerical model of sea breeze over South Florida*, Mon. Wea. Rev. *102*, 115–139.

PIELKE, R.A., *Mesoscale Meteorological Modeling*, 2nd Edition (Academic Press, San Diego 2002), 676 pp.

POTTY, K.V.J., MOHANTY, U.C., NANDI, B., and RAMESH, K.J., *Planetary boundary layer over monsoon trough region in a high resolution primitive equation model*. In *The Monsoon Trough Boundary Layer* (R. Narasimha *et al.*, eds.) (Indian Academy of Sciences, Bangalore 1997), PP. 339–358.

PRESS, W.H., FLANNERY, B.P., TEUKOLSKY, S.A., and VETTERLING, W.T., *Numerical Recipes* (Cambridge University Press, Cambridge, 1986).

RAMA KRISHNA, T.V.B.P.S., SHARAN, M., GOPALAKRISHNAN, S.G., and ADITI, (2003), *Mean structure of the nocturnal boundary layer under weak and strong wind conditions: EPRI case study*, J. Appl. Meteor. *42*, 952–969.

RAMAN, S., TEMPLEMAN, B., TEMPLEMAN, S., HOLT, T., MURTHY, A.B., SINGH, M.P., AGRAWAL, P., NIGAM, S., PRABHU, A., and AMEENULAH, S. (1990), *Structure of the Indian southwesterly pre-monsoon and monsoon boundary layers: Observations and numerical simulation*, Atmos. Environ. *24*, 723–734.

RAO, K.G., *Roughness length and drag coefficient at two MONTBLEX-90 tower stations*. In *The Monsoon Through Boundary Layer* (R. Narasimha *et al.*, eds.) (Indian Academy of Sciences, Bangalore 1997), pp. 245–259.

RAO, K.G., LYKOSSOV, V.N., PRABHU, A., SRIDHAR, S., and TONKACHEYEV, E., *The mean and turbulent structure simulation of the monsoon trough boundary layer using a one-dimensional model with e-ℓ and e-ε closures*. In *The Monsoon Trough Boundary Layer* (R. Narasimha *et al.*, eds.) (Indian Academy of Sciences, Bangalore 1997), pp. 359–392.

RUDRA KUMAR, S., AMEENULLA, S., and PRABHU, A., *MONTBLEX tower observations: Instrumentation, data acquisition and data quality*. In *The Monsoon Trough Boundary Layer* (R. Narasimha *et al.*, eds.) (Indian Academy of Sciences, Bangalore 1997), pp. 97–124.

SASAMORI, T. (1972), *A linear harmonic analysis of atmospheric motion with radiative dissipation*, J. Meteor. Soc. Japan. *50*, 505–518.

SATYANARAYANA, A.N.V., LYKOSSOV, V.N., and MOHANTY, U.C. (2000), *A study on atmospheric boundary layer characteristics at Anand, India using LSP experimental data Sets*, Boundary-Layer Meteor. *96*, 433–452.

SHARAN, M., GOPALAKRISHNAN, S.G., MCNIDER, R.T., and SINGH, M.P. (1996), *Bhopal gas leak: a numerical investigation of the prevailing meteorological conditions*, J. Appl. Meteor. *35*, 1637–1657.

SHARAN, M. and GOPALAKRISHNAN, S.G. (1997), *Comparative evaluation of eddy exchange coefficients for strong and weak wind stable boundary layer modeling*, J. Appl. Meteor. *36*, 545–559.

SHARAN, M., RAMA KRISHNA, T.V.B.P.S., and ADITI (2003), *Surface layer characteristics in a weak wind stable boundary layer*, Boundary- Layer Meteor. *108*, 257–288.

SIKKA, D.R. and NARASIMHA, R., *Genesis of the monsoon trough boundary layer experiment (MONTBLEX)*. In *The Monsoon Trough Boundary Layer* (R. Narasimha *et al.*, eds.) (Indian Academy of Sciences Bangalore 1997), pp. 1–32.

SINGAL, S.P., GERA, B.S., and AGARWAL, S.K. (1982), *Determination of structure parameter using sodar*, Boundary-Layer Meteor. *23*, 105.

SINGAL, S.P., GERA, B.S., and OJHA, V.K. (1993), *Sodar studies of the monsoon trough boundary layer at Jodhpur (India)*, Mausam *44*, 9–14.

STULL, R., *An Introduction to Boundary Layer Meteorology* (Kluwer Academic Publishers, The Netherlands 1988), 666 pp.

TJEMKES, S.A. and DUYNKERKE, P.G. (1989), *The nocturnal boundary layer: Model calculations compared with observations*, J. Appl. Meteor. *28*, 161–175.

VERNEKAR, K.G, SADANI, L.K., MOHAN, B., SAXENA, S., DEBAJE, S.B., PILLAI, J. S., MURTHY, B.S., and PATIL, M.N. (1991), *Structure and growth of Atmospheric boundary layer as observed by tethered balloon payload*, Indian J. Radio Space Phys. *20*, 312–315.

YADAV, A.K., RAMAN, S., and SHARAN, M. (1996), *Surface layer turbulence spectra and eddy dissipation during low winds in tropics*, Boundary-Layer Meteor. *79*, 205–223.

(Received May 22, 2006, accepted November 11, 2006)
Published Online First: June 19, 2007

Pure appl. geophys. 164 (2007) 1861–1880
0033–4553/07/091861–20
DOI 10.1007/s00024-007-0233-4

Pure and Applied Geophysics

Evolutionary Features of Marine Atmospheric Boundary Layer (MABL) over the Arabian Sea and the Onset of Monsoon over Kerala during ARMEX-2003

S. G. NAGAR,[1] A. R. DHAKATE,[1] and P. SEETARAMAYYA[2]

Abstract—In this study an attempt has been made to examine the evolutionary features of the dynamic and thermodynamic characteristics of the marine atmosphere over the South-East Arabian Sea near 9.22°N, 74.51°E just two to three days prior to the onset of southwest monsoon over Kerala during 2003 and seek the linkages with the large-scale flow in the lower and middle troposphere at that time over the region. The marine meteorological data collected onboard ORV Sagarkanya as part of the experiment ARMEX-2003 for 4–8 June, 2003 are used. The monsoon onset over Kerala occurred on 8 June, 2003. The observed changes in the marine atmospheric boundary layer (MABL) characteristics just two days prior to the onset are discussed. It is observed that the MABL increased in height up to 4 km on 6 June from an initial height 2.8 km on 5 June. The top of the MABL dried up (Relative Humidity RH ~ 30–40%) with weak and variable winds throughout the day on 6 June while the air at 850 hPa is relatively humid (RH ~ 50–80%) but not saturated. A sequential increase in RH is associated with a change in the winds from southwesterly to westerly from 6 June onwards until the onset date. The changes in the lower and middle troposphere flow patterns over the Arabian Sea and Indian region are highlighted.

Key words: ARMEX, onset of monsoon, conserved variables, deep convection.

1. Introduction

The Indian summer monsoon is the main rainy season for India. The 'burst' or onset of the summer monsoon over Kerala (at the southern tip of peninsular India) is a spectacular event in monsoon meteorology (ANANTHAKRISHNAN *et al.*, 1983). Several experiments have been conducted in the past, such as the IIOE (International Indian Ocean Experiment, 1963–1966), ISMEX-73 (Indo-Soviet Monsoon Experiment, 1973), MONSOON-77 (International Monsoon Experiment, 1977) and MONEX-79 (Monsoon Experiment, 1979) to understand the evolutionary processes that are prevailing over the Indian Ocean in general and the South-East Arabian Sea (SEAS) in particular. These experiments have led to the discovery of a Low Level Jet (FINDLATER, 1971), the incipient formation and development of an onset vortex over

[1] Indian Institute of Tropical Meteorology, Dr. Homi Bhabha Road, Pune 411 008, India.
E-mail: nagar@tropmet.res.in
[2] D4, Niranjan Complex, Pashan, Pune 411 021, India.

the East Central Arabian Sea (HUI, 1980; KRISHNAMURTI *et al.*, 1981; KERSHAW, 1985, 88) and a mini warm pool (SST $\geq$ 30.5°C) over the East Central Arabian Sea (SEETARAMAYYA and MASTER, 1984). Nevertheless, there are vagaries in the monsoon onset and its advance along the west coast of India (ANANTHAKRISHNAN *et al.*, 1983). The dynamical and the thermodynamical processes of the marine atmosphere over the Arabian Sea are yet to be understood properly. For better forecasting of the onset of southwest monsoon over Kerala from time to time it is needed to study these processes either by conventional or by numerical modeling technique.

The field experiment ARMEX (Arabian Sea Monsoon EXperiment), was the second large-scale land-ocean-atmosphere observational study implemented under the Indian Climate Research Program (ICRP), coordinated by the Department of Science and Technology as nodal agency. ARMEX is aimed at elucidating the linkages between the *in situ* atmospheric and oceanic processes in the Arabian Sea region and the variations in the monsoon climate over the Indian subcontinent (RAO, 2005) i.e., to understand regional scale ocean-atmosphere coupled processes over the Arabian Sea that are very important for predicting the vagaries of the monsoon. It was carried out in two phases during the period 15 June to 16 August, 2002 (Phase I) and 15 March to 10 April and 15 May to 15 June, 2003 (Phase II) to investigate the prevailing features both in the atmosphere and the ocean over the East Central Arabian Sea over the region 9–15°N, 70–75°E. The broad objectives of the ARMEX were (i) to study the Arabian Sea convection associated with intense rainfall events on the west coast of India during the monsoon period and (ii) to study the evolution, maintenance and the collapse of the Arabian Sea Warm Pool and pre-onset phase of the monsoon. The year 2003 was characterized by the late onset of monsoon over Kerala, which took place on 8 June, delayed by one week from its normal date 1 June. There was no intense low pressure system during the onset phase, but the monsoon was vigorous over Kerala on 8 June.

The primary objective of the present study is to understand the dynamical and thermodynamical stability of the marine atmosphere over the SEAS just prior to the onset of monsoon over Kerala on a sub-synoptic (2–3 days) time scale. The conserved variable analysis, mixing line theory and saturation point concept have been used to understand thermodynamical stability. The vertical profiles of wind as well as surface fluxes are used to study the dynamical stability. The large-scale evolution of the onset phase of monsoon is examined in terms of changes in SST over the SEAS, circulation features of the lower and middle troposphere and changes in the cloud types and rainfall over the ship and coastal stations along Kerala.

2. *Experimental Set up, Data and Quality Check*

During ARMEX-2003 the ship ORV Sagarkanya (cruise No. SK193) was stationary at 9.22°N, 74.51°E in the SEAS from 24 May 00 UTC to 7 June 0800

UTC. It has moved northward to 12.31°N, 73.22°E on 8 June 1725 UTC. This specific location was selected due to the existence of the Arabian Sea Warm Pool that generally originates in the months of March and April. Figure 1 shows the location of the ship. High resolution (at an interval of 10 to 25 hPa) GPS radiosondes (Vaisälä model RS80-15G) were launched (2–4 per day) at the ship by the Indian Institute of Science (IISc), Bangalore, India, to monitor upper air conditions.

Considerable efforts have been made to ensure the collection of high quality data during ARMEX. The relative humidity (humicap) and platinum resistance thermometer (PRT) sensors were calibrated in the laboratory before and after the field experiments using the same cables as those used onboard the ship. The measurement accuracies achieved onboard the ORV Sagarkanya met the same accuracy levels of 0.25°C and 0.2 m s^{-1} (or 2%, whichever is larger) sought by the World Ocean Circulation Experiment (WOCE) program for air temperature and wind speed, respectively, for observations over the oceans. The uncertainty achieved in the relative humidity is 2% (~ 0.5 g kg^{-1} in mixing ratio), which is marginally higher than the WOCE requirement of 1.7% (BHAT *et al.*, 2001).

The radiosonde data collected during 4 to 8 June onboard the ship contain 17 consecutive ascents (two each on 4 and 5 June, five on 6 June and four each on 7 and 8 June) generally close to 00, 0530, 0830, 1130 and 1730 UTC. These ascents have been used in the present study. The primary reason for selecting 6 June for detail

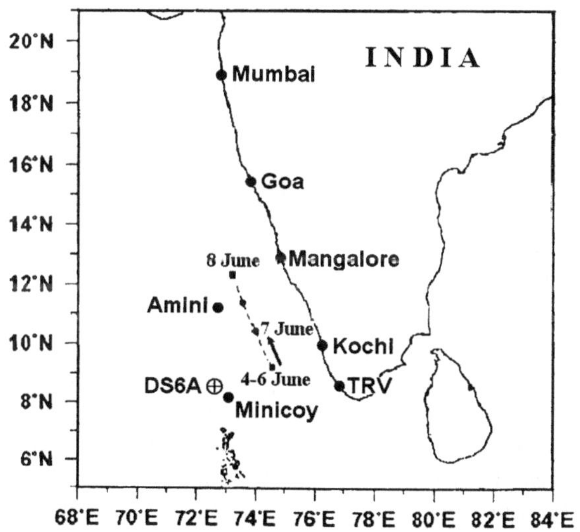

Figure 1
Areal map showing the location of the ship ORV Sagarkanya during 4–8 June, 2003. The figure also shows the locations of the Islands Amini and Minicoy, the coastal stations Thiruvananthapuam (TRV), Kochi and Mangalore along Kerala and the data buoy DS6A.

study is due to the fact that on this day five ascents at the stationary position are available. On 4 and 5 June only two ascents (00 and 1130 UTC) are available at the stationary position and from 7 June 0800 UTC onwards the ship started moving northward. At 1130 UTC it was at 9.57°N, 74.37°E, slightly northward from its stationary position. The one hourly surface meteorological data from the ship ORV Sagarkanya for temperatures (dry bulb, wet bulb, and sea surface), wind (speed and direction), humidity, mean sea-level pressure and precipitation have also been used along with the above profiles to understand the synoptic weather conditions. The SST observations made by the data buoy DS6A deployed at 8.35°N, 72.72°E at a depth of 2120 m are examined to delineate the response of the SEAS to the onset of monsoon over Kerala. The NCEP-NCAR Reanalysis wind data at 2.5° × 2.5° resolution over the domain 5°S–20°N, 40°E–85°E for 925 and 700 hPa levels are used to examine the changes in the lower and middle tropospheric flow patterns, respectively. The cloud and rainfall data over Minicoy and Amini Islands as well as over Thiruvananthapuram (TRV), Kochi and Mangalore situated along the Kerala coast are used to examine the changes in weather conditions. The locations of these stations are shown in Figure 1.

3. *Methodology*

The one-hourly surface meteorological data are used to compute the surface fluxes of sensible heat (Q_H), moisture (Q_E) and momentum (Tau) by using standard bulk aerodynamic method following Kara *et al.* (2002). The heat fluxes (Q_H and Q_E) are important for determining the thermal properties and stability at the ocean-atmosphere interface and the momentum flux is a major driving force for the upper ocean circulation. These fluxes contribute to the turbulent mixing within the oceanic mixed layer and in the atmosphere above the ocean surface.

The radiosonde data are recorded at an interval of 10 hPa upto 850 hPa and of 25 hPa above this level. These data have been linearly interpolated to an interval of 10 hPa above 850 hPa to make a uniform data set. Several thermodynamic parameters such as lifting condensation level (LCL), virtual potential temperature (θ_v), equivalent potential temperature (θ_e), saturated equivalent potential temperature (θ_{es}) and specific humidity (q) were computed with standard formulae (Bolton, 1980) using the 17 profiles of air temperature and humidity up to 500 hPa only to study the thermal structure and local stability of the MABL. The conserved variable analysis approach, as suggested by Betts (1985) and Betts and Albrecht (1987), is used to address the convective properties of the marine atmosphere over the SEAS. The boundary layer structure is defined from the gradients of θ_v, θ_e, θ_{es} and q. The θ_e represents the total energy as well as the entropy of the moist air. As the tropical atmosphere is conditionally unstable in nature, the θ_e is the most relevant attributor of the surface air and is an important parameter in the boundary layer for tropical convection. The deep convection is favored when θ_e at the surface is above a certain threshold value of 345 K over the tropical oceans (BETTS and RIDGWAY, 1989; BHAT *et al.*, 2000).

The subcloud layer is characterized by a layer of constant θ_v where unsaturated air mass is well mixed. This subcloud layer is also known as mixed layer (ML), which in turn plays an important role as a link between the surface layer and the free atmosphere above it. The ML carries the surface layer properties from the surface and transports them to the free atmosphere. The top of a constant θ_v layer gives an estimate of unsaturated ML (EMANUEL, 1994) and it generally coincides with the base of the cloud. The base of a cloud layer is defined as the level of condensation of an air parcel rising adiabatically from the surface to the level of its saturation (i.e., LCL of the surface). The cloud layer is generally capped by an inversion layer where θ_{es} shows an increase with height.

The MABL plays an important role in regulating convective activity and perhaps inhibits the deep convection over the Central and Western Arabian Sea during the height of the southwest monsoon. Its height is estimated by using the criteria of minimum θ_e and maximum θ_{es}. The MABL top is a layer which always interacts with the free atmosphere and forms a buffer layer in between the MABL and the free atmosphere. This buffer layer is the layer where θ_e and q inversions are observed.

The saturation point analysis, following BETTS (1982), is used to study the mixing processes in the boundary layer. When two air parcels of different characteristics mix together, the characteristics of the new parcel can be shown on the (θ_e - q) plot in which the mixing lines are straight lines. The MABL top appears as a marked kink in this plot and this kink marks the top of the layer that is thermodynamically coupled with the surface. The double or multiple mixing line structures with q reversal are observed due to precipitation-evaporation processes once the MABL (and the deepest clouds in it) reaches a sufficient depth. The low level stability is assessed from the profiles of θ_e and θ_{es} following KLOESEL and ALBRECHT (1989). The vertical profiles of wind field (direction and speed) are examined to study the dynamics of the mid-troposphere and advective processes.

4. *Results and Discussion*

4.1. *Evolution over the Location of the Ship*

Figure 2 shows the time series of the surface meteorological parameters from the ship during 4 June to 8 June, 2003. In this figure the onset date of monsoon over Kerala (i.e., 8 June) is marked by the vertical line. It is seen that the SST is more or less constant at 30.5°C while the air temperature (Ta) undulates near the SST. The Ta varies between 26°C (8 June) and 32°C (4, 6 and 7 June during daytime). The RH varies between 65 and 95%, with an average of 75%. Mean sea-level pressure

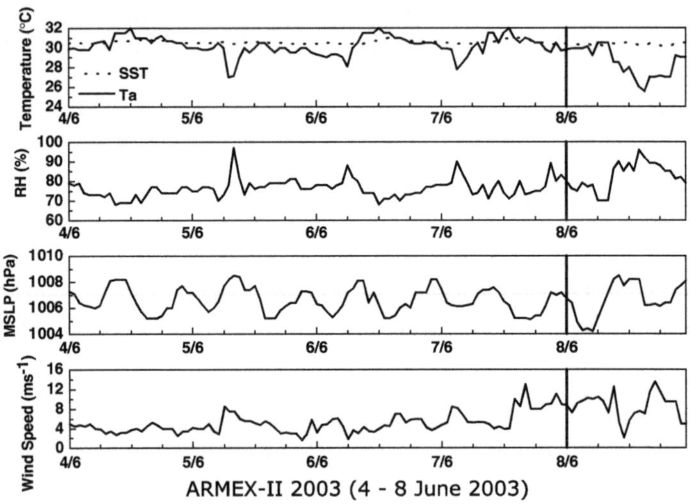

Figure 2
Time series of the surface meteorological parameters: Sea-Surface Temperature (SST) and Air Temperature (Ta), Relative Humidity (RH), Mean-Sea Level Pressure (MSLP), Wind Speed (ff) during 4–8 June, 2003 at the location of the ship. The date of onset of the monsoon over Kerala is marked by vertical thick line.

(MSLP) varies from 1004 to 1009 hPa with a semi-diurnal amplitude of 2 hPa. The wind speed varies from 2 to 12 m s^{-1}. The highest value of 12 m s^{-1} is observed on 7 June, just one day prior to the onset of the monsoon over Kerala.

Figure 3 shows the time series of Q_H, Q_E, Bowen ratio ($= Q_H/Q_E$) and Tau. It is seen that Q_H undulates around zero with a low value of -10 Wm^{-2} except to some events on 5, 7 and 8 June. The highest values on these days coincide with the lowest values of Ta and the highest values of RH. As the primary source of heat and water vapor is the ocean surface, it is plausible to assume that initial cooling and subsequent drying extends through the entire mixed layer (FITZJARRALD and GARSTANG, 1981a). Thus we can interpret that when Ta $>$ SST (i.e., $Q_H < 0$), the transfer of heat is towards the ocean surface which leads to the stable boundary layer conditions and complete lack of convective activity. Similarly for SST $>$ Ta (i.e., $Q_H > 0$), the transfer of heat is towards the atmosphere leading instability to the surface boundary layer i.e., increase in convection. In case of SST $=$ Ta, the condition of neutral buoyancy exists at the air-sea interface. Thus based on Q_H distribution we may assess that the marine atmosphere is at near-neutral conditions up to 7 June and then turns to unstable conditions following the onset of monsoon on 8 June. The Q_E varies between 20 and 380 Wm^{-2}. It shows an increase with an increase in wind speed as well as an increase in the difference in q between SST and Ta. The Bowen ratio follows a similar trend as that of Q_H with high values of

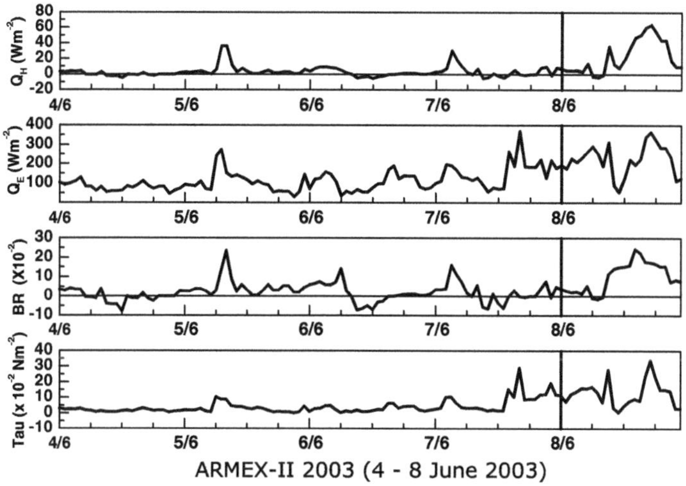

Figure 3
Time series of surface fluxes: Sensible heat flux (Q_H), moisture flux (Q_E), Bowen Ratio (BR) and momentum flux (Tau) during 4–8 June, 2003 at the location of the ship. The date of onset of the monsoon over Kerala is marked by vertical thick line.

20×10^{-2} coinciding with high values of Q_H on 5 June. Higher values of the Bowen Ratio indicate higher rate of sensible heat exchange. The Tau varies between 0.01 and 0.2 Nm^{-2}. The high value corresponds to the high wind speed on 7 June prior to the onset of monsoon over Kerala.

In general all the observed (and hence computed) surface parameters vary smoothly in regular intervals of time depending upon the diurnal forcings (tidal and solar radiation) from 4 June to mid-day of 7 June. There was an increase in all fluxes on 5 June at 0600 IST (Indian Standard Time) that lasted for about six hours. The fluctuations in the above parameters increase rapidly following the major changes in the atmospheric variables (Ta, MSLP and wind speed) from mid-day 7 June to the end of 8 June with the onset of monsoon. These rapid fluctuations are mainly due to the development of deep clouds along with the burst of the monsoon over Kerala.

The sudden increase of 4°C in Ta from 28°C to 32°C, the rapid decrease of 20% in RH from 85% to 65% and a corresponding decrease of 3 hPa in MSLP and 4 m s^{-1} in wind speed on 6 June has prompted us to examine the plausible reasons for these sudden changes on mesoscale/local scale basis. Fortunately, there are five consecutive radiosonde ascents taken at an interval of three to six hours on this day from early morning (00 UTC) to midnight (1730 UTC). Using these ascents the evolutionary features of the mixing processes are studied on this day.

Figures 4(a–e) show the vertical profiles of θ_v, θ_e, θ_{es} and q from the surface to 6000 m (500 hPa) for 6 June, 2003 at the ship's location for five consecutive hours

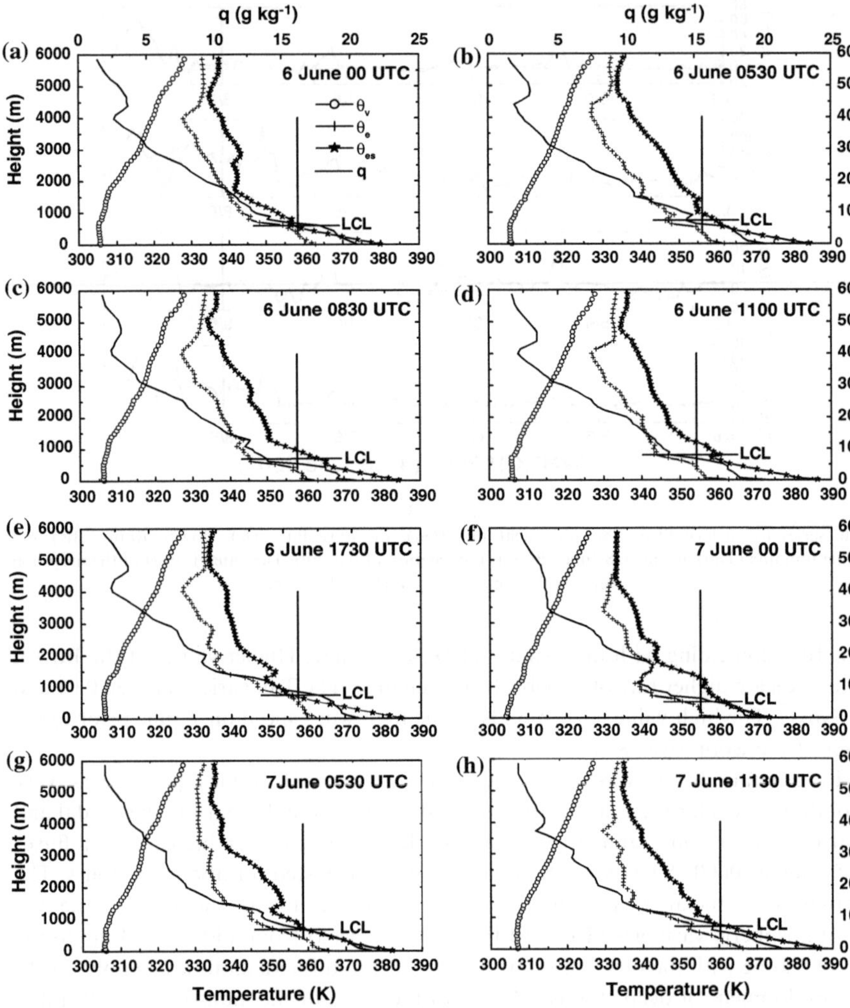

Figure 4
The vertical profiles of virtual potential temperature (θ_v), equivalent potential temperature (θ_e), saturated equivalent potential temperature (θ_{es}) and specific humidity (q) during 6 June and 7 June, 2003 at the location of the ship. The vertical thick line indicates the θ_e path for a non-entraining parcel originating from 970 hPa (i.e., mean sub-cloud layer). The lifting condensation level (LCL) of this parcel is also indicated.

viz. 00, 0530, 0830, 1130 and 1730 UTC and Figures 4(f–h) show the profiles for 7 June for 00, 0530 and 1130 UTC. As mentioned in section 2 the ship has moved northwards on 7 June 1130 UTC. In this diagram, the θ_e path for a nonentraining parcel originating from the mean of the subcloud layer (i.e., 970 hPa) is shown by the vertical thick line. The LCL of this parcel is also indicated in the Figure. Above the LCL of this parcel, the difference between this constant θ_e path of the rising parcel and the θ_{es} of the sounding is proportional to the temperature difference between a (non-entraining) parcel originating from the sub-cloud layer and the temperature of the environment. In case of inversion soundings, the θ_e path of the parcel intersects the θ_{es} profile of the environment below 600 hPa (~4000 m). In case of deep convection, the rising air parcel is always buoyant positively right up to 600 hPa, leading to an unstable condition. For the shallow or suppressed convection, the rising air parcel is negatively buoyant above its LCL at least up to 600 hPa leading to stable condition (KLOESEL and ALBRECHT, 1989). It is seen from the figure that all ascents are favorable for deep convection. The height of mixed layer (constant θ_v) varies between 600 and 630 m (940 hPa) on 6 June. It reduces to 335 m (970 hPa) on 7 June 00 UTC and again increases to ~ 630 m at 0530 UTC of the same day. The MABL (θ_e minimum) observed at the height of 4060 m (~ 630–625 hPa) on 6 June corresponds to the lower value of q (~2 g kg^{-1}). It reduces to 3500 m on 7 June 00 UTC.

In the early morning (Fig. 4a) the θ_e and θ_{es} are close to each other in the layer between 600 and 1400 m and they depart from each other up to the top of the MABL which shows the dryness. The θ_{es} curve shows a weak inversion layer between 2500–2700 m. This inversion is associated with subsidence from above this level. Due to this subsidence and insufficient moisture (RH < 80%) formation of deep clouds is prevented. The stratiform clouds of thickness ~200 m are seen in the lower part (500–700 m) of the MABL. Above the MABL, both the curves come close to each other in the free atmosphere. Two cloud layers are anticipated (i) from the ML to 1500 m and (ii) from 1600 m to 2700 m. A weak stable layer of nearly constant $\theta_{es} \approx 342.5$ K is noted between 1500 m and 2500 m. This type of conservation in θ_{es} is a conspicuous feature beneath the inversion layer in the tropics. BETTS and ALBRECHT (1987) also have observed similar weak stable layer of nearly constant $\theta_{es} \approx 343.5$ K over the Pacific ocean.

At 0530 UTC (Fig. 4b) the increase in solar insolation causes warming (as evident from an increase of ~3–5 K in θ_{es} from 00 UTC) and hence drying (~1–2 g kg^{-1}) of the subcloud layer below 600 m. The increase in temperature difference (θ_{es} - θ_e) shows an increase in subsaturation. Further, the LCL of the nonentraining parcel also increases from 560 m (Fig. 4a) to 730 m. A slight increase in q (~1 g kg^{-1}) around 720–900 m may be due to evaporation of falling rain (FITZJARRALD and GARSTANG, 1981b) or differential horizontal advection of moisture from surrounding environment (BETTS and ALBRECHT, 1987). The inversion level has come down to 1500 m from its previous position at 2700 m. At 0830 UTC (Fig. 4c), the atmospheric conditions are similar to that at 0530 UTC, except the cloud layer has come down to 400–500 m. There is a slight increase in moisture (1–2 g kg^{-1}) in the lower layers. At 1130 UTC (Fig. 4d), the mixed layer height has increased to

700 m. The lower layer cloud has small thickness ($\sim$ 50 m) in which a sharp decrease in moisture and an increase in θ_{es} is observed. The atmosphere is moist and neutrally stable (i.e., θ_e is conserved in the layer 700–2000 m). Another cloud layer in which q and θ_e slowly decrease with height and θ_{es} is nearly conserved is seen between 2000 m and 2200 m.

The 1730 UTC sounding (Fig. 4e) represents the cooling phase after the sunset (2300 IST). It is interesting to note from this figure that the cloud layer (600–1300 m) below the inversion is considerably closer to the saturation (as seen from the closeness of θ_e and θ_{es} curves). Multiple cloud layers are clearly seen between the ML and MABL with an increase in moisture. Both the curves (θ_e and θ_{es}) have exhibited more or less same value (333–334 K) above $\sim$ 4800 m. An inversion is seen in the layer 1400 m to 1600 m. The θ_e curve shows a sharp inversion in coherence with the q inversion in the layer 4150 m–4830 m.

Figure 4f shows multiple cloud layers with small inversions in the layers 1080 m–1280 m and 2000 m–2200 m. Above the MABL the sharp increase in moisture has now disappeared and the moisture has increased by 1–2 g kg^{-1} so that the atmosphere is completely saturated above 4500 m. This moisture increase in the middle levels may be attributed to the horizontal advection. At 7 June 0530 UTC (Fig. 4g) the distinct θ_e minimum at the top of the MABL (3500 m) is absent. Above the MABL, the atmosphere is neutrally stratified (constant θ_e). Figure 4h also shows the increase in moisture by 1 g kg^{-1} in the upper levels when compared with Figure 4d.

Figures 5(a–e) show the conserved parameter (θ_e – q) plots for 6 June and Figures 5(f–h) show similar plots for 7 June, 2003 at the ship's location. The top of the MABL and the ML are marked by arrows in these figures. The q reversals are observed in all the figures except at 6 June 00 UTC (Fig. 5a) and 7 June 0530 UTC (Fig. 5g), suggesting the presence of multiple cloud layers as observed in Figure 4. The saturation points between ML and MABL are quite close to each other suggesting well mixing, whereas, the distinct points below ML show moderate mixing. The ML dries up as the day progresses due to solar insolation so that its location has moved upward along the mixing line where a decrease in q signifies. It slowly decreases with an increase in moisture as well as nocturnal cooling. Further, above the MABL the q reversal and θ_e increase due to radiational warming and isothermal layers are observed in all the figures except at 7 June 0530 UTC. Though morning ascents (Figs. 5a and 5f) show cooling and moistening on 7 June, comparison of Figure 5b with Figure 5g and 5d with 5h clearly shows warming and an increase in moisture from the surface to the MABL on 7 June just prior to the onset of monsoon.

Figure 6 shows the time series of the vertical profiles of wind vectors from 4 June to 8 June, 2003. In this figure the horizontal thick line shows the top of the MABL on 6 June. It is seen that most of the time of observations, the surface winds are weak ($\sim$ 2 m s^{-1}) and variable except on 7 June. It has suddenly increased to 8–10 m s^{-1} on

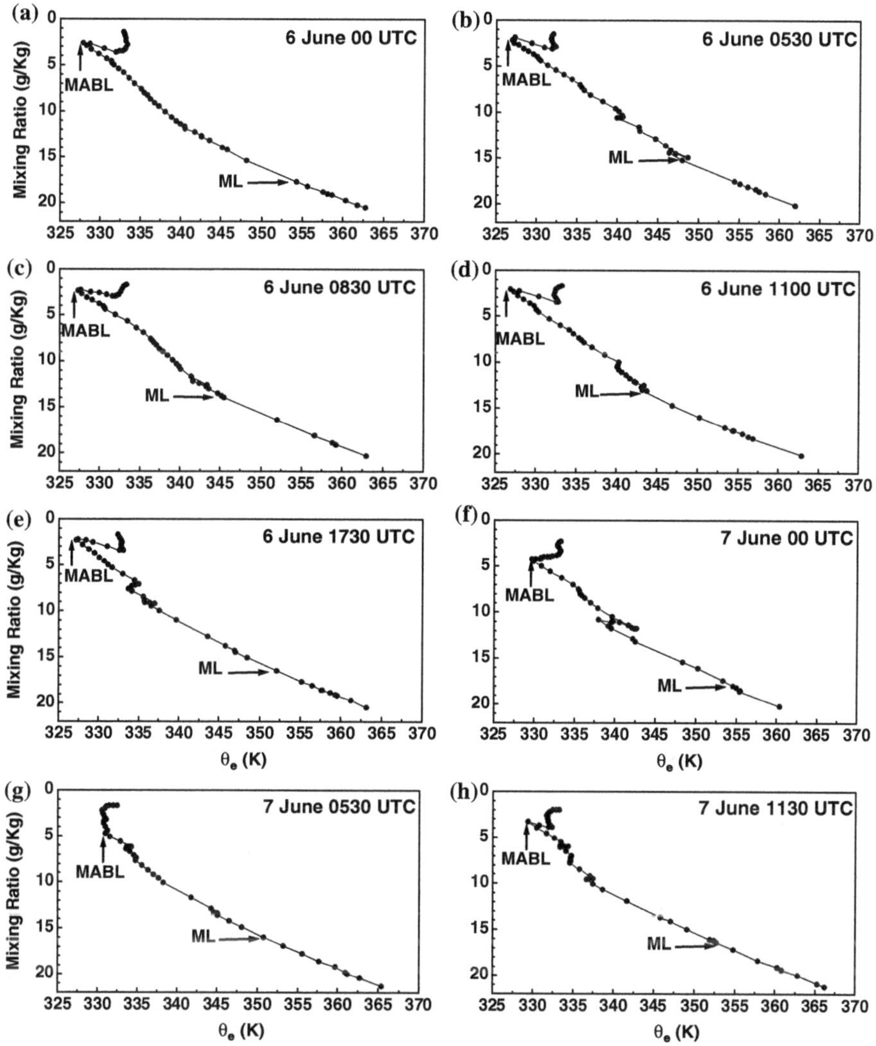

Figure 5
$(\theta_e - q)$ plots for 6 June and 7 June, 2003 at the location of the ship.

7 June and 12 m s^{-1} on mid-day of 8 June. These surface winds, which are northwesterly on 4 June, show backing to westerly on 5 June, further backing to southwesterly on 6 June, and then remain southwesterly up to 8 June.

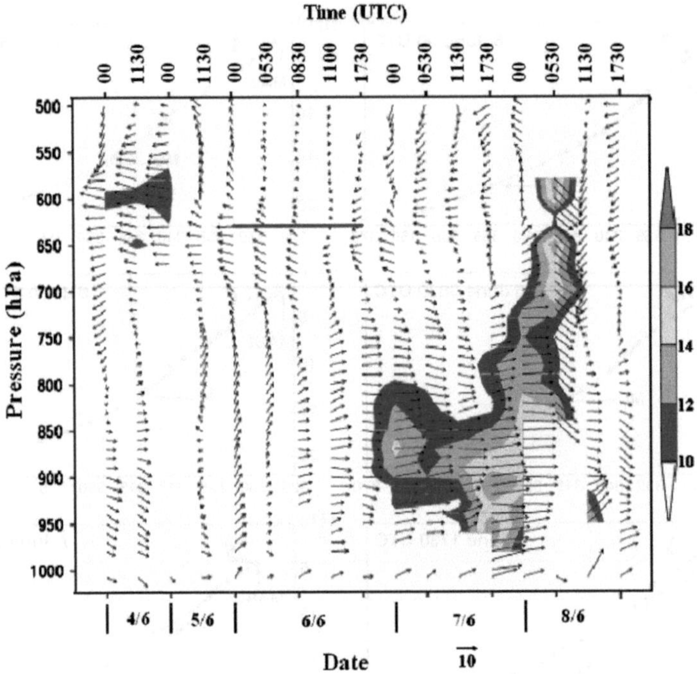

Figure 6
Time series of vertical profiles of wind vectors during 4–8, June 2003 at the location of the ship. The top of the MABL at 630 hPa on 6 June is marked by thick line.

On 4 and 5 June the winds are dry northwesterlies below 800 hPa above which northeasterlies originating from the Indian land mass are observed. The intensity of northeasterlies is a maximum (10–12 m s^{-1}) around 630–580 hPa. At the level of the MABL top ($\sim$ 630 hPa) a decrease in wind speed from 5 to 6 June with veering from northeast to southwest between 800 hPa and 700 hPa on 6 June 1100 UTC is noticeable. The northwesterlies have weakened on 5 June 1130 UTC. The winds back to southwesterlies at 6 June 00 UTC, paving a way for the onset of the monsoon over Kerala within a day or two. However, southweterlies have weakened at 6 June 0530 UTC and started strengthening between 950 and 800 hPa from 1730 UTC on the same day. The depth and intensity of southwesterlies have increased from 7 June 1730 UTC. On 8 June 00 UTC the southwesterly shows veering to westerly upto $\sim$ 700 hPa. The westerlies back to southwesterlies in the lower layers below 900 hPa (950 hPa) at 1130 UTC (1730 UTC).

Figure 7a shows the time series of the LCL of the surface during 4–8 June, 2003 at the ship's location. The LCL varies between 322 m (8 June) and 716 m (on 4 and 6 June i.e., 2 to 3 days prior to the onset). The increase in LCL on 6 June implies drying

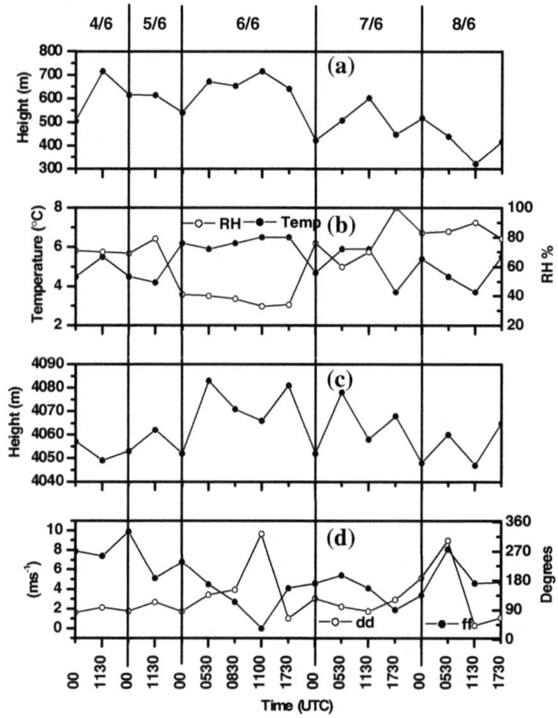

Figure 7
(a) Time series of LCL during 4–8 June, 2003. (b) Time series of air temperature (Temp) and relative humidity (RH) (c) geopotential height and (d) wind speed (ff) and direction (dd) at 630 hPa during 4–8 June, 2003 at the location of the ship.

and warming of the marine atmosphere associated with strong subsidence from above the boundary layer. The LCL height remains constant at 650 m on 6 June and 7 June 1130 UTC when a rise in surface pressure is noticed (Fig. 2). The rise in surface pressure over the Arabian Sea is a very conspicuous feature just 2–3 days prior to the onset over Kerala (KRISHNAMURTI *et al.*, 1981). This LCL height is more or less same as that observed by GROSSMAN and DURRAN (1984) and HOLT and SETHURAMAN (1985) over the East Central Arabian Sea during the advance process of the monsoon during MONEX-79 on 20 June. The LCL decreases on 7 June with increase in moisture in the lower layers. Figures 7(b–d) show the time series of temperature, RH, geopotential height and wind speed and direction at top of the MABL (630 hPa) from 4 to 8 June, 2003. It is clear from Figure 7b that the air temperature at the top of MABL is ~6.5°C on 6 June (2°C higher than the temperature on 5 June) with a sharp decrease in RH to ~ 30–40% (from 70–80% on

5 June). This temperature corresponds to a minimum $\theta_e = 327$ K of 6 June (Fig. 4). The geopotential height (Fig. 7c) has increased from 4052 m at 00 UTC of 6 June to 4083 m at 530 UTC of the same day. The winds at the MABL top decrease from 5 (7 m s^{-1}) to 6 June (1 m s^{-1}) with veering from northeast to southwest (Fig. 7d).

Figure 8 shows the three hourly observations of the total cloud amount (N, solid squares) and the amount of low cloud (N_L, solid circles) over the ship during 4–8 June, 2003. In this figure the genus of clouds low (C_L), medium (C_M) and high (C_H) are shown symbolically following WMO (World Meteorological Organization) code. The ship has observed the rainfall 10.4 mm and 0.5 mm on 7 and 8 June, respectively. It is seen that N increased from 2 oktas on 7 June 03 UTC to 6–8 oktas on 8 June 09 UTC and N_L increased to 5 oktas at the same time. The fair weather cumulus and stratocumulus observed up to 6 June are replaced by stratocumulus of bad weather (i.e., conditions which generally exist before, during and after precipitation) and nimbostratus from 7 June 15 UTC to 8 June 15 UTC.

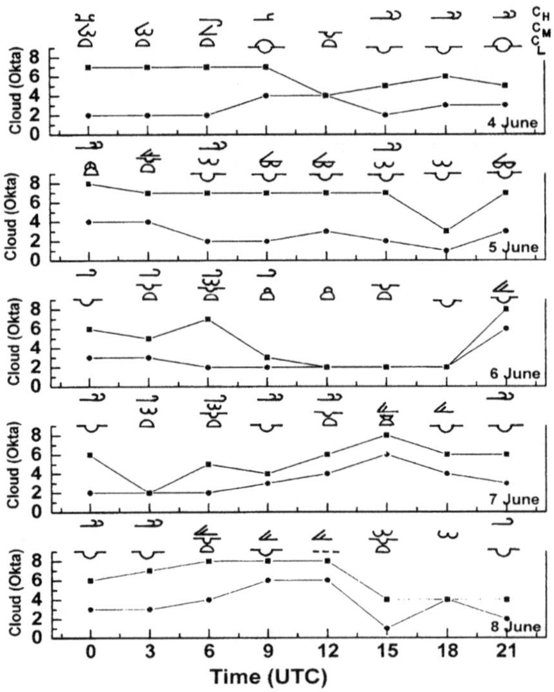

Figure 8
Time series of three hourly total cloud amount (N, solid squares) and low cloud amount (N_L, solid circles) over the ship during 4–8 June, 2003. The genus of low (C_L), medium (C_M) and high cloud (C_H) are plotted following WMO code.

4.2. Large Scale Evolution of the Onset Process of the Monsoon

4.2.1. Response of SST over the SEAS

Continuous monitoring of SST was made in the SEAS by the deep water data buoys network to study the changes in the warm pool region over the Lakshadweep area during the onset phase of summer monsoon 2003. Figure 9 shows the time series of SST at DS6A during 4–8 June, 2003. It is seen that SST decreases from 30.75°C (4 June) to 30.09°C (8 June). A sharp decrease in SST is followed by the monsoon onset at the buoy site on 6 June, 2003, two days prior to the onset over the Kerala coast. The high amplitude (~ 0.45°C) SST on 4 June prior to the onset has reduced to ~ 0.18°C on 7 and 8 June once the monsoon conditions set in over the buoy.

4.2.2. Pulsation in the strength of the southwesterly flow in the boundary layer

The monsoon bursts with the strong cross-equatorial flow. To highlight the increase in the strength of southwesterly flow in the boundary layer, the distribution of zonal component of wind (*u*) at 925 hPa over the domain 5°S–20°N, 40–85°E from NCEP/NCAR Reanalysis data for the period 4–8 June, 2003, 12 UTC is shown in Figure 10. It is clearly seen from the Figure that the strength of the westerlies increases from 4 m s^{-1} during 4–6 June along the SEAS and the Kerala coast to 6–8 m s^{-1} on 8 June. Figure 10e shows the time series of zonal wind at 9°N, 74°E for 925 hPa during 4–8 June. It shows a sharp increase from 3.8 m s^{-1} on 6 June to 6.3 m s^{-1} on 7 June and then further increase to 6.5 m s^{-1} (8.2 m s^{-1}) on 8 June (9 June).

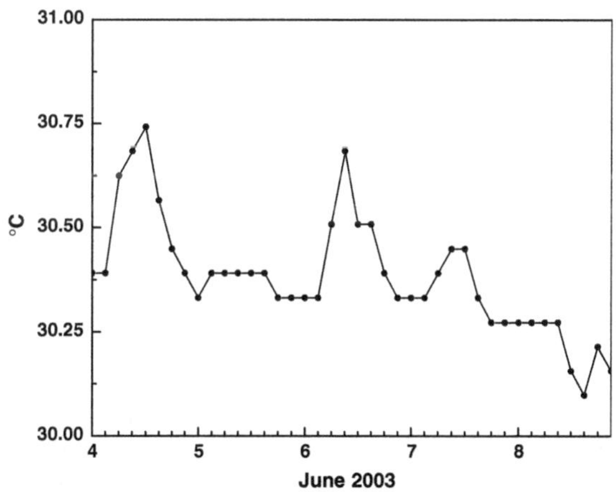

Figure 9
Time series of SST over the data buoy DS6A during 4–8 June, 2003.

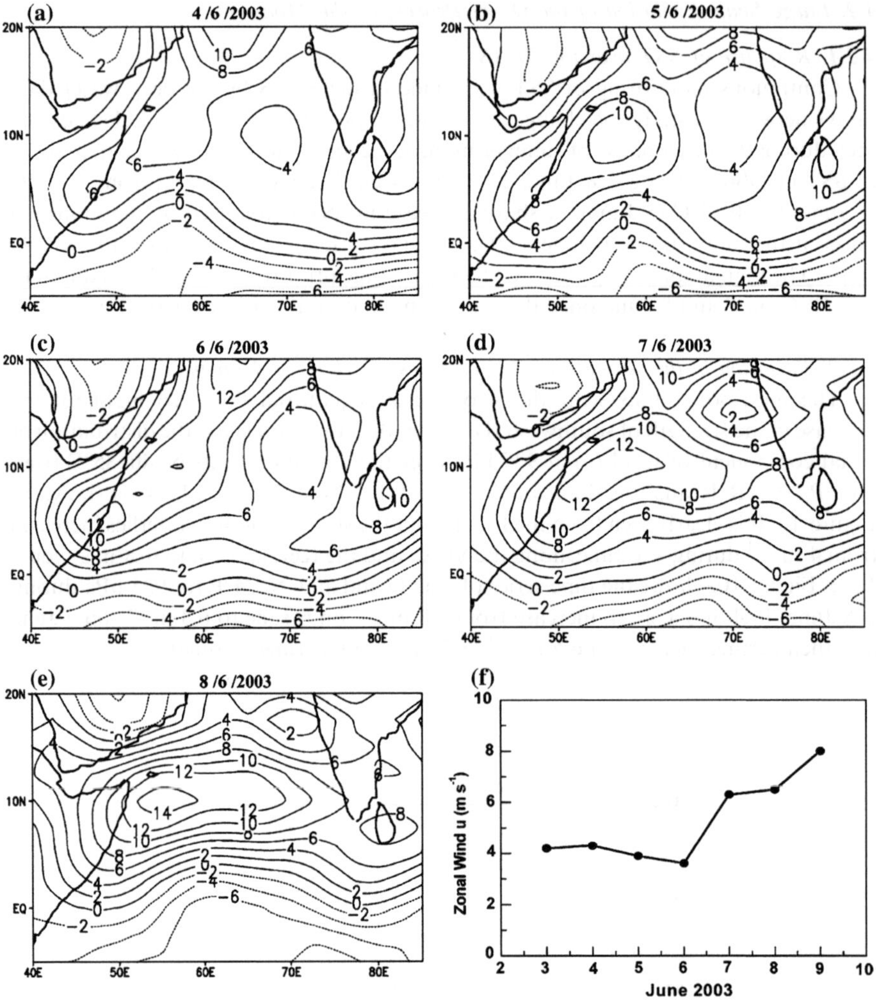

Figure 10
(a–e) The distribution of zonal wind at 925 hPa at 12 UTC during 4–8 June, 2003 over the domain 5°S–20°N, 40–85°E. (f) Time series of zonal wind over the ship during 3–9 June, 2003.

4.2.3. Changes in cloud types and rainfall along the Kerala coast

Table 1 illustrates the changes in N, C_L, C_M and rainfall over Amini and Minicoy islands, the ship ORV Sagarkanya and the coastal stations of Kerala, i.e., Thiruvananthapuram (TRV), Kochi and Mangalore. Minicoy and Kochi recorded rainfall during 6–8 June. The rainfall over Kochi has substantially increased to

Table 1

Variation in Total cloud amount (N, okta), Genus of clouds (C_L-Low cloud, C_M-Medium cloud) and Rainfall (mm) along the Kerala coast during 4–8 June, 2003

Date	Minicoy				Amini				Ship			
	N	C_L	C_M	Rain	N	C_L	C_M	Rain	N	C_L	C_M	Rain
4/6/03	7	4	0	0	6	8	3	0	7	2	5	0
5/6/03	8	3	2	8	5	8	0	0	7	8	2	0
6/6/03	4	4	3	Tr	6	8	3	18	5	8	0	0
7/6/03	6	3	3	7	3	8	0	0	2	1	3	10
8/6/03	7	8	3	4	8	8	7	0	7	5	0	1
Date	Trivandrum				Kochi				Mangalore			
	N	C_L	C_M	Rain	N	C_L	C_M	Rain	N	C_L	C_M	Rain
4/6/03	7	8	0	0	4	8	0	4	5	8	3	0
5/6/03	7	9	7	0	6	2	3	0	5	8	3	0
6/6/03	5	8	7	0	6	2	3	10	6	8	0	0
7/6/03	5	9	0	0	5	2	3	2	4	8	0	5
8/6/03	5	5	3	1	7	8	7	80	6	8	3	3

80 mm on the onset date. TRV did not record any rainfall. Although cloud amount is 5–7 oktas, only thunderstorm activity was recorded over this station. Similarly Mangalore has recorded a small amount of rainfall (5 mm and 3 mm) on 7 and 8 June. The monsoon had not arrived over this station at this point. The ship recorded the rainfall from 7 June onwards up to 12 June. With the arrival of the monsoon, rain producing clouds ($C_L = 3, 4, 7$ and 8: $C_M = 2, 3$ and 7) of increasing amounts are observed over Minicoy and Kochi.

4.2.4. Evolutionary changes in the mid-troposphere

The onset of monsoon conditions over the extreme south peninsula (south of 15°N) is associated with the formation of the monsoon trough at 700 hPa extending from the south Bay of Bengal and the establishment of the steady westerly winds over Kerala. From the analysis of the 5-day mean 700 hPa contour charts for the years 1957–1962, PANT (1964) has shown that the large-scale changes in the 5-day mean 700 hPa contour pattern were associated with the onset of the monsoon over different parts of the country. Figure 11 shows the 700 hPa patterns obtained from the NCEP/NCAR Reanalysis data for 4–8 June, 2003, 12 UTC over the domain 5°S–20°N, 40°E–85°E. The mid-tropospheric east-west shear line (marked by a thick line) along 55–70°E south of 10°N on 4 June propagates northwards along its western edge and spreads westwards on 5 June. This shear line shifts further northwards and is located along 13–15°N on 8 June in association with the onset of the monsoon over Kerala. Similar results were obtained by PANT (1981) during 1977 using MONSOON-77 data set.

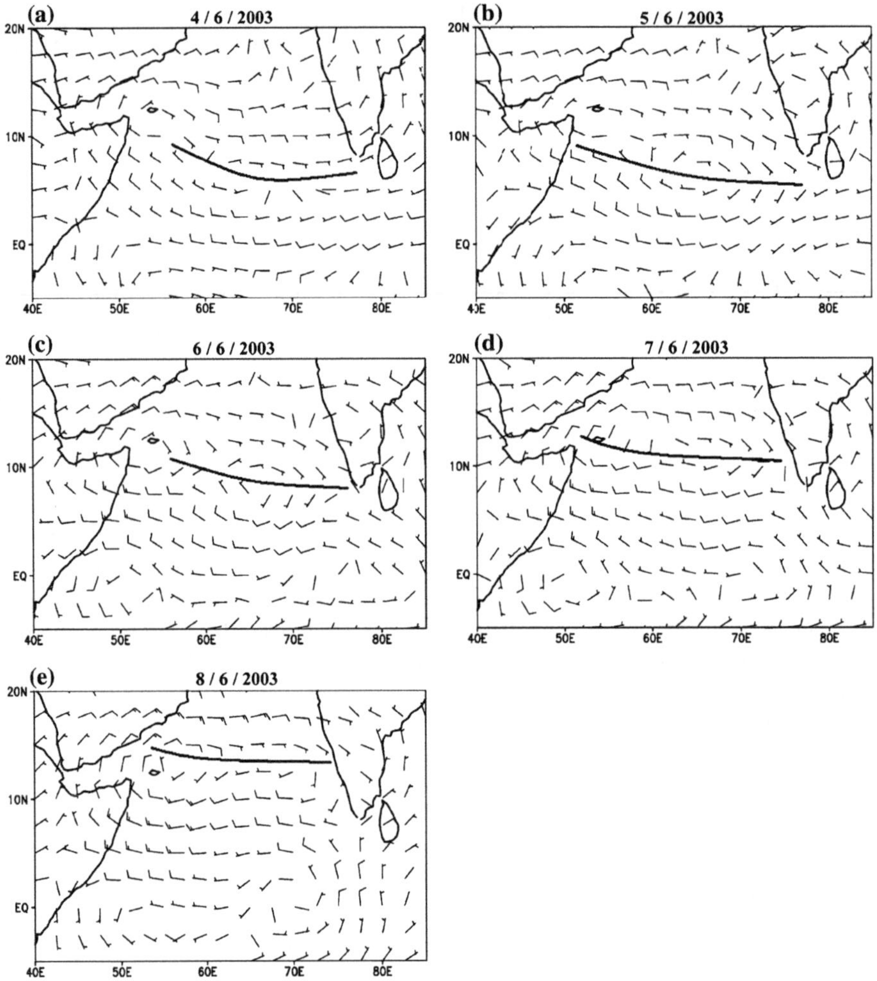

Figure 11
The flow patterns at 700 hPa at 12 UTC during 4–8 June, 2003 over the domain defined in Figure 10. The mid-tropospheric shear line is indicated by thick line.

5. *Conclusions*

The analyses of the surface marine meteorological parameters during 4–8 June, 2003 show smooth variations from 4 June to mid-day of 7 June. These parameters show rapid fluctuations due to deep cloud formation along with the onset of monsoon over Kerala on 8 June. The analysis of the vertical profiles of radiosonde data from 4 to 8 June ARMEX-2003 (just two to three days prior to the onset of monsoon over Kerala) has shown that top of the MABL depth has increased to 630 hPa (4060 m) on 6 June, i.e., two days prior to the onset. The MABL is humid and cloudy from the LCL to 1500 m and dry between 1500 and 4000 m. The θ_e and θ_{es} curves of 6 June have revealed a tendency of subsidence and cloud topped inversion in the layer between 2700 and 3000 m. The distinct θ_e minimum is not present on 7 June 0530 UTC. The winds between 800 hPa and 700 hPa veer from northeast (5 June) to southwest (6 June). The depth and intensity of southwesterlies has shown an increase from 6 June 1100 UTC onwards with the onset process on 8 June. The large-scale evolution of the monsoon is reflected in sharp fall in SST over the SEAS with a reduction in diurnal oscillations, strengthening of the zonal winds at the ship's location as well as over the coastal stations along Kerala, change of fair weather cumulus clouds to rain producing stratocumulus, nimbostrarus and altostratus and the northward movement of the mid-troposperic shear line at 700 hPa.

Acknowledgements

We are grateful to the Director, Indian Institute of Tropical Meteorology, Pune, for his constant encouragement and keen interest in the present study. We gratefully acknowledge the helpful criticisms offered by the reviewers.

References

Ananthkrishnan, R., Pathan, J. M., and Aralikatti, S. S. (1983), *The onset phase of the summer monsoon*, Current Science *52*, 755–764.

Betts, A. K. (1982), *Saturation point analysis of moist convective overturning*, J. Atmos. Sci. *39*, 1484–1505.

Betts, A. K. (1985), *Mixing line analysis of clouds and cloudy boundary layers*, J. Atmos. Sci. *42*, 2751–2763.

Betts, A. K. and Albrecht, B. A. (1987), *Conserved variable analysis of boundary layer thermodynamic structure over the tropical oceans*, J. Atmos. Sci. *44*, 83–99.

Betts, A. K. and Ridgway, W. (1989), *Climatic equilibrium of the atmospheric convective boundary layer over a tropical ocean*, J. Atmos. Sci. *46*, 2621–2641.

Bhat, G. S., Amennula, S., Venkataramana, M., and Sengupta, K. (2000), *Atmospheric boundary layer characteristics during the BOBMEX-pilot experiment*, Proc. Indian Acad. Sci. (Earth Planet Sci.) *109*, 229–237.

Bhat, G. S. and 16 other Authors (2001), *BOBMEX: The Bay of Bengal Monsoon EXperiment*, Bull. American. Met. Soc. *82(10)*, 2217–2243.

Bolton, D. (1980), *The computation of equivalent temperature*, Mon. Wea. Rev. *108*, 1046–1053.

Emanuel, K. A., *Atmospheric Convection* (Oxford University Press, New York 1994) 579 pp.

Findlater, J. (1971), *Mean monthly air flow at low levels over the western Indian Ocean*, Geophysics Memoirs *115*, 53 pp.

Fitzjarrald, D. R. and Garstang, M. (1981a), *Boundary-Layer growth over the Tropical Ocean*, Mon. Wea. Rev. *109*, 1762–1772.

FITZJARRALD, D. R. and GARSTANG, M. (1981b), *Vertical structure of the tropical boundary layer*, Mon. Wea. Rev. *109*, 1512–1526.
GROSSMAN, R. L. and DURRAN, D. R. (1984), *Interaction of low-level flow with the Western Ghat mountains and offshore convection in the summer monsoon*, Mon. Wea. Rev. *112*, 652–672.
HOLT, T. and SETHURAMAN, S. (1985), *Aircraft and ship observations of the mean structure of the marine boundary layer over the Arabian Sea during MONEX-79*, Boundary-Layer Meteor. *33*, 259–282.
HUI, D. (1980), *Results of summer MONEX field phase experiments*, GARP, FGGE Operation Report *A9*, 92–113.
KARA, A. B., ROCHFORD, P. A., and HURLBURT, H. E. (2002), *Air-sea flux estimates and the 1997–1998 ENSO event*, Boundary-Layer Meteor. *103*, 439–458.
KERSHAW, R. (1985), *Onset of the southwest monsoon and SST anomalies in the Arabian Sea*, Nature *315*, 561–583.
KERSHAW, R. (1988), *Effect of a sea-surface temperature anomaly on a prediction of the southwest monsoon over India*, Quart. J. Roy. Meteor. Soc. *114*, 325–345.
KLOESEL, K. A. and ALBRECHT, B. A. (1989), *Low-level inversions over the tropical-Pacific-thermodynamic structure of the boundary layer and the above-inversion moisture structure*, Mon. Wea. Rev. *117*, 87–101.
KRISHNAMURTI, T. N., ARDANUY, P., RAMANATHAN, Y., and PASCH, R. (1981), *On the onset vortex of the summer monsoon*, Mon. Wea. Rev. *109*, 341–363.
PANT, P. S. (1964), *Onset of monsoon over India*, Indian J. Meteor. and Geophys. *15*, 375–380.
PANT, P. S., *Medium-range forecasting of monsoon rains*. In *Monsoon Dynamics* (eds. Lighthills J. and Pearce R.) (Cambridge University Press, Cambridge, 1981) pp. 220–235.
SEETARAMAYYA, P. and MASTER, A. (1984), *Observed air-sea interface conditions and a monsoon depression during MONEX-79*, Arch. Met. Geoph. Biomet. Ser. *A-33*, 61–67.
RAO, P. SANJEEVA (2005), *Arabian sea monsoon experiment: an overview*, Mausam *56*(1), 1–6.

(Received May 12, 2006, accepted November 13, 2006)
Published Online First: July 14, 2007

MIX
Papier aus verantwortungsvollen Quellen
Paper from responsible sources
FSC® C105338

If you have any concerns about our products,
you can contact us on
ProductSafety@springernature.com

In case Publisher is established outside the EU,
the EU authorized representative is:
Springer Nature Customer Service Center GmbH
Europaplatz 3, 69115 Heidelberg, Germany

Printed by Libri Plureos GmbH
in Hamburg, Germany